D0710290

CONCRETE STRUCTURES

MEHDI SETAREH, PHD, PE

Virginia Polytechnic Institute and State University
Blacksburg, Virginia

ROBERT DARVAS, SE, PE

The University of Michigan
Ann Arbor, Michigan

PEARSON
Prentice
Hall

Upper Saddle River, New Jersey
Columbus, Ohio

i129253366

TA
682.4
.S48
2007   # 69734533

(+ 1 CD-ROM)

Library of Congress Cataloging-in-Publication Data

Setareh, Mehdi.
    Concrete structures / Mehdi Setareh, Robert Darvas.
        p. cm.
    Includes index.
    ISBN 0-13-198827-1
    1. Concrete construction.   I. Darvas, Robert.   II. Title.
    TA682.4.S48 2007
    624.1'834—dc22

                                    2006017887

Editor-in-Chief: Vernon Anthony
Senior Acquisitions Editor: Tim Peyton
Editorial Assistant: Nancy Kesterson
Production Coordination: Penny Walker, Techbooks
Production Editor: Holly Shufeldt
Design Coordinator: Diane Ernsberger
Cover Designer: Bryan Huber
Cover and Chapter Opener Photos: Corbis
Production Manager: Deidra Schwartz
Executive Marketing Manager: Derril Trakalo
Senior Marketing Coordinator: Liz Farrell
Marketing Assistant: Les Roberts

This book was set in Times Roman by Techbooks. It was printed and bound R.R. Donnelley & Sons Company. The cover was printed by Phoenix Color Corp.

Copyright © 2007 by Pearson Education, Inc., Upper Saddle River, New Jersey 07458.
Pearson Prentice Hall. All rights reserved. Printed in the United States of America. This publication is protected by Copyright and permission should be obtained from the publisher prior to any prohibited reproduction, storage in a retrieval system, or transmission in any form or by any means, electronic, mechanical, photocopying, recording, or likewise. For information regarding permission(s), write to: Rights and Permissions Department.

Pearson Prentice Hall™ is a trademark of Pearson Education, Inc.
Pearson® is a registered trademark of Pearson plc
Prentice Hall® is a registered trademark of Pearson Education, Inc.

Pearson Education Ltd.   Pearson Education Australia Pty. Limited
Pearson Education Singapore Pte. Ltd.   Pearson Education North Asia Ltd.
Pearson Education Canada, Ltd.   Pearson Educación de Mexico, S.A. de C.V.
Pearson Education—Japan   Pearson Education Malaysia Pte. Ltd.

PEARSON
Prentice
Hall

10 9 8 7 6 5 4 3 2 1
ISBN 0-13-198827-1

To Roufia, Ali, and Kamran with love,
Mehdi

To Eva with love,
Robert

PREFACE

The intended audience of this book is architectural engineering, undergraduate civil engineering, building construction, and architecture students. The manuscript complies with the provisions of the ACI Code 318-05. The easy to follow style of the text makes it valuable to engineering and non-engineering students. Furthermore, educators and practitioners interested in the analysis and design of concrete structures based on the latest ACI Code provisions may also benefit from it.

Chapter 1 covers the topic of concrete technology. It discusses the most important properties of the main components of reinforced concrete. This technology is essential for both architecture and engineering students.

Chapter 2 discusses the analysis and design of rectangular beams and one-way slabs, including a complete treatment of the Unified Design Method as recommended by the ACI 318-05. Several examples demonstrate the provisions of the latest changes in the ACI Code. It is written to benefit architecture and engineering students as well. Depending on the main objectives of the course and class time constraints, the instructor can select the specific topics and their details to be included for the intended audience.

Chapter 3: Special Topics in Flexure covers T-beams, doubly-reinforced beams, and a discussion of the deflection of reinforced concrete beams and slabs. These topics are more complex, but indispensable in the design of concrete structures. The detailed technical information presented is essential for engineering students. We recommend that only a brief discussion of each topic be used in courses for architecture students.

Chapter 4: Shear in Reinforced Concrete Beams covers the design of shear reinforcements in reinforced concrete beams. We consider this chapter to be important in both engineering and architecture courses. The depth of coverage may be left to the discretion of the instructor.

Chapter 5 covers the analysis and design of reinforced concrete columns. It includes a complete treatment of "short" columns with small and large eccentricities. Because most reinforced concrete columns are short and a complete treatment of slender columns is usually only covered in advanced engineering courses, we decided to cover that topic generally. We recommend this chapter be covered in engineering and architecture courses.

Chapter 6 is a treatise on the different floor systems typically used in reinforced concrete buildings. A simplified approach appropriate for both architecture and engineering students is used.

Chapter 7 discusses foundations, and earth-retaining walls. The chapter starts with a background on some aspects of soil mechanics and geotechnical investigations for building design. These topics are not usually covered in reinforced concrete structures textbooks. However, we are aware that many engineering students do not take a soil mechanics course as a prerequisite for a reinforced concrete class. Furthermore, soil mechanics and foundations courses are unavailable in nearly all architecture curriculums. The treatment of the subjects of foundations and earth-retaining walls are well-suited for both architecture and engineering students.

Chapter 8 is an introduction to pre-stressed concrete for both architecture and engineering students.

Chapter 9 discusses the use of the SI System in reinforced concrete design and construction. We decided against the use of the equivalent SI System within the main body of the book, as is done in many other textbooks. We felt that this resulted in a clearer text. Several examples on different topics covered in other chapters are again presented using the equivalent SI System.

Two unique features of this book are the "Self-Experiments" and an accompanying CD with images of concrete structures. From our experience we know that some engineering students and nearly all architecture students do not have access to a testing laboratory. Therefore, we included these simple-to-do sets of experiments that students can perform to learn about reinforced concrete from their own experiences. We believe these experiments may also help students gain a better understanding of concrete as a building material. The accompanying CD has a number of high-quality images of reinforced concrete structures, so that students can develop an appreciation of the potential this building material offers.

There are numerous problems at the ends of each chapter to be used as homework assignments. A complete Instructor's Solutions Manual is available upon request.

A step-by-step approach was adopted throughout the text. Most of the procedures for design or analysis are summarized in flowcharts, where all steps are numbered, and the example solutions follow these steps. In our experience this approach helps students try to follow the numerical solutions of various problems.

We would like to thank Professors Jay Stoeckel, Jack Davis, and Mr. Gerry Martin from the Ceco Concrete Construction, LLC for providing some of the images in the accompanying CD-ROM. The continued educational support by the Northeast Cement Shippers Association, and in particular Kim Frankin are greatly appreciated. We are also grateful to students at the School of Architecture + Design of Virginia Tech for their help and comments during the development of this book, in particular Mr. Amir Abu-Jaber for his assistance in typing and editing the manuscript and the solutions manual.

Finally, we wish to thank the Pearson Education editorial and production staff for their support and assistance. Many thanks to Bret Workman, who did a great job with text editing. In particular, the assistance of Penny Walker from Techbooks is greatly appreciated.

CONTENTS

CHAPTER 2   RECTANGULAR BEAMS AND ONE-WAY SLABS   33

1

REINFORCED CONCRETE TECHNOLOGY

1.1 INTRODUCTION

Concrete and reinforced concrete are extremely versatile building materials. Concrete, which essentially is a man-made stone, can take virtually any shape and form the designer envisions. Because concrete is like a heavy liquid when produced, it is poured into a mold and, when hardened, will take the shape of the pre-built form.

The skillful use of reinforced concrete opens unlimited vistas for the designer. Working with reinforced concrete is an experience in sculpting. Any sculptor working on the creation of an art object must be fully familiar with the possibilities and limitations of the material be it clay, metal, glass, or something else. Likewise, the designer must be fully cognizant of the nature of reinforced concrete. How is it made? How does it work? How will it serve in different environments?

Reinforced concrete is not a homogeneous material. It is a combination of two materials: concrete and reinforcing, which is most often steel. Concrete, while strong in compression, is relatively weak in tension. This weakness in tension must be corrected by adding steel reinforcing. The successful combination of these two distinctly different materials into one successful hybrid makes reinforced concrete the most widely used building and structural material in the world.

To gain a better understanding of the complexity underlying the construction of a reinforced concrete structure, consider Figure 1–1, which outlines the process and the "players" involved. Of course there are many more players (like those who manufacture materials such as cement, steel reinforcing rods, timber products used in form making, and so on) but their inclusion would unnecessarily complicate an already intricate web of involvements.

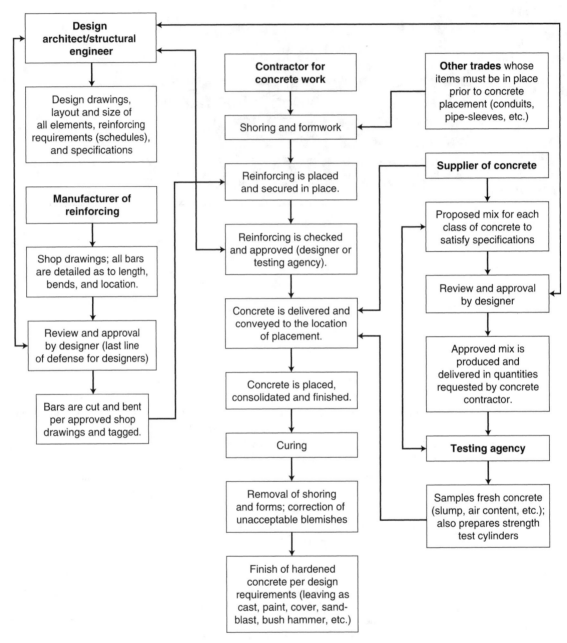

FIGURE 1–1   Concrete construction overview.

1.2 THE ACI CODE

The rules and regulations governing any construction in the United States are set by different model codes. The rules of a model code become the law in a given municipality when the legislative body (state legislature, city council, etc.) adopts them as the governing standard for building construction under its jurisdiction. There are many model codes, among them the *Uniform Building Code (UBC)* and the *International Building Code (IBC)*. The UBC is traditionally used in the western states of the United States. The IBC first appeared in the year 2000 and is a joint effort of the *BOCA (Building Officials and Code Administrators),* the *ICBO (International Conference of Building Officials),* and the *SBCCI (Southern Building Code Conference International).*

For the design of reinforced concrete structures, these model codes usually adapt the requirements set forth by the American Concrete Institute (ACI), headquartered in Farmington Hills, Michigan. Engineers, architects, concrete producers, contractors, chemists, and cement manufacturers from all over the world belong to this organization, but membership is available to anyone wishing to join. Student membership is also available.

There are several hundred technical committees within the ACI that deal with any aspects of producing and designing with concrete. These technical committees collect the vast existing (and continually forthcoming) research information and publish it in the *ACI Journal* as recommended standards. The updated collection of standards are also published every other year or so in a large six-volume set, *The ACI Manual of Standard Practice.*

One of the committees (ACI Committee 318) compiles and publishes a document, *Building Code Requirements for Structural Concrete,* that contains the most up-to-date rules recommended by the collective knowledge in the Institute. In the past a new updated edition was published about every 6 or 7 years. Now the Institute appears to have adopted a 3-year cycle to update the Code and incorporate the latest and best available research information. The latest edition, *ACI 318-05,* appeared in 2005. This book has already been updated to contain the latest changes in the Code. Proposed changes in any new edition are first published in the *ACI Journal* for review and comments from the membership; then the final revised document is submitted to the Institute's membership for approval. Upon approval, it becomes an ACI Standard that governs the design of concrete and reinforced concrete structures.

1.3 CONCRETE INGREDIENTS

Concrete is a mixture composed of a filler material (aggregate) bound together by a hardened paste. One might call it man-made stone. The hardened paste is the result of a chemical reaction, called *hydration,* between cement and water. In addition, admixtures—various chemicals, usually in liquid form—are often used to impart desirable qualities to the freshly mixed or hardened concrete. The paste fills the voids between the aggregate particles, gravel or crushed stone and sand, and binds them together. The aggregate size distribution is carefully controlled to minimize the resulting voids that must be filled with the paste. Minimizing the amount of paste helps to minimize the amount of cement, which is the most expensive ingredient of the mixture, because it requires a large amount of energy in its manufacture. The usual proportion of the aggregate in normal-weight concrete is about 65%–75% by volume, while the paste makes up about 33%–23%. The remaining volume is air.

Thus the four ingredients of concrete (see Figure 1–2) are (1) cement (i.e., the binder); (2) fine and coarse aggregates, which fill the bulk of the volume; (3) water and air; and (4) admixtures, which are used to impart certain desirable properties. These ingredients, carefully proportioned, are combined in a mixer.

FIGURE 1–2 Concrete ingredients (Courtesy of Portland Cement Association).

Portland Cement

Mankind has used natural cements since ancient times. The magnificent stone structures built by the Romans all used finely ground cementitious materials (pozzolans) in the mortar. Other materials may be used to bind aggregate particles together (asphalts are used in making asphaltic concrete for road construction), but in the making of structural concrete, hydraulic cements are used without exception. Hydraulic cements harden by reacting with water. The most important hydraulic cement is the one first made by an English mason, Joseph Aspdin, who received an English patent in 1824 for the composition and the process. He called it *portland cement* because the concrete made with the cement had the color of natural limestone quarried on the Portland peninsula. Portland cement is a fine powdery material, composed mainly of calcium silicates and aluminum silicates. The materials needed to make cement are found in virtually every part of the world:

1. Limestone, which provides *calcium oxide* (CaO)

2. Clays, shales, and so on, which supply *silicon dioxide* (SiO_2) and *aluminum trioxide* (Al_2O_3)

The materials are pulverized, mixed in the right proportions, then baked in a rotary kiln at very high (about 2,300°F) temperatures. The product from the kiln, a glassy-looking ceramic, is called *clinker*. During the baking, chemical changes occur in the original materials, which form four important compounds (among others of somewhat lesser importance). These are *dicalcium silicate, tricalcium silicate, tricalcium aluminate,* and *tetracalcium aluminoferrite.* The relative proportions of these compounds influence the characteristics of different cements.

The clinker is then ground into a fine powder. The average size of the particles is only 0.0004 in. This is just an average; there are many smaller size particles. Actually there are about 7 trillion particles per pound of cement. The particles' combined surface area is about 2,000 ft^2/lb. Usually, small amounts of gypsum and various other minerals are mixed with the ground clinker to adjust the setting time of the cement or to impart some desirable properties to the final product. Different types of cement are used for various jobs and conditions. For building structures in most cases, Type I—normal portland cement, or Type III—high-early-strength portland cement are used.

While the basic raw materials of cement (limestone, clay, shale, etc.) are relatively cheap, the making of cement, chiefly the previously described baking process, requires large amounts of energy (e.g., natural gas). Thus the cement is by far the most expensive component of concrete. To save cement, other materials that have hydraulic properties can be substituted for some part of the cement. Substitution of up to 20% by weight may be permitted. Fly ash, a by-product of coal-fired power plants, and ground blast furnace slag are two such commonly used substitutes.

Fine and Coarse Aggregates

Aggregates, such as the filler material, make up the bulk of the volume in concrete. Thus it is important that the aggregates be of good quality, strong and resistant to the environmental forces (physical and chemical) that will affect the concrete throughout its intended life. Aggregates should not contain chemicals or materials that might lead to the destruction of the inner structure of the concrete.

As mentioned before, approximately 65%–75% of the total volume of concrete is aggregates. In a somewhat arbitrary way they are divided into two classes. The particles that pass a #4 sieve, that is, less than 0.25 in. are called *fine aggregates* or sand. *Coarse aggregates*—natural gravel or crushed stone—are particles that are larger than 0.25 in. Aggregates are mostly dug or dredged from a pit, river, lake, or seabed. They are also produced by crushing rocks (limestone, dolomite, etc.) and boulders.

Producers usually wish to fill most of the volume with the cheaper ingredients, that is, the aggregates, so they first carefully separate the different grain sizes (the fine and the coarse), then mix them in desirable proportions. In the resulting particle distribution, the successively smaller particles fill the voids between the larger parts. This is referred to as *good gradation*.

At a certain point however, the smallness of a fine aggregate particle becomes deleterious to the quality. The cement paste must coat all the particles to bind them together. The smaller the particle, the larger is its surface-to-volume ratio. Thus particles less than 0.006 in. are undesirable. If necessary, aggregates are carefully washed to rid them of adhered clay or mud particles.

On the other end of the scale, the maximum size of the coarse aggregate must be controlled as well. The gravel in the concrete mix must pass between closely spaced reinforcing bars, and the concrete must smoothly fill often-intricate forms. In general, the maximum-size aggregate should be no larger than one-fifth of the narrowest dimension of the concrete form. Furthermore, in building structures, where the minimum allowable clear spacing between reinforcing bars is 1 in., the maximum size of coarse aggregate particles is usually limited to about $^3/_4$ in. in the concrete mix.

The unit weight of concrete made with gravel (or crushed stone) and sand aggregates varies between 140 to 150 pounds per cubic foot (pcf). In calculations an average weight of 145 pcf is used for unreinforced concrete, while for the weight of reinforced concrete structures a value of 150 pcf is used. The difference between these two values tends to account for the greater unit weight of the embedded steel reinforcement.

The last 50 years has also seen a growing development in the use of lightweight aggregates. In concrete structures, because the self-weight of the structure is a much larger component of the total loads than in steel or wood-framed structures, it is often desirable to use lighter aggregates than gravel or stone. Concretes made with lightweight aggregates also have better insulating properties. Most of these aggregates are artificially produced. For structural purposes expanded shales and clays are used almost exclusively. Their use enables the production of structural (as opposed to insulating) concretes with only 110 to 115 pcf unit weight. Lightweight structural concrete is more expensive than normal weight concrete, but its lighter weight often reduces the overall cost of the structure.

Water and Air

Water   Water is an important and necessary part of making concrete. The water used to make concrete has to be free of chemicals and unwanted elements. In general, if the water is drinkable, it can be used to make concrete, although some waters that are not fit for drinking may also be suitable for concrete. Two important aspects about the role of water in concrete need to be discussed: the hydration process and the water/cementitious materials ratio.

The Hydration Process   When water is mixed with cement, a chemical reaction starts between them. This is called *hydration,* which creates the binding quality of the paste. The two calcium silicates that make up about 75% of portland cement react with water to form two new compounds: *calcium hydroxide* and the more important *calcium silicate hydrate*. The latter compound first appears as a gel, which later turns into a solid. The surface area of the calcium silicate hydrate is enormous. Its crystals can be discerned only in a scanning electron microscope. These crystals adhere to each other, as well as to the grains of sand and gravel, cementing (gluing) all parts together.

The hydration process develops in three stages. These are *setting, hardening,* and *strength development*. They all are related to the rate of reaction between the cement and the water. This rate of reaction must be carefully determined and regulated to allow sufficient time for the concrete to be transported, placed, and finished. When the hydration advances to a certain stage of setting, the concrete becomes difficult or impossible to handle. Thus the concrete must be placed and consolidated in the forms, usually within 2 hours after batching (the mixing of ingredients with the water). Temperature also has a major influence on the rate of the hydration, so various chemical admixtures may be added to either retard or accelerate the process.

From the age of about 2 hours to about 6 to 8 hours, the hardening stage takes place. After hardening, one may step on the concrete without leaving an imprint on the surface. The concrete is far from being strong at this stage, however. Thereafter begins the third stage, that is, the strength development that is quite rapid in the early days and gradually becomes slower.

Hydration continues throughout the life of a concrete structure as long as free moisture is available to react with unhydrated parts of cement particles.

Water/Cementitious Materials Ratio The water/cementitious (w/cm) materials ratio is of paramount importance. It greatly influences the quality of the paste, hence the quality of the concrete. It is defined as the weight of the total water to the weight of the cement (or cementitious products) in the mix. The total water must also account for the water contained by moist aggregates. The free water adhering to the aggregates can be significant, so it must be carefully determined, and the weight of the additional water into the mix must be adjusted accordingly.

For complete hydration only an approximate w/cm ratio of 0.25 (25 lb of water for every 100 lb of cementitious material) is needed. This is a theoretical value only. Evaporation of water from the mix cannot be prevented, thus reducing the amount available for the hydration process. Furthermore, concrete made with such a small w/cm ratio is too dry and unworkable. More water (higher w/cm ratio) is needed to produce a concrete that is *workable*. Therefore, a minimum w/cm ratio between 0.35 and 0.40 is usually required.

Workability is not a scientifically definable term. It refers to the ease of placing, consolidating, and finishing fresh concrete. It is true that more water in the mix tends to increase the workability, but excess water creates all sorts of problems. Practically all desirable properties of concrete, such as strength and durability, are adversely affected by high w/cm ratios.

To begin with, concrete with excess water has a tendency to segregate. When the fresh concrete is too fluid the heavier particles (coarse aggregate) settle on the bottom of the form, that is, they segregate from the ideal distribution of particles. Then the excess water migrates to the surface in a process called *bleeding,* producing a weak top layer. Large amounts of bleed water also make it difficult to properly finish the top surface of floors. In the later stages of hardening and strength gain, the excess water, that is, the water that was not used during hydration, will evaporate from the concrete through tiny capillaries. This results in voids that weaken the concrete. Figure 1–3 illustrates the dramatic strength loss with increasing w/cm ratio while all other parameters, such as total cement content, are kept constant in a given mix.

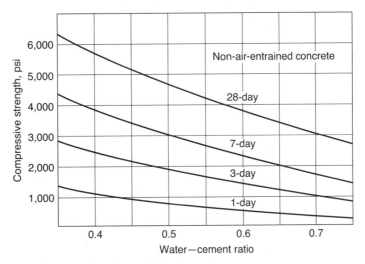

FIGURE 1–3 Changes of concrete compressive strength with w/cm ratio and age.

A balance must be struck between having too little water and an unworkable mix, and having too much water that results in loss of strength and durability. Thus an optimum water content must be used. Optimum water content is the minimum amount necessary in a mix to maintain good workability. As will be discussed in the section on admixtures, there are certain chemical compounds that, when added to the fresh concrete, temporarily increase its fluidity. These are called *water reducing agents (plasticizers)* or *high-range water reducing agents (superplasticizers)*. The former agents reduce the water requirement by 5%–10%, while the latter ones reduce it by as much as 20%–30% without the loss of workability.

Air   All concrete, even after the most careful consolidation, contains some air. Two types of air may be present in a concrete mix: unwanted or "bad air," and wanted or "good air." Bad air is basically large bubbles of air entrapped inside the mix. These bubbles create discontinuity in the concrete's texture and weaken its strength. Every effort is made to minimize this type of air in hardened concrete. At the time of placement the aim is to consolidate the concrete to the maximum possible density and bring these unwanted air bubbles to the surface. Unfortunately, even with the best consolidation efforts, these large air bubbles (well visible to the naked eye) get stuck in the concrete by adhering to aggregates, reinforcing bars, and most often to the inner surface of the formwork, especially in columns and sides of deep beams. There they become visible on the surface of the hardened concrete and are commonly referred to as "bug-holes." In well-consolidated concrete, these may represent about 1% or less of the total volume. Their presence is not a source of major weakness.

The second type of air, good air, is deliberately introduced into the concrete. This process is called *air entrainment*. Properly used air-entraining agents distribute tiny (microscopic) air bubbles uniformly in the concrete. The size of the bubbles ranges between 0.0004 to 0.004 in., in excess of 3 billion air bubbles per cubic foot of concrete. The chemicals used to create them are added to the concrete using special admixtures. Air-entrainment makes the concrete mix more workable (thus requiring less water), slightly decreases the weight of the concrete, and, most importantly, increases the durability of the concrete.

Admixtures

Admixtures are chemicals added to the concrete batch during mixing or just prior to placement to enhance properties such as rate of setting, hydration, workability, strength, and so on. Four main types of admixtures are discussed here.

Air-entraining Admixtures   Air entraining agents are *hydrophobic,* that is, they repel water. Thus a film (e.g., soap film) forms on the surface of the bubbles that prevents them from collapsing or coalescing. The film also keeps water out of the bubbles. These bubbles are finely dispersed throughout the concrete during mixing. They do increase the workability of the concrete, but their most important role is to increase concrete durability. *Durability* in this context refers to concrete's resistance to the destructive process of freeze and thaw cycles that occur in certain climates.

Hardened concrete contains fine capillaries that enable moisture to penetrate. As free water in moist concrete freezes, it expands. The expansion is significant. Ice takes up about

9% more volume than unfrozen water. This expansion exerts hydraulic pressure on the yet unfrozen water, which in turn exerts pressure on the surrounding paste structure. If this pressure is too great for the tensile strength of the paste to withstand, the paste structure will rupture and collapse to provide the excess room needed for the ice. This creates more volume the next time around for the penetrating water, thus even more room is needed to accommodate the expanding ice. This cyclic phenomenon continues, resulting in scaling and crumbling of the concrete.

Entrained air voids act as relief reservoirs in the paste structure. The expanding water in the capillaries can enter the storage space provided by these well-dispersed tiny bubbles by overcoming the air pressure existing within the bubbles. On thawing, the water, driven out by the compressed air, returns to the capillaries.

To impart proper freeze/thaw resistance to building structures the accepted range is to have about 5%–7% entrained air in the hardened concrete volume. Air content of the freshly mixed concrete can be measured right at the job site prior to placement. This ensures that the hardened concrete will have an appropriate amount of entrained air. The actual air content can also be established on core samples taken from the already hardened concrete. The two different tests may give different results, because some air inevitably will be lost during placement and consolidation.

Accelerating Admixtures (Accelerators)   This admixture hastens the setting of concrete by speeding up the hydration process, which in turn makes the concrete gain strength faster, especially at an early age. Similar results may be obtained by using Type III, or high-early-strength portland cement, by lowering the w/cm ratio, or by curing the concrete at higher temperatures. Accelerators are traditionally used in cold weather construction. Cement hydration is an *exothermic* process, that is, it generates heat. The use of accelerators reduces the setting and hardening times. The accelerated hydration produces a larger amount of heat, which helps to prevent the concrete from freezing. In the past, calcium chloride was used as an accelerator, and some products on the market still contain calcium chloride. This chemical, however, has many potentially dangerous side effects (e.g., chloride ions in the the presence of moisture enhance corrosion of the reinforcing), so its use is strongly discouraged. Several, non-chloride–based and noncorrosive accelerators are available for use.

Superplasticizers   This admixture reduces the water needed to create a flowing concrete as well as the water that otherwise would be needed for proper workability. The reduced w/cm ratio results in a higher strength concrete with the same amount of cement. Superplasticizers are indispensable when concrete is pumped between the point of discharge from the delivery truck and the point of placement. Normal structural concrete is said to be workable when the slump is about 3 in. (see Section 1.5 for the slump test that is used to check consistency and workability). Such concrete is too stiff to flow through a 5-in. or 6-in. diameter hose. Adding a superplasticizer will temporarily increase a 3 in. slump to 8 in. or 9 in.; thus the concrete behaves like a liquid for a short time.

Retarding Admixtures (Retarders)   As the name implies, retarders have an effect opposite to that of accelerators. They slow down concrete hydration and increase the setting time. Retarders are used for hot weather construction because the hydration process is much faster at elevated temperatures. Their use enables the contractor to place and finish the concrete before advancing hydration makes the concrete difficult to handle. Retarders are also used to make exposed aggregate elements in the precast concrete industry. A layer of retarder paste is smeared on the inside of the form prior to the placement of the concrete. In about 12 to 24 hours (depending on the curing technique used), the precast concrete element is removed from the form and the retarder paste is washed away, exposing the surface of the underlying aggregate structure. (Note: For additional information, refer to *ACI 212.3R: Chemical Admixtures for Concrete,* reported by ACI Committee 212.)

1.4   CURING

Freshly placed concrete is consolidated to bring large entrapped air bubbles and excess water to its surface. This is usually done by high-speed vibrators. The vibrating action reduces the friction between the particles and makes the concrete behave like a thick fluid. At the same time, entrapped air bubbles and excess water are forced to rise to the surface. Vibrators are elongated cylinders with an unbalanced weight rotating inside at a high frequency. Vibrators are of different diameters, from about $3/4$ up to 6 in.; the most frequently used ones are 2 to $2^1/_2$ in. in diameter. Frequencies may vary from 5,000 to 15,000 cycles per minute. Vibrators should be rapidly lowered into the concrete and then slowly withdrawn for best effect.

After the freshly placed concrete is finished, it is necessary to create the best possible environment for the concrete to harden and gain strength. This process is called *curing*. Hydration and strength gain will continue as long as unhydrated cement particles and adequate moisture are present for the chemical reaction. Thus the moisture in the concrete after the consolidation and finishing processes must remain in the concrete. If the concrete dries out (i.e., the relative humidity *inside* drops below 80%), the hydration stops. Similarly, if the moisture in the concrete freezes, the hydration will stop and the expansion of ice will destroy the paste matrix, which is at its early stages of formation.

So the concrete should be kept moist and comfortably warm. Concrete is kept moist by covering it to prevent evaporation from the surface, or sprinkling it several times daily. Chemical curing compounds also are available. These are sprayed on the concrete to form a film that prevents moisture from escaping.

In the wintertime, freshly placed concrete is covered with insulation blankets. It is also a usual practice to enclose the space below the fresh concrete and heat the space with propane space heaters. This process not only prevents the freshly placed concrete from freezing but enhances the speed of the hydration. (Note: Detailed information may be found in the ACI Standards, *ACI 305—Hot Weather Concreting* and *ACI 306—Cold Weather Concreting*. These contain state-of-the-art recommendations regarding the topics.)

1.5   TESTING CONCRETE

Testing of concrete aims (1) to ensure that it has the required properties called for in the design documents and specifications, and (2) to determine the properties of concrete in an existing structure.

Many tests can be performed to evaluate certain properties of fresh or hardened concrete. The three most commonly used tests are as follows.

Slump Test

The slump test measures the consistency and workability of concrete. This test is performed on fresh concrete either as it is discharged from the truck (known as *testing at the point of delivery*), or after it has been conveyed to the point of placement. The distinction is sometimes important, for significant slump loss may occur during conveyance. The device used in this test is a 12 in. high truncated metal cone, 4 in. wide at the top and 8 in. wide at the base (see Figure 1–4a). The method of sampling the fresh concrete, and of filling and consolidating the concrete inside the slump cone is standardized in the ASTM (American Society for Testing and Materials) C143 standard.

      The cone is filled with concrete in three equal-volume layers, and each layer is consolidated within the cone by 25 strokes of a $5/8$-in. diameter rod with a rounded end. After

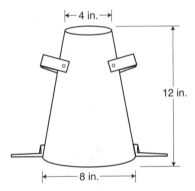

FIGURE 1–4a Slump cone.

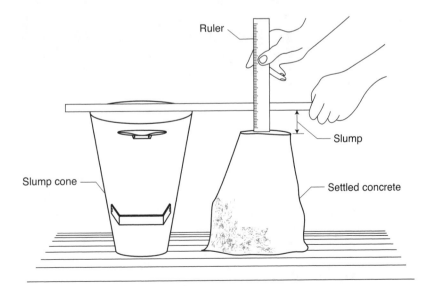

FIGURE 1–4b Slump test.

FIGURE 1–5   Slump test (Courtesy of Portland Cement Association).

the third layer is filled, the excess concrete is struck off with the steel rod, and the cone is carefully lifted off. The cone is then placed upside down next to the concrete, and the steel rod is placed across its top. The distance measured from the bottom edge of the rod to the original center of the slumped concrete mass is the *slump* (Figures 1–4b and 1–5). The slump recommended for good workability and an acceptable w/cm ratio depends on the type of construction. The common range of slump in building structures is 3 to 4 in. (unless super-plasticizers are used).

Cylinder Test

The most important property of hardened concrete is its compressive strength, f'_c. This value refers to the cylinder compressive strength of the concrete at the age of 28 days and

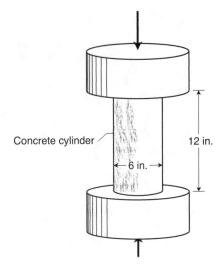

FIGURE 1–6   Cylinder test.

forms the basis of the design of a structure. In the United States compressive strength of concrete is measured on 6 in. diameter by 12 in. high cylinders. (Note: The sampling of the fresh concrete is governed by ASTM C172, "Method of Sampling Freshly Mixed Concrete." The making and curing of the cylinders are governed by ASTM C31, "Practice for Making and Curing Concrete Cylinder Test Specimens").

The architect or design engineer specifies the number of cylinders cast for testing. Typically one set of cylinders is made from about every 50 to 100 yd^3 of concrete, but not less than one set from each day's pour. Usually three cylinders comprise one set. After the concrete is hardened, the cylinders are transported to a testing laboratory where they are placed in a curing chamber. The temperature inside the curing chamber is kept at 72°F (room temperature) with 100% relative humidity. These cylinders thus treated are called *lab-cured cylinders*. They indicate how good the concrete mix was, not how good the concrete is in the structure, as the contractor may not maintain ideal curing conditions on the site. To determine the strength development of the concrete in the field, extra cylinders may be cast and kept in the field to cure under the same conditions as those of the structure. These are known as *field-cured cylinders*. Comparing the strength of field-cured cylinders to that of lab-cured cylinders helps to determine how successful the contractor's efforts were in providing good curing.

The strength test is performed in accordance with ASTM C39, "Test Method for Compressive Strength of Cylindrical Concrete Specimens." Compression force is applied to the prepared concrete cylinder by a hydraulic jack (Figures 1–6 and 1–7). The load is increased progressively at a rate of 35 ± 5 psi (pounds per square inch) per second until the concrete cylinder fails. The load required to break the cylinder is noted, then divided by the cross-sectional area of the cylinder. The result gives the breaking stress, or cylinder strength. A strength test is the average strength of two cylinders cast from the same sample.

The acceptance of the concrete (from the strength point of view) is regulated by the ACI Code. The strength of the concrete is considered satisfactory when:

- The arithmetic average of any three consecutive strength tests equals or exceeds f'_c, (i.e., the specified design strength)

FIGURE 1–7   Cylinder test (Courtesy of Portland Cement Association).

■ No individual strength test (the average of two cylinders) falls below f'_c by more than 500 psi when f'_c is 5,000 psi or less; or by more than 0.10 f'_c when f'_c is more than 5,000 psi

Often extra cylinders are cast and tested at an earlier age (7 days) to evaluate the development of strength. Although different cements may gain strength at somewhat different rates depending on the relative proportions of the main chemical compounds, the 28-day strength can be estimated by extrapolating early test data.

The problem with the 28-day strength test is that if the results are unsatisfactory, the remedy is usually difficult and expensive. In any major project, construction progresses far in 28 days, often resulting in two or three additional floors. Thus, removing the weak concrete and replacing it is rarely an option. The various strengthening methods of the structure are generally very expensive. So it is of paramount importance to have good quality control throughout the process from mixing (making sure that all the required ingredients are there in the right proportions), to transporting, placing, finishing, and curing.

Core-Cylinder Test and In Situ Tests

Core-cylinder tests are used to evaluate the strength of the concrete in an existing structure. The sample is obtained by coring the hardened concrete. The size of the sample is typically 2 in. in diameter and 4 in. high, and its compressive strength is determined in a manner similar to that for a normal cylinder test. Larger-diameter and longer cylinders may be cored; however, the ratio of height to diameter should preferably equal 2.

Nondestructive tests exist, that is, tests that do not require the removal of a sample. The most popular is the *rebound hammer test*. This test uses a calibrated spring-loaded device that shoots a rod against the concrete surface. A dial gage measures the rebound that is correlated to the concrete's modulus of elasticity, from which an estimate of the compressive strength can be made.

1.6 MECHANICAL BEHAVIOR OF CONCRETE

Concrete in Compression

Concrete is very strong in compression. In the United States, cylinder tests are used to study the behavior of concrete in compression. In other parts of the world, compression testing is typically done on 20-cm cube samples. Results obtained from the two different tests are different for samples made of the same concrete. This is due to the shape and proportions of the samples.

The deformation of the sample under load during testing may also be measured to establish the stress-strain diagram. Axial compression stress is defined as the force divided by the cross-sectional area $\left(f = \dfrac{P}{A} \right)$, which has units of psi (pounds per square inch) or ksi (kip per square inch). In SI (International System) units, the stress may be measured in KPa (kilopascal) or MPa (megapascal) units. Strain is the deformation of a unit length of the member and is defined as $\left(\epsilon = \dfrac{\Delta \ell}{\ell} \right)$, where $\Delta \ell$ is the change in the length, and ℓ is the original length. Strain is a dimensionless number, for example, inch/inch.

It must be emphasized that the cylinder test used to evaluate the strength of concrete is only a representative sample and provides only an indicative and correlative value of how the concrete may behave in the structure. The cylinder, when tested, is free to expand laterally. The concrete in a structure may be confined by its surrounding. A confined sample of concrete is much stronger. Note also that the cylinder test determines the strength of the concrete under *short-term* loading. Research indicates that under *long-term* loading (e.g., within building structures), the strength of the concrete is less than that exhibited by the cylinder testing. Different concrete mixes exhibit not only different strengths, but very different deformation characters. Figure 1–8 shows the stress-strain diagrams of typical concrete mixes.

A study of the stress-strain curves leads to important observations. At small strain levels there seems to be a straight-line relationship between strain and stress, that is, concrete

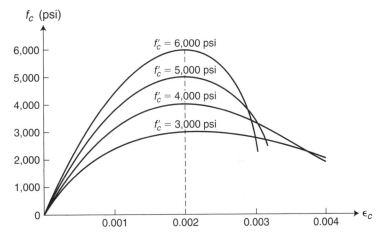

FIGURE 1–8   Stress-strain diagram of concrete in compression.

seems to follow Hooke's law (stress is linearly proportional to strain). This almost elastic relationship is valid up to about 30%–50% of the ultimate strength. The relationship then starts to deviate from this reasonably assumed straight line, that is, with increasing strain the stress grows, but at a slower rate. The peak stress level in typical concretes used in construction occurs near a strain value of 0.002. Then the stress in the cylinder starts to decrease with increasing strain until an *ultimate strain* value is reached, at which point the sample fails. The ultimate strain is different for different strength concrete mixes, as shown in Figure 1–8. In general, weaker concrete has greater ultimate strain. Note that in reinforced concrete design an ultimate useful strain of 0.003 is assumed. This will be further discussed in following chapters where we deal with the design of reinforced concrete members.

Most building projects utilize concrete with f'_c in the range of 3,000 to 6,000 psi. The last couple of decades have seen the industrial development and utilization of ultra-high-strength concretes. Concretes with 10,000 to 12,000 psi compressive strength are routinely available from many suppliers in larger cities, and even higher-strength concretes, some exceeding 20,000 psi, can be manufactured. Ultra-high-strength concretes are used mainly in columns of high-rise buildings.

The compressive strength of concrete, f'_c, varies with time. Figure 1–9 shows this variation. Well cured concrete gains most of its potential compressive strength within the first 28 days. After that, the strength gain proceeds at a much slower rate, although the hydration process between cement and water may continue in the presence of available free water.

Modulus of Elasticity of Concrete   In elastic materials (or materials that behave in an elastic way up to a certain stress level), a definite linear relationship exists between stress and strain. The coefficient in the relationship is called the *modulus of elasticity*. The capital letter E is used to denote this modulus, and the relationship is defined as *stress = modulus of elasticity × strain*.

The behavior of concrete, as described by the typical stress-strain curves in Figure 1–8, is not this simple. The diagrams are not linear; thus, the E value (i.e., the slope of the tangent to the curve at any point) is changing continuously. To simplify the matter and establish a value that can be used in calculations, a substitute E value is used. The E value

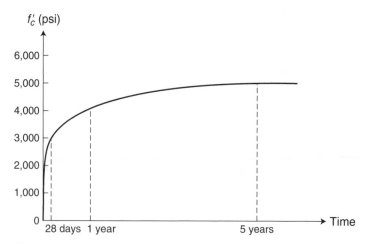

FIGURE 1–9   Compressive strength versus time for concrete.

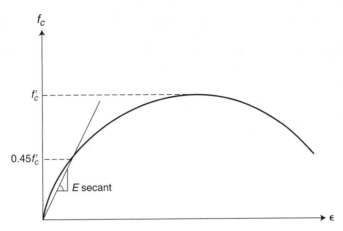

FIGURE 1–10   Concrete secant modulus.

used is the *secant modulus,* which is the slope of a line connecting the point of zero stress and zero strain to the stress point of $0.45 f_c'$ and its corresponding strain (Figure 1–10). By definition, this value is the modulus of elasticity of concrete.

This value is different for different strength concretes: Stronger concretes have greater E. Furthermore, concretes made with different aggregates (normal-weight concrete, lightweight structural concrete, etc.) also exhibit different moduli of elasticity.

The value for the modulus of elasticity is needed when calculating instantaneous (also called elastic) deformations of structures under load, such as the deflection of a beam. This is justified, for at stress levels that exist during normal use of structures, the concrete responds in a quasi-elastic manner to short-term loads. After studying the results of hundreds of tests and applying statistical analysis (fitting a mean curve to the values), researchers have determined that Equation 1–1 provides a reasonable approximation for the modulus of elasticity of concrete, E_c (ACI Section 8.5.1):

$$E_c = 33 \, w_c^{1.5} \sqrt{f_c'} \qquad (1\text{–}1)$$

where w_c = the weight of the concrete in pounds per cubic foot and f_c' = the ultimate cylinder strength, or specified compressive strength of concrete in pounds per square inch. The resulting unit for E_c is psi, and substitution into the equation must be made using the units as defined.

Unreinforced normal-weight concrete is about 145 pcf. When w_c = 145 pcf is substituted in Equation 1–1, the result, after rounding, is Equation 1–2 (ACI Code, Section 8.5.1):

$$E_c = 57{,}000 \sqrt{f_c'} \qquad (1\text{–}2)$$

As discussed above, there are two ways of determining the modulus of elasticity: (1) by testing, and (2) by using the approximate equation provided by the ACI Code. Because the concrete that will go into the structure has not been made, placed, and cured at the time of design, the designer is invariably forced to use the accepted approximate equation.

EXAMPLE 1–1

Find the modulus of elasticity of a concrete mix with the compression strength, $f_c' = 3,500$ psi. Assume the mix is lightweight structural concrete with a unit weight of 110 pcf.

Solution

The ACI approximate equation for E_c is:

$$E_c = 33w_c^{1.5}\sqrt{f_c'}$$

Substituting $w_c = 110$ pcf and $f_c' = 3,500$ psi:

$$E_c = 33(110)^{1.5}\sqrt{3,500}$$
$$E_c = 2,252,356 \text{ psi}$$

or

$$E_c = 2,252 \text{ ksi}$$

Concrete in Tension

The strength of concrete in tension is only about 8%–12% of its compressive strength, f_c', that is, it is a very weak material in tension. The ratio of tensile to compressive strength is greater in low-compressive-strength concrete than it is in high-compressive-strength concrete. In fact the tensile strength of concrete is completely disregarded when designing reinforced concrete structures in flexure (bending). It is somewhat cumbersome to make reliable concrete samples that could be tested in pure tension, so substitute tests are often used. One such test determines the tensile strength of the concrete in an unreinforced beam by testing it in flexure. Because the tensile strength of concrete is much less than its compressive strength, the beam will fail on the tension side of the cross-section. If we know the load, span, and cross-section of the beam, we can calculate the maximum moment on the beam and, consequently, the ultimate tensile stress at failure. This tensile stress value is called the *modulus of rupture*, or f_r.

From statistical analysis of data, an empirical formula (Equation 1–3) evolved and has been adopted by the ACI Code (ACI Equation 9–10):

$$f_r = 7.5\sqrt{f_c'} \tag{1–3}$$

In this equation f_c' and f_r are in psi units. (Substitution of f_c' must be made in psi, otherwise the formula will produce erroneous results). This formula is simple to use, but in most cases, it overestimates the true tensile strength of a concrete element.

The other test is the *splitting tensile strength test* (or *diametral compression test*). In this test a concrete cylinder lying on its side is tested diametrically in compression. The load generates tensile stresses perpendicular to the direction of the applied load. The load is increased until the cylinder splits. The tensile strength can be calculated from the maximum compression load that was applied. Tensile strength established by this method, split tensile strength (f_{ct}), is a more reliable value than the modulus of rupture.

For structural lightweight concretes, the formula for the modulus of rupture is modified. If the value of the split tensile strength (f_{ct}) is specified, the modulus of rupture for lightweight concrete is determined by Equation 1–3a (ACI Code, Section 9.5.2.3).

$$f_r = 1.12 f_{ct} \leq 7.5 \sqrt{f_c'} \qquad (1\text{–}3a)$$

If f_{ct} is not specified, the value obtained from Equation 1–3 must be multiplied by 0.75 for all-lightweight concrete and by 0.85 for sand-lightweight concrete.

EXAMPLE 1–2

A test was performed to determine the modulus of rupture of a concrete. A concrete beam 3 in. × 6 in. in cross-section and 9'-0" long was cast and supported at the ends on masonry blocks. The beam was loaded at the one-third points of the span with "concentrated" loads. The beam failed when it cracked on the bottom face at a load of 150 lb at each location (which led to an immediate collapse). The applied load and test setup is shown in Figure 1–11. The compressive strength of the concrete was determined as 4,000 psi via a cylinder test. The concrete weight was $w_c = 150$ pcf. Calculate the modulus of rupture of the concrete using (a) the results of the test, and (b) the ACI approximate equation.

Solution

a. *Test Results* The beam is subjected to two loads: its weight, and the two concentrated loads as shown in Figure 1–11. The beam self-weight is a uniformly distributed load, with a magnitude of

$$w = (150)\frac{(3)(6)}{(12)(12)} = 18.75 \text{ lb/ft}$$

Conversion factor for in^2 to ft^2

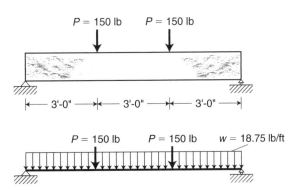

FIGURE 1–11 Beam loading for Example 1–2.

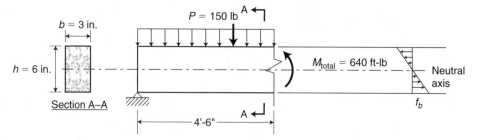

FIGURE 1–12   Internal forces and stresses for Example 1–2.

The maximum moment for the beam occurs at the midspan. The equations for the maximum moments are as follows:

$$M_{max} = \frac{w\ell^2}{8} \text{ (for the beam with uniform load)}$$

$$M_{max} = \frac{P\ell}{3} \text{ (for the beam with concentrated loads)}$$

$$M_{total} = \frac{w\ell^2}{8} + \frac{P\ell}{3}$$

$$M_{total} = \frac{18.75(9)^2}{8} + \frac{150(9)}{3}$$

$$M_{total} = 190 + 450 = 640 \text{ ft-lb}$$

M_{total} is the internal moment at the mid-span of the beam. This moment creates a set of compression stresses at the top, and tensile stresses at the bottom of the beam as shown in Figure 1–12. The maximum bending stress occurs at the top (compressive) and the bottom (tensile) of the cross section. The equation for the maximum bending stress is:

$$f_b = \frac{Mc}{I} = \frac{M}{S_m}$$

where c is the distance from the neutral axis (where stress is zero) to the top or bottom of the beam, I is the moment of inertia of the section about its neutral axis, and S_m is the elastic section modulus. For a rectangular shape, S_m is:

$$S_m = \frac{bh^2}{6}$$

Therefore,

$$S_m = \frac{(3)(6)^2}{6} = 18 \text{ in}^3$$

Conversion factor for feet to inches

$$f_r = f_b = \frac{M}{S_m} = \frac{(640)(12)}{18} = 427 \text{ psi}$$

b. *ACI Approximate Equation*

$$f_r = 7.5\sqrt{f_c'} = 7.5\sqrt{4{,}000}$$
$$f_r = 474 \text{ psi}$$

1.7 VOLUME CHANGES IN CONCRETE

Concrete is not an inert material, so its dimensions change in response to environmental influences. The most important ones are temperature change, concrete shrinkage, and concrete creep.

Temperature Change

Concrete, like most other materials, expands with rising temperature and contracts with falling temperature. Suppose a concrete element with the length ℓ is restrained at only one end (A) (see Figure 1–13a). Under an increase in temperature of ΔT (degrees of Fahrenheit), the element expands and has an increase in length equal to $\Delta\ell$. This increase in the length can be calculated using Equation 1–4.

$$\Delta\ell = \alpha\,\Delta T \ell \qquad (1\text{–}4)$$

where α is the coefficient of thermal expansion, which depends on the type of material. For normal weight concrete, α is about 5.5×10^{-6} to 6×10^{-6} in./in./°F.

The length change caused by thermal expansion/contraction in a concrete element can be calculated using Equation 1–4. For example, due to a 100°F temperature change, a 200-ft-long building will change its length by $\Delta\ell = \alpha\,\Delta T\ell = (6 \times 10^{-6})(100)$ $(200 \times 12) = 1.44$ in., a very significant length change indeed.

Now, if both ends of the concrete element (Figure 1–13b) are restrained, the length cannot grow at any increase in temperature, and the restraint causes longitudinal compression stresses.

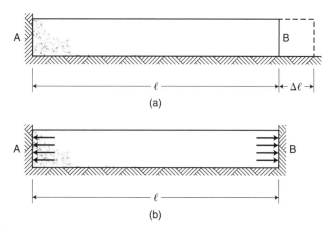

FIGURE 1–13 Effects of temperature on concrete: (a) free to move, (b) restrained.

Because $f = E_c\varepsilon$ and $\varepsilon = \Delta\ell/\ell$, Equation 1–5 can be used to calculate the change in length in terms of the stress f.

$$\Delta\ell = \varepsilon\ell = \frac{f}{E_c}\ell \tag{1–5}$$

Combining Equations 1–4 and 1–5 yields Equation 1–6.

$$\frac{f}{E_c}\ell = \alpha\,\Delta T\ell$$

$$f = E_c\alpha\,\Delta T \tag{1–6}$$

where E_c is the modulus of elasticity of the material in psi (or ksi), and f is the resulting stress in psi (or ksi) developed in the restrained element due to a change in temperature equal to ΔT.

As shown by Equation 1–6, large stresses can build up if the length change is restrained. The buckling of pavements often seen on hot days is the result of two neighboring pavement slabs pressing each other (in the absence of a wide enough expansion joint) while trying to expand. The buckling relieves the prevented expansion. On the other hand, tensile stresses will build up when concrete tries to shorten with dropping temperatures if the free contraction is somehow hindered. For example, for an $f'_c = 4{,}000$ psi concrete:

$$f = E_c\alpha\,\Delta T$$
$$f = (57{,}000\sqrt{4{,}000})(6 \times 10^{-6})(1.0) = 21.6\ \text{psi}$$

for each degree of temperature change ($\Delta T = 1°F$), if the length change is fully prevented. If the concrete in the above example has an ultimate tensile strength of $7.5\sqrt{4000} = 474$ psi, the theoretical value of the temperature drop that will crack this concrete is only $474/21.6 = 21.9°F$, a rather small temperature change.

Admittedly it is very rare that concrete is *fully* restrained against movement due to temperature change. But the unsightly cracking of concrete structures all around us provides ample testimony to the results of restrained volumetric changes.

The value of α for concrete is quite similar to that of steel (6.5×10^{-6} in./in./°F). Thus, the reinforcing steel inside the concrete will expand or contract at about the same rate as the surrounding concrete, without significant stresses resulting from expanding or contracting at a different rate. Aluminum, for example, has a coefficient of expansion roughly twice that of steel. Thus, the use of aluminum as reinforcement for concrete is not a good idea; for when the temperature rises, the aluminum rod expands at twice the rate of the surrounding concrete at the interface between the two materials. The conflicting expansion rates cause all kinds of "weird" stresses at the interface, breaking down the necessary bond between the two materials.

Concrete Shrinkage

Shrinkage means that the concrete becomes smaller in volume. There are many causes of shrinkage, but the most significant contributor to this phenomenon is the loss of water.

As previously discussed, more water is needed in a concrete mix than the cement uses for hydration. Some of this excess water bleeds and evaporates during and immediately following consolidation while the concrete is still plastic. The heavier parts in the still-fluid concrete tend to settle, causing what is known as *setting shrinkage* or *plastic shrinkage*. Reinforcing bars or large aggregates near the surface obstruct the uniform settlement of the concrete, thus enhancing the formation of thin hairline cracks on the surface. These hairline cracks look like cobwebs: lots of relatively short, thin cracks in all directions. Their depth is usually limited to small fractions of an inch.

After the concrete hardens, it still contains free water in the capillaries and water adsorbed on the surface of particles. As this water slowly evaporates, the concrete continues to shrink, not unlike a sponge shrinks as it dries. This causes what is known as *drying shrinkage*. The rate of the drying shrinkage is tied to the speed of the evaporation, which in turn depends on the porosity of the concrete and the environment, that is, temperature and relative humidity. Concrete in highly humid climates shrinks less than corresponding concrete does in arid climates.

More than 90% of the drying shrinkage happens within the first few weeks after casting. Drying shrinkage, however, is partially reversible. Thus, if the concrete gets soaked it swells, and when it dries out again it shrinks. If the drying shrinkage could take place without any restraint whatsoever (a theoretical proposition rather than what really occurs), no stress buildup would result. Because, however, free shrinkage is usually restrained (i.e., something prevents the concrete from shortening in any direction), tensile stresses start to develop and build up. In moderate climates the average dimensional change is about 300 millionths (300×10^{-6} in./in.). Compared to the length change due to a decrease in temperature, the effect of the average shrinkage value is similar to that of a 50°F temperature drop. If the developing tensile stress is greater than the tensile strength of the concrete at any point, the concrete will crack. The crack should be thought of as a relief from tension caused by the prevention of free movement.

An example is a long wall that has been cast on top of its footing. The footing has already cured and hardened. When the wall tries to shrink, the footing restrains its bottom edge from moving. The top of the wall, however, is free to shrink lengthwise. Thus, a tug of war results between the top and the bottom of the wall, resulting in one or more cracks with diminishing width from top to bottom (Figure 1–14a).

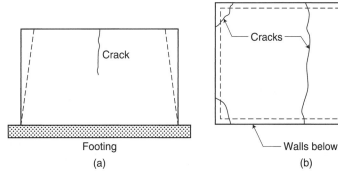

Crack

Footing

(a)

Cracks

Walls below

(b)

FIGURE 1–14 Shrinkage cracks.

Another example is a floor that is cast over walls placed earlier, thus hardened. As the floor shrinks and tries to change its long dimensions, it cannot because the walls restrain it. The buildup of tensile stresses results in cracks, especially in the corners, where the edges of the slab try to move in two different directions. The relief comes as diagonal cracks in the corners (Figure 1–14b).

Plastic shrinkage cracks are characterized as random surface cracks, that is, they do not penetrate the full thickness of the concrete element. Drying shrinkage cracks, on the other hand, are usually full depth and quite wide ($^1/_8$ in. or more is not unusual).

The mitigation of the effects of shrinkage requires good design and construction practices. The following actions help to minimize cracking in slabs and walls due to shrinkage:

- *Use the minimum amount of water in the concrete mix.* The concrete should have not only the smallest w/cm ratio, but also the smallest amount of water in absolute terms. This also means using the smallest amount of cement necessary to achieve the desired concrete strength, because more cement introduces more excess water in the mix. Such a tactic is also good for keeping costs down.
- *Use good curing technique.* Moist curing helps keep the excess water from evaporating too soon (i.e., before the concrete has a chance to develop its tensile strength).
- *Limit the size of the pour to about 60 to 80 ft maximum length in any direction.* The construction is broken up into segments by the use of *construction joints* (Figure 1–15a). If the second pour is three to four days after the first pour, some shrinkage has already taken place in the first pour. On some projects the pouring sequence may follow a checkerboard pattern. Other construction techniques leave a gap between two neighboring pours (12 to 24 in.) that is filled in when the larger pours have undergone most of their shrinkage.
- *Provide reinforcing steel (shrinkage reinforcement).* Because steel bars are bonded to the concrete, they restrain and limit the change of length of the concrete.
- *Use shrinkage compensating cement (Type K).* This particular cement type expands during the early stages of hydration, before any drying shrinkage occurs due to

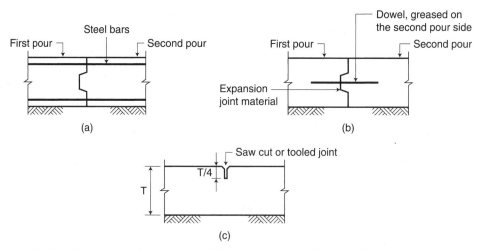

FIGURE 1–15 (a) Construction joint, (b) expansion joint, (c) control joint.

moisture loss. Reinforcing is also provided in both directions in a wall or a slab, and the expansion of the concrete at the early stage of hardening induces tension in the steel (i.e., the expanding concrete tries to elongate the reinforcing bars). If we recall Newton's law on action and opposite and equal reaction, it is easy to understand that the steel in turn will compress the concrete. When shrinkage sets in and causes tension in the concrete, it first will have to overcome the precompression in the concrete. Thus, the forces will either completely cancel each other out, or at least the resulting tensile stresses will be greatly reduced.

- *Provide **expansion/contraction** or **control joints** (Figures 1–15b and 1–15c, respectively).* At an *expansion joint* the longitudinal reinforcing is interrupted. The joint is filled with an elastomeric material that can be compressed when the concrete expands, and permits the free moving of the two parts relative to each other when the concrete shrinks. A key-way (or a dowel that is greased on one side of the joint to prevent bonding) forces the two parts to stay together in the out-of-plane sense while still allowing them to move freely longitudinally. This prevents one side from moving higher or lower than the other and thus creating a trip hazard or a step. The role of the *control joint* is different. A weakening groove, usually $1/8$ in. wide and about one-fourth to one-fifth of the slab thickness, is either tooled into the freshly finished concrete, or cut with a saw into the concrete as soon as it hardens enough so as not to leave an imprint on the surface. This allows the shrinking concrete to crack along that straight line where the section is weakened. Control joints essentially locate shrinkage cracks along predetermined paths instead of letting them naturally meander all over the slab or wall.

Creep of Concrete

A structure deforms when it is subjected to loads. For example, beams and slabs deflect, columns become shorter, and so on. For every stress level, there is a corresponding strain. Strain is nothing else than the deformation of a unit length of the material.

Concrete structural elements experience two types of deformations under loads: (1) instantaneous or elastic deformation, and (2) long-term deformation, or *creep*.

Instantaneous deformations occur as soon as the member is subjected to load. This is similar to what happens in other construction materials such as steel.

Creep, on the other hand, is the gradual long-term deformation of concrete under a sustained load. Nearly 75% of the total creep happens during the first year, and the total creep can be two to three times the instantaneous deformation (Figure 1–16).

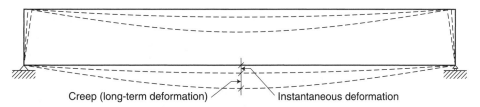

Creep (long-term deformation)   Instantaneous deformation

FIGURE 1–16   Instantaneous and long-term deformation in a concrete beam.

The causes of creep are complex. Interestingly, one contributor is the loss of adsorbed water. In drying shrinkage the loss occurs due to the lower relative humidity of the ambient atmosphere, and this loss leads to the shrinkage. In the case of creep, the sustained compression on the concrete squeezes some of the moisture out of the concrete. This in turn lets the solids consolidate even more.

The second major cause of creep is thought to be microfracturing in the hardened paste around sharp edges of aggregates under the effect of compression.

Creep deformations can be very significant. They are caused mainly by dead loads or sustained loads, because the self weight and some permanently attached superimposed dead loads are dominant in most concrete structures, whereas the transitory (live) loads are less significant. Creep could account for an additional 100%–300% of the instantaneous deformations. Thus, a beam's original deflection of 1 in. may grow to anywhere between 2 in. to 4 in. If this additional deformation is not accounted for in the detailing of attached items, such as partitions, it may cause serious distress in them.

1.8   REINFORCING STEEL

Reinforcing steel is used to overcome the weakness of concrete in tension. The role of the reinforcing is to resist the tension in structures. Thus, a hybrid structural composite called *reinforced concrete* is created, where each material does the work it is well suited for. Concrete takes care of the compression, while the steel takes care of the tension.

Behavior of Steel Under Stress

To better understand the material that will be discussed in the following chapters, we review the behavior of steel under stress. This review will also help to familiarize the reader with the terminology that will be used later.

Consider Figure 1–17, which is a typical stress-strain diagram for steel in tension. There are four distinct zones in the stress-strain diagram for steel. First is the *elastic*

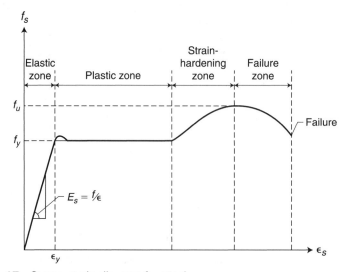

FIGURE 1–17   Stress-strain diagram for steel.

zone, where steel under stress will go back to its original length if it is released. In this zone the stress in the material is linearly proportional to the strain. (Robert Hooke formulated this relationship, so we refer to it as *Hooke's law.* Hooke worked with Christopher Wren on the construction of St. Paul's Cathedral in London, England.) When steel is pulled beyond the elastic zone (elastic or proportional limit) it *yields. Yielding* is an elongation of the steel with no appreciable change in stress. The onset of yielding (elastic or proportional limit) is the beginning of the *plastic zone.* When the steel is pulled beyond the proportional limit it will not return to its original length, but remain permanently deformed.

The stress at which steel yields is called *yield stress* and it is noted as f_y. The corresponding strain is called *yield strain* or ε_y. By the time yielding ends, the corresponding strain is about 8 to 10 times the strain at the proportional limit. After yielding, the steel's stress/strain curve starts to "climb again" in a curvilinear mode until it reaches a plateau called the *ultimate strength* (f_u). This curvilinear zone of the stress/strain curve is called *strain hardening*.

Another important mechanical property of steel is its *modulus of elasticity* (E_s). For steel the modulus of elasticity corresponds to the slope of the stress-strain diagram in the elastic zone (see Figure 1–17). E_s is about 29,000 ksi.

Three forms of reinforcements are commonly used in concrete structures: (1) *steel bars,* (2) *welded wire reinforcements* (WWR), and (3) *prestressing steel.* (A fourth is short steel, glass, or plastic fibers mixed into the fresh concrete. Bars manufactured from advanced composite materials, such as fiberglass and carbon fibers, are also used in special cases. Discussion of these reinforcing methods, however, falls beyond the scope of this text.)

Steel Bars   Modern reinforcing bars are round, rolled sections. In the past, square bars were also used and may be encountered in old buildings built before World War I or shortly thereafter.

Round reinforcing steel comes in two different variations: *deformed* and *plain.*

Deformed bars have a pattern of ribs, or *deformation,* rolled on them. These deformations provide better relative slip resistance between the steel bar and concrete. In addition to the chemical bond that exists between the cement paste and the steel surface, these ridges provide a mechanical anchorage as well. Figure 1–18 shows a few examples of deformed bars. The ACI Code mandates the use of deformed bars in all new reinforced concrete structures.

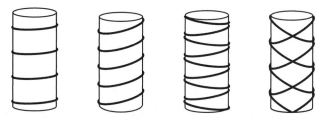

FIGURE 1–18   Examples of deformed bars.

Plain bars do not have any deformations and rely on surface bonding only to prevent relative slippage. These are no longer in use, although they may be encountered in old structures.

Different grades of steel are used, typically made from either new steel, scrap metal, or a mixture of both. *Grades* of steel represent their guaranteed minimum yield stress in ksi units. For example, Grade 60 steel refers to reinforcing steel with a guaranteed minimum yield stress of $f_y = 60$ ksi. Table A1–1 in Appendix A lists the different types of steel used as reinforcing bars along with their mechanical properties. Each type of steel in Table A1–1 has an ASTM designation such as A615. Different types of steel in Table A1–1 are:

- *Billet steel (A615)*—This is new steel and the most common type.
- *Rail steel (A996)*—This is made of recycled railroad track.
- *Axle steel (A996)*—Similar to rail steel, but made from axle as scrap metal. Axle and rail steels are not as readily available as billet steel.
- *Low-alloy steel (A706)*—This type of steel provides enhanced weldability.

Of all these reinforcing steels, most construction uses A615 Grade 60 (billet) steel. Grade 75 is sometimes used in columns. Grade 50 has not been around for quite a while. Grade 40 is almost never used, for it has only two-thirds of the strength of Grade 60 steel, and its cost in place per lb is the same.

Bar Sizes Steel bars are made in different sizes. Bar size, in general, represents the diameter of the steel bar in inches. From #3 to #8 (#1 or #2 bars do not exist), each number represents the diameter of a bar in fractions of $^1/_8$ in. For example, #3 bar means that the diameter of the bar is $^3/_8$ in., and #8 is $^8/_8$ in. = 1 in. diameter.

The heavier (larger-diameter) #9, #10, and #11 bars do not precisely follow the $^1/_8$ in. rule, but they are close.

In addition, there are #14 and #18 bars, which are very large, heavy bars. They are used mostly in large columns in high-rise construction and are available on special order.

Table A1–2 includes the diameters and areas of the available steel bars.

Identification of Steel Bars Steel bars used in concrete construction have special identification marks rolled on them. These marks provide information such as where the bars were produced, the bar size, type of steel, and their grade (see Figure 1–19).

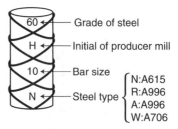

FIGURE 1–19 Identification marks for steel reinforcement.

Epoxy-Coated Bars The highly alkaline environment that the concrete provides for the embedded reinforcing usually protects it from corrosion. Some structures, however, such as bridge decks, parking structures, coastal structures, and so on are often exposed to moisture containing chloride salts. The deicing salts (road salts) that are used on roads and bridges, and carried into parking structures by the automobiles, contain large amounts of soluble chlorides. When such solutions get into contact with the reinforcing, the result is corrosion (oxidization or rusting) of the steel.

The rust (ferrous oxide) grows to about eight to tenfold the volume of the original steel. Thus, as the rust tries to create "elbow room" for itself, the internal pressure starts to crack, split, and spall the concrete around it. This in turn provides more access to the dangerous chloride-laden moisture.

One way to protect reinforcing in this kind of environment is the use of *epoxy-coated bars*. Epoxy resin is an excellent adhesive and protects the steel from chloride attacks. A note of caution is in order, however: Such bars must be handled carefully to prevent nicks or cracks in the coating. Such places are especially attractive to chloride ions and often become nodes of violent and rapid corrosion in the reinforcement.

Welded Wire Reinforcements (WWR) In certain situations it is more economical to use *welded wire reinforcements* (WWR) in lieu of a series of small-diameter bars. WWR are thin wires spaced at certain distances in two orthogonal directions and fabricated in either large sheets, or in long rolls in the case of light-gage wires. They are welded together at intersection points, usually by the electric resistance welding method. The chief advantage of using WWR is the labor saving. Individual reinforcing bars are placed one by one and are secured by tying them together at every intersection. This ensures that they will remain at the desired location throughout the concrete placement, consolidation, and finishing process.

WWR are available in commonly standardized wire sizes and spacing. Table 1–1 lists some of the commonly used styles of WWR. The standard designation of the reinforcement represents the spacing and the wire sizes. In the modern designation system the W-number represents the approximate cross-sectional area of the wire in multiples of 0.01 in^2. Thus, the cross-sectional area of a W4.0 wire is about 0.04 in^2. As an example, $6 \times 12 - W4.0 \times W2.1$ represents wires with cross-sectional areas of 0.04 in^2 and 0.021 in^2 in a rectangular grid of 6 in. $\times$ 12 in. as shown in Figure 1–20.

TABLE 1–1 Some Commonly Stocked Styles of Welded Wire Reinforcements

Reinforcement Designation	Steel Area (in²/ft)	
	Longitudinal	Transverse
6×6 − W1.4 × W1.4	0.029	0.029
4×12 − W2.1 × W0.9	0.062	0.009
6×6 − W2.1 × W2.1	0.041	0.041
4×4 − W1.4 × W1.4	0.043	0.043
6×6 − W2.9 × W2.9	0.058	0.058
6×6 − W4.0 × W4.0	0.080	0.080
4×4 − W4.0 × W4.0	0.120	0.120

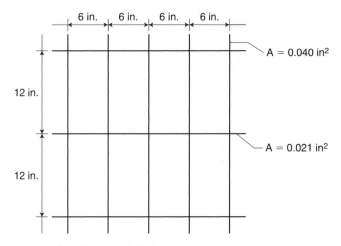

FIGURE 1–20 $6 \times 12 - W4.0 \times W2.1$.

Prestressing Steel   When we discussed the mechanical properties of concrete, we noted that the tensile strength of concrete is small. Early users of reinforced concrete soon realized that if compressive stresses were induced into regions where the loads (dead and live) caused tension, this tension would have to overcome the pre-existing compressive stresses before inducing tensile stresses that could result in cracking or failure. Hence the concept of prestressing was developed.

Two different techniques are used to achieve prestressing. One is known as *pretensioning,* the other is *posttensioning.* (These will be discussed in detail in Chapter 8.)

A much stronger steel product than ordinary reinforcing steels is needed for prestressing purposes. Most of the time, seven-wire strands are used (six wires wrapped around a core wire in a helical form; see Figure 1–21). The wires are cold-drawn (i.e., the wires are pulled through a series of smaller and smaller round openings without any preheating). The cold working increases the toughness and the strength of the steel. Because the wires are stretched way beyond yield during manufacturing, the strands manufactured from them have no yield levels comparable to those of ordinary reinforcing bars. The most commonly used prestressing strands have a nominal ultimate strength of 270 ksi.

Some special applications use either smooth or deformed prestressing bars of varying diameters from $3/4$ in. to $2^{1}/_{2}$ in. These are available with ultimate strengths of up to 160 ksi.

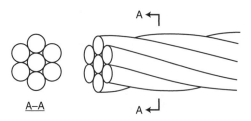

FIGURE 1–21   Seven-wire strand for prestressing.

PROBLEMS

1–1   What is hydration in concrete?

1–2   What is the significance of compression strength of concrete, and how is it measured?

1–3   What are the applications of air-entraining admixtures?

1–4   What is the modulus of elasticity of concrete (E_c) and how is it determined?

1–5   Define the modulus of rupture, f_r, for concrete.

1–6   What are the differences between deformed bars and welded wire reinforcements?

1–7   Draw the bending moment and shear force diagrams for a 12 in. $\times$ 24 in. concrete beam made of lightweight concrete with the unit weight of 110 pcf, subjected to a uniformly distributed load of 1.0 kip/ft. Assume the beam is simply supported and has a 10'-0" span.

1–8   Determine the modulus of elasticity, E_c, and the modulus of rupture, f_r, for a normal-weight concrete (w_c = 145 pcf) with a specified compressive strength, f'_c, of 3,500 psi.

1–9   Determine the maximum concentrated load that can be applied at the center of a 6 in. $\times$ 6 in. simply-supported plain concrete beam before it cracks in tension. The beam has a 6'-0" span and is constructed of sand-lightweight concrete with a unit weight of 120 pcf. The specified compressive strength is 3,000 psi. Use the ACI Code recommended value for the modulus of rupture.

1–10   Determine the maximum span for an 8 in. $\times$ 12 in. simply-supported plain concrete beam constructed of normal-weight concrete and loaded by a uniformly distributed load of 2 kip/ft just before it fails. The specified compressive strength of the concrete is 4,000 psi. Use the ACI Code–recommended value for the modulus of rupture.

SELF-EXPERIMENTS

In the self-experiments of this chapter, you learn about the different aspects of making concrete by using simple tools.

Experiment 1 (Making a Concrete Sample)

The following materials are needed:

1. Three 20-oz empty tin cans (cylinder shape)
2. Three large bowls
3. Cement (can be obtained from a local hardware store)
4. Sand and gravel
5. Tap water
6. A $^3/_8$ in. or $^1/_2$ in. diameter wood dowel, about 12 in. to 15 in. long

    Make three samples: (1) cement sample, (2) concrete with a w/cm ratio of 1.0, and (3) concrete with a w/cm ratio of 0.5.

(1) Cement Sample

Pour 10 oz of cement with 6 oz of water into a bowl and mix them thoroughly. Note how much effort is used to mix the cement with water. Then place the mix in can number 1. Consolidate the mix in the can by prodding it with the dowel about 12–15 times.

(2) w/cm = 1.0 Concrete Sample

Pour 3 oz of cement, 10 oz of sand, 10 oz of gravel, and 3 oz of water into a bowl, and mix them thoroughly. Again note how much effort is needed to mix the materials. Then place the concrete in can number 2. Consolidate the mix in the can by prodding it with the dowel about 12–15 times.

(3) w/cm = 0.5 Concrete

Pour 4 oz of cement, 10 oz of sand, 10 oz of gravel, and 2 oz of water into a bowl, and mix them thoroughly. As in the first two cases, pay attention to the amount of effort needed to make the mix. Then place the concrete in can number 3. Consolidate the mix in the can by prodding it with the dowel about 12–15 times.

Leave the three samples for approximately 6 hours at room temperature. Check them every 6 hours for 3 days. Record any observations. Answer the following questions:

- Which mix was easiest to make (i.e., which one was most workable)?
- Which mix resulted in the most bleeding?
- Was any sign of hydration observed?

At the end of the 3 days cut the tin cans to completely expose the samples. Answer the following questions:

- What are the differences in the textures of the three samples?
- Which sample has the most uniformity of material?

2

RECTANGULAR BEAMS AND ONE-WAY SLABS

2.1 INTRODUCTION

This chapter covers the analysis (checking the strength) and the design (sizing the concrete and steel) of reinforced concrete beams and slabs that span primarily one way.

The previous chapter emphasized that concrete is very weak in tension, but strong in compression. As a result, reinforcements are used to supply tensile strength in concrete members (most commonly in the form of round reinforcing bars or *rebars*). Like any other building system, reinforced concrete structures have advantages and disadvantages.

2.2 ADVANTAGES OF REINFORCED CONCRETE

1. *Can be cast into any shape* This is the main advantage of reinforced concrete compared to other building materials. Concrete members can be made into any desired shape by using forms.

2. *Has great resistance to fire and water* Concrete loses its structural integrity much more slowly than wood or steel when subjected to high temperature. In fact, concrete is often used as fireproofing material. Concrete also better resists exposure to water, does not corrode like steel, and does not lose strength as wood does. Certain chemicals in water, however, can harm concrete.

3. *Is a low-maintenance material* Concrete does not corrode, so it does not need to be painted and regularly maintained when exposed in the environment.

4. *Has very long service life* Reinforced concrete structures that are well designed and built last a very long time.

2.3 DISADVANTAGES OF REINFORCED CONCRETE

1. *Has very low tensile strength* Concrete has a very low tensile strength in comparison to its compressive strength. Consequently, reinforcing steel bars are needed to counteract the development of tensions in concrete structures.

2. *Requires shoring and forms* This is a major disadvantage of concrete because it raises the cost of concrete structures, especially in countries such as the United States where labor costs are high. Shoring and formwork often constitute more than half the total cost of the structure.

3. *Has variations in properties* The mechanical and physical properties of concrete are sensitive and require careful proportioning, mixing, curing, and so on. Eliminating large variation in these properties demands carefully monitored procedures.

4. *Results in heavy structural members* Reinforced concrete structures are heavier than similar steel or wood structures. This results in larger building dead loads, which in turn result in larger foundations. Concrete structures are also more sensitive to differential settlements. Thus, concrete structures require relatively good soil conditions.

2.4 ON THE NATURE OF THE DESIGN PROCESS

Before attending to the main topic of this chapter, which is the analysis and design of bending members, a discussion on the concept of *design* is appropriate.

Ask ten people about the meaning of the word "design" and you probably will get ten different answers. Design also has very different meanings to architects and to engineers. And to top it all off, design is often viewed as synonymous with sizing of members. So we hope that readers will forgive the rather loose usage of the term *design.*

Structural design of reinforced concrete structures is an iterative process. It begins with the layout of the structure or, in other words, with the selection of the structural system. Any practitioner will admit that this initial step is by far the hardest part of the process. It requires the designer to come up with a synthesized whole for the building, laying out all the component elements (columns, girders, beams (or joists), and slabs). Furthermore, the designer must also estimate the sizes of the elements within the space in order to go to the next step, that is, to *analysis.*

The flowchart of Figure 2–1 presents a somewhat simplified picture of the process. Oddly enough, it begins with a step in synthesis, or the conception of the structure. This step is nonmathematical, for the aim of the study at this point is to look at what the building structure should do. What spaces are required? What is the minimum column spacing required to fit the architectural program?

But before we reach the part designated as "Analysis" or "Design," we must complete another exercise: identifying the loads that the structure may be subjected to in its life span.

Loads generally fall into two major categories: gravity loads and lateral loads. Gravity loads are further divided into two major groups: *dead loads* and *live loads.* One can only guess how this nomenclature came into usage. Perhaps people originally identified loads

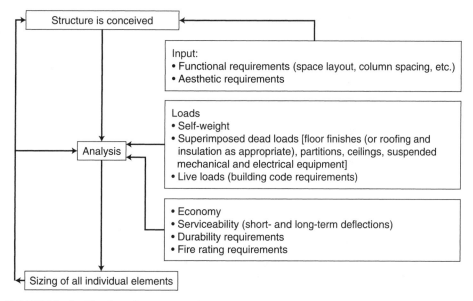

FIGURE 2–1 The iterative nature of structural design.

that were stationary as "dead," and loads that moved as "live." Today, we make a somewhat different distinction between these two loads. Dead loads are those that remain permanently attached to the structure, while other loads that are transitory in nature are referred to as live loads. Thus, furniture and stored items as well as loads from people's activities are in the latter category. For example, most of the weight in a library's stack area is from the stored books with only a very small part of the floor loads coming from the visitors; nevertheless, the stacks and the books are considered live loads. In addition, environmental effects such as moisture or temperature changes may create stresses in the structure, so they also may be loosely defined as loads that the structure must safely withstand.

Before any meaningful analysis can be performed to calculate and appropriately size any component element within a structure, designers must establish the loads that such an element can safely support, or at least must reasonably approximate them.

In a concrete structure, the *self-weight* is a very significant part of the dead loads. Because self-weight depends on the size of the particular member, a reasonable estimate must be made on the size. After the designer estimates the size, he or she can calculate the loads from the self-weight, assuming that reinforced concrete weighs about 150 lb/ft^3. At this point we do not want to tax the student's attention with detailed discussion on the selection of an appropriately sized beam or slab, and all of the reasons thereof. This subject will be discussed later in this chapter. In any case, if during the design process the designer determines that an initial estimate of the member's size, and thus the self-weight, was significantly in error, he or she has to re-analyze the member, taking into account the newly adjusted size; thus, the *iterative* nature of the design and sizing.

Superimposed dead loads (SDL) are somewhat ambiguous. Often these items and their precise location in space are not completely known at this stage of the design (see Figure 2–1). Partition layouts have not been decided yet, or may change in the future. Ductwork, piping, and light fixtures may go anywhere. So the designer is forced to make a blanket estimate on these. Most practitioners estimate that the combination of these items will exert about 15 to 20 lb/ft^2 of floor area. (The only areas that need more careful attention are those where some special flooring, such as stone or terrazzo, is planned. These items exert about 12 to 13 lb/ft^2/in. thickness. Thus, a 2 in. terrazzo flooring weighs about 25 psf.)

Live loads (LL) are prescribed by building codes for the particular usage of a space. These loads are listed as uniformly distributed minimum loads and represent the current professional wisdom. Because live loads are not uniformly distributed except in very isolated cases, they have very little, if anything, to do with the real loads that may occur on structures. Actual surveys show that total loads, uniformly averaged out over the whole floor area, amount to only about 15%–20% of the codes' mandated minimums in spaces like hotels, residential buildings, and offices. These minimums, however, represent a statistical probability of the loads that the structure may experience in a projected lifetime of 50 or 100 years. Furthermore, these code-prescribed live loads also try to account for the dynamic nature of many loads by treating them as equivalent static loads.

This discussion of loads should suffice to show that any calculation made during the load analysis phase will contain unavoidable inaccuracies and uncertainties. These errors are inevitable no matter how carefully the designer tries to evaluate the currently envisioned, but essentially future loads.

EXAMPLE 2–1

In this simple floor plan, beams 12 in. wide and 20 in. deep are spanning 30 ft. The beams are located 9′-0″ center to center. A 5-in. thick slab spans from beam to beam. (See Figure 2–2.) The floor structure will be used in a general office building, thus (per Code) the minimum uniformly distributed live load is 50 lb/ft^2. Calculate the dead and live loads that one interior beam has to carry. Assume 20 psf for the superimposed dead load for the partitions, mechanical and electrical systems, and so on.

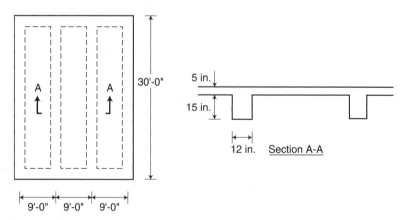

FIGURE 2–2   Floor plan and section.

Solution

The beams are 9 ft apart, so each beam is assumed to be responsible for the loads that occur 4.5 ft from either side of the beam's centerline. Thus, each linear foot of beam will support loads from 9 ft^2 of floor in addition to the weight of the stem.

Loads from the slab:

5 in. slab self-weight ($^5/_{12}$) $\times$ 150	62.5 psf
Superimposed dead loads, estimated	20.0 psf
Total dead load on slab	82.5 psf
Dead loads on beam from slab: 9 ft $\times$ 82.5 =	742.5 lb/ft
Volume of stem per foot: (12 $\times$ 15)/144 $\times$ 1 ft =	
1.25 ft^3/ft of beam	
Weight of stem: 1.25 $\times$ 150 =	187.5 lb/ft
TOTAL DEAD LOADS: w_D =	930 lb/ft

In addition, the beam will support live loads from 9 ft^2 of floor area on each linear foot of beam. Thus:

TOTAL LIVE LOADS: w_L = 9 $\times$ 50 psf = 450 lb/ft

Summary: See Figure 2–3.

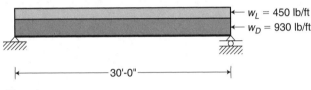

w_L = 450 lb/ft
w_D = 930 lb/ft

|———————————— 30'-0" ————————————|

FIGURE 2–3   Floor beam.

2.5 LIVE LOAD REDUCTION FACTORS

We complete this discussion of loads by dealing with the concept of *live load reduction factors*. These are derived from statistical analyses of the probability of having the maximum amount of live loads everywhere on a floor of a building. Studies indicate that the larger the floor area that contributes loads to a particular member, the less likely it is that every square foot of that area will bear the maximum amount of live loads.

Different codes deal with this concept somewhat differently. Some codes relate the live load reduction to the *tributary area* (A_T), or the area directly loading the particular element under investigation. Other codes relate the live load reduction to the so-called *influence area* (A_I), the area in which a part, however small, of any load may contribute to the loading of a particular element under investigation. In other words, the influence area for a structural member is the part of the building structure that may fail if that member is removed.

As an example consider Figure 2–4, which shows the floor framing plan for a reinforced concrete building. To determine the influence area for beam B-1, assume that this beam is removed. This will cause the slabs supported by B-1 to fail. As a result, the influence area for B-1 is $(A_I)_{B-1}$, the area between column lines 1, 2, A, and B. Following this logic, if we remove girder G-1, the beams it supports will fail, and consequently the slabs supported by the beams. Thus, the area between column lines 1, 2, B, and D $(A_I)_{G-1}$ will collapse. A similar study will show that the influence area for column C-1 is the area between column lines 1, 3, D, and F.

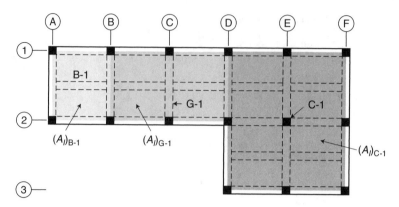

FIGURE 2–4   Influence areas for different structural members.

One variation of the live load reduction formula is given in Equation 2–1:

$$L_{\text{red}} = L_0\left(0.25 + \frac{15}{\sqrt{A_I}}\right)$$
(2–1)

where
  L_{red} = the reduced design live load per square foot of area supported by the member
  L_0 = the unreduced design live load per square foot of area supported by the member
  A_I = the influence area of the member in square feet

Equation 2–1 is applicable whenever $A_I > 400$ ft². The usage of live load reduction is limited in that the reduction cannot exceed 50% ($L_{\text{red}} \geq 0.5L_0$) for members supporting one floor and cannot exceed 60% ($L_{\text{red}} \geq 0.4L_0$) for members supporting two or more floors. Live load reductions do not apply for live loads in excess of 100 psf, except for members supporting two or more floors, in which case the live load can only be reduced up to 20%.

EXAMPLE 2–2

For the interior beam of Example 2–1, determine the reduced live loads.

Solution

The influence area, A_I, for the beam is:

$$A_I = 2 \times 9 \times 30 = 540 \text{ ft}^2$$

Because this area is larger than 400 ft², a reduced live load may be used in the design of the beam. The reduced design live load is:

$$L = 50\left[0.25 + \frac{15}{\sqrt{540}}\right] = 50 \times 0.895 = 44.8 \text{ psf}$$

Thus, the reduced design live load on this beam is:

$$w_L = 44.8 \times 9 = 403 \text{ lb/ft}$$

rather than the previously calculated load of 450 lb/ft.

2.6 CONTINUITY IN REINFORCED CONCRETE CONSTRUCTION

Many readers may have encountered only statically determined structural elements. These are simply supported beams (with or without cantilevers at their ends), cantilevers fixed at one end and free to move at the other, simple posts, and so on. These elements are all characterized by needing only the equations representing static equilibrium ($\Sigma H = 0, \Sigma V = 0, \Sigma M = 0$) to solve for the reactions.

A review of what "reactions" means may be needed here. A building element does not exist in a stand-alone vacuum. It is connected to other elements. At a point of connection the free relative displacement between the element under study and the rest of the structure is denied. This denial of free movement results in the transmission of a force (or moment) at the connection between the supporting and the supported elements. Look at Figure 2–5a for example. Here a beam end is supported on a wall. Elsewhere within the span the beam is free to deflect, or move vertically. But this ability to displace vertically is denied at the place of the support.

Figures 2–5b and 2–5c show the symbols of a *hinge* type of support and a *roller*. In the hinge support, the two relative displacement components (vertical and horizontal) are denied between the beam (the member under investigation) and the support below it. Thus, vertical and horizontal forces could be transmitted at the point between the beam and the support. (The forces coming from the support to the supported member are called *reaction forces*.) At a roller support (Figure 2–5c) only relative vertical displacement is denied; the beam could still freely roll horizontally without resistance. Correspondingly only a vertical force could be transmitted between the beam and the support. Figure 2–5d shows a beam end built into a large mass. The beam end cannot move horizontally or vertically, and it cannot rotate with respect to the mass. This condition is called *fixity*. The usual symbol of fixity is shown in Figure 2–5e. In this condition, horizontal force, vertical force, and a moment may be transmitted between the member and the support at that location.

All of these support conditions are quite familiar to students who have had a first course in structures. These support conditions represent what may be called *absolute conditions*: The displacement (vertical, horizontal, or rotational) is either freely available, or

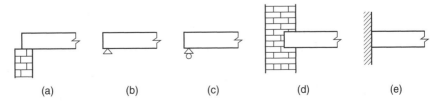

(a)   (b)   (c)   (d)   (e)

FIGURE 2–5   The meaning of the different support conditions.

FIGURE 2–6   Joist before and after deformation.

completely denied. As will be pointed out later, there is an infinite number of conditions in between, especially as related to rotations. Consider, for example, a flexible joist supported by a wall or beam at its ends (Figure 2–6). The mere supporting certainly precludes vertical displacement of the joist, thus a force transfer occurs. An action force is transmitted from the joist to the wall or beam, and an equal but opposite reaction force is transmitted from the supporting element to the joist. As the joist deflects under load, its supported ends can rotate freely; thus, the moments at the ends are zero.

Reinforced concrete construction is monolithic, which means that members are intimately built together with neighboring members. Slabs are continuous over supporting beams and girders; beams and girders are continuous over supporting interior columns, and so on.

Figure 2–7 illustrates the point. The slab in the beam and slab structure is continuous in both horizontal directions over the beams. The beams are continuous over other beams or columns.

A simple problem is presented here to clarify the concept. Admittedly, this problem does not occur in reinforced concrete structures, but it serves to illustrate the concept. A continuous structural member is represented by an imaginary center line (see Figure 2–8). On this two-span beam, Span 2 is larger than Span 1. If the loads are about the same, Span 2 will deflect more. Consequently this deflection will try to force Span 1 to curve upward slightly near the center support to follow Span 2. (The tangent to the deformation curve will rotate toward Span 2.) Study of the deformation curve shows that the beam bends into an upward curvature, that is, tension develops at the top of the beam, between the two points of inflection (where the moment in the beam is zero), whereas elsewhere the beam bends downward, resulting in tensions at the bottom. The moment diagram is shown below the

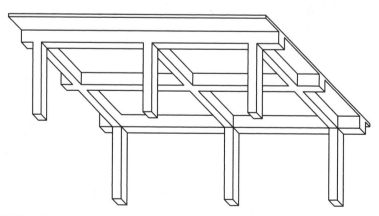

FIGURE 2–7   Beam and slab floor framing.

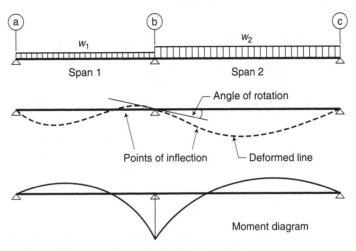

FIGURE 2–8 Deformations and moments in a two-span beam.

deformation line of the beam. The moments are referred to as positive when tension is on the bottom, and negative when tension is on the top.

The deformation line in Figure 2–8 shows that the longer span (Span 2) will force the beam to rotate toward itself at the center support. The resistance against this rotation comes from the bending stiffness of the member in Span 1. Stiffness is the ability of a member to resist deformation. There are several different types of stiffness, such as flexural, shear, axial, and torsional. Each type refers to a specific ability to resist a certain type of deformation. The greater the stiffness, the more is the effort required to bring about the specific deformation.

The flexural stiffness of a member is linearly related to the moment of inertia, I, which is a cross-sectional property, and to the modulus of elasticity, E, the ease of extendibility or compressibility of the material; and is inversely related to the length, ℓ, of the member. Thus, if K represents the flexural stiffness, $K = k \dfrac{EI}{\ell}$, where k is a numerical constant that depends on the support conditions of the other end of the member.

In the simple beam shown in Figure 2–8, if the flexural stiffness of Span 1 is infinitely large, it will resist any attempt by Span 2 to rotate the section over the center support toward itself. Hence the condition for Span 2 will approach that of full fixity at its left end. On the other hand, if the stiffness of Span 1 is very small, it will offer very little resistance against the efforts of Span 2 to rotate freely at the center support. Thus, as far as Span 2 is concerned, such a condition might be a "simple support," regardless of the continuity.

2.7 PROPAGATION OF INTERNAL FORCES

The free-body diagrams that resulted from the continuity are shown in Figure 2–9. Double subscripts identify the locations of shears and moments. Thus, if the first span is from a to b then V_{ab} represents the shear in that span at end a, and so on.

The two-span continuous beam is dissected to show the propagation of loads and moments. Each "cut" shows every force and every moment as they act on the part under

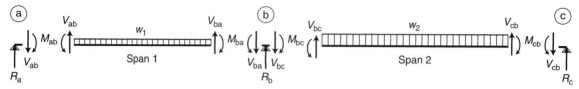

FIGURE 2–9 Propagation of internal forces on a two-span beam.

consideration. For example, M_{ba} is shown as a clockwise arrow on Span 1, whereas it is shown as a counterclockwise arrow on the small part over the b support. These are two manifestations of the same moment, a concept well known from Newtonian physics (action and reaction). Similarly, V_{ba} is shown at the same cut as an upward force on Span 1 that comes from the support to the beam, as well as a downward force that comes from the beam to the support.

Consider now the following self-evident statement: When a structure is in equilibrium, every part must be in equilibrium. Thus the well known equilibrium conditions of $\Sigma H = 0$, $\Sigma V = 0$, and $\Sigma M = 0$ apply for each individual part that is arbitrarily cut out of the structure. For example, the reaction force on the left-hand support, R_a, must equal the shear force, V_{ab}, transferred by the beam to that support. If we consider that $\Sigma M = 0$ on the same piece, we conclude that M_{ab} must equal zero, for there is no other moment on the piece to maintain equilibrium. On the small piece just above the b support, the reaction force from the support R_b must equal the sum of V_{ba} and V_{bc}. Note also that $M_{ba} = M_{bc}$ in order to satisfy equilibrium conditions.

Figure 2–10 shows a three-story-high, three-bay-wide reinforced concrete frame with all the joints numbered. The two outer bays are shown as somewhat wider than the inner bay. Thus, when they are all loaded in an approximately uniform way, the larger spans will try to

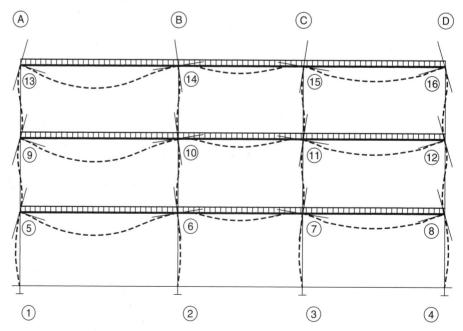

FIGURE 2–10 Deformations of a three-bay and three-story monolithic structure.

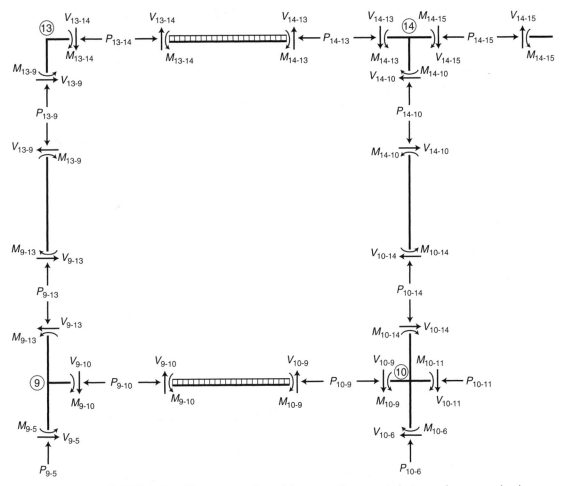

FIGURE 2–11   The propagation of forces and moments between beams and columns.

rotate the ends of the inner bay (between column lines B and C) toward themselves. Thus, the joints on line B will rotate counterclockwise, and the joints on line C will rotate clockwise. At the exterior ends, the loads on the beams will rotate the joints on line A clockwise, and the joints on line D counterclockwise.

From the study of the deformation lines, we can draw some important general conclusions. The beams will have two curvature reversals (inflection points or points of counterflexure). They curve downward in their midspans, resulting in tensions at the bottom (positive moment region). They will curve upward near their ends, resulting in tensions at the top (negative moment region).

The columns on the two upper floors, due to the forced rotations of their ends, will bend into a double curve (S curve). Depending on the amount of fixity available at the footing level, the lower columns will bend either into a double curve when the fixity at the base is significant, or into a single curve when the resistance against rotation at the base approaches that of a hinge.

Figure 2–11 shows free-body diagrams for part of the frame. Again $\Sigma H = 0$, $\Sigma V = 0$, and $\Sigma M = 0$ apply for each individual part. Thus, the axial force in beam 13-14 must equal the shear at the top of column 9-13 for Node 13 to be in equilibrium. The axial force in the column equals the shear at the left end of beam 13-14. And the moment at the end of column 9-13 must maintain equilibrium with the moment at the left end of beam 13-14. Mathematically:

For $\Sigma H = 0$ $V_{13\text{-}9} - P_{13\text{-}14} = 0$
For $\Sigma V = 0$ $P_{13\text{-}9} - V_{13\text{-}14} = 0$
For $\Sigma M = 0$ $M_{13\text{-}14} - M_{13\text{-}9} = 0$

The reader may want to study and write out the equilibrium equations for other free-body parts.

2.8 ON THE "FICKLENESS" OF LIVE LOADS

As stated earlier, loads permanently attached to the structure are referred to as *dead loads*, and transitory loads are referred to as *live loads*. The nature of live loads is that sometimes they are there and sometimes they are not, so it is entirely possible that the live loads are fully present in one bay, while completely missing in other bays. Figures 2–12a through 2–12d show the effects of loading one span at a time on a four-bay continuous beam. In each case the deformation and the moment diagram are shown schematically under different *live loading* conditions. Deformations are shown as dashed lines.

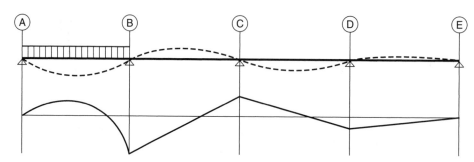

FIGURE 2–12a The effects of live loads on span A-B.

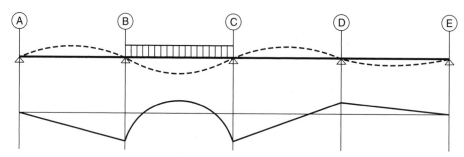

FIGURE 2–12b The effects of live loads on span B-C.

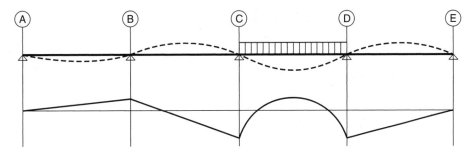

FIGURE 2–12c The effects of live loads on span C-D.

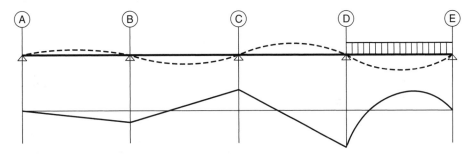

FIGURE 2–12d The effects of live loads on span D-E.

A study of the deformation lines and the moment diagrams of these four different cases leads to the following conclusions:

1. The largest positive moments due to *live loads* in a given span occur when live loads are on that span and on every second span on either side. This is known as a *checkerboard* pattern loading. See Figures 2–13a and 2–13b.

2. The largest negative moments due to *live loads* near a support occur when live loads are on neighboring spans and on every other span on either side. See Figures 2–13c through 2–13e.

Thus, on a continuous beam the number of live loading patterns that result in maximum moment effects equals the number of supports. For example, in a four-span beam with five supports, five different live loading patterns need to be considered to find the possible absolute maximums in each of the positive and negative moment zones.

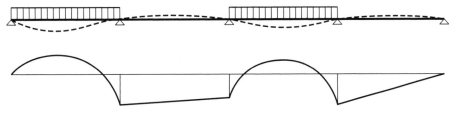

FIGURE 2–13a Live loads in the first and third bays. Largest positive moments in first and third spans.

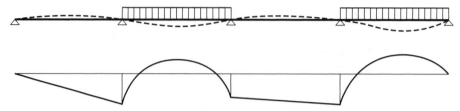

FIGURE 2–13b Live loads in the second and fourth bays. Largest positive moments in second and fourth spans

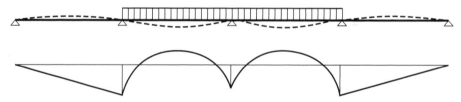

FIGURE 2–13c Live loads in the first, second, and fourth bays. Largest negative moments at second support.

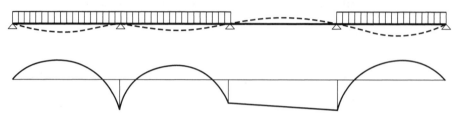

FIGURE 2–13d Live loads in the second and third bays. Largest negative moment at third support.

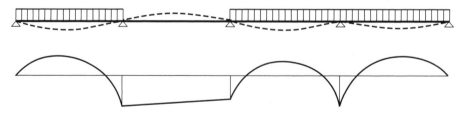

FIGURE 2–13e Live loads in the first, third, and fourth bays. Largest negative moment at fourth support.

These are only the moments that are due to the effects of the live loads. The cases, shown in Figures 2–13a through 2–13e must be combined with the moments resulting from the dead loads, that is, the loads that are permanently present on the structure, whose effects are not variable. The combinations of the dead load moments and the live load moments will result in a maximum possible moment at every location along the beam. The live and dead loads, when plotted into a graph such as the one shown in Figure 2–14, produce a

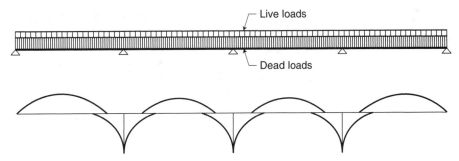

FIGURE 2–14 Maximum moments due to dead loads and different combinations of live loads.

diagram that represents all these combinations. This is called *the diagram of maximum moments* or *the moment envelope*.

Two important points must be noted here. Figure 2–14 shows that in some portions of each span, only positive moments occur, and in others, only negative moments, regardless of the distribution of the live loads. There are portions of each span, however, where either positive or negative moments may occur. This fact is significant in that it affects how a continuous beam must be reinforced.

The second point is that so far we have assumed that the continuous beam is similar to a mathematical line supported on knife-edge supports. The result of such a simplified assumption is that the reactions appear as concentrated forces and the moment diagram has a sharp peak (cusp) at those points. This result, however, is not in conformance with the physical reality. Supports (columns) have a width over which the reactions are distributed. This modifies the moment diagram within the width of the support to something similar to the sketch shown in Figure 2–15. The exact shape of the moment diagram at this location is quite immaterial, for both theoretical studies and numerous test results clearly show that the critical negative moments in the beam occur at the faces of the supports. (Refer to ACI Code, Section 8.7.3.)

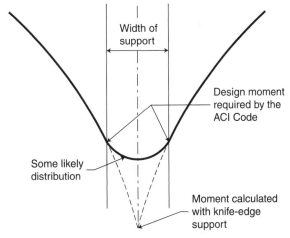

FIGURE 2–15 The true moments in beams at columns.

2.9 THE ACI CODE MOMENT AND SHEAR COEFFICIENTS

The complexities involved in the design of a very simple continuous beam may seem quite bewildering. In practice, however, a vastly simplified procedure is available in most cases.
Any moment along a span may be expressed as follows:

$$M_u = \text{coefficient} \cdot w_u \, \ell_n^2 \qquad\qquad (2\text{--}2)$$

where
w_u is the intensity of the total factored load (see Section 2.10), or the load per unit length. This variable should be evaluated and applied separately for each span if the live loads are different in each one.

ℓ_n is the net (clear) span for positive moment or shear, or the average of adjacent net (clear) spans for negative moment.

When certain conditions are satisfied, the ACI Code permits the use of approximate moments and shears in the design of continuous beams and one-way slabs in lieu of the detailed analysis for maximum moments outlined in the previous section. Approximate moments and shears usually provide reasonable and sufficiently conservative values for the design of these horizontal flexural elements.

ACI Code 318, Section 8.3.3 requires the following conditions for the use of these coefficients:

- *There are two or more spans* The beam or slab is continuous; that is, the approximation does not apply to a single span only.
- *Spans are approximately equal, with the longer of two adjacent spans not greater than the shorter by more than 20 percent* The larger span tends to pull the shorter neighboring span upward if there are significant differences between adjacent spans.
- *Loads are uniformly distributed.*
- *Unit live load does not exceed three times the unit dead load* This is usually the case with reinforced concrete structures.
- *Members are prismatic* This means that the cross section is constant along the length of the span.

The ACI Code design moments and shears are applicable when these preconditions are satisfied. Table A2–1 and the accompanying figure list the coefficients for the moments and shears according to the end conditions and number of spans. In the authors' experience, the ACI coefficients are somewhat more conservative than values obtained from detailed computerized analysis; thus, their use will result in additional safety for the structure.

In actual practice the use of simplified methods to find the design moments and shears is in decline. Many proprietary computer programs are available that not only help evaluate all the most critical loading combinations, but also aid in the design of the required reinforcing. These programs require the sizes of the members as input, for the analysis of an indeterminate structure requires them. (The result, or the output, depends on the relative stiffnesses of the members.) Thus, the application of these coefficients is still very useful for obtaining quick results that can be used in preliminary sizing of the members, which in turn enables the development of input data for a more detailed computerized analysis.

2.10 THE CONCEPT OF STRENGTH DESIGN

The first design theory of reinforced concrete, developed near the end of the nineteenth century, simply borrowed its approach from the prevailing theory of elasticity. The method assumed that reinforced concrete elements at usual actual loads will have stress levels that might be considered to fall within the elastic zone. Figure 1–8 indicates that concrete in compression may follow an approximately linear stress/strain relationship as long as the stress level does not exceed 50% of its ultimate strength level. Steel reinforcing behaves elastically below its yield point. So the concept of *working stress design* (WSD) was not an unreasonable methodology, and the underlying calculation technique is still used when estimating deformations (deflections) in structural elements. (See Section 3.3 for a more detailed discussion.)

The WSD method, however, has many conceptual drawbacks. First and foremost, it does not account for differences between dead and live loads. Rather, it simply lumps them together and assigns a "collective" margin of safety, regardless of the origin of the load. Dead loads can be estimated much more accurately than can live (transitory) loads; thus, logic dictates that the part of the load that comes from dead loads could use a much smaller safety factor against failure. On the other hand, the magnitude and the distribution of the live loads are much more uncertain.

Another, and equally important, drawback of the WSD method is it inaccurately assumes that concrete behaves in a linear fashion with increasing stress levels. Merely knowing a stress level does not ensure a correct prediction of an undesirable level of stress (i.e., failure), because steel has a linear stress response to strain whereas concrete has a nonlinear one.

The third, and perhaps the most significant, drawback of the WSD method is that it is unimportant to know the stress level in a structure at a given loading. What is important is to know how much overload the structure can take before it fails.

Strength is needed to have a *safe design,* or *adequate strength,* so that the structure does not fail whether the actually occurring loads were underestimated or excess load is placed on the structure. Thus, *load factors* (i.e., values used to magnify the actual loads [called *working* or *service* loads]), or moments or shears therefrom, are used so as to create a *demand* on the strength. The concept of demand states, for example, that the structure (or, more precisely, a given element under investigation) must have an *ultimate strength* (i.e., before it fails) not less than those given by Equation 2–3a (ACI Code, Section 9.2.1).

$$U = 1.4(D + F)$$

or $$U = 1.2(D + F + T) + 1.6(L + H) + 0.5(L_r \text{ or } S \text{ or } R)$$

or $$U = 1.2D + 1.6(L_r \text{ or } S \text{ or } R) + (1.0L \text{ or } 0.8W)$$

or $$U = 1.2D + 1.6W + 1.0L + 0.5(L_r \text{ or } S \text{ or } R)$$

or $$U = 1.2D + 1.0E + 1.0L + 0.2S$$

or $$U = 0.9D + 1.6W + 1.6H$$

or $$U = 0.9D + 1.0E + 1.6H \tag{2–3a}$$

where
U = required (ultimate) strength
D = effect from dead loads
L = effect from live loads
W = effect from wind loads
E = effect from seismic (earthquake) loads
L_r = effect from roof live loads
S = effect from snow loads
F = effect from fluid loads
H = effect from soil loads
R = effect from rain loads

The multipliers applied to the effects in the various load combinations are the *load factors*. These guard against accidental overloading of the structure. They also take account of the imprecision in establishing the magnitude, or the distribution, of the loads. Thus, for example, greater load factors are assigned to live loads (or wind loads, or earthquake loads) than to dead loads to account for greater uncertainty.

Also, dead loads sometimes actually help to counteract the effect of wind or earthquake loads. For these conditions a more conservative approach is to presume that calculated dead loads are somewhat less than assumed. Such a concept is accounted for by the sixth and seventh load combinations in Equation 2–3a. These load combinations can be simplified by combining all live loads as L and using the larger load factor. In addition, for $U = 1.4(D + F)$ to govern the design, the condition of $D > 8L$ must exist, which is not very probable in most cases. Therefore, the load combination given in Equation 2–3b will be used throughout this book:

$$U = 1.2D + 1.6L \qquad\qquad (2\text{–}3b)$$

where D includes the effects from all the dead loads and L is due to all the live loads.

2.11 DESIGN (ULTIMATE) STRENGTH

The ultimate strength of a section within a structure (as discussed in detail later for separate and combined cases of bending moment, shear, torsion, and axial load) is calculated from the sizes (dimensions) of the section, the materials (steel and concrete) employed, and the amount of reinforcing used. This calculation gives us the *supply,* or the resisting strength furnished by the section. In flexural design, for example, this calculated quantity is designated as M_n, which is called *nominal moment strength* or *nominal resisting moment*. Nominal strength is the calculated strength, provided that everything goes according to plan; that is, the concrete is at least as strong as assumed in the design, the dimensions of the beam, slab, or any designed element is exactly as shown on the plans, the required reinforcing is placed exactly where it was assumed in the calculations, and so on. But experience shows that there is no such thing as perfectly executed plans, even in the best circumstances. ACI 117-90, "Standard Tolerances for Concrete Construction and Materials" lists tolerances that are reasonable to expect when good workmanship is provided. Furthermore, the calculation processes employ simplified mathematical models that should be considered as only reasonable approximations of reality. The design methodology also tries to reflect the relative importance of different structural components. The failure of columns, for example, may result in collapse of an entire building, but the failure of a beam typically causes only limited local damage.

In light of all these possible detrimental effects to the assumed strength, a *strength reduction factor* (φ-factor), sometimes referred to as an *under-strength* factor, is introduced to the above defined *nominal strength*. This factor accounts for the fact that the section's strength may be less than assumed in the analysis.

Thus, we arrive at the concept of *useable strength* (or supply), which is the product of the nominal strength and the strength reduction factor.

Different φ factors are used for different types of effects. Equation (2–4) gives some φ factors.

$$\text{Flexure} \quad \phi = 0.90$$
$$\text{Shear and torsion} \quad \phi = 0.75$$
$$\text{Axial compression (columns)} \quad \phi = 0.65 \tag{2–4}$$

Hence the *ultimate strength design* (USD) method can be stated as the following inequality:

$$\text{Demand} \leq \text{Supply}$$
or $$\text{required ultimate strength} \leq \text{useable design strength}$$
or $$\text{effects of loads} \leq \text{resisting capacity of member}$$

And so for a beam subjected to gravity (dead and live) loads, for example, Equations 2–6 through 2–8 represent this concept.

$$M_u = 1.2M_D + 1.6M_L \tag{2–5}$$

and

$$M_u \leq \phi M_n \tag{2–6}$$

Defining the design resisting moment, M_R, as

$$M_R = \phi M_n \tag{2–7}$$

the following must hold for the beam to be safe:

$$M_u \leq M_R \tag{2–8}$$

On the left side of Equation 2–8 is the demand. The demand depends only on the span, the type of support (e.g., simply supported, cantilevered, etc.), and the loads. All this information comes from the static analysis.

On the right side of Equation 2–8 stands the supplied strength of the section (design resisting moment, M_R), which depends on the size and shape of the cross section, the quality of the materials employed (f_c' and f_y), and the amount of reinforcing furnished. Thus, the left side of the inequality is unique, but the right side is undefined. An infinite number of different sizes, shapes, and reinforcing combinations could satisfy a given problem. *The only rule is that the supplied useable strength be larger than (or at least equal to) the required strength.*

EXAMPLE 2–3

Assume that the beam in Example 2–1 is simply supported. Calculate the required ultimate flexural strength (factored moment from the loads). Use the permitted reduced live load.

Solution

$$M_D = 930 \times 30^2/8 = 104{,}625 \text{ lb-ft}$$
$$M_L = 403 \times 30^2/8 = 45{,}338 \text{ lb-ft}$$

Thus:

$$M_u = 1.2 \times 104{,}625 + 1.6 \times 45{,}338 = 198{,}091 \text{ lb-ft} \quad \text{(or 198.1 kip-ft)}$$

The same result could be obtained by using factored loads (the loads multiplied by their respective load factors).

$$w_u = 1.2 \times 930 + 1.6 \times 403 = 1{,}761 \text{ lb/ft} = 1.761 \text{ kip/ft}$$

and

$$M_u = 1.761 \times 30^2/8 = 198.1 \text{ kip-ft}$$

Notice that when finding factored loads from service or working loads, the nature of the loads does not change; only their magnitudes are multiplied by the corresponding load factors. If a service load is distributed, its factored value is also distributed; if the service load is concentrated, its corresponding factored load is also concentrated. The following example clarifies this point.

EXAMPLE 2–4

Determine factored loads for the beam shown in Figure 2–16.

Solution

For the left half of the beam:

$$w_{u1} = 1.2w_D + 1.6w_L$$
$$w_{u1} = 1.2 \times 1.0 + 1.6 \times 2.0 = 4.4 \text{ kip/ft}$$

For the right half of the beam:

$$w_{u2} = 1.2w_D + 1.6w_L$$
$$w_{u2} = 1.2 \times 1.0 + 1.6 \times 0 = 1.2 \text{ kip/ft}$$

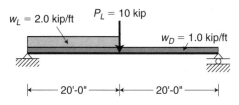

FIGURE 2–16   Example 2–4 (service loads).

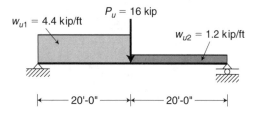

FIGURE 2–17   Example 2–4 (factored loads).

The concentrated load is a live load only:

$$P_u = 1.2P_D + 1.6P_L$$
$$P_u = 1.2 \times 0 + 1.6 \times 10 = 16 \, \text{kip}$$

The factored loads on the beam are shown in Figure 2–17.

2.12   ASSUMPTIONS FOR THE FLEXURAL DESIGN OF REINFORCED CONCRETE BEAMS

To this point we have discussed the calculations for the left side of the design Equation 2–8 (demand) in some detail. In this section we develop the right side of the design equation. To establish the *supply,* or the ultimate flexural strength, of a reinforced concrete section, we must discuss the stages of stress that a reinforced concrete section experiences before reaching failure. This discussion of these different stages of stress under increasing bending moments will also illuminate the assumptions made in developing expressions for calculating the ultimate strength of the section. To keep the discussion simple, we will examine a beam with a rectangular cross-section like the one shown in Figure 2–18.

The symbols in Figure 2–18 will be used throughout this book. They are the standard ACI symbols used with reinforced concrete. Thus:

b = width of the section

h = the overall depth of a section

d = the effective depth of a section, or the depth from the centroid of the tension reinforcement to the compression face

A_s = the sum of the cross-sectional areas of the reinforcing bars

Notice that the reinforcement is not placed at the very bottom of the beam. The first and foremost reason for this placement is to provide corrosion protection to the reinforcement. The inner environment of concrete is highly alkaline (high pH value) and helps to protect the reinforcement. The concrete cover also provides fire protection to the reinforcement. Furthermore, the concrete surrounds the reinforcing steel, which enables intimate bonding and allows the concrete and the steel, two individual materials, to work together. The required minimum concrete cover is given in Section 7.7 of the ACI Code. For beams it is 1.5 in. to the stirrups. (The stirrups, usually made out of #3 or #4 bars, will be discussed in Chapter 4.)

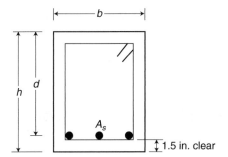

FIGURE 2–18 Definition of symbols used in a rectangular beam section.

Figure 2–19 shows a simply supported beam that has a simple rectangular cross section made of plain concrete (homogeneous material). This type of beam is almost never used in an actual building, but it will give us insight into the behavior of concrete beams.

The uniformly distributed load (Figure 2–19a) represents the self weight plus some superimposed load. The slightly exaggerated deflected shape is shown in Figure 2–19b, and the moment diagram in Figure 2–19c. Attention will be directed to the section where the bending moment is the greatest. This location is where the stresses and the strains are also the largest.

Figure 2–20 shows the cross section of the beam and the distribution of strains and stresses if the beam is unreinforced. Figure 2–21 illustrates the distribution of the strains and stresses in a 3-D form. As long as the bending moments are small, that is, the resulting tensile stresses at the bottom are less than the ultimate tensile strength of the concrete, the section will behave as if it were made of a homogeneous, quasi-elastic material. The bottom is in tension, and the top is in compression.

Direct your attention to the *strain* diagram first. Strain represents changes in length. The strain distribution is linear from bottom to top.

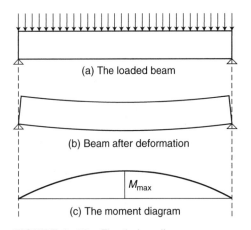

FIGURE 2–19 Elastic bending.

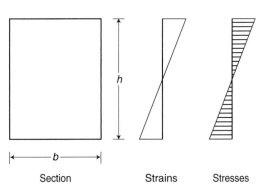

FIGURE 2–20 Linear distribution of strains and stresses.

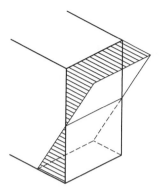

FIGURE 2–21 3-D representation of linear strain or stress distribution.

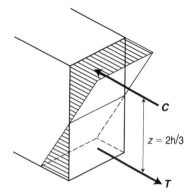

FIGURE 2–22 The internal couple in homogeneous beams.

The farther up or down a point is from the imaginary center, the greater the strain in the beam. The largest tensile strains are at the bottom, whereas the largest compressive strains are at the top. There is a line across the section where the strain is zero. This is called the *neutral axis*. The straight-line distribution of strains is known as the *Bernoulli–Navier hypothesis*. This distribution is called a "hypothesis" because it results not from mathematical derivation, but from careful measurements made on countless tests of many different materials, including concrete. The distribution of *stresses* is also linear when the material follows Hooke's law, as steel does below the so-called *proportional limit*. Stresses are forces acting on a unit area. Thus, it is possible to determine the resultant for these forces. *The resultant, which is a tensile (T) or compressive (C) force, is equal to the volume of the stress block,* For example, if the largest compressive stress is f_{cmax}, then the sum of all the compressive forces is given by Equation 2–9.

$$C = \tfrac{1}{2}[f_{cmax} \times (h/_2) \times b] \qquad (2\text{--}9)$$

Similarly, the sum of all tensile forces is given by Equation 2–10.

$$T = \tfrac{1}{2}[f_{tmax} \times (h/_2) \times b] \qquad (2\text{--}10)$$

These resultants will be located at the centroid of the wedge-shaped stress blocks, as shown in Figure 2–22. Equilibrium requires that these resultants be equal in magnitude, and together they form an *internal couple*. The internal couple is equivalent to the bending moment at the section.

EXAMPLE 2–5

For the beam of Figure 2–22, assume $b = 12$ in., $h = 24$ in. and $M_{max} = 38.4$ kip-ft. Determine the bending stresses and the equivalent tensile and compression forces acting on the section.

Solution

The section modulus is:

$$S = b \times h^2/6 = 1{,}152\,\text{in}^3$$

Thus the maximum stresses are:

$$f_{max} = M_{max}/S = 38.4 \times 12/1152 = 0.400\,\text{ksi}$$

Then

$$C = T = 1/2 \times [0.400 \times (24/12) \times 12] = 28.8\,\text{k}$$

The moment arm between the maximum stresses is $z = 2 \times 24/3 = 16$ in.
The moment equivalent of this couple is:

$$C \times z = T \times z = 28.8 \times 16 = \frac{460.8\,\text{kip-in}}{12} = 38.4\,\text{kip-ft}$$

which agrees with the given moment, $M_{max} = 38.4$ kip-ft.

The concept of the internal couple will become a very important tool in considering a reinforced concrete beam. If the beam in Example 2–5 has enough tensile strength to withstand the applied 0.400 ksi (400 psi) tensile stress, the beam will not fail. As discussed earlier, concrete has a rather limited tensile strength. The modulus of rupture, which was said to represent the ultimate tensile strength of concrete in flexure, is given in Equation 1–3.

As mentioned previously, the modulus of rupture is a statistical average (with a considerable coefficient of variation) that is empirically derived from many laboratory tests. At increasing loads, a magnitude very soon is applied at which the beam's tensile strength is exhausted. At that point, somewhere near the maximum moment, the beam will crack. Without reinforcement, the crack will instantly travel upward and the beam will collapse, as shown in Figure 2–23.

In the following discussion the beam is assumed to have flexural reinforcement. Such a beam is shown in Figure 2–24. As long as the tensile stresses in the concrete at the bottom of the section are less than the modulus of rupture, there will be no cracks. At the location of the reinforcing steel, the concrete and the steel have identical strains. The steel

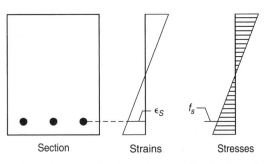

Section   Strains   Stresses

FIGURE 2–23 Bending failure of an unreinforced concrete beam.

FIGURE 2–24 Strain and stress distribution of a reinforced concrete beam prior to cracking.

is bonded to the concrete, thus they must deform together. But the two different materials respond differently to deformation because they have a different modulus of elasticity, so the stresses will be different. In this particular case the stress in the steel will be much larger than that in the concrete.

For example, assume a concrete with $f'_c = 3,000$ psi. Then $E_c = 57,000\sqrt{3,000} = 3,122,000$ psi $= 3,122$ ksi. The modulus of elasticity of the reinforcing steel is $E_s = 29,000$ ksi. According to Hooke's law the stress equals the product of the modulus of elasticity and the strain. So it follows that the stress in the steel will be about 9 times higher (the ratio of the two moduli of elasticity values) than the stress in the concrete in the immediate vicinity. This ratio is usually designated as $n = E_s/E_c$ and is called the *modular ratio*.

The concrete cracks under increasing applied forces, and it is the reinforcement that carries the tension across the crack. The crack travels up to a height, then stops somewhere below the neutral axis as seen in Figure 2–25. The shaded area represents the uncracked part of the section. Where the strains are still small near the neutral axis, the concrete is still able to transfer some tensile stresses (albeit very small), even in the cracked section; *however, the amount of tensile force represented by the still un-cracked tensile stress volume is so small that it is simply neglected.*

Assuming, therefore, that the concrete does not carry any tension after cracking, the bending moment in the section is transferred across from one side of the crack to the other via the tension in the steel and the compression in the concrete, as seen in Figure 2–26. This assumption simplifies the development of an appropriate formula for the internal couple. The tensile component of this couple is at the centroid of the reinforcing steel, while the compressive component is at the centroid of the wedge-shaped compression block. Comparing Figures 2–22 and 2–26 indicates that the T force now is concentrated at the centroid of the reinforcing.

In Figure 2–26 the compression stress block is represented as a triangular wedge shape. This representation is more or less accurate as long as the compressive stresses in the concrete remain quite low. Figure 1–8 shows the generic shapes of the stress-strain curve of concrete in compression, and the assumption of linear distribution of stresses may be justified up to approximately $0.5f'_c$.

As the applied loads increase, there is a corresponding increase in bending moments throughout the beam. Thus, many more sections away from the location of the maximum moment will develop tensile stresses that exceed the concrete's ultimate tensile strength, resulting in the development of more cracks. While theoretically the spacing between

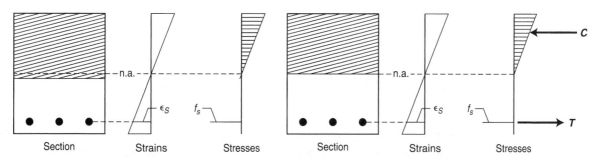

FIGURE 2–25 Strains and stresses after cracking.   **FIGURE 2–26** The internal couple after cracking.

cracks is very small, it does not happen that way, because the formation of a crack relieves tensile strains in the concrete in its immediate neighborhood. Initially the cracks are very fine hairline cracks, and a magnifying glass may be needed to locate them. These hairline cracks do not indicate that there is anything wrong with the beam: They occur naturally in reinforced concrete beams subjected to flexure under normal working load conditions. In fact, the reinforcement does not even do much work until after the concrete has cracked.

As the bending moment at the section increases, the magnitude of T and C, the tension and compression components of the internal couple, must also increase. In the reinforcement this is simply reflected as an increase in stresses. Correspondingly, the steel also will experience greater strains and elongation. As long as the strains in the reinforcing are less than the yield strain, the relationship between stresses and strains remains linear.

In the concrete, however, the increased compression strains result in a nonlinear response of the stresses while maintaining the required increase in the volume of the stress block. The concrete stress block becomes more and more bounded by a curvilinear surface. Ultimately, the contour will resemble the one shown in Figure 2–27. This diagram is the same as the ones shown in Figure 1–8, except the axes are reversed. At the origin, the strains and stresses are zero, just like on the beam section at its neutral axis. At the top there is a strain value of 0.003, which is a value selected by the ACI Code (somewhat arbitrarily) as the *ultimate useful strain.* Somewhere between these two limits (in the neighborhood of 0.002) the peak stress (the *maximum compressive strength* or simply *compressive strength*) occurs. In calculations this value is designated as f'_c; it is the specified compression strength of the concrete.

On the tension side (i.e., at the reinforcement), Figure 1–17 shows the stress-strain curve of the reinforcing steel, or the response of the steel to increasing strain values. This

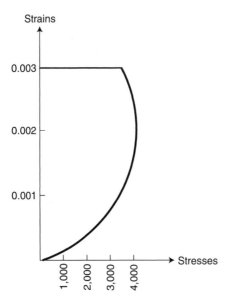

FIGURE 2–27 Typical curvilinear stress distribution in the concrete at ultimate strength.

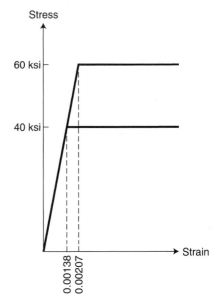

FIGURE 2–28 Assumed bilinear stress-strain diagram of reinforcing steel.

curve clearly shows that the steel has significant residual strength even after it has yielded, but this residual strength (the strength gained in the strain hardening zone) is neglected. Thus, we assume that the stresses will linearly increase with increasing strains up to yield, after which ever-increasing strains produce no corresponding increase in stresses. Scientifically, this curve is known as a *bilinear stress-strain diagram,* and the response of the steel as *elasto-plastic behavior*. Figure 2–28 shows the assumed stress-strain diagram for 40 and 60 ksi steel, respectively.

2.13 DIFFERENT FAILURE MODES

As a first case assume that a beam has a relatively *small* amount of reinforcing steel. Such a beam is shown in Figure 2–29. With increasing demand on the internal couple the stresses in the steel will reach yield *before* the demand on the concrete compression block reaches the ultimate concrete compressive strength. With increasing elongation in the steel, still prior to yield, the cracks will become wider and more visible. When the steel starts to yield (i.e., elongate rapidly), the relatively narrow crack at the bottom opens up. This forms a wedge that shifts the neutral axis upward, thus decreasing the area available for the compressive stress block, until the concrete crushes on the compressive side as a *secondary* failure. The *primary* cause of failure was due to the yielding of the reinforcement. In a somewhat misleading way such sections are sometimes referred to as *underreinforced* sections. This unfortunate expression implies that the section is underreinforced as compared to the capacity of the compression part of the section. (In Section 2.17 we will discover that the behavior of an under-reinforced section is classified as *tension-controlled* or *transition-controlled* depending on the level of tensile strain in the steel at the time of failure.)

As a second case consider a beam that has a relatively large amount of reinforcing. For such a beam the steel will be able to develop the T part of the internal couple without yielding. As demand on the compression stress block increases, however, the capacity to provide a sufficiently large volume of concrete stresses will be exhausted, reaching the state shown in Figure 2–27. In such a case the *primary* failure occurs in the concrete. These types of sections are referred to as *overreinforced,* that is, the beam has more reinforcing in the section than what could be used with the largest possible compressive stress block.

A casual observer may care little about what initiated the failure of the beam. But the two modes of failure vastly differ. The first mode, in which the primary failure happens due to the yielding of the reinforcing, is a *ductile* process and is preceded by significant cracking, fairly large deflections, and similar warning signs. The beam, in a way, tells you that something bad is about to happen.

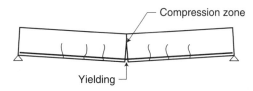

FIGURE 2–29 Tension-controlled failure of a reinforced concrete beam.

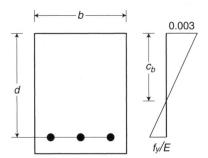

FIGURE 2–30   Strain distribution at "balanced" failure.

In the second mode there are no such obvious signs of impending failure. The reinforcement, in providing the tensile part of the internal couple, experiences relatively low strains, so the few hairline cracks do not serve as warning signs. Consequently when the failure occurs, it happens in a sudden, explosive way—the concrete failure in compression is very abrupt.

Between these two different failure modes is a special case, known in the literature as the *balanced-failure* condition. Balanced failure is a theoretical limit dividing the underreinforced and overreinforced failure modes. We feel that this is an unfortunate terminology, because the word *balance* (i.e., equilibrium) should not be used to describe a failure mode that is anything but the maintenance of balance. We would prefer to use the expression *simultaneous failure.* But whatever terminology is used, it refers to the amount of reinforcement in a section that causes the concrete at the compression side to fail at exactly the same time the steel begins to yield. So the strain in the steel will be the yield strain, and the strain at the extreme edge of the concrete will be 0.003. This balanced condition is depicted in Figure 2–30.

2.14   THE EQUIVALENT STRESS BLOCK

A quick look at Figure 2–27, or at its 3-D representation in Figure 2–31, should convince anyone that it would be impractical to calculate the value of C by figuring out the volume of the stress block. The calculation would require integral calculus, even if there was an easy way to express the shape of the curve mathematically. A reasonable approximation can be obtained by substituting a stress block whose volume is about the same as the true stress volume enclosed in Figure 2–31, and whose centroid is fairly close to that of the true stress volume. This is known as the *equivalent stress block,* and is shown in Figure 2–32.

The relationship between the true stress block and the equivalent stress block has been established by studying many concrete stress-strain curves. The simple rectangular block has been adopted for its simplicity and ease of calculation. If a uniform stress value of $0.85f_c'$ is adopted, then only the relationship between the depth of the equivalent stress block a and the distance of the neutral axis from the top c is needed. This relationship is given in Equation 2–11.

$$a = \beta_1 c \qquad (2\text{--}11)$$

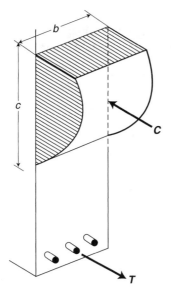

FIGURE 2–31   True stress
distribution in the concrete at
ultimate strength.

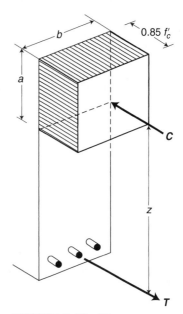

FIGURE 2–32   The
equivalent stress block.

To account for the somewhat different shapes of the stress-strain curves of different strengths of (refer to Figure 1–8) concrete, β_1 is given by the ACI Code (Section 10.2.7.3) as follows:

$\beta_1 = 0.85$ for concrete strength f'_c up to and including 4,000 psi. For strengths above 4,000 psi, β_1 shall be reduced at a rate of 0.05 for each 1,000 psi of strength in excess of 4,000 psi, but β_1 shall not be taken less than 0.65.

Equation 2–12 gives the expression to calculate β_1 for $f'_c > 4,000$ psi.

$$\beta_1 = 0.85 - 0.05\left(\frac{f'_c - 4,000}{1,000}\right) \geq 0.65 \tag{2–12}$$

The equivalent stress block makes it extremely easy to manipulate the expression to calculate the ultimate (design) resisting moment of a given section. The moment arm of the internal couple, z, can be calculated using Equation 2–13.

$$z = d - \frac{a}{2} \tag{2–13}$$

The numerical value of the internal couple can be expressed in two different ways, using the designation of M_n for the *nominal resisting moment* and M_R for the *design resisting moment*. These moments can be calculated using Equations 2–14 and 2–15, respectively.

$$M_n = Tz \quad \text{or} \quad M_n = Cz \tag{2–14}$$

$$M_R = \phi M_n = \phi Tz = \phi Cz \tag{2–15}$$

where

$T = A_s f_y$ (the area of the reinforcing multiplied by the yield stress of the steel)

$C = 0.85 f_c' ab$ (the volume of the equivalent stress block)

Equilibrium requires that T be equal to C, thus

$$A_s f_y = 0.85 f_c' ab \qquad (2\text{--}16)$$

Solving this equation for a gives Equation 2–17 for calculating the depth of the equivalent stress block.

$$a = \frac{A_s f_y}{0.85 f_c' b} \qquad (2\text{--}17)$$

Note that a will increase as larger amounts of reinforcement, or reinforcing steel with greater strength is used. On the other hand a will be smaller if a wider section, or stronger concrete is used. Note, however, that a is independent of the depth of the section.

2.15 THE STEEL RATIO (ρ)

Sometimes it is useful to express A_s as a fraction of the working cross section, which is the product of the width b and the effective depth (or *working depth*) d. The term *steel percentage* or, more accurately, *steel ratio* refers to the ratio between the area of the reinforcing steel and the area of the working concrete section.

The steel ratio is calculated using Equation 2–18.

$$\rho = \frac{A_s}{bd} \qquad (2\text{--}18)$$

Note that ρ is a nondimensional number, area divided by area, so it is not a percentage per se. But it can be made into a percentage by multiplying it by 100. For example, assume the following beam data: $b = 12$ in., $h = 24$ in., $A_s = 3$ #6 bars $= 3 \times 0.44 = 1.32$ in^2, and #3 stirrups in the beam.

Then $d = 24 - 1.5$ in. (concrete cover) $- 0.375$ in. (diameter of the stirrup) $- 0.75$ in. (diameter of #6 bar)$/2 = 21.75$ in. Thus, the steel ratio is $\rho = \dfrac{A_s}{bd} = \dfrac{1.32}{12(21.75)} = 0.00506$ (or 0.506%).

2.16 THE BALANCED STEEL RATIO

Section 2.13 discussed the two possible different failure modes of reinforced concrete beams in bending. The theoretical dividing point between them, the "balanced failure," was also discussed. In this case the steel in the outermost layer (if there is more than one layer)

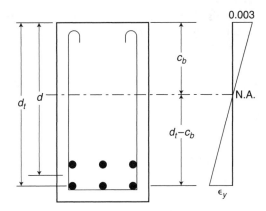

FIGURE 2–33   Strain distribution at balanced failure.

reaches its yield strain exactly when the maximum compressive strain in the concrete reaches the 0.003 value. The strain distribution at balanced failure resembles the one shown in Figure 2–33. In order to cover the more general (although not so frequent) case of multilayer reinforcing in the beam, a distinction is made between d, the working depth, and d_t, the depth to the outermost layer of reinforcing on the tension side. When there is only one layer of reinforcement, $d = d_t$.

From the similarity of the two triangles above and below the neutral axis, c_b, the depth of the neutral axis at balanced failure can be expressed as a function of d_t and f_y.

$$\frac{c_b}{d_t - c_b} = \frac{0.003}{\epsilon_y} \tag{2–19}$$

Solving for c_b

$$c_b = \frac{0.003 d_t}{0.003 + \epsilon_y} \tag{2–20}$$

because

$$\epsilon_y = \frac{f_y}{E_s} = \frac{f_y}{29{,}000{,}000} \tag{2–21}$$

We can substitute and rearrange to obtain

$$c_b = \frac{87{,}000}{87{,}000 + f_y} d_t \tag{2–22}$$

In these equations f_y is substituted in psi.

With this information the depth of the equivalent stress block at *balanced failure* can be calculated using Equation 2–23.

$$a_b = \beta_1 c_b = \frac{A_{sb} f_y}{0.85 f_c' b} \tag{2-23}$$

where A_{sb} is the theoretical amount of reinforcing needed to cause a balanced failure mode.

When c_b from Equation 2–22 and $A_{sb} = \rho_b bd$ from Equation 2–18 are substituted into Equation 2–23.

$$\beta_1 \frac{87,000 \, d_t}{87,000 + f_y} = \frac{\rho_b bd f_y}{0.85 f_c' b}$$

the steel ratio for balanced failure, ρ_b, can be calculated using Equation 2–24.

$$\rho_b = \frac{0.85 f_c'}{f_y} \beta_1 \frac{87,000}{87,000 + f_y} \frac{d_t}{d} \tag{2-24}$$

If $d_t = d$, which means there is only one layer of reinforcing steel (by far the most frequent case), then Equation 2–24 becomes Equation 2–25.

$$\rho_b = \frac{0.85 f_c'}{f_y} \beta_1 \frac{87,000}{87,000 + f_y} \tag{2-25}$$

Note that the value of ρ_b depends only on the selected materials (f_c' and f_y) and is independent of the size of the section. (The ratio $\dfrac{d_t}{d}$ becomes necessary only when there is more than one layer of reinforcement.)

2.17   ELABORATION ON THE NET TENSILE STRAIN IN STEEL (ϵ_t)

In an effort to generalize the approach for members subject to both bending and axial compressive forces, the ACI Code strives to treat these combination cases together. The different failure modes were discussed in Section 2.13. These modes are distinguished by whether the primary failure is due to yielding of the steel or to crushing of the concrete. The former is called *tension-controlled* failure, and the latter is *compression-controlled* failure. It was also previously noted that tension-controlled failure results in highly desirable ductility, whereas compression-controlled failure is abrupt and nonductile in nature. Unfortunately, as will be discussed later in Chapter 5, the desire to have only ductile tension-controlled failure modes cannot always be satisfied. But in flexural members, at least we can control the failure behavior by using no more steel than an amount that ensures the desirable ductility. In the past this was accomplished by limiting the reinforcement ratio, ρ, to $0.75\rho_b$ in flexural members. Since 2002 the ACI Code has adopted a new approach that is a better integration of dealing with members subject to axial stresses whether from flexure,

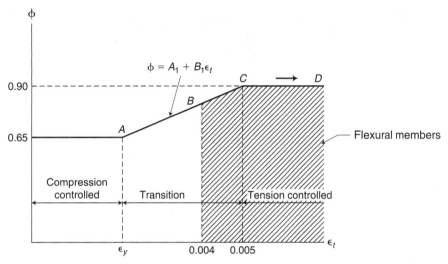

FIGURE 2–34 Variation of ϕ versus ϵ_t.

or axial compression, or both. If ductile failure mode cannot always be assured, then the use of a larger safety factor against a nonductile type of failure is warranted. This larger safety factor is obtained by regulating the ratio between the useful ultimate moment or design resisting moment ($M_R = \phi M_n$) and the nominal ultimate moment (M_n). This requires only an adjustment in the ϕ (strength reduction) factor.

The ACI Code (Section 9.3.2) defines three different types of section behavior: *tension-controlled, compression-controlled,* and a *transition zone,* which is the zone between the tension- and the compression-controlled failure zones. Figure 2–34 shows a graphical representation of these zones, and defines and separates the three regions. Theoretically the division between compression-controlled failure and tension-controlled failure is where $\epsilon_t = \epsilon_y$. In other words, the section is compression-controlled if the strain in the steel is less than the yield strain; and is tension-controlled if the strain in the steel is greater than the yield strain when the compression strain in the concrete reaches the limit of 0.003. For design purposes, however, the ACI Code requires a safely assured tension-controlled section; thus, it defines a section as tension-controlled only when the steel strain at ultimate strength is greater than 0.005. Between the two limits, yield strain and 0.005, the Code defines a *transition zone* with lowered ϕ values.

Note that the ACI Code allows ϵ_t for flexural members to be as small as 0.004 at ultimate strength. A somewhat diminished ϕ factor, however, is required in conjunction.

It may be helpful here to repeat what was discussed in Section 2.13 in a somewhat different format. Figure 2–35 defines graphically the behavior of reinforced concrete sections.

1. A compression-controlled section is a reinforced concrete section in which the strain in the concrete reaches 0.003 at ultimate strength, but the strain in the steel (ϵ_t) is less than the yield strain (ϵ_y). (See Figure 2–35a.) In other words, at the ultimate strength of the member, the concrete compressive strain reaches 0.003 before the steel in tension yields. This condition results in a brittle or sudden failure of beams and should be avoided. In

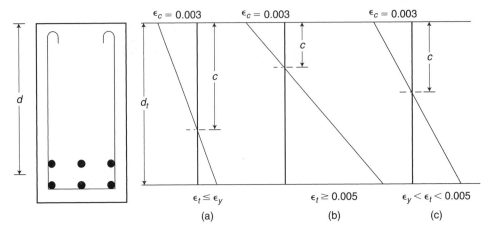

FIGURE 2–35 Strain distribution and net tensile strain (ϵ_t) at behavior limits:
(a) compression-controlled sections; (b) tension-controlled sections;
(c) transition-controlled sections.

reinforced concrete columns, however, a design based on compression-controlled failure behavior cannot be avoided. As shown in Figure 2–34, $\phi = 0.65$ is mandated for this case, which is considerably less than the $\phi = 0.90$ that is used for tension-controlled sections. The reasons for this additional factor of safety are: (1) compression-controlled sections have less ductility; (2) these sections are more sensitive to variations in concrete strength; and (3) the compression-controlled sections generally occur in members that support larger loaded areas than do members with tension-controlled sections.

 2. *A tension-controlled section* is a reinforced concrete section in which the tensile strain in steel (ϵ_t) is more than 0.005 when the compression strain in concrete reaches 0.003 (see Figure 2–35b). In other words, when a section is tension-controlled at ultimate strength, steel yields in tension well before the strain in the concrete reaches 0.003. Flexural members with tension-controlled sections have ductile behavior. As a result, these sections may give warning prior to failure by *excessive deflection* or *excessive cracking,* or both. Not all tension-controlled sections will give both types of warning, but most tension-controlled sections should give at least one type of warning. Both types of warnings, excessive deflection and cracking, are functions of the strain, particularly the strain on the tension side. Because tensile strains are larger than compressive strains in tension-controlled sections at failure, the ACI Code allows a larger ϕ factor (0.90) for these types of members.

 3. *A transition-controlled section* is a reinforced concrete section in which the net tensile strain in the steel (ϵ_t) is between yield strain (ϵ_y) and 0.005 when the compression strain in the concrete reaches 0.003. (See Figure 2–35c.) Some sections, such as those with a limited axial load and large bending moment, may have net tensile strain in the extreme steel (ϵ_t) between these limits. These sections are in a transition region between compression- and tension-controlled sections. In Figure 2–34, the line *AC* represents the Code-defined relationship between ϕ and ϵ_t in the transition-controlled zone. The value of ϕ in the transition zone can be calculated using Equation 2–26.

$$\phi = A_1 + B_1\epsilon_t \qquad (2\text{–}26)$$

where the coefficients A_1 and B_1 may be expressed as

$$A_1 = \frac{0.00325 - 0.9\,\epsilon_y}{0.005 - \epsilon_y}$$

$$B_1 = \frac{0.25}{0.005 - \epsilon_y}$$

Table A2–2a in Appendix A lists the values for the coefficients A_1 and B_1 for commonly used reinforcing steels.

2.18 THE LOCATION OF THE NEUTRAL AXIS AND LIMIT POSITIONS

Consider the strain diagram shown in Figure 2–36. The location of the neutral axis at ultimate strength (c) depends upon the net tensile strain of the steel. Observe the solid and the dotted lines. Because the strain at the compression face is constant (0.003), c becomes smaller as the steel strain increases. Using similar triangles of the strains above and below the neutral axis, an expression can be derived to calculate the depth of the neutral axis, c.

$$\frac{c}{d_t - c} = \frac{0.003}{\epsilon_t} \tag{2–27}$$

$$c\epsilon_t = 0.003(d_t - c)$$

Solving Equation 2–27 for c:

$$c = \frac{0.003}{0.003 + \epsilon_t}\, d_t \tag{2–28}$$

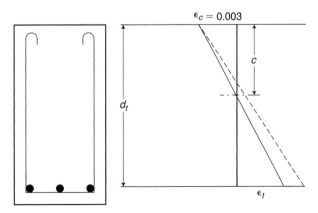

FIGURE 2–36 Variation of the location of the neutral axis (c) with the tensile stress in steel (ϵ_t).

The ratio of c/d_t, given in Equation 2–29, is often used to check if a section is tension-controlled.

$$\frac{c}{d_t} = \frac{0.003}{0.003 + \epsilon_t} \qquad\qquad (2\text{–}29)$$

Two values of ϵ_t are of special interest. The first one is $\epsilon_t = 0.004$. This is the absolute minimum steel strain permitted by the ACI Code for members in flexure. (Refer to Figure 2–34 and ACI Code, Section 9.3.2.) Substituting this ϵ_t value into Equation 2–29 gives us Equation 2–30.

$$\frac{c}{d_t} = \frac{3}{7} = 0.429 \qquad \text{or} \qquad c = 0.429 d_t \qquad\qquad (2\text{–}30)$$

Equation 2–30 gives the lowest permissible value of the neutral axis depth. In other words, this defines the largest permissible concrete area in compression ($c \leq 0.429 d_t$).

The second value of interest is $\epsilon_t = 0.005$. Solving Equation 2–29 for this case, we obtain Equation 2–31 for the lowest location of the neutral axis depth for tension-controlled sections.

$$\frac{c}{d_t} = \frac{0.003}{0.003 + 0.005} = \frac{3}{8} = 0.375 \qquad \text{or} \qquad c = 0.375 d_t \qquad\qquad (2\text{–}31)$$

2.19 RELATIONSHIP BETWEEN ϕ AND d_t/c

Equation 2–29 shows that the ratio of either c/d_t, or its inverse, d_t/c, are in direct relationship with the steel tensile strain ϵ_t. Then it is possible to modify Figure 2–34 to show the ACI Code–prescribed strength reduction factor's (the ϕ factor's) variation in terms of the d_t/c ratio. (For convenience of graphing, the relationship is shown in terms of d_t/c.) Figure 2–37 expresses the changing ϕ values with respect to the ratio d_t/c. Note that the ratio d_t/c_b is the ratio of d/c at the balanced failure point.

Table A2–2b in Appendix A of this text lists the values for the coefficients A_2 and B_2 that describe the variations in ϕ values through the transition zone. The limiting ratios between the depth of the member and the location of the neutral axis (d_t/c) and its inverse at the balanced failure point (i.e., d_t/c_b or c_b/d_t) are also included.

2.20 LIMITATIONS ON THE STEEL PERCENTAGE (ρ) FOR FLEXURAL MEMBERS

With the help of Equations 2–30 and 2–31, the corresponding largest ρ values (i.e., the steel percentages that satisfy those limiting conditions) can be determined. For $\epsilon_t = 0.004$ (lowest

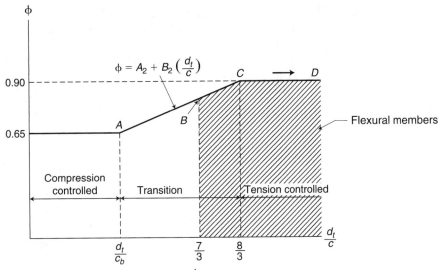

FIGURE 2–37 Variation of ϕ versus $\dfrac{d_t}{c}$.

permitted steel strain value at ultimate strength of flexural members), the maximum depth of the neutral axis is calculated using Equation 2–32.

$$c_{max} = \frac{3}{7} d_t \tag{2–32}$$

The corresponding depth of the equivalent stress block (refer to Equations 2–11 and 2–17) is given by Equation 2–33.

$$a_{max} = \frac{A_{s,max} f_y}{0.85 f'_c b} = \frac{3}{7} \beta_1 d_t \tag{2–33}$$

where $A_{s,max}$ is the amount of reinforcing steel necessary to have $\epsilon_t = 0.004$.

Substituting for $A_{s,max} = \rho_{max} bd$ in Equation 2–33, then rearranging, the largest ρ value can be determined.

$$\frac{\rho_{max} bd f_y}{0.85 f'_c b} = \frac{3}{7} \beta_1 d_t \tag{2–34}$$

$$\rho_{max} = \frac{3}{7} (0.85) \beta_1 \frac{f'_c}{f_y} \cdot \frac{d_t}{d}$$

or

$$\rho_{max} = 0.364 \beta_1 \frac{f'_c}{f_y} \cdot \frac{d_t}{d} \tag{2–35}$$

Equation 2–35 gives the maximum percentage of reinforcing steel permitted by the ACI Code in flexural members, unless the capacity is augmented by the use of compression reinforcing. (See more on that in Chapter 3.)

For sections with a single layer of reinforcing, $d_t/d = 1.0$ and Equation 2–35 is simplified as indicated in Equation 2–36.

$$\rho_{max} = 0.364 \, \beta_1 \frac{f_c'}{f_y} \qquad (2\text{–}36)$$

In a similar way, we can determine the value of ρ that will ensure an $\epsilon_t = 0.005$, the upper limit of ρ needed to ensure a tension-controlled (ductile) failure in beams at their ultimate strength. Designate this value of ρ as ρ_{tc}. After changing the right side of Equation 2–34 accordingly (see Equations 2–30 and 2–31), then the value of ρ_{tc} can be calculated using Equation 2–38 (or Equation 2–39 for the special case of a section with only one layer of reinforcement).

$$\frac{\rho_{tc} \, bdf_y}{0.85 f_c' b} = \frac{3}{8} \beta_1 d_t \qquad (2\text{–}37)$$

$$\rho_{tc} = 0.319 \, \beta_1 \frac{f_c'}{f_y} \cdot \frac{d_t}{d} \qquad (2\text{–}38)$$

$$\rho_{tc} = 0.319 \, \beta_1 \frac{f_c'}{f_y} \qquad (2\text{–}39)$$

Table A2–3 in Appendix A lists the values of ρ_{max} and ρ_{tc} for various grades of steel (f_y) and concrete strength (f_c') combinations. The value of the strength reduction factor (ϕ) is shown in the right column of the table. This value varies when $\rho_{tc} < \rho < \rho_{max}$, or the beam's failure mode is in the transition zone (see Section 2.17). Table A2–3 indicates that not much is gained in terms of useable moment capacity with the required reductions in the ϕ values and when the reinforcing percentage is increased from ρ_{tc} to ρ_{max}, especially when higher strength steels are used.

2.21 MINIMUM STEEL RATIO (ρ_{min}) FOR REINFORCED CONCRETE BEAMS

When a reinforced concrete beam, for architectural or other reasons, is relatively large in cross section, or carries little load, the calculations may require only a very small amount of reinforcing steel. Such a section, if accidentally overloaded, will fail in a sudden, brittle manner. The reason is that the ultimate moment strength provided by the reinforced section is actually less than the strength of the same section without any reinforcing. Thus, the stress in the reinforcement will immediately reach yield at the first crack, causing the section to fail suddenly.

To ensure that reinforced beam's ultimate strength is larger than that of the unreinforced beam, Section 10.5.1 of the ACI Code requires a minimum amount of flexural steel in reinforced concrete beams. This requirement is given in Equation 2–40.

$$A_{s,min} = \frac{3\sqrt{f_c'}}{f_y} bd \geq \frac{200}{f_y} bd \qquad (2\text{–}40)$$

This minimum amount of steel ($A_{s,min}$) provides enough reinforcement to ensure that the moment strength of the reinforced concrete section is more than that of an unreinforced concrete section, which can be calculated from its modulus of rupture.

In the past, the ACI Code required only an $A_{s,min} = \frac{200}{f_y} bd$. For concrete strength greater than about 4,440 psi, however, this is not sufficient to ensure the desired aim; $\frac{3\sqrt{f_c'}}{f_y} bd$ rectifies this condition. Because $A_{s,min} = \rho_{min} bd$, Equation 2–40 may be expressed mathematically in terms of ρ_{min} (minimum steel ratio) as shown in Equation 2–41.

$$\rho_{min} = \max\left\{\frac{3\sqrt{f_c'}}{f_y}, \frac{200}{f_y}\right\} \qquad (2\text{–}41)$$

Table A2–4 in Appendix A provides values of ρ_{min} for different grades of steel and compressive strengths of concrete.

2.22 ANALYSIS OF RECTANGULAR REINFORCED CONCRETE SECTIONS

Analysis of a section means finding the $M_R = \phi M_n$ value. This may be necessary when checking an existing structure or element to determine if the strength provided by the section (*supply*) is sufficient to satisfy M_u that is calculated from the loads (*demand*). Finding M_R also makes it possible to calculate the maximum live load that may be permitted on the element.

An analysis can be performed only when all parameters that influence the ultimate strength of a section are known. There are five of these parameters, namely:

The dimensions of the section	b and d
The materials used in the beam	f_c' and f_y
The tensile reinforcement in the beam	A_s

Next we show two methods for calculating the value of M_R.

M_R Calculation: Method I

This method closely follows the already discussed and established formulae. Figure 2–38 shows the stress and strain distributions for a reinforced concrete rectangular beam at ultimate strength. For the most general case, a beam section with multilayer reinforcing is shown.

The resisting moment can be calculated from the internal couple and using Equations 2–42 through 2–44.

$$T = A_s f_y \qquad C = 0.85 f_c' ba \qquad z = d - a/2$$
$$M_n = Tz = A_s f_y (d - a/2) \qquad (2\text{–}42)$$

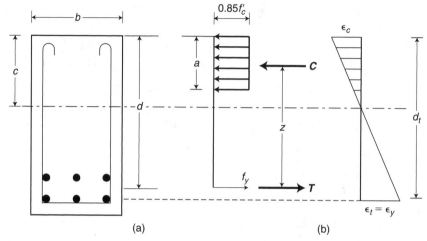

FIGURE 2–38 Stress and strain distributions on a reinforced concrete section.

$$M_n = Cz = 0.85f'_c\, ba\,(d - a/2) \tag{2-43}$$
$$M_R = \phi M_n \tag{2-44}$$

The calculation proceeds as follows:

Step 1 Calculate $\rho = \dfrac{A_s}{bd}$ and check if $\rho \geq \rho_{min}$ (from Table A2–4); if not, the beam does not satisfy the minimum requirements of the ACI Code, and its use for load carrying is not permitted. Determine whether $\rho \leq \rho_{max}$ (from Table A2–3); if not, the beam has too much reinforcing and does not satisfy the latest ACI Code's limitations. A practical solution for this is to disregard the excessive amount of reinforcement, assume that the section is in the transition zone, and continue the calculations with the maximum permissible amount of reinforcing.

Step 2 Calculate the depth of the equivalent stress block from Equation 2–17:

$$a = \frac{A_s f_y}{0.85f'_c\, b}$$

Step 3 Calculate the location of the neutral axis from Equation 2–11:

$$c = \frac{a}{\beta_1}$$

Step 4 Determine whether

$$\frac{c}{d_t} \leq \frac{3}{8}$$

If yes, the beam is in the tension-controlled failure zone; set $\phi = 0.90$ and go directly to step 5. If not $\left(\text{i.e.,} \dfrac{3}{8} \leq \dfrac{c}{d_t} \leq \dfrac{3}{7}\right)$, the ϕ factor must be adjusted accordingly. Therefore, calculate the reduced ϕ:

$$\phi = A_2 + \frac{B_2}{\dfrac{c}{d_t}} \qquad \text{(refer to Table A2–2b for } A_2 \text{ and } B_2\text{)}$$

Step 5   Calculate $M_R = \phi M_n = \phi A_s f_y \left(d - \dfrac{a}{2} \right)$ (refer to Equations 2–42 and 2–44)

Figure 2–39 summarizes the analysis steps.

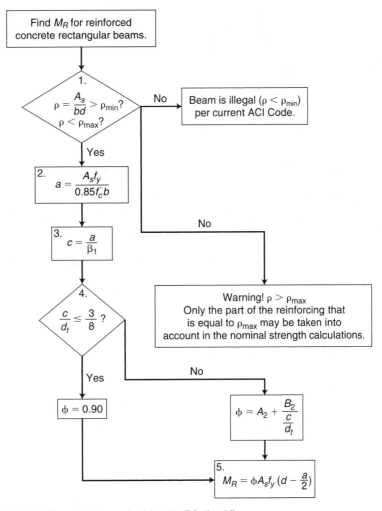

FIGURE 2–39 Flowchart to calculate M_R (Method I).

EXAMPLE 2–6

Use Method I to determine the design resisting moment, M_R, of the reinforced concrete beam section shown below. $f'_c = 4$ ksi, and $f_y = 40$ ksi. The reinforcement is 3 #9 bars, $A_s = 3.00$ in^2.

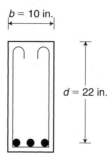

Solution

Using the steps of Figure 2–39:

Step 1   Find the steel ratio, ρ:

$$\rho = \frac{A_s}{bd} = \frac{3}{10 \times 22} = 0.0136$$

From Table A2–4 $\longrightarrow$ $\rho_{min} = 0.0050 < 0.0136 \therefore$ ok
From Table A2–3 $\longrightarrow$ $\rho_{max} = 0.0310 > 0.0136 \therefore$ ok

Step 2   Calculate the depth of the compression zone, a:

$$a = \frac{A_s f_y}{0.85 f'_c b} = \frac{3 \times 40}{0.85 \times 4 \times 10} = 3.53 \text{ in.}$$

Step 3   From the depth of the equivalent stress block, determine the location of the neutral axis, c:

$$c = \frac{a}{\beta_1} = \frac{3.53}{0.85} = 4.15 \text{ in.}$$

Step 4   Determine whether the section is tension-controlled or is in the transition zone:

$$\frac{c}{d_t} = \frac{4.15}{22} = 0.189 < 0.375 \therefore \text{ ok}$$

Therefore, the section is tension-controlled and the strength reduction factor $\phi = 0.90$.

Step 5   Calculate the resisting moment, M_R:

$$M_R = \phi A_s f_y \left(d - \frac{a}{2} \right)$$

$$M_R = \frac{0.90 \times 3 \times 40 \left(22 - \dfrac{3.53}{2} \right)}{12} = 182 \text{ ft-kip}$$

EXAMPLE 2–7

Repeat Example 2–6 for $f_y = 60\,\text{ksi}$, and $f_c' = 3\,\text{ksi}$.

Solution

Step 1

$$\rho = \frac{A_s}{bd} = 0.0136$$

Table A2–4 $\longrightarrow$ $\rho_{\min} = 0.0033 < 0.0136$ $\therefore$ ok
Table A2–3 $\longrightarrow$ $\rho_{\max} = 0.0155 > 0.0136$ $\therefore$ ok

Step 2

$$a = \frac{A_s f_y}{0.85 f_c' b} = \frac{3 \times 60}{0.85 \times 3 \times 10} = 7.06 \text{ in.}$$

Step 3

$$c = \frac{a}{\beta_1} = \frac{7.06}{0.85} = 8.30 \text{ in.}$$

Step 4

$$\frac{c}{d_t} = \frac{8.30}{22} = 0.377 > 0.375$$

$\therefore$ Section is in the transition zone (although just barely).

$$\phi = A_2 + \frac{B_2}{\dfrac{c}{d_t}}$$

Use Table A2–2b to determine A_2 and B_2; then

$$\phi = 0.233 + \frac{0.25}{0.377} = 0.90$$

Step 5

$$M_R = \frac{0.90 \times 3 \times 60 \left(22 - \dfrac{7.06}{2} \right)}{12} = 249 \text{ ft-kip}$$

EXAMPLE 2–8

Calculate M_R for the reinforced concrete beam section shown below. $f_y = 60\,\text{ksi}, f_c' = 4\,\text{ksi}$, and $A_s = 7.62\,\text{in}^2$.

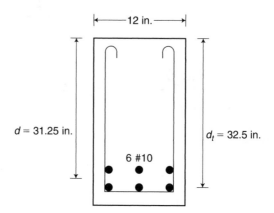

Solution

Step 1

$$\rho = \frac{A_s}{bd} = \frac{7.62}{12 \times 31.25} = 0.0203$$

From Table A2–4 $\rho_{\min} = 0.0033 < 0.0203$ ∴ ok
From Table A2–3 $\rho_{\max} = 0.0207 > 0.0203$ ∴ ok

Step 2

$$a = \frac{A_s f_y}{0.85 f_c' b} = \frac{7.62 \times 60}{0.85 \times 4 \times 12} = 11.21\,\text{in.}$$

Step 3

$$c = \frac{a}{\beta_1} = \frac{11.21}{0.85} = 13.19\,\text{in.}$$

Step 4

$$\frac{c}{d_t} = \frac{13.19}{32.5} = 0.406 > 0.375$$

∴ Section is in the transition zone. With the help of Table A2–2b:

$$\phi = A_2 + \frac{B_2}{\dfrac{c}{d_t}} = 0.233 + \frac{0.25}{0.406} = 0.85$$

Step 5

$$M_R = \frac{0.85 \times 7.62 \times 60 \left(31.25 - \dfrac{11.21}{2} \right)}{12} = 831 \text{ kip-ft}$$

M_R Calculation: Method II

This method results in the development of design aid tables, which are more user-friendly. The tables will also be useful when the aim is to design beam sections to satisfy a given M_u demand instead of analyzing.

The expressions for the components of the internal couple are

$$T = A_s f_y \qquad C = 0.85 f'_c\, ba \qquad z = d - a/2$$

Because $T = C$, the depth of the equivalent stress block is

$$a = \frac{A_s f_y}{0.85 f'_c b}$$

Substituting from Equation 2–18, $A_s = \rho bd$. Equation 2–45 can be used to calculate a.

$$a = \frac{\rho bd f_y}{0.85 f'_c b} = \frac{\rho d f_y}{0.85 f'_c} \tag{2–45}$$

Substituting from Equation 2–11, $a = \beta_1 c$, c can be determined using Equation 2–46.

$$\beta_1 c = \frac{\rho d f_y}{0.85 f'_c}$$

$$c = \frac{\rho f_y d}{0.85 f'_c \beta_1} \tag{2–46}$$

or

$$\frac{c}{d_t} = \frac{\rho f_y}{0.85 f'_c \beta_1} \cdot \frac{d}{d_t} \tag{2–47}$$

Equation 2–47 is usually the preferred equation to check if a section is tension-controlled.

If $^3/_8 \le c/d_t \le {}^3/_7$, the strength reduction factor, ϕ, must be adjusted accordingly.

$$\phi = A_2 + \frac{B_2}{\dfrac{c}{d_t}}$$

Substituting for c/d_t from Equation 2–47, Equation 2–48 can be used to calculate the adjusted strength reduction factor.

$$\phi = A_2 + B_2 \frac{0.85 f'_c \beta_1}{\rho f_y} \cdot \frac{d_t}{d} \tag{2–48}$$

Equation 2–48 provides the values of ϕ in the transition zone. In order to simplify the equation, introduce a new steel ratio, ρ_t:

$$\rho_t = \frac{A_s}{bd_t}$$

(Note that $d_t = d$ and $\rho_t = \rho$ when the beam has only a single layer of steel.)

Substituting $\rho = \rho_t \dfrac{d_t}{d}$, Equation 2–48 can be rewritten as Equation 2–48a.

$$\phi = A_2 + B_2 \frac{0.85f_c'\beta_1}{\rho_t f_y} \tag{2–48a}$$

From Equation 2–43 (see also Figure 2–38):

$$M_R = \phi M_n = \phi Cz$$

$$M_R = \phi(0.85f_c'ba)\left(d - \frac{a}{2}\right)$$

Substituting from Equation 2–45 for a:

$$M_R = \phi\left(0.85f_c'b\,\frac{\rho df_y}{0.85f_c'}\right)\left(d - \frac{\rho df_y}{1.7f_c'}\right)$$

Rearranging and simplifying:

$$M_R = bd^2\left[\phi\rho f_y\left(1 - \frac{\rho f_y}{1.7f_c'}\right)\right] \tag{2–49}$$

If the product in the bracket is designated as R (called the *resistance coefficient*, which has units of stress, psi or ksi) as shown in Equation 2–50,

$$R = \phi\rho f_y\left(1 - \frac{\rho f_y}{1.7f_c'}\right) \tag{2–50}$$

the expression for M_R is simplified to Equation 2–51.

$$M_R = bd^2 R \tag{2–51}$$

It is clear from Equation 2–50 that R depends on the materials used (i.e., f_c', f_y, and the steel ratio (ρ) in the beam), but it is independent of the dimensions of the section. Thus tables for R can be developed in terms of ρ for the various combinations of materials. Values of R can be found from Tables A2–5 through A2–7. In these tables the ρ_{min} value is printed in bold. Reinforcement ratio (ρ) values less than ρ_{min} may not be used in beams, but may be used in slabs and footings.

These tables were developed with R in psi. Using R in psi and beam dimensions b and d in inches results in lb-in. units for M_R. Because kip-ft are usually used in moment calculations, appropriate conversions must be made between lb-in. and kip-ft for the correct use of the tables.

$$M_R \text{ (ft-kip)} = b \text{ in. } (d \text{ in.})^2 \frac{R \text{ (psi)}}{12,000}$$

The tables must be used with care, especially when large ρ values result in the section being in the transition zone. The value of ϕ depends on f_c', f_y, ρ, and $\dfrac{d_t}{d}$, thus if

$$\rho \leq \rho_{tc} \longrightarrow \phi = 0.90$$

and if

$$\rho_{max} \geq \rho > \rho_{tc} \longrightarrow \phi = A_2 + B_2 \frac{0.85f_c'\beta_1}{\rho f_y} \cdot \frac{d_t}{d}$$

The values of ρ_{tc} and ρ_{max} for common grades of steel and concrete strength are listed in Table A2–3.

An important note here is that Tables A2–5 to A2–7 have been developed based on ρ (i.e., beams with a single layer of reinforcement). If the beam has multiple layers of reinforcement ($\rho_t \neq \rho$), the R value must be modified by adjusting it to an R' value based on ρ_t. This can be easily done by using Equation 2–51a.

$$R' = \frac{\phi'}{\phi} R \qquad (2\text{–}51\text{a})$$

The values of ϕ', which are listed in Tables A2–5 to A2–7, correspond to the values of ρ_t.

The use of Method II for analysis of reinforced concrete beam sections involves the following steps:

Step 1   Determine whether $\rho \geq \rho_{min}$; if not, then the beam does not satisfy the minimum requirements of the ACI Code and its use for load carrying is not permitted.

Determine whether $\rho \leq \rho_{max}$; if not, the beam has too much reinforcing and does not satisfy the latest ACI Code's limitations. A practical solution for this is to disregard the excessive amount of reinforcement, assume that the section is in the transition zone, and continue the calculations with the maximum permissible amount of reinforcing.

Step 2   Use ρ, f_c' and f_y to obtain R and ϕ values from the appropriate Tables A2–5 through A2–7. If the beam has a single layer of steel or $\phi = 0.90$, find M_R from Step 3. Otherwise move to Step 4.

Step 3   Calculate $M_R = \dfrac{bd^2R}{12,000}$ (b, d = in; R = psi; M_R = ft-kip)

Step 4   For beams with multiple layers of reinforcement, calculate $\rho_t = \dfrac{A_s}{bd_t}$ and obtain the corresponding strength reduction factor (ϕ') from Tables A2–5 through A2–7.

Step 5   Calculate the modified value of the coefficient of resistance $\left(R' = R\dfrac{\phi'}{\phi} \right)$.

Step 6   Calculate $M_R = \dfrac{bd^2R'}{12,000}$.

The flowchart for Method II is shown in Figure 2–40.

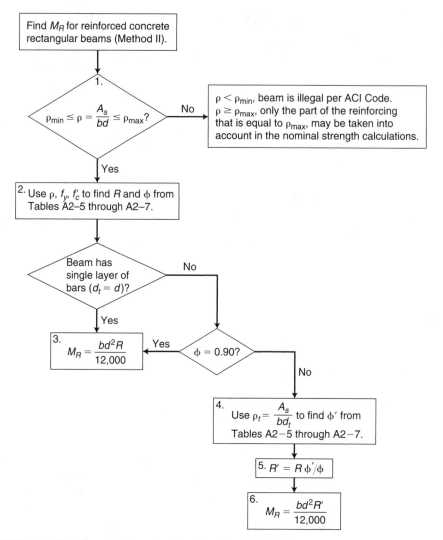

FIGURE 2–40 Flowchart to calculate M_R using Method II.

Method II has fewer steps to follow, so it is easier to use. Method I, however, is more general as it does not require the use of design tables (which may not be readily available) and it is adaptable to any grade of steel or compressive strength of concrete, not just the ones listed in the tables.

EXAMPLE 2–9

Solve Example 2–6 using Method II.

Solution

Step 1   From Example 2–6:

$$\rho = 0.0136 > \rho_{min} = 0.0050 \quad \therefore \text{ ok}$$

From Table A2–3 $\longrightarrow \rho_{max} = 0.0310 > 0.0136 \quad \therefore \text{ ok}$

Step 2   Using $\rho = 0.0136$, $f_y = 40\,\text{ksi}$, and $f_c' = 4\,\text{ksi}$, obtain the resistance coefficient, R, from Table A2–5b:

$$R = 450\,\text{psi}, \phi = 0.90$$

Step 3   Because the beam has a single layer of reinforcement:

$$M_R = \frac{bd^2R}{12,000} = \frac{10 \times 22^2 \times 450}{12,000}$$

$$M_R = 182\,\text{ft-k}$$

which is the same as determined in Example 2–6.

EXAMPLE 2–10

Solve Example 2–7 using Method II.

Solution

Step 1   From Example 2–7:

$$\rho_{max} = 0.0155 > \rho = 0.0136 > \rho_{min} = 0.0033 \quad \therefore \text{ ok}$$

Step 2 $\rho = 0.0136$, $f_c' = 3\,\text{ksi}$, and $f_y = 60\,\text{ksi}$. From Table A2–6a:

$$R = 615\,\text{psi}, \phi = 0.90$$

Step 3

$$M_R = \frac{bd^2R}{12,000}$$

$$M_R = \frac{10 \times 22^2 \times 615}{12,000} = 248\,\text{ft-kip}$$

which is about the same as the result determined in Example 2–7.

EXAMPLE 2–11

Solve Example 2–8 using Method II.

Solution

Step 1   From Example 2–8:

$$\rho = 0.0203 > 0.0033 \quad \therefore \text{ ok}$$

Because there are two layers of reinforcement, adjust ρ_{max} using Table A2–3:

$$\rho_{max} = 0.0207 \frac{d_t}{d} = 0.0207 \frac{32.5}{31.25} = 0.0215 > 0.0203 \quad \therefore \text{ ok}$$

Step 2   Use ρ, f_y, and f_c' to obtain R from Table A2–6b.

$$\rho = 0.0203$$
$$f_c' = 4 \text{ ksi} \longrightarrow \text{Table A2–6b} \longrightarrow R = 825 \text{ psi}$$
$$f_y = 60 \text{ ksi} \qquad\qquad\qquad\qquad \phi = 0.82$$

Step 3   Because the beam has two layers of reinforcement and ϕ is not equal to 0.90, determine ρ_t and ϕ' and adjust the resistance coefficient, R:

Step 4

$$\rho_t = \frac{A_s}{bd_t} = \frac{7.62}{12 \times 32.5} = 0.0195$$

From Table A2–6b $\longrightarrow \phi' = 0.85$

Step 5   Adjusted value of the resistance coefficient (R') is:

$$R' = R \frac{\phi'}{\phi} = 825 \times \frac{0.85}{0.82} = 855 \text{ psi}$$

Step 6 $$M_R = \frac{bd^2 R'}{12,000} = \frac{12 \times 31.25^2 \times 855}{12,000} = 835 \text{ ft-kip}$$

This result is about the same as that from using Method I. The difference is insignificant and is due to rounding errors in the calculations.

2.23   SELECTION OF APPROPRIATE DIMENSIONS FOR REINFORCED CONCRETE BEAMS AND ONE-WAY SLABS

Selection of Depth

The selection of a beam's depth is almost always a controversial issue. On the one hand, the building designer wants to minimize the depth of the structure in order to maximize the headroom without unduly increasing the height of the building. On the other hand, structural elements that are too shallow lead to increased short- and long-term deflections. These, in turn, may be detrimental to attached nonstructural building elements. Excessive deflections of concrete structures may result in cracked walls and partitions, non-functioning doors, and so on.

To guide in the design of well-functioning structures, the ACI Code (Section 9.5.2.1) recommends a set of span/depth ratios, with the comment that the designer does not have to calculate deflections (an involved and somewhat uncertain process) if the utilized depth is at least equal to the values provided in ACI Table 9–5(a). These values are summarized graphically in Figures 2–41 and 2–42.

Note from Figures 2–41 and 2–42 that the recommended minimum depth for simply supported beams is span/16, whereas for one-way slabs this value is span/20. These types of support conditions are quite rare in monolithic reinforced concrete construction, because

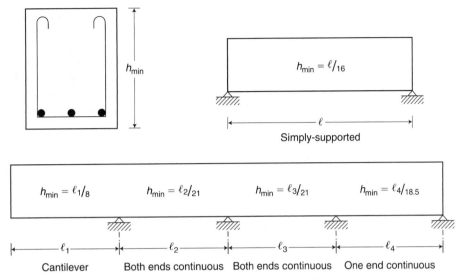

FIGURE 2–41   Minimum depth requirements for reinforced concrete beams.

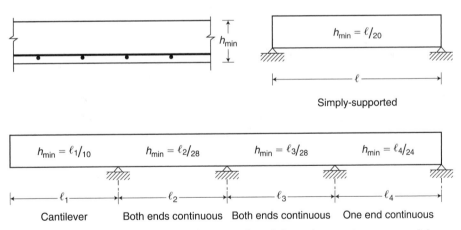

FIGURE 2–42   Minimum depth requirements for reinforced concrete one-way slabs.

in most cases either continuity or some other type of restraint is available at the supports. If the member is continuous at both ends, $h_{min} = $ span/21 for beams and $h_{min} = $ span/28 for one-way slabs. Finally, if the beam is continuous at only one end, the minimum depth is span/18.5, and for one-way slabs is span/24.

A cautionary note is in order here. Span 2 (ℓ_2) in Figures 2–41 and 2–42 is shown as "Both ends continuous." This assumption is valid only if the cantilever at the left end of ℓ_2 is long enough to develop a significant end moment. Experience shows that when the cantilever length is at least $\ell_2/3$, the span ℓ_2 may safely be assumed as "both ends continuous" from the point of view of satisfactory deflection control.

The values shown in Figures 2–41 and 2–42 are applicable only to normal-weight concrete (w_c = 145 lb/ft³) and Grade 60 reinforcement. For other conditions, the ACI Code recommends the following modifications:

a. For lightweight concrete in the range of 90–120 pcf, the values in Figures 2–41 and 2–42 need to be multiplied by $(1.65 - 0.005w_c)$ where w_c = unit weight of concrete in lb/ft³. This factor should not be less than 1.09. For a typical lightweight structural concrete, w_c = 115 pcf. Then the multiplier is 1.65−0.005 × 115 = 1.075 < 1.09. Use a multiplier equal to 1.09.

b. For f_y other than 60,000 psi, the values obtained from Figures 2–41 and 2–42 shall be multiplied by:

$$\left(0.4 + \frac{f_y}{100,000} \right) \tag{2–52}$$

If the selected beam depth is less than the recommended h_{min}, the beam deflection has to be calculated and checked against the ACI Code requirements. Therefore, if a beam does not satisfy the minimum depth requirements, it may still be acceptable if computation of deflection proves it to be satisfactory.

Selection of Width

Minimum Bar Spacing in Reinforced Concrete Beams In Section 2.12 we discussed the role of concrete cover over the reinforcement. Reinforcing bars also need space between them to ensure adequate bond surface at their interface with the concrete. The space should also be larger than the size of the largest aggregate particle in the concrete.

Sections 7.6.1, 7.6.2, and 3.3.2 of the ACI Code require a minimum clear space for single and multiple layers of bars as follows:

Minimum Space (s_{min}) for Single Layer of Bars The *minimum space (s_{min})* for a single layer of bars in beams (see Figure 2–43a) is the largest of the following: the diameter of bar (d_b), 1 in., and $^4/_3$ of maximum size aggregate used in the concrete mixture.

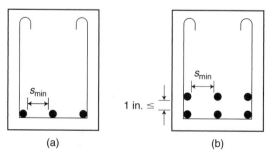

FIGURE 2–43 Minimum spacing between reinforcing bars: (a) single layer; (b) multiple layer.

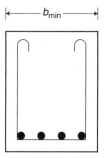

FIGURE 2-44   Minimum beam width (b_{min}).

Mathematically:

$$s_{min} = \max\{d_b, 1 \text{ in.}, {}^4\!/_3 \text{ max. aggregate size}\}$$

Note that in most building structure applications (save for footings and foundations) the usual concrete mix limits the size of the aggregate to $^3\!/_4$ in. Thus, a 1 in. minimum spacing satisfies the third of the spacing requirements.

Minimum Space for Multiple Layers of Bars   Where reinforcement is placed in two or more layers (see Figure 2–43b), bars in the upper layers shall be placed directly above bars in the lower layer with clear distance between layers not less than 1 in. In addition, the requirements of single-layer bars must also be satisfied.

Minimum Width (b_{min}) of Reinforced Concrete Beams   We use the minimum required space between bars in a single layer to calculate the *minimum beam width* needed to provide enough room for a specific number and size of bars. To compute b_{min}, consider Figure 2–44. Usually #3 or #4 bars are used for stirrups. Also, the minimum cover for bars in beams is 1.5 in. Therefore, we can calculate b_{min} by adding the minimum required spaces and the bar diameters.

As an example, suppose that the beam in Figure 2–44 is reinforced with 4 #8 bars. Assuming #4 stirrups, the minimum width for this beam is:

$$b_{min} = \underset{\underset{\text{Cover}}{\uparrow}}{2 \times 1.5 \text{ in.}} + \underset{\underset{\text{Stirrups}}{\uparrow}}{2 \times \frac{1}{2} \text{ in.}} + \underset{\underset{\text{Main bars}}{\uparrow}}{4 \times 1 \text{ in.}} + \underset{\underset{s_{min}}{\uparrow}}{3 \times 1 \text{ in.}} = 11 \text{ in.}$$

Note that $s_{min} = 1$ in. was used; this assumes that $^4\!/_3$ of the maximum aggregate size is less than or equal to 1 in. Table A2–8, based on the above example, shows b_{min} for different numbers and sizes of bars in a single layer.

2.24   CRACK CONTROL IN REINFORCED CONCRETE BEAMS AND ONE-WAY SLABS

It was previously mentioned that a reinforced concrete member will always crack when subjected to bending. In fact, the reinforcing really starts working only after the development of cracks. Nevertheless, designers try to minimize the size of the cracks. Limitation of crack width is desirable for three main reasons: (1) appearance; (2) limitation of corrosion of the reinforcement; and (3) water-tightness.

Laboratory experiments have shown that several parameters influence the width and spacing of flexural cracks. The first is the *concrete cover* over the reinforcing. The smaller the cover, the smaller the crack width will be. The cover cannot be reduced beyond a certain limit, however, because a minimum cover is needed for fire and corrosion protection. Thus, the Code requires a minimum cover of 1.5 in. over the stirrups for beams and $3/4$ in. for joists and slabs. The 1.5 in. cover over the stirrups results in a cover of $1^7/8$ in. to 2 in. over the main reinforcement.

The second important parameter is the *maximum stress in the reinforcement* (directly related to the strain, or the elongation of the steel) at *service load* levels. This value may be assumed to be roughly $0.66 f_y$. The higher the stress level is in the steel, the wider the cracks are expected to be. Thus, using more reinforcing than required to satisfy the ultimate strength capacity can reduce the width of cracks by reducing the stresses (and strains) at working load levels. This is not an economical choice, however. The same is true if steel with $f_y = 40,000$ psi is used instead of steel with $f_y = 60,000$ psi. The section would need 50% more steel, but the much lower levels of stress at service load levels would help limit the crack width.

Another important parameter is the *maximum spacing* of the reinforcing bars. For minimizing the width of cracks, placing more and smaller bars closer together is preferable to placing a few large bars farther apart. The ACI Code (Section 10.6.4) limits the maximum spacing of the tensile reinforcement in beams and one-way slabs. The empirical formula for maximum spacing, given in Equation 2–53, is based on the tensile stress in the steel and the concrete cover.

$$s = 15\left(\frac{40,000}{f_s}\right) - 2.5c_c \leq 12\left(\frac{40,000}{f_s}\right) \tag{2–53}$$

where s is the center-to-center spacing (in inches) of flexural tension reinforcement nearest to the extreme tension face; f_s is the calculated tensile stress (in psi) at service load in steel or $2/3f_y$; and c_c is the least distance (in inches) from the surface of the reinforcement to the tension face. Equation 2–53 cannot address the control of cracking for all the different causes discussed.

If $f_y = 60,000$ ksi, the right side of Equation 2–53 is limited to 12 in. (since $f_s = 2/3f_y$). The left side of the inequality relates the maximum spacing (s) to the concrete cover (c_c). To better comprehend Equation 2–53, consider Figure 2–45, which shows the

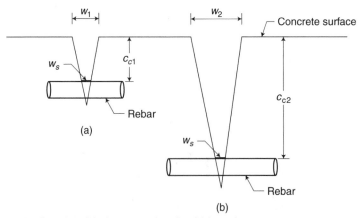

FIGURE 2–45   Relationship between crack width and concrete cover.

reinforcing bar with two different covers, c_{c1} and c_{c2}. If the concrete cover is *increased* from c_{c1} to c_{c2} and the crack width at the level of the reinforcement (w_s) is constant, the *surface crack width increases* from w_1 to w_2. Figure 2–45 clearly shows the relationship between surface crack width and amount of concrete cover.

We can use the maximum spacing limitation (s) given by Equation 2–53 to determine the maximum beam width (b_{max}) as a function of the number of bars placed in the section. For example, for 4 #4 main bars, #4 stirrups, and $f_y = 60,000$ psi, the maximum permissible spacing of bars (s) is:

$$s = 15\left(\frac{40,000}{f_s}\right) - 2.5c_c \leq 12\left(\frac{40,000}{f_s}\right)$$

$$s = 15\left(\frac{40,000}{{}^2/_3 \times 60,000}\right) - 2.5(1.5 + 0.5) \leq 12\left(\frac{40,000}{{}^2/_3 \times 60,000}\right)$$

$$s = 10 \text{ in.} \leq 12 \text{ in.} \longrightarrow s = 10 \text{ in.}$$

and the maximum beam width (b_{max}) is:

$$b_{max} = 2 \times 1.5 \text{ in.} + 2 \times {}^1/_2 \text{ in.} + {}^1/_2 \text{ in.} + 3 \times 10 \text{ in.} = 34.5 \text{ in.} \approx 34 \text{ in.}$$

$$\underset{\text{Cover}}{\uparrow} \qquad \underset{\text{#4 stirrups}}{\uparrow} \quad \underset{\text{#4 bar}}{\uparrow} \qquad \underset{s}{\uparrow}$$

Note that in the above calculation, s is the *center-to-center* distance of the reinforcing bars. Therefore, only the diameter of one bar was used to determine b_{max}. The last column of Table A2–8 lists b_{max} for different sizes and numbers of bars in a single layer. In practice, b_{max} is rarely a problem for beams; however, the maximum spacing limitation is an important issue when designing reinforcing layouts in slabs.

Table A2–9 shows the areas of reinforcing steel (A_s) for different sizes and numbers of bars.

2.25 DESIGN OF BEAMS

The ultimate strength of a beam depends on five parameters. These are the materials (f'_c and f_y), the dimensions of the section (b and d), and the amount of reinforcement (A_s). The last three parameters may be expressed in the form of the steel ratio $\rho = A_s/bd$.

Whichever way these parameters are expressed, they are always five in number. There is only one equation (or, more precisely, one inequality), however, that expresses the problem:

$$M_u \leq M_R$$

The left side of this inequality depends only on the applied loads. The right side of the inequality, on the other hand, depends on all five of the variables listed above. Thus, this problem has an infinite number of solutions. But if four out of the five parameters are pre-selected or assumed, the inequality can be readily solved.

As an example, contemplate the following considerations. In a floor of a given structure, it would be quite impractical to vary the quality of the concrete. Consequently, every beam and slab of the floors of the structure is usually cast with the same quality concrete (same f'_c) throughout. (In columns, the use of a different quality concrete may be warranted; but even then all columns in a given floor level would have the same concrete mix.) So preselecting the concrete quality for the slabs and beams throughout a building is standard practice.

The same is true with the reinforcement. Labor is the dominant factor in the price of the "in-place" reinforcing steel. And the basic cost per ton of reinforcing steel with $f_y = 40$ ksi and $f_y = 60$ ksi is very near the same, so there is no economic incentive to use the former. In fact, 60 ksi steel provides 50% more strength than 40 ksi steel, thus making it cheaper to use.

Of the three remaining variables, b (the width of the section), d (the working depth of the section), and A_s (the amount of reinforcement), two must still be preselected in order to solve for the remaining unknown quantity. Generally speaking, practitioners select a concrete section (b and h) and then solve for a minimum required amount of reinforcement to satisfy the demanded factored moment requirements. Often all beams have the same depth and width to enable the contractor to reuse the forms. In other cases keeping the depth of all beams uniform satisfies the minimum headroom requirement throughout the structure.

In general, two types of problems arise: (1) The beam's sizes (b and h) are set using the considerations stated above and the designer needs only to determine the required area of steel (A_s); this is by far the most common problem. (2) The beam's sizes (b and h) and area of steel (A_s) are all unknown and determined by the designer during the process; this problem is more academic than practical.

b, h = known, A_s = unknown

The flowchart in Figure 2–46 shows the steps for the design process.

Step 1   Find the maximum factored bending moment, M_u.

Step 2   Because the bar sizes are not yet known, assume the distance from the edge of concrete in tension to the center of steel ($\bar{y}$) is 2.5 in. This is a reasonable assumption if the cover is 1.5 in., the stirrup diameter is $3/8$ in. (#3) or $1/2$ in. (#4), the main reinforcement is #8 to #10 bars or smaller, and there is only one layer of reinforcement.

Step 3   Use the assumed value of d to calculate the required resistance coefficient (R).

$$M_R = bd^2R \qquad \text{(refer to Equation 2–51)}$$

If b and d are in inches, and R in psi, M_u will need to be converted to in.-lb from its usual ft-kip units.

$$M_R = \frac{bd^2R}{12{,}000}$$

Set $M_u = M_R$:

$$M_u = M_R = \frac{bd^2R}{12{,}000}$$

$$R = \frac{12{,}000M_u}{bd^2}$$

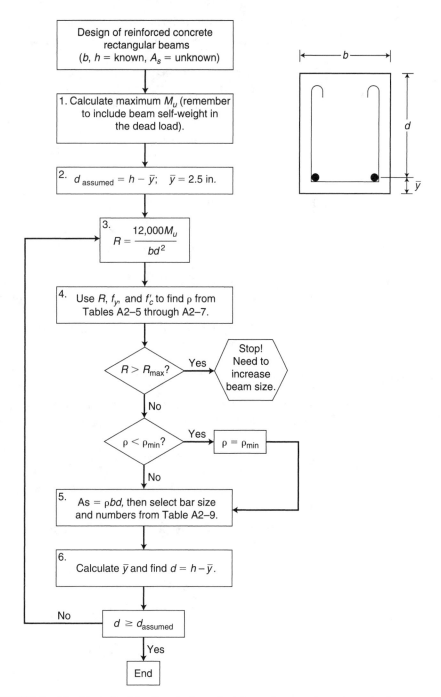

FIGURE 2–46 Flowchart for the design of reinforced concrete rectangular beams (b, h = known, A_s = unknown).

Step 4   Use R, f_y, and f_c' to determine ρ from Tables A2–5 through A2–7. If R is greater than the maximum R value (R_{max}) to be found in the tables, it means that the selected sizes are too small and must be increased.

If the value obtained is less than ρ_{min}, it means that the beam sizes b and h are larger than needed to carry the loads with minimum reinforcement. This may happen when other considerations dictate the beam sizes. In this case use $\rho = \rho_{min}$ from Table A2–4, because the beam *must* always have the required minimum reinforcement.

Step 5   Determine how much steel is needed and select bars using Table A2–9. It is also helpful to use Table A2–8 here, because it lists how many of a certain size of bar may be fitted into the selected b in a single layer.

Step 6   Once the bar sizes are known, the exact effective depth (d) can be calculated. If this depth is greater than what was assumed at the beginning of process, the design will be conservative as it will have more moment capacity than what was demanded. If the effective depth is less than the assumed value (e.g., the section needs multiple layers of reinforcements), then the process needs to be repeated with a new value of d. Insignificant differences in the assumed and recalculated values in d (less than $3/8$ in. in slabs and $1/2$ in. in beams) may be neglected and the reinforcing need not be redesigned.

Note that having multiple layers of reinforcing bars may influence the value of the strength reduction factor, ϕ.

EXAMPLE 2–12

Figure 2–47a shows the partial framing plan of a beam-girder reinforced concrete floor system. The slab is 6 in. thick, and is subjected to a superimposed dead load of 30 psf. The floor live load is 100 psf. Beam B-2 has a width of 12 in. ($b = 12$ in.), and a total depth of 30 in. (including the slab thickness). Determine the steel required at Section 1.1. Use the ACI Code coefficients to calculate moments. Assume that the beam end is integral with the column. Use $f_c' = 4$ ksi, $f_y = 60$ ksi, and assume that the unit weight of concrete is 150 pcf. Stirrups are #3 bars.

Solution

Step 1   Before calculating the moments at the selected location, we must determine the floor loads:

$$\text{Weight of slab} = 150 \times (6/12) = 75 \text{ psf}$$
$$\text{Superimposed dead load} = 30 \text{ psf}$$
$$\text{Total dead load} = 105 \text{ psf}$$
$$\text{Live load} = 100 \text{ psf}$$

The tributary width for beam B-2 is 15'-0"; therefore, the uniform dead and live loads are:

$$w_D = \frac{105 \times 15}{1000} + \frac{150\left(\dfrac{12}{12} \times \dfrac{24}{12}\right)}{1000} = 1.88 \text{ kip/ft}$$

— Beam weight

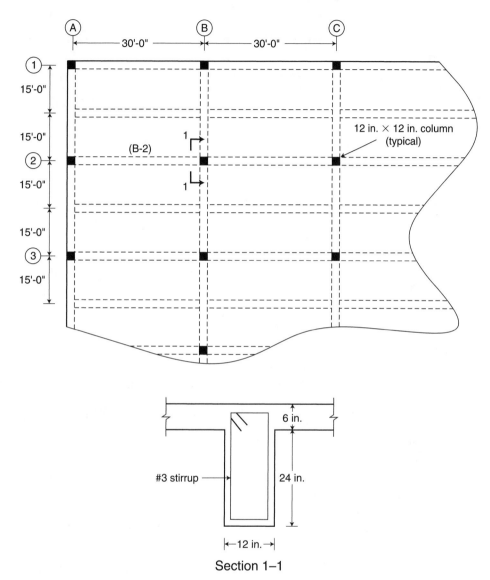

Section 1–1

FIGURE 2–47a Framing plan and section for Example 2–12.

$$w_L = \frac{100 \times 15}{1000} = 1.5 \text{ kip/ft} \qquad \text{Note : Reduction of live load is neglected here.}$$

$$w_u = 1.2w_D + 1.6w_L = 1.2 \times 1.88 + 1.6 \times 1.5 = 4.65 \text{ kip/ft}$$

The beam clear span $\ell n = 30 \text{ ft} - (0.5 \text{ ft} + 0.5 \text{ ft}) = 29 \text{ ft}$

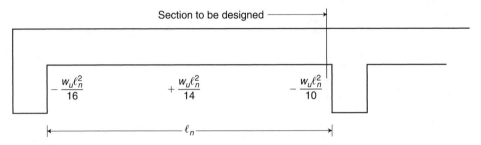

FIGURE 2–47b   Moments using the ACI coefficients (Example 2–12).

Figure 2–47b shows the moments using the ACI coefficients from Table A2–1 for an exterior beam. Because the problem requires designing the reinforcement at section 1-1:

$$(M_u)^- = \frac{w_u\,\ell_n^2}{10} = \frac{4.65(29)^2}{10} = 391 \text{ ft-kip}$$

Step 2   Assuming the distance ($\bar{y}$) from the edge of the beam in tension to the center of tensile steel is 2.5 in.:

$$d = h - \bar{y} = 30 \text{ in.} - 2.5 \text{ in.} = 27.5 \text{ in.}$$

Step 3   The required resistance coefficient, R, is:

$$R = \frac{12{,}000 M_u}{bd^2} = \frac{12{,}000 \times 391}{12(27.5)^2}$$
$$R = 517 \text{ psi}$$

Step 4

$$R = 517 \text{ psi}$$
$$f_c' = 4 \text{ ksi} \longrightarrow \text{Table A2–6b} \longrightarrow \rho = 0.0106$$
$$f_y = 60 \text{ ksi}$$

Note that $\rho = 0.0106$ corresponding to $R = 519$ psi was conservatively selected.

$$\text{Table A2–4} \longrightarrow \rho_{\min} = 0.0033 < \rho = 0.0106 \quad \therefore \text{ ok}$$

Step 5   Find the required amount of steel:

$$A_s = \rho bd = 0.0106(12)(27.5) = 3.50 \text{ in}^2$$

From Table A2–9 $\longrightarrow$ Try 3 #10 ($A_s = 3.81$ in^2)

The reinforcement is placed at the top of the beam, because the moment is negative at the section under investigation, which causes tension at the top. Figure 2–47c shows a sketch of the beam.

$$\text{Table A2–8} \longrightarrow b_{\min} = 10.5 \text{ in.} < 12 \text{ in.} < b_{\max} = 24 \text{ in.} \quad \therefore \text{ ok}$$

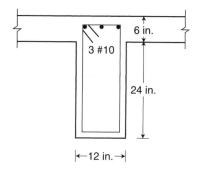

FIGURE 2–47c   Sketch of beam for Example 2–12.

Step 6   Check for the actual effective depth, d:

$$\bar{y} = 1.5 \text{ in.} + {}^{3}\!/_{8} \text{ in.} + {}^{1.27}\!/_{2} = 2.51 \text{ in.}$$

Cover   Stirrup   Bar diameter

$$d = h - y = 30 \text{ in.} - 2.51 \text{ in.} = 27.49 \text{ in.} \approx d_{\text{assumed}} = 27.5 \text{ in.} \quad \therefore \text{ ok}$$

b, h, A_s = unknown

There is still only one design equation, but the problem now is formulated differently. It is somewhat more "contorted" than the previous one, for if the designer does not like the results obtained with the assumed cross section and the corresponding reinforcement, he or she can just change the width or the depth (or both) and recalculate the reinforcement until satisfied with the design.

A first assumption may be an arbitrary selection of the steel ratio ρ. When ratios close to the ρ_{max} value are chosen, the amount of steel required creates a rather congested layout, especially in the positive moment regions (steel is placed in the bottom of the beam). On the other hand, an unnecessarily large concrete section may result if the section's moment requirement can be satisfied with ρ_{min}. Most practical designs have steel ratios somewhere between ρ_{max} and ρ_{min}.

Generally speaking, if ρ is assumed to be about $0.6\rho_{\text{max}}$ or less the beam proportions will likely be such that excessive deflection will not be a problem. Therefore, Table 2–1 is provided as an aid for the designer. In this table, ρ_{des} was calculated as $0.6\rho_{\text{max}}$ as a starting point.

TABLE 2–1   Design Steel Ratio (ρ_{des})

f_y (psi)	ρ_{des}		
	f'_c = 3,000 psi	f'_c = 4,000 psi	f'_c = 5,000 psi
40,000	0.0139	0.0186	0.0218
60,000	0.0093	0.0124	0.0146
75,000	0.0074	0.0099	0.0116

Then the corresponding R value may be obtained from Tables A2–5 through A2–7. The value bd^2 can be determined using M_u:

$$M_u = Rbd^2 \longrightarrow bd^2 = \frac{M_u}{R}$$

Two unknowns remain, however: b and d. There are no ACI Code requirements on the *geometrical proportioning* of beams. But it is more economical to design beams as deep and narrow rather than wide and shallow sections. This means that the effective depth, d, should be larger than the width, b. Generally speaking, the most economical beam sections for spans up to 25 ft usually have a d/b ratio between 1.5 and 2.5. For longer spans, a d/b ratio of 3 to 4 may be more suitable. Economy for a specific beam (or set of beams) is not the same as economy for the overall building. In fact, sometimes it is more economical to design wide and shallow beam sections due to the savings in the floor-to-floor height, even though this design will require more reinforcing steel.

Figure 2–48 summarizes the steps of the design process:

Step 1   Find the factored loads and moments.
Step 2   Use f_y and f_c' to select a ρ_{des} value from Table 2–1. Then find the corresponding R value from the appropriate design table (Tables A2–5 through A2–7).
Step 3   The formula for M_R is:

$$M_R = \frac{bd^2R}{12,000}$$

and the design of the beam requires that $M_R \geq M_u$. For the most economical case, $M_u = M_R$; therefore

$$\frac{bd^2R}{12,000} = M_u$$

Solving for bd^2:

$$bd^2 = \frac{12,000M_u}{R}$$

Now we must preselect one dimension or the other: We either assume b and solve for d, or the other way around. A third possibility is to assume a certain proportion between d and b, for example, $d/b = 2$; then the problem again becomes straightforward.

Step 4   Use the values of b and d from above to find the required area of reinforcement (A_s):

$$A_s = \rho bd$$

and select the size and number of bars using Tables A2–8 and A2–9.

Step 5   Now find the beam's total depth (h) using the effective depth (d) from Step 3 and size of bars:

$$h = d + \bar{y}$$

Then round h up to the nearest 1 in.

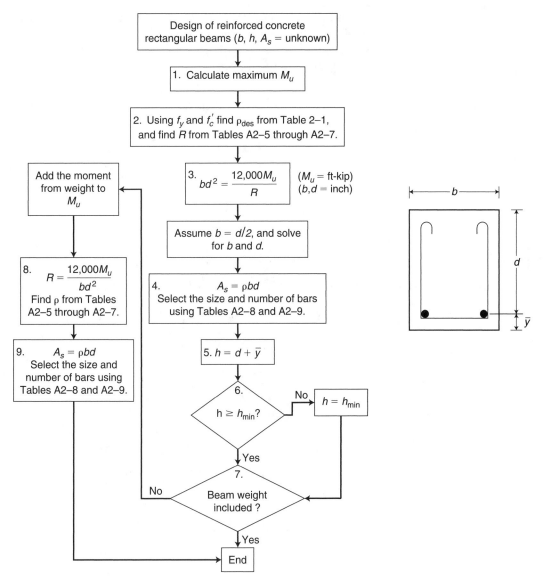

FIGURE 2-48 Flowchart for the design of reinforced concrete rectangular beams (b, h, A_S = unknown).

Step 6   Check the beam depth for expected deformation performance by comparing it with h_{min} as recommended by the ACI Code (see Figure 2–42). If $h < h_{min}$, use h_{min}. In this case you may want to go back and recalculate A_s.

Step 7   Because the beam sizes were not known when the loads were calculated, the beam's self-weight could only be estimated. Experienced designers usually use their own rule of thumb for this purpose. For example, some engineers assume the beam's self-weight to be about 10%–20% of the loads it carries.

Others estimate the total depth (h) to be roughly 6%–8% of the span, and $b \cong 0.5h$, and find a preliminary estimate for the beam's weight. But if we desire a more accurate value of the beam's weight, we can estimate it now and make corrections to the dead load and the total M_u.

Step 8   Find a new R value:

$$R = \frac{12,000M_u}{bd^2}$$

and find the corresponding steel ratio (ρ) using Tables A2–5 through A2–7.

Step 9   Find the required area of steel:

$$A_s = \rho bd$$

and select the numbers and sizes of bars from Tables A2–8 and A2–9.

EXAMPLE 2–13

Determine the required area of steel for a reinforced concrete rectangular beam subject to a total factored moment, $M_u = 400$ ft-kip, that already includes the estimated weight of the beam. $f_c' = 4,000$ psi and $f_y = 60,000$ psi and use $\rho_{des} = 0.0124$ from Table 2–1.

Solution

Step 1

$$M_u = 400 \text{ ft-kip}$$

Step 2

$$\text{For } f_c' = 4 \text{ ksi}, f_y = 60 \text{ ksi and } \rho = 0.0124$$
$$\text{using Table A2–6b} \longrightarrow R = 596 \text{ psi}$$

Steps 3 & 4   Search now for the beam's sizes:

$$bd^2 = \frac{12,000M_u}{R} = \frac{12,000 \times 400}{596}$$
$$bd^2 = 8,054 \text{ in}^3$$

There are an infinite number of solutions, that is, an infinite number of concrete cross sections that will satisfy the design problem, even with the provision that $\rho = 0.0124$ (1.24%). The table below lists a few solutions. Take your pick!

b	10 in.	12 in.	14 in.	16 in.	18 in.	20 in.
d	28.4 in.	26.0 in.	24.0 in.	22.5 in.	21.2 in.	20.1 in.
$A_{s,required}$	3.52 in²	3.87 in²	4.17 in²	4.46 in²	4.73 in²	4.98 in²
$h_{practical}$	32 in.	30 in.	28 in.	26 in.	24 in.	24 in.

A couple of important observations must be made here. All of these sections have approximately 1.24% reinforcement, but the quantity of reinforcing grows as the beam becomes wider and shallower. Furthermore, the concrete cross-sectional area (and, consequently, the self-weight of the beam) also increase.

Another way to solve this same problem is to select a d/b ratio. For example, suppose that after determining that

$$bd^2 = 8{,}054 \text{ in}^3$$

the designer selects a $d/b = 2.0$ ratio. Then:

$$b = \frac{d}{2}$$

$$\frac{d}{2}(d^2) = \frac{d^3}{2} = 8{,}054$$

$$d = 3\sqrt{2 \times 8{,}054} = 25.3 \text{ in.}$$

$$b = \frac{d}{2} = \frac{25.3}{2} = 12.65 \text{ in.} \longrightarrow \text{Select } b = 13 \text{ in.}$$

$$h = 25.3 + 2.5 = 27.8 \text{ in.} \longrightarrow \text{Select } h = 28 \text{ in.}$$

$$A_s = \rho bd = 0.0124 \times 12.65 \times 25.3 = 3.97 \text{ in}^2$$

EXAMPLE 2–14

Use the floor framing plan and loadings of Example 2–12 (Figure 2–47a) to design the reinforced concrete rectangular beam along grid line 2. Assuming that the beam width $b = 12$ in., determine the beam depth, h, and required steel for the location of the maximum bending moment. Use ACI Code coefficients for calculation of moments. Assume that the beam end is integral with the column, $f_y = 60$ ksi, $f'_c = 4$ ksi, and the unit weight of the concrete is 150 pcf. The stirrups are #3 bars.

Solution

Step 1   Find the maximum ultimate moment, M_u.

From Example 2–12:

$$w_D = \frac{105 \times 15}{1000} = 1.58 \text{ kip/ft} \quad \text{(without the weight of the beam's stem)}$$

$$w_L = \frac{100 \times 15}{1000} = 1.5 \text{ kip/ft} \quad \text{(without the use of live load reduction)}$$

$$w_u = 1.2w_D + 1.6w_L = 1.2 \times 1.58 + 1.6 \times 1.5 = 4.3 \text{ kip/ft}$$

$$\ell_n = 30 \text{ ft} - \left(\frac{1}{2} + \frac{1}{2}\right) = 29 \text{ ft}$$

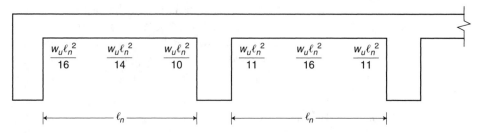

FIGURE 2–49a Moments using the ACI coefficients (Example 2–14).

Using the ACI coefficients (Table A2–1) to calculate moments (Figure 2–49a), we determine that the maximum bending moment for the beam along line 2 is at the first interior column (negative moment):

$$M_u = \frac{w_u \ell_n^2}{10} = \frac{4.3(29)^2}{10} = 362 \text{ ft-kip}$$

Step 2 From Table 2–1 $\longrightarrow$ $f_c' = 4\,\text{ksi}$, $f_y = 60\,\text{ksi}$ $\longrightarrow$ $\rho_{des} = 0.0124$
From Table A2–6b $\longrightarrow$ $R = 596\,\text{psi}$

Step 3 Determine the beam's sizes:

$$bd^2 = \frac{12{,}000 M_u}{R} = \frac{12{,}000 \times 362}{596}$$

$$bd^2 = 7{,}289 \text{ in}^3$$

$$b = 12 \text{ in.} \longrightarrow 12d^2 = 7{,}289$$

$$d^2 = 607 \longrightarrow d = 24.7 \text{ in.}$$

Step 4 Calculate the required area of steel, and select the number and size of the reinforcing bars:

$$A_s = \rho bd = (0.0124)(12)(24.7) = 3.68 \text{ in}^2$$

From Table A2–9 $\longrightarrow$ Try 4 #9 ($A_s = 4\,\text{in}^2$)

Table A2–8 $\longrightarrow$ $b_{min} = 12 \text{ in.} = 12 \text{ in.}$ ∴ ok

Table A2–8 $\longrightarrow$ $b_{max} = 34 \text{ in.} > 12 \text{ in.}$ ∴ ok

Step 5 Use the selected bar sizes and the effective depth (d) to calculate the total beam depth (h):

$$\bar{y} = 1\frac{1}{2} + \frac{3}{8} + \frac{1.128}{2} = 2.44 \text{ in.}$$

$$h = d + \bar{y} = 24.7 + 2.44 = 27.14 \text{ in.}$$

This value is usually rounded up to the nearest 1 in. Thus:

$$h = 28 \text{ in.}$$

Step 6 Check to see if the beam depth is more than the recommended minimum for deflection control. The case for the beam with one end continuous results in the largest required depth (see Figure 2–41):

$$h_{min} = \frac{\ell}{18.5} = \frac{30 \times 12}{18.5} = 19.5 \text{ in.} < 28 \text{ in.}   \therefore \text{ ok}$$

Step 7   Calculate the correct beam weight. The total beam depth is 28 in. The concrete slab, however, is 6 in. thick; therefore, the beam depth (the stem) below the slab is 28 in. − 6 in. = 22 in.

$$\text{Stem weight} = \frac{150 \left(\dfrac{12}{12} \times \dfrac{22}{12} \right)}{1,000} = 0.28 \text{ kip/ft}$$

The total uniform dead load acting on the beam (w_D):

$$w_D = 1.58 + 0.28 = 1.86 \text{ kip/ft}$$
$$w_u = 1.2 \times 1.86 + 1.6 \times 1.5 = 4.63 \text{ kip/ft}$$
$$(M_u)^- = \frac{w_u \ell_n^2}{10} = \frac{4.63(29)^2}{10} = 390 \text{ ft-kip}$$

Step 8

$$R = \frac{12,000 M_u}{bd^2}$$

$$R = \frac{12,000 \times 390}{12(24.7)^2}$$

$$R = 639 \text{ psi}$$

From Table A2–6b $\longrightarrow$ $\rho = 0.0134$ (this corresponds to $R = 638$ psi, which is very close).

Step 9

$$A_s = \rho bd = (0.0134)(12)(24.7) = 3.97 \text{ in}^2$$
From Table A2–9 $\longrightarrow$ Use 4 #9 bars.

The selected reinforcement is the same as it was for the previous design cycle. Figure 2–49b shows the sketch of the beam.

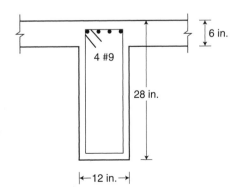

FIGURE 2–49b   Final design of Example 2–14.

2.26   SLABS

Slabs or plates are very important components of reinforced concrete structures. The elements we have studied until now, could be described abstractly by a line: Bending of that line in a vertical plane by the loads described their behavior. These elements are called linear elements, because one of their three dimensions, the length, is much greater than the other two, i.e. the dimensions of the cross section.

Slabs (plates), on the other hand, cannot be described by a line. They have two dimensions, length and width, that are significantly larger than the third one, the thickness. Mathematically plates are described as planes. A mathematically exact analysis of slabs is not provided here but a discussion of their behavior is in order.

A slab can bend in two directions, so its bent shape is described not by the shape of a single line, but rather by the bent shape of a surface. A slab must carry the loads to the supports, hence it will bend accordingly. The behavior of a slab depends on the support conditions, that is, on how the designer chose to support it. The types of supports are:

a. *Line supports* (beams, girders, walls) Slabs that are supported by these types of building elements are referred to as *one-* or *two-way slabs*. In this chapter we discuss only one-way slabs, although an attempt is made to explain the difference between one-way and two-way slabs. Chapter 6 discusses the different types of two-way slabs used as floor systems.

b. *Point supports* (columns, posts, suspension points, etc.) Slabs supported by these types of supports are referred to as *flat slabs* or *flat plates*. We will discuss these in more detail in Chapter 6.

c. *Continuous media* (slabs on grade)

The simple sketch in Figure 2–50 illustrates the behavior of a *one-way* slab. The beams that support the slab are poured together with the slab. Slabs are often not just single span, as shown here, but continuous over several spans defined by the beams' spacing. In the case of uniformly distributed loads, the most common for slabs (for it is quite rare to place large concentrated loads on slabs), every one-foot-wide strip of the slab is loaded identically; hence, the design is limited to only a one-foot-wide strip and the selection of the reinforcing for that strip. Then it is assumed that all the other strips behave the same way, that is, they need the same amount of reinforcing. Figure 2–50 also illustrates that if only one imaginary strip is loaded, the adjacent slab strips will have to help. This is

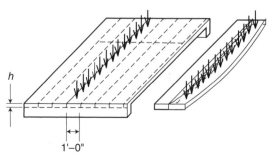

FIGURE 2–50   One-way slab behavior.

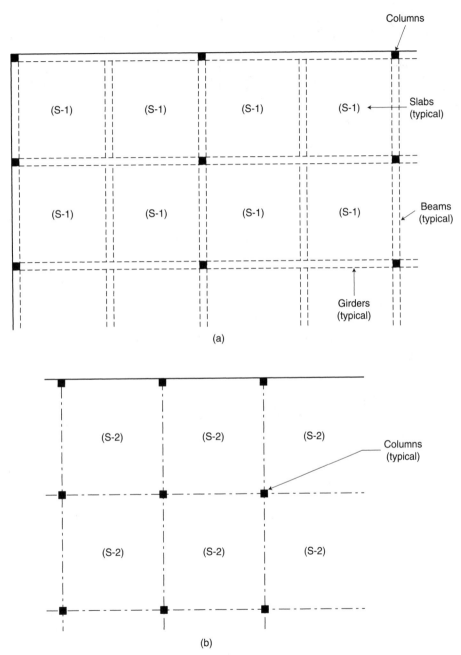

FIGURE 2–51   (a) Slabs in beam girder floor system; (b) flat plate slab.

because it is impossible for a monolithic structure to get the deformation diagram shown on the right of the figure.

Figures 2–51 and 2–52 show the framing plan of different reinforced concrete floor/roof systems. In Figure 2–51a, slab S-1 is supported by the surrounding beams and girders. In Figure 2–51b slab S-2 is part of a flat plate floor system, in which slabs are

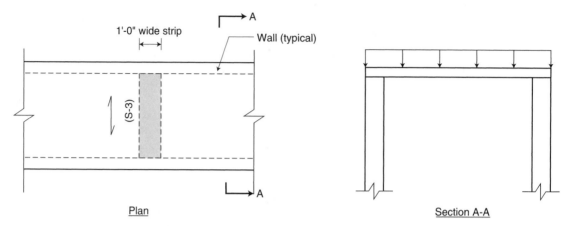

FIGURE 2–52 Slab supported by walls.

directly supported by columns. In Figure 2–52 slab S-3 is supported by two parallel walls, which can be made of concrete or masonry.

2.27 BEHAVIOR OF REINFORCED CONCRETE SLABS UNDER LOADS

Depending on the geometry and location of the supports, most slabs are divided into two groups: one-way slabs, and two-way slabs.

One-way slabs bend mainly in one direction. If the supporting elements of the slab are only two parallel members such as beams or walls, the slab is forced to bend in a perpendicular direction. Figure 2–52 shows the plan view of a slab supported by two parallel walls. Because every 1 ft wide strip can be considered to be the same as all the others, only a single 1 ft wide strip of slab needs to be considered in analysis and design.

The slab's geometry is an important factor that affects its behavior under loads. Figure 2–53a shows a slab supported by edge beams B-1 and B-2. Determining the distribution of loads from the slab to the supporting beams can be simplified by assuming that the load is transferred to the nearest beam. Such an assumption is represented by drawing 45-degree lines from each slab corner. The enclosed areas show the tributary loads to be carried by each beam. Beam B-1 will carry large trapezoidal loads compared to the triangular loads that will be carried by beam B-2. As the ratio of longer span (ℓ_ℓ) to shorter span (ℓ_s) increases, B-1 carries more loads than does B-2, that is, more loads are transferred in the shorter span of the slab.

In fact, if the ratio ℓ_ℓ/ℓ_s is greater than or equal to 2.0 ($\ell_\ell/\ell_s \geq 2.0$), the load carried by B-2 is quite small, and it can be neglected altogether. Therefore, if $\ell_\ell/\ell_s \geq 2.0$, the slab behaves as a one-way slab for all practical purposes, even though the slab is supported on all four edges.

To better understand this assumption, consider Figure 2–53b, in which two 1 ft wide strips of slab in the long (ℓ) and short (s) directions are shown at midspan for both. The load

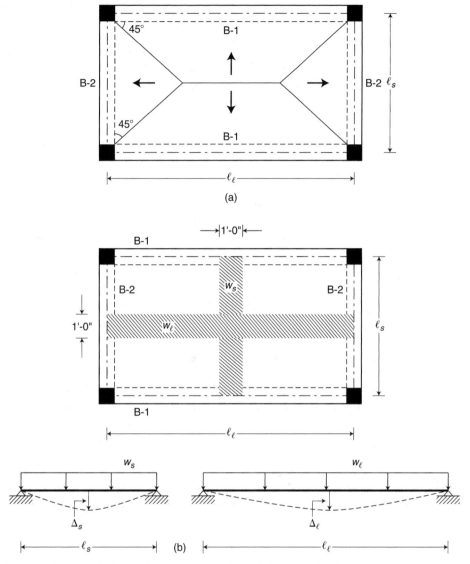

FIGURE 2–53 (a) Slab (edge supported); (b) slab load distribution.

carried by the short 1 ft wide strip is w_s, and the load carried by the long 1 ft wide strip is w_ℓ. If we assume that the slab is simply-supported along all edges, we can calculate the maximum mid-span deflections for the short (Δ_s) and long (Δ_ℓ) 1 ft wide strips from Equations 2–54 and 2–55.

$$\Delta_s = \frac{5w_s\ell_s^4}{384EI} \tag{2-54}$$

$$\Delta_\ell = \frac{5w_l\ell_l^4}{384EI} \tag{2-55}$$

The two deflections must be equal. Thus, an expression may be developed that relates the loads and spans, as shown in Equation 2–56.

$$\Delta_s = \Delta_\ell$$

$$\frac{5w_s \ell_s^4}{384EI} = \frac{5w_\ell \ell_\ell^4}{384EI}$$

$$w_s \ell_s^4 = w_\ell \ell_\ell^4$$

$$\frac{w_s}{w_\ell} = \frac{\ell_\ell^4}{\ell_s^4} = \left(\frac{\ell_\ell}{\ell_s}\right)^4 \tag{2–56}$$

The assumption for one-way behavior is $\ell_\ell/\ell_s \geq 2.0$. If $\ell_\ell/\ell_s = 2.0$ is substituted into Equation 2–56, w_s is equal to $16w_\ell$. Thus, the load transferred in the shorter direction (w_s) is 16 times larger than that transferred in the long direction (w_ℓ), when $\ell_\ell/\ell_s \geq 2.0$. Therefore, it is reasonable to assume that the loads are transferred mainly in the shorter direction.

Despite all the foregoing reasoning, structural engineers often design slabs as one-way slabs, even when the slabs' proportions do not satisfy the $\ell_\ell/\ell_s \geq 2.0$ requirement. The reason is that the shrinkage and temperature reinforcing needed in the long direction is usually quite enough to satisfy the small moment's requirements. Design and analysis of floor systems with two-way slabs are discussed in Chapter 6.

2.28   REINFORCEMENT IN ONE-WAY SLABS

In general, two types of reinforcement are used in one-way slabs: main reinforcement, and shrinkage and temperature reinforcement.

Main Reinforcement

The *main reinforcement* resists the bending moments. It is designed to act in the direction of the one-way slab's bending, which is along the shorter span length. Figure 2–54 shows the main reinforcement in a one-way slab supported by two parallel walls. The slab is assumed to be simply supported by the walls. In other words, no moment is transferred from the slab to the walls. Because the bottom portion of slab is in tension, the main reinforcement is placed in the bottom. Similarly, the main reinforcement is placed in a continuous construction where tension develops. For this case, as shown in Figure 2–55, the main reinforcement is at the bottom of the slab in the midspan region (positive moment) and at the top of the slab over the supports (negative moment). Typically, #4 bars or larger are used as main reinforcement, #3 bars are susceptible to permanent distortion caused by the construction crew walking over them. This is more critical for the top (negative moment) bars as the slab effective depth (d) may be reduced.

Shrinkage and Temperature (S & T) Reinforcement

As discussed in Chapter 1, fresh concrete loses water and shrinks soon after placement. In addition, variations in temperature cause the concrete to expand and contract. These volume

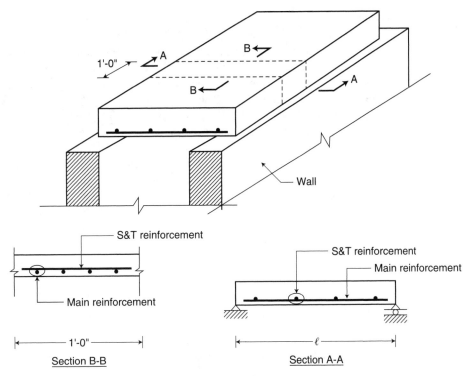

FIGURE 2–54 One-way slab reinforcement (simple span).

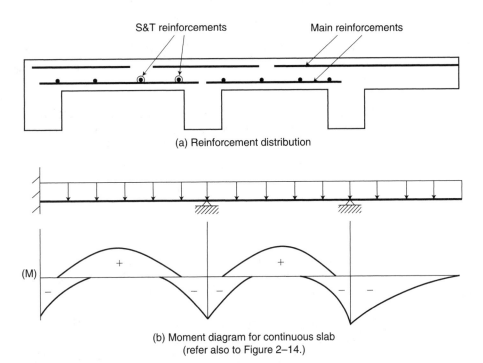

(a) Reinforcement distribution

(b) Moment diagram for continuous slab
(refer also to Figure 2–14.)

FIGURE 2–55 One-way slab reinforcement (continuous construction).

changes, when restrained, may result in cracking of concrete, especially in the early stages of strength development. Reinforcing bars are used to resist developing tensions in order to minimize cracks in concrete caused by shrinkage and temperature. The main longitudinal reinforcement in beams plays that role as well. Because the cross-sectional dimensions of beams are relatively small and beams may freely change their cross-sectional dimensions without restraint, shrinkage and temperature reinforcement are not needed perpendicular to the main bars.

This is not the case in reinforced concrete slabs. Slabs typically have large dimensions in two directions, thus they need shrinkage and temperature reinforcement, which is placed in the direction perpendicular to the main reinforcement. Figures 2–54 and 2–55 show such reinforcement for simple-span one-way slabs and continuous one-way slabs, respectively. In addition, temperature and shrinkage reinforcement helps distribute concentrated loads to a wide zone transversely to the one-way direction. (This is necessary in bridges, for example, to distribute large wheel loads onto a much wider strip than the one directly affected by the concentrated load.)

Minimum Reinforcements for One-Way Slabs

As discussed above, two types of reinforcement are used in one-way slabs. The ACI Code sets the following minimum reinforcement criteria for both the main and the shrinkage and temperature reinforcements.

Minimum Main Reinforcement   The minimum main reinforcement for slabs is equal to that required for shrinkage and temperature reinforcements (ACI Code, Section 10.5.4):

$$A_{s,min} = A_{s(S\&T)} \qquad (2\text{--}57)$$

In other words, if the calculated main reinforcement is less than that required for shrinkage and temperature reinforcement, the designer must use at least the latter amount.

Minimum Shrinkage and Temperature Reinforcement   The ACI Code (Section 7.12.2.1) requires shrinkage and temperature reinforcement based on the grade of steel, as given in Equations 2–58 through 2–60.

$$\text{For } f_y = 40 \text{ or } 50 \text{ ksi} \longrightarrow A_{s(S\&T)} = 0.002bh \qquad (2\text{--}58)$$

$$\text{For } f_y = 60 \text{ ksi} \longrightarrow A_{s(S\&T)} = 0.0018bh \qquad (2\text{--}59)$$

$$\text{For } f_y > 60 \text{ ksi} \longrightarrow A_{s(S\&T)} = \frac{0.0018 \times 60}{f_y}bh \geq 0.0014bh \qquad (2\text{--}60)$$

In Equations 2–58 through 2–60, $b = 12$ in. (slab width), which corresponds to the width of the 1 ft wide strip, h is the overall thickness of the slab in inches, and $A_{s(S\&T)}$ is the area of steel in square inches per foot of width.

Minimum Concrete Cover for the Reinforcement in Slabs   A minimum concrete cover is needed for the reinforcement to prevent various detrimental effects of the environment on reinforcing bars. Concrete cover is always measured from the closest

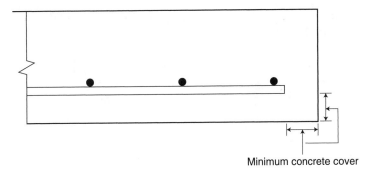

FIGURE 2–56   Minimum cover for slabs.

concrete surface to the first layer of reinforcing. This is shown in Figure 2–56. Section 7.7.1 of the ACI Code requires a minimum concrete cover of $^3/_4$ in. for #11 and smaller bars, and 1.5 in. for #14 and #18 bars, provided that the concrete slab is not exposed to weather or not in contact with the ground.

Bar Spacing in Reinforced Concrete Slabs   No specific minimum spacing of bars is required in slabs other than what was already discussed for beams. For practical reasons, however, bars are not placed closer than 3 in. to 4 in.

The ACI Code has different maximum spacing requirements for the main and the shrinkage and temperature reinforcements. These are as follows:

Maximum Spacing of Main Reinforcement Bars   ACI 318-05 has two sets of requirements regarding maximum bar spacing for the main reinforcement in one-way slabs: (1) Section 10.5.4 requires that the maximum spacing of bars be limited to three times the slab thickness or 18 in., whichever is smaller; and (2) Section 10.6.4 limits the maximum main reinforcement spacing (s) of one-way slabs, as calculated by Equation 2–53, in order to control the width and spacing of flexural cracks.

We can use the required minimum cover of $^3/_4$ in. for one-way slabs ($c_c = 0.75$ in.) and $f_s = {^2/_3}f_y = {^2/_3}(60{,}000) = 40{,}000$ psi to determine the maximum spacing for $f_y = 60$ ksi reinforcement. Substituting into Equation 2–53:

$$s = 15\left(\frac{40{,}000}{40{,}000}\right) - 2.5(0.75) \le 12\left(\frac{40{,}000}{40{,}000}\right)$$

$$s = 13.1 \text{ in.} \le 12 \text{ in.}$$

$$s = 12 \text{ in.}$$

Therefore, the maximum main reinforcement spacing with $f_y = 60$ ksi steel for one-way slabs is given by Equation 2–61a.

$$s_{\text{max, main}} = \min\{3h, 12 \text{ in.}\} \qquad (2\text{–}61\text{a})$$

Similarly, when using $f_y = 40$ ksi steel as main reinforcement, Equation 2–53 will simplify to Equation 2–61b.

$$s_{max,\,main} = \min\{3h, 18 \text{ in.}\} \tag{2–61b}$$

Maximum Bar Spacing of Shrinkage and Temperature Reinforcement ACI Code, Section 7.12.2.2 limits the spacing of the shrinkage and temperature reinforcements to five times the slab thickness or, 18 in., whichever is smaller:

$$s_{max,\,(S\&T)} = \min\{5h, 18 \text{ in.}\} \tag{2–62}$$

Minimum Thickness of Slab for Deflection Control The minimum recommended thickness for one-way slabs required to adequately control excessive deflections is based on Table 9–5(a) of the ACI Code, which is summarized graphically in Figure 2–42. Lesser thicknesses are permitted if the designer can show through a detailed deflection analysis that the Code's serviceability requirements are met.

2.29   AREAS OF REINFORCING BARS IN SLABS

A 1 ft (12 in.) wide strip of slab is typically used for the analysis and design of one-way slabs. Thus, it is advantageous to define the amount of steel in a 1 ft wide strip as a function of the bar size and the spacing.

Table A2–10 lists spacing and bar sizes for slabs. The table provides the areas of reinforcement averaged out to 1 ft width for different sizes and spacing of bars. (One can interpolate for $1/2$ in. spacing increments, if so desired.)

For example, with #5 @ 8 in. o.c. (#5 bar at 8 in. on-center spacing), the table, under #5 bars spaced at 8 in., provides the area of steel per foot of section 0.47 in^2. In other words, $0.47 \text{ in}^2/\text{ft}$ is equivalent to one #5 bar every 8 in.

Another example: If 0.50 in^2 of reinforcement is required for a 1 ft wide strip of a slab, the table offers several options, including #4 @ 4 in. ($A_s = 0.60 \text{ in}^2$), #5 @ 7 in. ($A_s = 0.53 \text{ in}^2$), #6 @ 10 in. ($A_s = 0.53 \text{ in}^2$), and so on.

2.30   ANALYSIS OF REINFORCED CONCRETE ONE-WAY SLABS

In general, one-way slabs and reinforced concrete beams are analyzed very similarly. There are a few differences, however. These are listed below:

1. For the analysis of one-way slabs, b is always 12 in.
2. Slabs require a different amount of concrete cover over the reinforcement.
3. Slabs require shrinkage and temperature reinforcement.
4. The Code-specified minimum amounts of reinforcing steel for slabs and beams are different.
5. Minimum required depth/span ratios for adequate control of deflection are different.
6. Bar spacing requirements are different.

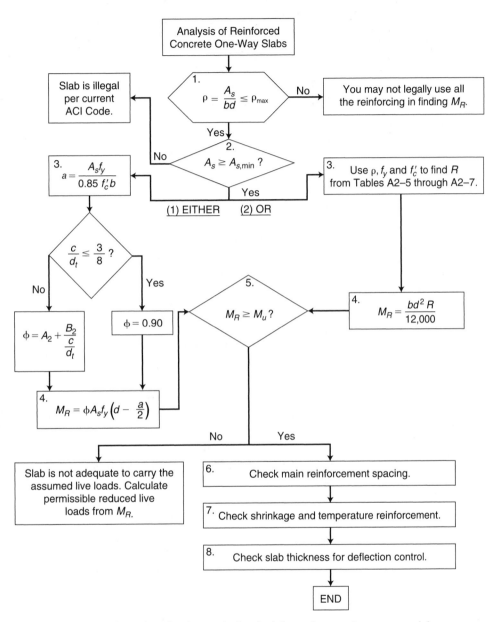

FIGURE 2-57 Flowchart for the analysis of reinforced concrete one-way slabs.

Figure 2–57 summarizes the steps for the analysis of reinforced concrete one-way slabs. They are as follows:

Step 1   Calculate the steel ratio, $\left(\rho = \dfrac{A_s}{bd}\right)$. A_s is the area of steel in a 1 ft wide strip of slab from Table A2–10. Compare ρ with ρ_{max} from Table A2–3. The maximum permitted steel ratio is the same for beams and slabs.

Step 2   Compare A_s with $A_{s,min}$, which is the minimum required area of steel for the control of shrinkage and temperature-induced volumetric changes. If $A_s \leq A_{s,min}$, the proportioning of steel and concrete is not acceptable according to the current ACI Code and the slab's use is illegal. If $A_s \geq A_{s,min}$, however, then one of the following methods can be used to check the adequacy of the slab:

Method I

Step 3   Calculate the depth of the compression zone:

$$a = \frac{A_s f_y}{0.85 f'_c b}$$

Determine the location of the neutral axis (c):

$$c = \frac{a}{\beta_1}$$

If $\dfrac{c}{d_t} \leq \dfrac{3}{8}$, the section is tension-controlled and $\phi = 0.90$. Otherwise, the section will be in the transition zone. Calculate the strength reduction factor, ϕ:

$$\phi = A_2 + \frac{B_2}{\dfrac{c}{d_t}}$$

A_2 and B_2 are listed in Table A2–2b.

Step 4   Calculate the section's resisting moment (M_R):

$$M_R = \phi A_s f_y \left(d - \frac{a}{2} \right)$$

(If A_s is in^2, f_y ksi, d and a in. units are used, then M_R will have kip-in unit. Divide the result by 12 to obtain M_R in the customary units of kip-ft).

Step 5   Compare M_R with the maximum factored moment from the applied loads. If $M_R < M_u$, the slab is not adequate to carry the *assumed* loads. Proceed to calculate a new permissible live load that the slab may legally support. If $M_R \geq M_u$, the section can take the assumed loads, but the reinforcing still needs to be checked for conformance with other Code requirements.

Step 6   Check spacing requirements. The maximum allowable spacing of main reinforcement is min{$3h$, 12 in.}, or min{$3h$, 18 in.} for $f_y = 60$ ksi and $f_y = 40$ ksi steel, respectively.

$$3 \text{ in.} \leq s \leq \min\{3h, 12 \text{ in.}\} \qquad \text{for} f_y = 60 \text{ ksi}$$
$$3 \text{ in.} \leq s \leq \min\{3h, 18 \text{ in.}\} \qquad \text{for} f_y = 40 \text{ ksi}$$

Step 7   Check the amount and spacing of shrinkage and temperature reinforcement, $(A_s)_{S\&T}$. (Refer to Equations 2–58 through 2–60.)

$$3 \text{ in.} \leq s_{S\&T} \leq \min\{5h, 18 \text{ in.}\}$$

Step 8   Check the thickness of the slab against the minimum thickness of one-way slabs for desirable deformation control (see Figure 2–42).

$$h_{min} = \ell/20 \text{ for simply-supported slabs}$$
$$h_{min} = \ell/10 \text{ for cantilevered slabs}$$
$$h_{min} = \ell/28 \text{ for both ends continuous slabs}$$
$$h_{min} = \ell/24 \text{ for one end continuous slabs}$$

If the slab thickness is less than the above limits, calculate the deflection and check it against the Code's serviceability requirements.

Method II

Steps 1 and 2 are the same as in Method I.

Step 3   Use f_y, f_c', and the calculated steel ratio (ρ) to obtain the resistance coefficient, R, from Tables A2–5 through A2–7.

Step 4   Use the R value to calculate the section's resisting moment.

$$M_R = \frac{bd^2 R}{12,000}$$

R is in psi, $b = 12$ in., and d in inches. M_R will be in units of ft-kip. Steps 5, 6, 7, and 8 are the same as in Method I.

EXAMPLE 2–15

Figure 2–58 shows a section through a reinforced concrete simply-supported one-way slab of an existing building. The maximum moment from dead loads, including the slab weight, is 3.0 (ft-kip)/ft, and that from live loads is 2.0 (ft-kip)/ft. Check the adequacy of the slab, including the shrinkage and temperature reinforcements, using (a) Method I, and (b) Method II.

Use a concrete cover of $^3/_4$ in., $f_c' = 3.0$ ksi, and $f_y = 40.0$ ksi.

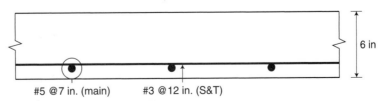

#5 @ 7 in. (main)   #3 @ 12 in. (S&T)

FIGURE 2–58   Sketch of one-way slab for Example 2–15.

Solution

Step 1   Check the reinforcement ratio in the slab:

Diameter of #5 bars

$$\bar{y} = \frac{3}{4} + \frac{\frac{5}{8}}{2} = 1.06 \text{ in.}$$

Cover

$$d = h - \bar{y} = 6 \text{ in.} - 1.06 \text{ in.} = 4.94 \text{ in.}$$

#5 @ 7 in. (main reinforcement) $\longrightarrow$ Table A2–10 $\longrightarrow A_s = 0.53 \text{ in}^2/\text{ft}$

$$\rho = \frac{A_s}{bd} = \frac{0.53}{12 \times 4.94} = 0.00894$$

$f_c' = 3 \text{ ksi} \longrightarrow$ Table A2–3 $\longrightarrow \rho_{max} = 0.0232 > 0.00894 \qquad \therefore \text{ ok}$

$f_y = 40 \text{ ksi}$

Step 2   Check the minimum area of main reinforcement. For slabs, this area is the same as the requirement for shrinkage and temperature reinforcement:

$$A_{s,min} = A_{s(S\&T)} = 0.002bh \qquad (f_y = 40 \text{ ksi})$$
$$A_{s,min} = (0.002)(12)(6) = 0.14 \text{ in}^2/\text{ft}$$
$$A_s = 0.53 \text{ in}^2/\text{ft} > 0.14 \text{ in}^2/\text{ft} \qquad \therefore \text{ ok}$$

(a) Method I

Step 3   Calculate the depth of the compression zone:

$$a = \frac{A_s f_y}{0.85 f_c' b} = \frac{0.53 \times 40}{0.85 \times 3 \times 12}$$
$$a = 0.69 \text{ in.}$$

The neutral axis is located at c:

$$c = \frac{a}{\beta_1} = \frac{0.69}{0.85} = 0.81 \text{ in.}$$
$$d_t = d = 4.94 \text{ in.}$$
$$\frac{c}{d_t} = \frac{0.81}{4.94} = 0.164 < 0.375 \qquad \therefore \phi = 0.90$$

Step 4

$$M_R = \phi M_n = \phi A_s f_y \left(d - \frac{a}{2} \right)$$

$$M_R = (0.9)(0.53)(40) \left(4.94 - \frac{0.69}{2} \right)$$

$$M_R = \frac{87.7 \text{ in.-kip}}{12 \text{ in./ft}} = 7.31 \text{ ft-kip}$$

Step 5   Calculate the factored applied moment on the slab:

$$M_u = 1.2M_D + 1.6M_L$$

$$M_u = 1.2 \times 3.0 + 1.6 \times 2.0 = 6.8 \text{ ft-kip} < 7.31 \text{ ft-kip} \qquad \therefore \text{ ok}$$

Step 6   Check the main reinforcement spacing:

$$3 \text{ in.} \leq s \leq \min\{3h, 18 \text{ in.}\}$$

The main reinforcement is #5 @ 7 in.:

$$3 \text{ in.} < 7 \text{ in.} < \min\{3 \times 6 \text{ in.}, 18 \text{ in.}\}$$

$$3 \text{ in.} < 7 \text{ in.} < 18 \text{ in.} \qquad \therefore \text{ ok}$$

Slab is ok

Step 7   Check the shrinkage and temperature reinforcements:

$$A_{s(S\&T)} = 0.002bh = (0.002)(12)(6) = 0.14 \text{in}^2/\text{ft}$$

From Table A2–10 $\longrightarrow$ #3 @ 12 in. $\longrightarrow$ $A_s = 0.11 \text{ in}^2/\text{ft} < 0.14 \text{ in}^2/\text{ft}$
$$\therefore \text{ N.G.}$$

Therefore, the shrinkage and temperature reinforcement in the slab does not satisfy the current ACI Code's minimum requirement.

(b) Method II

$$\rho = 0.00894$$

Step 3 $f_c' = 3 \text{ ksi} \longrightarrow$ Table A2–5a $\longrightarrow R = 299 \text{ psi (by interpolation)}$
$f_y = 40 \text{ ksi}$

Step 4

$$M_R = \frac{bd^2 R}{12,000}$$

$$M_R = \frac{(12)(4.94)^2(299)}{12,000}$$

$$M_R = 7.3 \text{ ft-kip/ft}$$

This value is the same as the resisting moment we calculated in Step 4 using Method I. Steps 5, 6, and 7 are the same as those of Method I.

EXAMPLE 2–16

Figure 2–59 shows the partial floor framing plan and section of a reinforced concrete floor system. The weight of the ceiling and floor finishing is 5 psf, the mechanical and electrical systems are 5 psf, and the partitions are 15 psf. The floor live load is 150 psf. The concrete is normal weight, $f_c' = 4 \text{ ksi}$, and $f_y = 60 \text{ ksi}$. Check the adequacy of slab S-1 in the exterior bay at (a) midspan, and (b) over the interior supporting beam. Assume the slab is cast integrally with the supporting beams and use ACI code coefficients to calculate moments. Use $^3/_4$ in. cover for the slab.

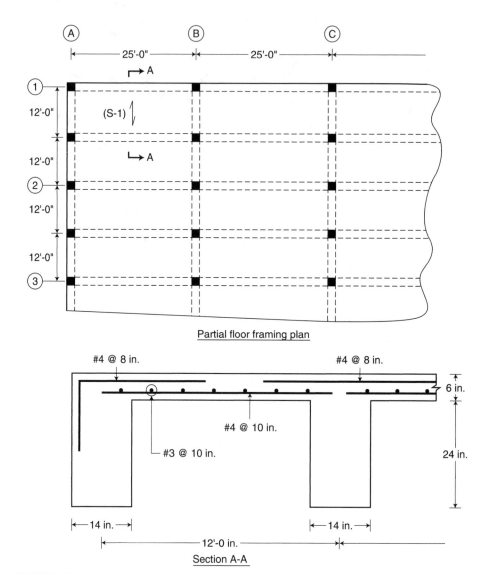

FIGURE 2–59   Framing plan and section for Example 2–16.

Solution

a.   *Check the Slab at the Midspan*

Step 1   The main reinforcement at the midspan (positive moment) is #4 @ 10 in.

$$\text{\#4 @ 10 in.} \longrightarrow \text{Table A2–10} \longrightarrow A_s = 0.24 \text{ in}^2/\text{ft}$$

$$\bar{y} = \frac{3}{4} + \frac{\frac{4}{8}}{2} = 1.0 \text{ in.}$$

$$d = h - \bar{y} = 6 \text{ in.} - 1 \text{ in.} = 5 \text{ in.}$$

$$\rho = \frac{A_s}{bd} = \frac{0.24}{(12)(5)} = 0.0040$$

$f'_c = 4\,\text{ksi} \longrightarrow \text{Table A2–3} \longrightarrow \rho_{max} = 0.0207 > 0.004 \qquad \therefore \text{ ok}$

$f_y = 60\,\text{ksi}$

Step 2

$$A_{s,min} = A_{s(S\&T)} = 0.0018bh \qquad (f_y = 60\,\text{ksi})$$

$$A_{s,min} = (0.0018)(12)(6) = 0.13\,\text{in}^2/\text{ft}$$

$$A_s = 0.24\,\text{in}^2/\text{ft} > 0.13\,\text{in}^2/\text{ft} \qquad \therefore \text{ ok}$$

Method II is followed for the rest of the solution, as it requires fewer steps.

Step 3 $\rho = 0.0040$

$f'_c = 4\,\text{ksi} \longrightarrow \text{Table A2–6b} \longrightarrow R = 208\,\text{psi}$

$f_y = 60\,\text{ksi}$

Step 4

$$M_R = \frac{bd^2 R}{12{,}000}$$

$$M_R = \frac{(12)(5)^2(208)}{12{,}000}$$

$$M_R = 5.2\,\text{ft-kip}$$

Step 5   The slab's dead and live loads are:

$$\text{Weight of slab} = 150\left(\frac{6}{12}\right) = 75\,\text{psf}$$

$$\text{Ceiling and floor finishing} = 5\,\text{psf}$$

$$\text{Mechanical and electrical} = 5\,\text{psf}$$

$$\text{Partitions} = 15\,\text{psf}$$

$$\text{Total dead load} = 100\,\text{psf}$$

$$\text{Total live load} = 150\,\text{psf}$$

The slab's tributary width is 1'-0":

$$w_D = \frac{100 \times 1}{1000} = 0.10\,\text{kip/ft}$$

$$w_L = \frac{150 \times 1}{1000} = 0.15\,\text{kip/ft}$$

$$w_u = 1.2w_D + 1.6w_L = 1.2 \times 0.10 + 1.6 \times 0.15$$

$$w_u = 0.36\,\text{kip/ft}$$

$$\ell_n = 12\,\text{ft} - \frac{14\,\text{in.}}{12} = 10.83\,\text{ft}$$

The maximum factored moment at the midspan of the exterior bay of the slab is:

$$M_u = \frac{w_u \ell_n^2}{14}$$

$$M_u = \frac{(0.36)(10.83)^2}{14}$$

$$M_u = 3.0 \text{ ft-kip} < M_R = 5.2 \text{ ft-kip} \quad \therefore \text{ ok}$$

Because M_R is much larger than M_u, the slab is overdesigned for positive moment.

Step 6 Check the spacing requirements for the main reinforcement:

$$3 \text{ in.} \le s \le \min\{3h, 12 \text{ in.}\}$$
$$3 \text{ in.} < 10 \text{ in.} < \min\{3 \times 6 \text{ in.}, 12 \text{ in.}\}$$
$$3 \text{ in.} < 10 \text{ in.} < 12 \text{ in.} \quad \therefore \text{ ok}$$

Step 7 Check shrinkage and temperature reinforcement:

$$A_{s(\text{S\&T})} = 0.0018bh \quad (f_y = 60 \text{ ksi})$$
$$A_{s(\text{S\&T})} = 0.0018(12)(6) = 0.13 \text{ in}^2/\text{ft}$$

$$\#3 @ 10 \text{ in.} \longrightarrow \text{Table A2–10} \longrightarrow A_s = 0.13 \text{ in}^2/\text{ft} \quad \therefore \text{ ok}$$

Check the spacing of the shrinkage and temperature reinforcement:

$$3 \text{ in.} \le s \le \min\{5h, 18 \text{ in.}\}$$
$$3 \text{ in.} < 10 \text{ in.} < \min\{5 \times 6 \text{ in.}, 18 \text{ in.}\}$$
$$3 \text{ in.} < 10 \text{ in.} < 18 \text{ in.} \quad \therefore \text{ ok}$$

Step 8 For deflection control, the minimum recommended thickness (without calculating deflections) for the one-end-continuous slab is:

$$h_{\min} = \frac{\ell}{24} = \frac{12 \times 12}{24} = 6 \text{ in.} = 6 \text{ in.} \quad \therefore \text{ ok}$$

Slab is ok at mid-span.

b. *Check the Slab at Supports*

Step 1 The main reinforcement at the supports (negative moment) is #4 @ 8 in.

$$\#4 @ 8 \text{ in.} \longrightarrow \text{Table A2–10} \longrightarrow A_s = 0.30 \text{ in}^2/\text{ft}$$

$$\bar{y} = \frac{3}{4} + \frac{\frac{4}{8}}{2} = 1.0 \text{ in}$$

$$d = h - \bar{y} = 6 \text{ in.} - 1 \text{ in.} = 5 \text{ in.}$$

$$\rho = \frac{A_s}{bd} = \frac{0.30}{(12)(5)} = 0.005$$

$$f_c' = 4\,\text{ksi} \longrightarrow \text{Table A2--3} \longrightarrow \rho_{max} = 0.0207 > 0.005 \quad \therefore \text{ ok}$$
$$f_y = 60\,\text{ksi}$$

Step 2

$$A_{s,min} = A_{s(S\&T)} = 0.0018bh\,(f_y = 60\,\text{ksi})$$
$$A_{s,min} = (0.0018)(12)(6) = 0.13\,\text{in}^2/\text{ft}$$
$$A_s = 0.30\,\text{in}^2/\text{ft} > 0.13\,\text{in}^2/\text{ft} \quad \therefore \text{ ok}$$

$$\rho = 0.005$$

Step 3

$$f_c' = 4\,\text{ksi} \longrightarrow \text{Table A2--6b} \longrightarrow R = 258\,\text{psi}$$
$$f_y = 60\,\text{ksi}$$

Step 4

$$M_R = \frac{bd^2R}{12,000}$$
$$M_R = \frac{(12)(5)^2(258)}{12,000}$$
$$M_R = 6.5\text{ft-kip}$$

Step 5 The dead and live loads from part a are:

$$w_u = 0.36\,\text{kip/ft (from part a)}$$
$$\ell_n = 10.83\,\text{ft (from part a)}$$

The maximum factored moment at the first interior support for an exterior bay of the slab is (Table A2–1):

$$M_u = \frac{w_u\ell_n^2}{10}$$
$$M_u = \frac{(0.36)(10.83)^2}{10}$$
$$M_u = 4.2\text{ft-kip} < M_R = 6.5\,\text{ft-kip} \quad \therefore \text{ ok}$$

Step 6 Check the spacing of the main reinforcement:

$$3\,\text{in.} \leq s \leq \min\{3h, 12\,\text{in.}\}$$
$$3\,\text{in.} \leq s \leq \min\{3 \times 6\,\text{in.}, 12\,\text{in.}\}$$
$$3\,\text{in.} < 8\,\text{in.} < 12\,\text{in.} \quad \therefore \text{ ok}$$

The shrinkage and temperature reinforcement and the minimum depth for deflection were checked in part a.

Slab is ok at the support.

2.31 DESIGN OF REINFORCED CONCRETE ONE-WAY SLABS

The design process of one-way slabs is similar to that of reinforced concrete rectangular beams. Figure 2–60 summarizes the steps for the design of reinforced concrete one-way slabs. They are as follows:

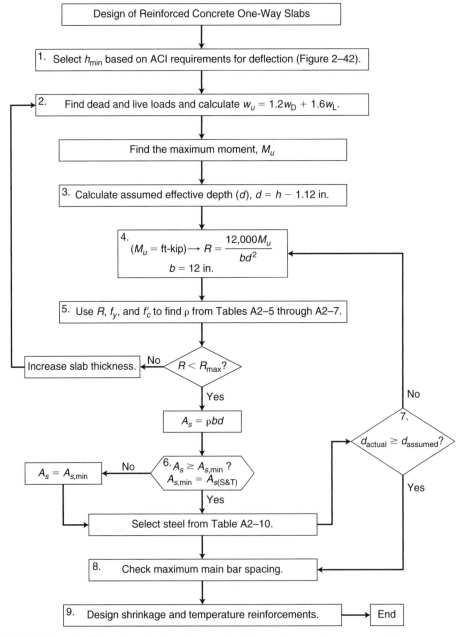

FIGURE 2–60 Flowchart for the design of reinforced concrete one-way slabs.

Step 1   Select the slab thickness. The slab thickness is generally based on the minimum ACI requirements for deflection control (see Figure 2–42). This is usually rounded up to the nearest $\frac{1}{2}$ in. for slabs with $h \le 6$ in. and to the nearest 1 in. for those with $h > 6$ in.

Step 2   Calculate the factored loads (w_u), and then determine the maximum factored moment, M_u.

Step 3   Determine the slab's effective depth, d. Because the bar sizes are not yet known, assume #6 bars with $\frac{3}{4}$ in. cover.

$$\bar{y} = 1.12 \text{ in. to } 1.25 \text{ in.}$$

Therefore, the assumed effective depth:

$$d = h - 1.12 \text{ in.}$$

Step 4   Determine the required resistance coefficient (R):

$$R \text{ (psi)} = \frac{12{,}000 M_u}{bd^2}$$

$b = 12$ in., and d is in inches. M_u is in ft-kip and R in psi.

Step 5   Using R, f_y, and f_c' select ρ (steel ratio) from Tables A2–5 through A2–7. If the value of R is more than the maximum value shown in these tables ($R > R_{max}$), the selected slab thickness is not adequate for the loads and needs to be increased. (Note that in most cases this does not happen. The required thickness for deflection control is usually more than what is required to carry the loads.)

$$A_s = \rho bd$$

Step 6   Check the minimum reinforcement requirement. The minimum area of steel for the main reinforcement must not be less than that required for shrinkage and temperature reinforcement:

$$A_{s,min} = A_{s(S\&T)}$$

If $A_s < A_{s,min}$, the slab requires only a small amount of reinforcing steel, A_s. Use at least $A_{s,min}$, however. Select the bar size and spacing from Table A2–10.

Step 7   Check for actual depth (d_{actual}) based on the bar selected. If $d_{actual} < d_{assumed}$, go back to Step 4 and revise. Repeat if the difference is too large (larger than $\frac{1}{8}$ in. for slabs $h < 6$ in. and $\frac{1}{4}$ in. for $h > 6$ in.).

Step 8   Check bar spacing. The spacing of bars selected in Step 6 has to be checked against the ACI Code requirements for maximum allowable spacing.

Step 9   Design the shrinkage and temperature reinforcements according to the ACI Code requirements.

EXAMPLE 2–17

Design the one-way slab (S-1) of Example 2–16. Determine the reinforcement at (a) the midspan and (b) the supports.

Solution

a. *Slab Design at the Midspan*

Step 1 Because S-1 is one end continuous, the minimum slab thickness (h_{min}) is:

$$h_{min} = \frac{\ell}{24} = \frac{12 \times 12}{24} = 6 \text{ in.}$$

Step 2 Determine the loads on the slab:

$$\text{Weight of slab} = 150(^6/_{12}) = 75 \text{ psf}$$
$$\text{Ceiling and floor finishing} = 5 \text{ psf}$$
$$\text{Mechanical and electrical} = 5 \text{ psf}$$
$$\text{Partitions} = 15 \text{ psf}$$

$$\text{Total dead load} = 100 \text{ psf}$$
$$\text{Total live load} = 150 \text{ psf}$$

On a 1 ft wide strip

$$w_D = \frac{100 \times 1}{1000} = 0.10 \text{ kip/ft}$$

$$w_L = \frac{150 \times 1}{1000} = 0.15 \text{ kip/ft}$$

$$w_u = 1.2w_D + 1.6w_L = 1.2 \times 0.10 + 1.6 \times 0.15$$
$$w_u = 0.36 \text{ kip/ft}$$

$$\ell_n = 12 \text{ ft} - \frac{14 \text{ in.}}{12} = 10.83 \text{ ft}$$

The maximum factored moment at the midspan of S-1 (see Figure 2–61) is:

$$M_u = \frac{w_u \ell_n^2}{14}$$

$$M_u = \frac{(0.36)(10.83)^2}{14}$$

$$M_u = 3.0 \text{ ft-kip}$$

Step 3 Assuming $^3/_4$ in. cover, calculate the slab's effective depth:

$$d = h - 1.12 \text{ in.} = 6 \text{ in.} - 1.12 \text{ in.} = 4.88 \text{ in.}$$

Step 4 Calculate the required resistance coefficient, R:

$$R = \frac{12,000 M_u}{bd^2}$$

$$R = \frac{12,000 \times 3.0}{(12)(4.88)^2} = 126 \text{ psi}$$

Step 5   Find ρ from Tables A2–5 through A2–7:

$$R = 126 \, psi$$
$$f_c' = 4 \, ksi \longrightarrow \text{Table A2–6b} \longrightarrow \rho = 0.0024$$
$$f_y = 60 \, ksi$$

Therefore, the required area of main reinforcement (A_s) is:

$$A_s = \rho bd = (0.0024)(12)(4.88)$$
$$A_s = 0.14 \, in^2/ft$$

Step 6   The minimum amount of reinforcement for slabs cannot be less than the required shrinkage and temperature reinforcement steel:

$$A_{s,min} = A_{s(S\&T)} = 0.0018bh \quad \text{for } f_y = 60 \, ksi$$
$$A_{s,min} = (0.0018)(12)(6) = 0.13 \, in^2/ft < 0.14 \, in^2/ft \quad \therefore \text{ ok}$$
$$A_s = 0.14 \, in^2/ft$$

From Table A2–10 $\longrightarrow$ use #4 @ 17 in. $(A_s = 0.14 \, in^2/ft)$

Note that according to Section 2.28, the smallest size bar for main reinforcement is #4.

Step 7   Check for the actual effective depth.

$$d_{actual} = 6 - \frac{3}{4} - \frac{\frac{4}{8}}{2} = 5.0 \, in. > d_{assumed} = 4.88 \, in. \quad \therefore \text{ ok}$$

Step 8   Check the main reinforcement spacing, s, $(f_y = 60 \, ksi)$.

$$3 \, in. \leq s \leq \min\{3h, 12 \, in.\}$$
$$3 \, in. < 17 \, in. < \min\{3 \times 6 \, in., 12 \, in.\}$$
$$3 \, in. < 17 \, in. > 12 \, in. \quad \therefore \text{ N.G.}$$

Therefore:

Use # 4 @ 12 in. for the main reinforcement at the midspan.

Step 9   Calculate the required shrinkage and temperature reinforcement.

$$A_{s(S\&T)} = 0.0018bh = 0.13 \, in^2/ft$$

From Table A2–10 $\longrightarrow$ use #3 @ 10 in.

The shrinkage and temperature reinforcement spacing (s) has to be within the following range:

$$3 \, in. \leq s \leq \min\{5h, 18 \, in.\}$$
$$3 \, in. < 10 \, in. < \min\{5 \times 6 \, in., 18 \, in.\}$$
$$3 \, in. < 10 \, in. < 18 \, in. \quad \therefore \text{ ok}$$

Therefore,

Use #3 @ 10 in. for the shrinkage and temperature reinforcement.

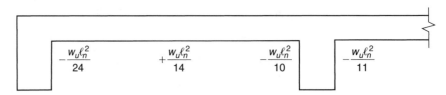

FIGURE 2–61 Design factored moments for slab S-1 of Example 2–17 using ACI Code coefficients from Table A2–1.

b. *Slab Design at the Supports*

Step 1 From Step 1 of part a:

$$h_{min} = 6 \text{ in.}$$

Step 2 The factored uniformly distributed load on the slab (w_u) from Step 2 of part a is:

$$w_u = 0.36 \text{ kip/ft}$$

and the clear span (ℓ_n) is:

$$\ell_n = 10.83 \text{ ft}$$

From Figure 2–61, the moments at the exterior and interior supports are:

$$M_u^- = \frac{w_u \ell_n^2}{24} = \frac{(0.36)(10.83)^2}{24} = 1.76 \text{ ft-kip (exterior support)}$$

$$M_u^- = \frac{w_u \ell_n^2}{10} = \frac{(0.36)(10.83)^2}{10} = 4.22 \text{ ft-kip (interior support)}$$

Step 3

$$\text{Assume } d = h - 1.12 \text{ in.} = 6 - 1.12 = 4.88 \text{ in.}$$

Step 4

$$R = \frac{12,000 \, M_u}{bd^2} = \frac{12,000 \times 1.76}{(12)(4.88)^2} = 74 \text{ psi (exterior support)}$$

$$R = \frac{12,000 \, M_u}{bd^2} = \frac{12,000 \times 4.22}{(12)(4.88)^2} = 177 \text{ psi (interior support)}$$

Step 5

$$\text{For exterior support} \begin{cases} R = 74 \text{ psi} \\ f'_c = 4 \text{ ksi} \\ f_y = 60 \text{ ksi} \end{cases} \longrightarrow \text{Table A2–6b} \longrightarrow \rho_{ext.} = 0.0014$$

$$\text{For interior support} \begin{cases} R = 177 \text{ psi} \\ f'_c = 4 \text{ ksi} \\ f_y = 60 \text{ ksi} \end{cases} \longrightarrow \text{Table A2–6b} \longrightarrow \rho_{int.} = 0.0034$$

Therefore:

$$(A_s)_{\text{ext.}} = \rho bd = (0.0014)(12)(4.88) = 0.082 \text{ in}^2/\text{ft}$$
$$(A_s)_{\text{int.}} = \rho bd = (0.0034)(12)(4.88) = 0.20 \text{ in}^2/\text{ft}$$

Step 6 From Step 6 of part a:

$$A_{s,\text{min}} = A_{s(\text{S\&T})} = 0.13 \text{ in}^2/\text{ft}$$
$$(A_s)_{\text{ext}} = 0.082 \text{ in}^2/\text{ft} < 0.13 \text{ in}^2/\text{ft} \quad \therefore \text{ N.G.}$$

Therefore, use

$$(A_s)_{\text{ext.}} = 0.13 \text{ in}^2/\text{ft}$$
$$(A_s)_{\text{int.}} = 0.20 \text{ in}^2/\text{ft} > 0.13 \text{ in}^2/\text{ft} \quad \therefore \text{ ok}$$

From Table A2–10 $\longrightarrow$ Try #4 @ 12 in. (exterior supports)

$$(A_s)_{\text{int.}} = 0.20 \text{ in}^2/\text{ft}$$

From Table A2–10 $\longrightarrow$ Try #4 @ 12 in. (interior supports)

Step 7 This is the same as in part a.

Step 8 Check the main reinforcement spacing:

$$3 \text{ in.} \leq s \leq \min\{3h, 12 \text{ in.}\}$$
$$3 \text{ in.} \leq s \leq \min\{3 \times 6 \text{ in.}, 12 \text{ in.}\}$$
$$3 \text{ in.} \leq s \leq 12 \text{ in.}$$
$$s_{\text{int.}} = s_{\text{ext.}} = 12 \text{ in.} = 12 \text{ in.} \quad \therefore \text{ ok}$$

$\therefore$ Use #4 @ 12 in. for the exterior and interior supports.

Step 9 The shrinkage and temperature reinforcement was designed in part a. Figure 2–62 shows the slab as designed.

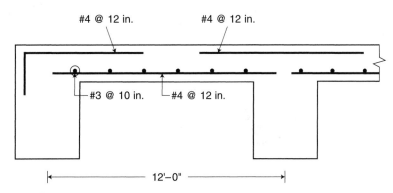

FIGURE 2–62 Slab S-1 designed in Example 2–17.

PROBLEMS

In the following problems, unless noted otherwise, use 150 pcf for the unit weight of concrete, 1.5 in. for beam clear concrete cover, and 0.75 in. for slab clear concrete cover.

2–1 Consider a section with a width (b) of 14 in. and reinforced with 4 #9 bars in a single layer. $f'_c = 4,000$ psi, and $f_y = 60,000$ psi. Determine the moment capacity of the section, M_R, using Method I or II, for the following cases:

(a) $d = 28$ in.   (c) $d = 36$ in.
(b) $d = 32$ in.   (d) $d = 40$ in.

Show the changes in M_R with respect to the section's effective depth. Calculate the percentages of increase in M_R versus d.

2–2 Consider a rectangular reinforced concrete beam with an effective depth of 36 in. reinforced with 4 #9 bars. $f'_c = 4,000$ psi, and $f_y = 60,000$ psi. Determine M_R using Method I or II for the following cases:

(a) $b = 14$ in.   (c) $b = 18$ in.
(b) $b = 16$ in.   (d) $b = 20$ in.

Show the variation in M_R with b. For each case calculate the percentage of increase in M_R versus b.

2–3 Consider a reinforced concrete beam with a width (b) of 14 in. and an effective depth (d) equal to 36 in.. $f'_c = 4,000$ psi, and $f_y = 60,000$ psi. Determine the moment capacity of this beam, M_R, for the following reinforcements:

(a) 4 #6 bars   (c) 4 #8 bars
(b) 4 # 7 bars   (d) 4 #9 bars

Show the variation of M_R with respect to the area of reinforcements (A_s). For each case calculate the percentage of increase in M_R versus A_s.

2–4 Consider a reinforced concrete beam with a width (b) of 14 in., and an effective depth (d) of 36 in. reinforced with 4 #8 bars. Use $f_y = 60,000$ psi. Determine the moment capacity, M_R, of this beam for the following cases:

(a) $f'_c = 3,000$ psi
(b) $f'_c = 4,000$ psi
(c) $f'_c = 5,000$ psi

2–5 Rework Problem 2–4 for $f'_c = 4,000$ psi and for the following steel yield strengths:

(a) $f_y = 40,000$ psi
(b) $f_y = 60,000$ psi
(c) $f_y = 75,000$ psi

2–6 Determine the useful moment strength of the section shown below in accordance with the ACI Code. Use $f'_c = 4,000$ psi, $f_y = 60,000$ psi, and #3 stirrups and follow Method II in the calculations.

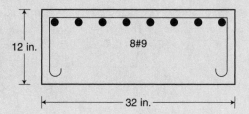

2–7 The rectangular reinforced concrete beam shown below is subjected to a dead load moment of 180 ft-kip and live load moment of 90 ft-kip. Determine whether the beam is adequate for moment capacity. $f'_c = 4,000$ psi, and $f_y = 60,000$ psi. The stirrups are #3 bars.

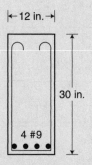

2–8 The beam below supports 500 lb/ft service dead loads and 600 lb/ft service live loads in addition to its self-weight. Calculate the maximum simply-supported span ($\ell = ?$) for the beam. Use Method II in the calculations. Use $f'_c = 5,000$ psi and $f_y = 60,000$ psi.

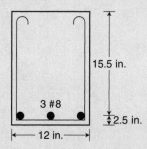

2–9 A rectangular beam carries uniformly distributed service (unfactored) dead loads of 3.0 kip/ft, including its own self-weight and 1.5 kip/ft service live loads. Based on the beam's moment capacity, calculate the largest factored concentrated loads, P_u, that may be placed as shown on the span in addition to the given distributed loads. The beam width is 18 in., and has a total depth of 30 in. with 5 #11 bars. Use $f'_c = 5,000$ psi, $f_y = 60,000$ psi, and #3 stirrups.

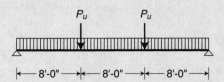

2–10 The beam shown below is part of a beam-girder floor system. It is subjected to a superimposed dead load of 4.0 kip/ft (excluding the beam weight) and a live load of 2.0 kip/ft. Check the adequacy of this beam. Use $f'_c = 4,000$ psi, $f_y = 60,000$ psi, and #3 stirrups. Assume knife edge type supports at the centers of the walls.

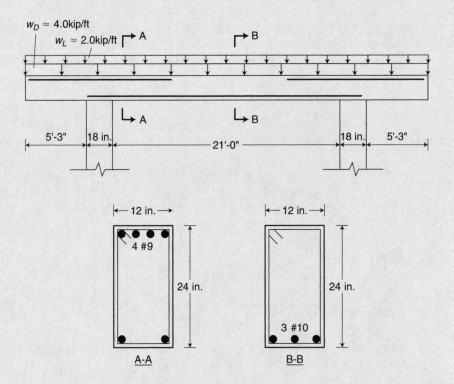

Note: Check both sections A-A and B-B. Neglect the reinforcement in the bottom of the beam at section A-A.

2–11 Determine the moment capacity, M_R, of the reinforced concrete section shown below if subjected to a negative moment. The stirrups are #3 bars. Use $f'_c = 4,000$ psi and $f_y = 60,000$ psi.

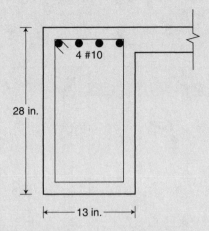

2–12 The figure below shows the cross section of a floor system consisting of a reinforced concrete beam supporting precast concrete planks. The beam span is 20'-0" with 16'-0" spacing. Calculate the maximum service live load per square foot of floor area. Use $f'_c = 4,000$ psi and $f_y = 60,000$ psi. The unit weight of lightweight (LW) concrete used is 108 pcf. Assume the beam is simply-supported.

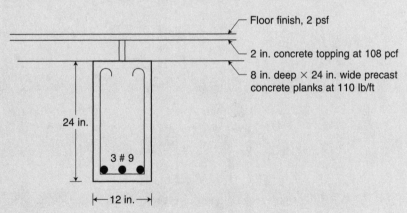

2–13 The 16 in. × 27 in. rectangular reinforced concrete beam shown below is reinforced with 4 #10 bars in the positive moment region and 3 #11 bars in the negative moment region. Determine the maximum factored uniformly distributed load, w_u, for this beam. Stirrups are #4, $f'_c = 5,000$ psi, and $f_y = 60,000$ psi.

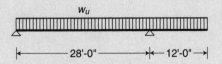

2–14 The beam of Problem 2–11 is part of a beam-girder floor system shown below (beam B-1). The floor slab is 6 in. thick concrete, and the weight of the mechanical/electrical systems is 5 psf. Assume 15 psf for partition loads, and miscellaneous dead loads of

5 psf. What is the maximum allowable live load for this floor? Consider only the *negative* moment capacity of the section. (Note: Use the ACI moment coefficients. Live load is not to be reduced.)

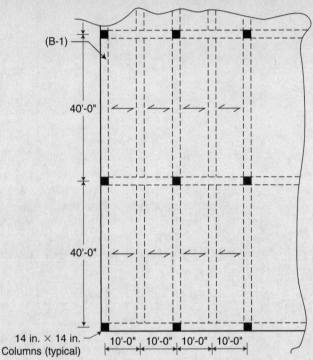

(B-1)

40'-0"

40'-0"

14 in. × 14 in.
Columns (typical)

10'-0" 10'-0" 10'-0" 10'-0"

2–15 Calculate the required areas of reinforcement for the following beams. Use $f_c' = 4{,}000$ psi and $f_y = 60{,}000$ psi.

(a) $b = 10$ in., $d = 20$ in., $M_u = 200$ ft-kip
(b) $b = 12$ in., $d = 24$ in., $M_u = 300$ ft-kip
(c) $b = 18$ in., $d = 36$ in., $M_u = 500$ ft-kip

2–16 Design a rectangular reinforced concrete beam subjected to a factored load moment, $M_u = 250$ ft-kip. The architect has specified width $b = 10$ in. and total depth $h = 24$ in.. Use $f_c' = 4{,}000$ psi, $f_y = 60{,}000$ psi, and #3 stirrups.

2–17 Redesign the beam in Problem 2–16, assuming that the clear height for the building requires the total beam depth to be limited to 20 in. Determine the beam width (b) and the area of steel (A_s) in such a way that the section will be in the tension-controlled failure zone.

2–18 Design a rectangular beam for $M_u = 300$ kip-ft. Use $f_c' = 3{,}000$ psi, $f_y = 60{,}000$ psi, and #3 stirrups. Size the beam for $\rho = 0.01$ and $b/d = 0.5$ (approximate). Do not consider the beam's self-weight.

2–19 The 16 in. × 27 in. rectangular reinforced concrete beam shown below is subjected to concentrated loads of $P_D = 12.0$ kip and $P_L = 8.0$ kip. The uniformly distributed dead load, w_D, is 1.6 kip/ft (including the beam's self-weight), and the live load, w_L, is 1.0 kip/ft. Determine the required reinforcements. Sketch the section and show the selected bars. Use $f_c' = 5{,}000$ psi and $f_y = 60{,}000$ psi.

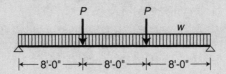

2–20 An artist is designing a sculpture that is to be supported by a rectangular reinforced concrete beam. The sculpture's weight is estimated to be 400 lb/ft (assumed as a live load). The beam section must be limited to $b = 8$ in. and $h = 12$ in. The artist wants to make his sculpture as long as possible. What is the maximum possible length of this cantilever beam without the use of compression reinforcement? Use $f'_c = 4{,}000$ psi, $f_y = 60{,}000$ psi, and #3 stirrups.

2–21 A 14 in. $\times$ 24 in. rectangular precast reinforced concrete beam supports a factored uniform load, $w_u = 4.0$ kip/ft, including the beam's self-weight. Determine the reinforcements required at the supports and the midspan. Use $f'_c = 4{,}000$ psi and $f_y = 60{,}000$ psi.

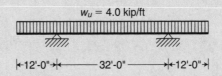

2–22 An 8 in. thick simply-supported reinforced concrete one-way slab is subjected to a live load of 150 psf. It has a 12 ft span and is reinforced with #4 @ 8 in. as the main reinforcement and #4 @ 12 in. as shrinkage and temperature reinforcement. Determine whether the slab is adequate. Use $f'_c = 4{,}000$ psi and $f_y = 60{,}000$ psi.

2–23 A 5 in.-thick simply-supported reinforced concrete one-way slab is part of a roof system. It is supported by two masonry block walls, as shown below. Assume a superimposed dead load (roofing, insulation, ceiling, etc.) of 15 psf and a roof snow load of 30 psf. Check the adequacy of the slab, including the required shrinkage and temperature reinforcement. Use $f'_c = 4{,}000$ psi and $f_y = 60{,}000$ psi. The bearing length of the slab on the wall is 6 in.

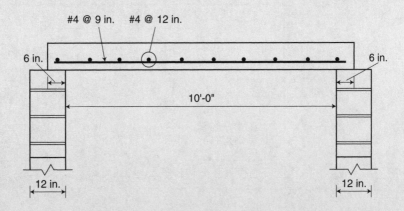

2–24 The figures below show the framing plan and section of a reinforced concrete floor system. The weight of the ceiling and floor finishing is 5 psf, that of the mechanical and electrical systems is 5 psf, and the weight of the partitions is 20 psf. The floor live load is 80 psf. The 6 in.-thick slab exterior bay (S-1) is reinforced with #6 @ 9 in. as the main

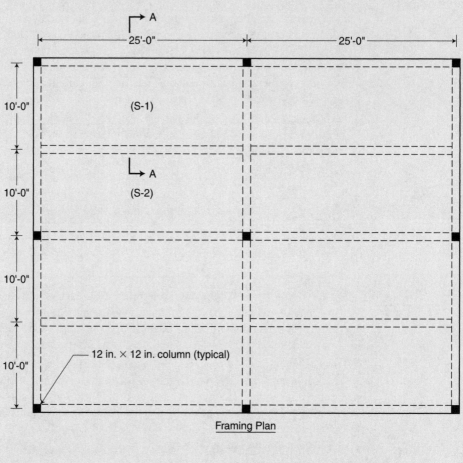

Framing Plan

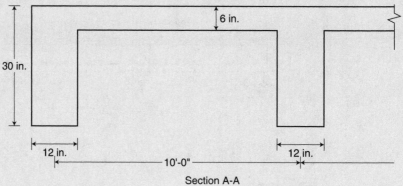

Section A-A

reinforcement at the midspan and #4 @ 12 in. for the shrinkage and temperature reinforcement. Check the adequacy of the slab. Use the ACI moment coefficients. Use $f'_c = 4,000$ psi and $f_y = 60,000$ psi.

2–25 Design a 6 in.-thick one-way slab for a factored moment, $M_u = 10$ ft-kip. Use $f'_c = 4,000$ psi and $f_y = 60,000$ psi.

2–26 Find the reinforcements for the midspan and supports for an interior 6 in.-thick slab (S-2) of the floor of Problem 2–24. Sketch the slab and show the reinforcements including the shrinkage and temperature reinforcement steel.

SELF-EXPERIMENTS

The main objective of these self-experiments is to understand the behavior of beams in bending (tension and compression) and changes in concrete strength with time, finding the modulus of rupture, and understanding the behavior of reinforced concrete beams under loading. The other objective is to understand the different aspects of concrete slabs. Remember to include all the details of the tests (sizes, time of day concrete was poured, amounts of water/cement/aggregate, problems encountered, etc.) with images showing the steps (making concrete, placing, forming, performing tests, etc.).

Experiment 1

In this experiment you learn about the behavior of beams in bending. Obtain a rectangular-shaped piece of Styrofoam with the proportions of a beam. Make slots on the top and bottom of the beam, as shown in Figure SE 2–1.

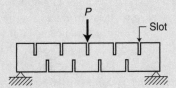

FIGURE SE 2–1   Styrofoam beam with slots.

Place the beam on two supports and add a load at the center as shown in Figure SE 2–1. Answer the following questions:

1. What happened to the slots at the top and bottom of the beam?
2. Did the slots stay straight after adding the load?
3. Any other observations?

Experiment 2

You must start and perform Experiments 2 and 3 at the same time. In this experiment, you find the modulus of rupture for a plain concrete beam and learn about concrete curing and gaining strength with time.

For this experiment you will build four beams using concrete with w/cm ratio = 0.5. Size the beams as you wish, but do not make them excessively small or large (for practical reasons). After forming the beams (you can use cardboard or wood for your forms, depending on the beam size), spray water on two of the beams while keeping the other two dry. Keep your concrete beams indoors, as the concrete may freeze and stop the hydration process. After 2 days, test two of your test beams (one kept dry and one kept wet) by placing loads on them, as shown in Figure SE 2–2.

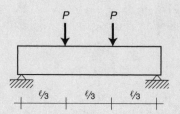

FIGURE SE 2–2   Plain concrete beam test.

Increase the loads until the beams fail. Record the loads at which the two specimens fail.

After seven days, repeat the tests with the remaining two beams and record the loads at which they fail.

Experiment 3

In this experiment, you will learn about the importance of reinforcing steel in concrete beams and compare the results with those of Experiment 2.

When you pour the four plain concrete beams for Experiment 2, build two reinforced concrete beams with the same dimensions as those of the plain concrete beams. You can use steel wires for the reinforcement (depending on your beam size). Place these wires on only one side of the beam (singly-reinforced beam).

After 2 days, place one of the beams on two supports and apply loads as shown in Figure SE 2–3a. Increase the load, and record your observations.

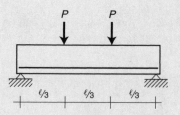

FIGURE SE 2–3a   Reinforced concrete beam test 1.

Repeat this test for the remaining reinforced concrete beam after seven days. (Perform these tests at the same time as Experiment 2.) DO NOT TRY TO FAIL THE REINFORCED CONCRETE BEAMS! Turn the beams upside down (Figure SE 2–3b) and repeat the tests. Add loads until the beams fail. Record your observations.

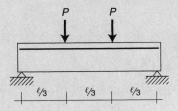

FIGURE SE 2–3b   Reinforced concrete beam test 2.

Answer the following questions regarding Experiments 2 and 3:

1. Which of the samples (dry or wet) had more strength? Why?
2. Was the 7-day-old sample stronger than the 2-day-old one? Why?
3. Find the modulus of rupture for the 7-day-old plain concrete beams.
4. How did the reinforcement affect the concrete beam strength?
5. What happened when you turned the beam upside down and tested it?

Experiment 4

This experiment demonstrates the behavior of one-way and two-way slabs, and the reinforcing of one-way slabs.

Test 1

Use two Styrofoam pieces to represent one-way and two-way slabs. For the two-way slab, cut the Styrofoam into a square piece, and for the one-way slab make it such that length/width ≥ 2. Place the square Styrofoam on two parallel supports and apply a load as shown in Figure SE 2–4a. Support the same model on four edges and repeat the test as shown Figure SE 2–4b. Make notes on how the two models deform and their differences.

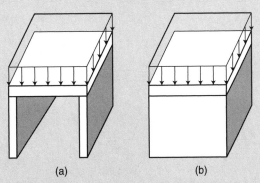

(a)   (b)

FIGURE SE 2–4   Slabs under loads: (a) two parallel supports; (b) supports along all edges.

Test 2

Repeat Test 1 using the one-way slab model. Record your observations.

Experiment 5

This experiment deals with the reinforcement in slabs.

Cast two slab models with a thickness of approximately 1 in. and a width of at least 12 in. Make one from plain concrete and the other from concrete reinforced with a grid of thin wires (provide about $1/4$ in. cover).

One week after making the samples, compare the two slabs in terms of crack formation. Which one has more surface cracks?

3

SPECIAL TOPICS
IN FLEXURE

3.1 T-BEAMS

Introduction

In cast-in-place reinforced concrete systems, the concrete for beams and slabs is poured at the same time. As a result, a monolithic system is obtained, that is, beams and slabs working together to carry the loads.

There are several different types of reinforced concrete floor systems, as we will discuss in detail later in Chapter 6. Here we will use a beam-girder floor system to study T-beams. Figure 3–1 shows the floor framing plan and the section of a typical beam-girder floor system. The floor beams (B-1) support the one-way slab (S-1).

The slab transfers the load to the beams (B-1); then the girders (G-1) carry the loads from the beams. The girders are supported by columns (C-1). Because the one-way slab is continuously supported by the beams, the load on the beams is a uniformly distributed load. The girders, however, support the beams at their ends, so the loads on the girders are concentrated. Thus, the flow of the gravity loads is from the slab to the beams, from the beams to the girders, from the girders to the columns, from the columns to the footings, and from the footings to the ground.

In cast-in-place concrete construction, concrete is poured in the forms after the formwork is built and the rebars are placed, creating a monolithic system of slabs, beams, and girders. There is no physical separation between beams and slabs as in steel construction. So when a beam bends, part of the slab attached to the beam works with the beam and helps the beam carry the load. At the midspan the top part of the beam is in compression. As a result the slab, which is attached to the top of the beam, is subjected to compression stress. But at the support,

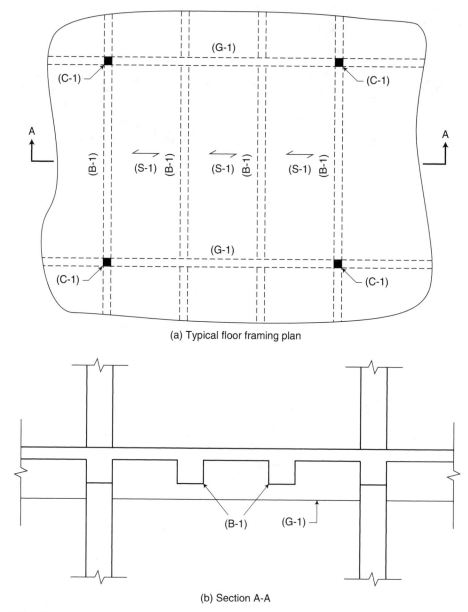

(a) Typical floor framing plan

(b) Section A-A

FIGURE 3–1 Beam-girder floor system.

the top portion of beam, including the neighboring slab, is in tension. Therefore, the slab does not help carry the beam load because the concrete does not take any tensile stresses.

Figure 3–2 shows cross sections and moments for a typical beam (B-1). At the midspan the moment is positive, so steel reinforcement is needed at the bottom of the beam (A_s^+). In this case the concrete slab and part of the beam web are in compression. The shape

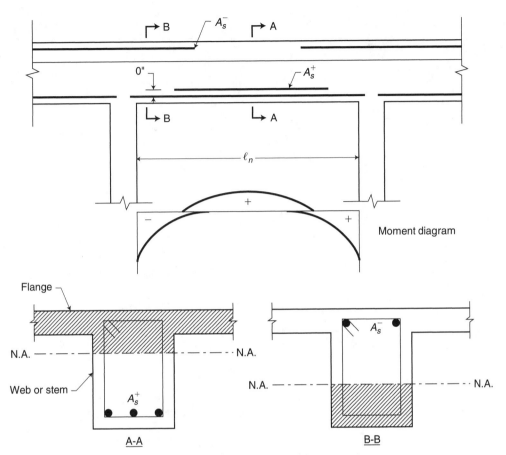

FIGURE 3–2 Beam behavior at midspan (T-beam) and over the support (rectangular beam).

of the compression zone looks like a T-shape, so it is called a *T-beam*. Over the supports, however, there are negative moments. This requires steel reinforcement at the top of the beam (A_s^-). In certain special cases, the ACI Code requires part of the positive reinforcements (A_s^+) to be extended over the supports. In these cases reinforcing is used in the compression zone, resulting in a *doubly-reinforced* beam (see Section 3.2).

Effective Flange Width (b_{eff})

The attached slab zone of a T-beam is referred to as the *flange* of the beam. The portion below the flange is called the *web*. How much of the slab width acts as part of the beam is a rather complex matter. It depends on many parameters that define how much of a slab's width is "dragged" into compression by the beam. The phenomenon that dissipates the compression in the slab that lies farther away from the beam's web is known as "shear lag."

The ACI Code simplifies the matter by defining an effective flange width (b_{eff}), in which the stresses due to bending are assumed to be uniform. Figure 3–3, which shows the floor framing plan and a section through the midspan of a reinforced concrete floor system,

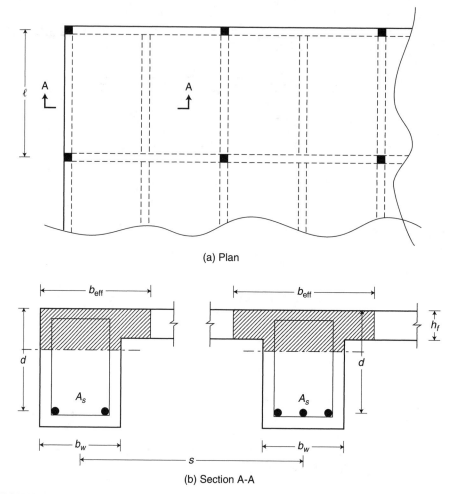

(a) Plan

(b) Section A-A

FIGURE 3–3 Effective flange widths for T- and L-beams.

also shows the effective width for an edge beam and for an interior beam. The edge beam is called an *L-beam* because the compression zone has an L shape. The interior beam is a T-beam. The beams' span is ℓ, and their center-to-center distance is designated by s. The slab or flange thickness is designated by h_f.

The effective flange widths of T- and L-beams are based on Sections 8.10.2 and 8.10.3 of the ACI Code and are given in Equations 3–1 and 3–2.

a. b_{eff} for T-beams:

$$b_{\text{eff}} \leq \min\left\{\frac{\ell}{4}, b_w + 16h_f, s\right\} \tag{3–1}$$

b. b_{eff} for L-beams:

$$b_{\text{eff}} \leq \min\left\{b_w + \frac{\ell}{12}, b_w + 6h_f, \frac{(b_w + s)}{2}\right\} \tag{3–2}$$

Minimum Steel for T-beams

The minimum amount of steel for a T-beam is the same as that for a rectangular beam having working dimensions of b_w (width of web) and d. Equation 3–3 gives the minimum amount of steel required.

$$A_{s,\text{min}} = \max\left\{\frac{3\sqrt{f_c'}}{f_y}b_w\,d, \frac{200}{f_y}b_w\,d\right\} \tag{3-3}$$

Equation 3–4 gives the requirement in terms of minimum steel ratio.

$$\rho_{\text{min}} = \max\left\{\frac{3\sqrt{f_c'}}{f_y}, \frac{200}{f_y}\right\} \tag{3-4}$$

Table A2–4 lists the values for ρ_{min}.

The compression zone in the negative moment regions (near the columns) is at the bottom of the web, where there is no flange attached. The section, therefore, is simply rectangular. The analysis and design of these sections were discussed in Chapter 2.

Analysis of T-beams

The behavior of the T-beam (or L-beam) depends on the shape of the compression zone. The depth of the equivalent stress block (a) may be above or below the bottom of the flange, depending on the proportioning of the beam and the slab and the amount of reinforcement used. Figures 3–4a and 3–4b show these two cases, respectively.

When the neutral axis is within the flange's depth, the T-beam (or L-beam) acts like a wide rectangular beam with a rectangular compression zone of size $b_{\text{eff}} \times a$. In the rare cases when a small b_{eff} is coupled with relatively large positive moments, the $b_{\text{eff}} \times h_f$ zone is not adequate to develop the compression part of the internal couple. Then a part of the web becomes in compression to aid the compression zone. The analysis and design of such beams are somewhat different from those of rectangular beams.

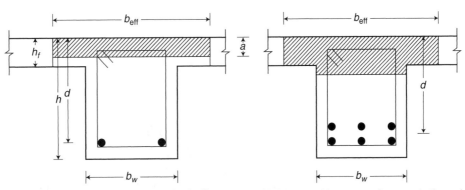

(a) T-beam with compression zone in the flange   (b) T-beam with compression zone in the web

FIGURE 3–4   Different types of T-beams.

Thus, two slightly different sets of procedures are used for the analysis of T- (or L-) beams based on the shape of the compression zone. The flowchart in Figure 3–6 summarizes the different steps of analysis of T- (or L-) beams. They are as follows:

Step 1   Calculate the effective flange width (b_{eff}).

Step 2   Check the minimum area of steel $A_{s,\min}$ or the minimum steel ratio from Table A2–4. Note if the areas of reinforcing satisfy the current ACI Code's requirements.

Step 3   Assume that the steel yields in tension before the concrete crushes in compression (i.e., $f_s = f_y$). Then calculate the total tensile force, T:

$$T = A_s f_y$$

Step 4   Calculate the compression force if the entire flange is in compression, C_f:

$$C_f = 0.85 f_c' b_{eff} h_f$$

The internal couple requires that $T = C$, that is, the compression force and the tensile force must be equal.

If $T < C_f$ (or $a < h_f$), the full depth of the flange thickness is not needed to develop the compression part of the internal couple. In that case the depth of the equivalent stress block is less than the thickness of the flange, and *case a* below is applicable; otherwise, use *case b*.

Case a: The compression zone is within the flange ($a \leq h_f$); the beam behaves like a rectangular beam.

Step 5   Determine the steel ratio, ρ:

$$\rho = \frac{A_s}{b_{eff} d}$$

Step 6   Use f_y, f_c', and ρ to obtain the resistance coefficient, R, from Tables A2–5 through A2–7.

The resistance coefficient obtained, R, is only applicable for beams with a single layer of reinforcement ($d_t = d$). If the beam has multiple layers of reinforcement, R may need to be revised. If the value of the strength reduction factor, ϕ, in the last step is 0.90, no change in the value of R is necessary.

If $\phi < 0.90$, however, then compute $\rho_t = \dfrac{A_s}{b_{eff} d_t}$ and obtain the corresponding value of ϕ' from Tables A2–5 through A2–7. Then calculate R' ($R' = R\phi'/\phi$).

Step 7   Calculate M_R:

$$M_R = \phi M_n = b_{eff} d^2 R / 12{,}000$$

or

$$M_R = \phi M_n = b_{eff} d^2 R' / 12{,}000$$

M_R is in ft-kip, b and d are in inches, and R and R' are in psi.

Step 8   After calculating M_R, check to ensure the beam can safely carry the loads by comparing M_R with the maximum factored moment (M_u). Also, check the depth of the beam to determine if deflection calculations are required according to the ACI Code (see Figure 2–41).

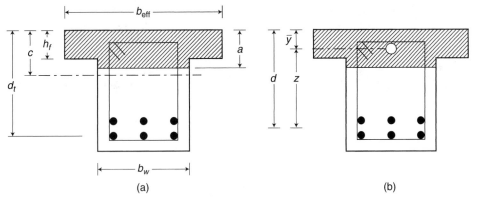

FIGURE 3–5 T-beam with neutral axis below the flange.

Case b: The compression zone extends below the flange ($a > h_f$); compression zone is T-shaped.

Figure 3–5 shows the T-shaped compression zone and the corresponding definition of symbols used below.

Step 5   Determine the depth of the compression zone (a) by equating the tensile force to the compression forces in the flange and the web:

$$T = C_f + 0.85\,f_c'\,b_w\,(a - h_f)$$
$$T - C_f + 0.85\,f_c'b_wh_f = 0.85\,f_c'b_wa$$
$$a = \frac{T - C_f + 0.85f_c'b_wh_f}{0.85f_c'b_w}$$
$$a = \frac{T - C_f}{0.85\,f_c'b_w} + h_f \qquad (3\text{–}5)$$

Step 6   Locate the neutral axis (c) and check to ensure the section satisfies the ACI Code's requirements for being in the tension-controlled or transition zones. The neutral axis is located at:

$$c = \frac{a}{\beta_1}$$

If $\dfrac{c}{d_t} > \dfrac{3}{7}$ the section does not satisfy the ductile failure requirements, as $\epsilon_t < 0.004$ when $\epsilon_c = 0.003$.

If $\dfrac{c}{d_t} \leq \dfrac{3}{7}$, determine the strength reduction factor, ϕ, using the relationships below:

$$\text{if } \quad \frac{c}{d_t} \leq \frac{3}{8} \quad \longrightarrow \quad \phi = 0.90$$

$$\text{if } \quad \frac{c}{d_t} > \frac{3}{8} \quad \longrightarrow \quad \phi = A_2 + \frac{B_2}{c/d_t}$$

A_2 and B_2 are obtained from Table A2–2b.

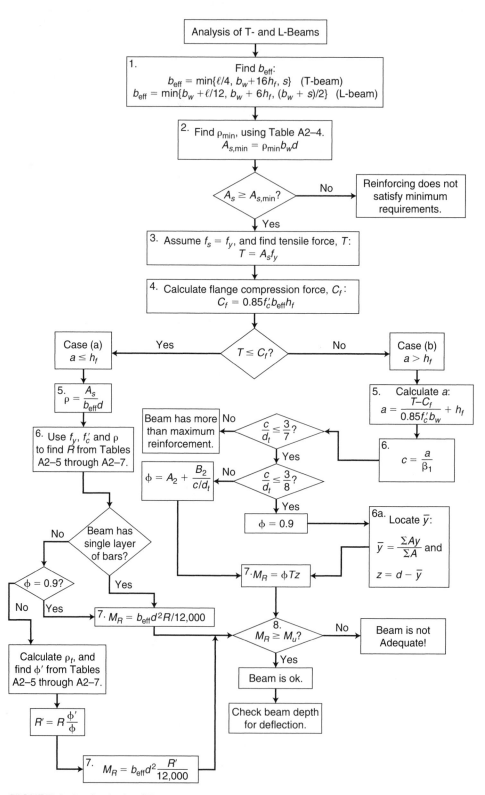

FIGURE 3–6 Analysis of T- and L-beams.

Step 6a   Determine the centroid of the compression zone by dividing it into rectangular parts and using Equation 3–6:

$$\bar{y} = \frac{\Sigma Ay}{\Sigma A} \tag{3–6}$$

$\bar{y}$ is the distance from the top of the beam to the centroid of the compression zone.

The moment arm (z), which is the distance between the tensile and compression forces, is:

$$z = d - \bar{y}$$

Step 7   Calculate the design resisting moment, M_R:

$$M_R = \phi M_n = \phi T z \tag{3–7}$$

Step 8   After computing M_R, check to ensure the beam is adequate. Also check the depth of the beam for deflection (Figure 2–41).

EXAMPLE 3–1

Figure 3–7 shows the partial floor framing plan and sections of a reinforced concrete floor system. The slab is 4 in. thick, and the weight of mechanical/electrical systems, ceiling, and floor finishing is 24 psf. The floor live load is 200 psf. The beam ends are integral with their support, $f'_c = 3$ ksi, $f_y = 60$ ksi, and a unit weight of concrete of 150 pcf. Stirrups are #4 bars. Use ACI coefficients for calculation of bending moments.

   a.   Check the adequacy of the edge beam (B-1) at midspan.

   b.   Check the adequacy of the interior beam (B-2) at midspan.

Solution   *Use the flowchart of Figure 3–6.*

(a) *Edge Beam (B-1)* B-1 is an L-beam for positive moment (midspan):

Step 1   Calculate the effective flange width:

$$b_{\text{eff}} = \min\left\{ b_w + \frac{\ell}{12}, b_w + 6h_f, \frac{(b_w + s)}{2} \right\}$$

where
$$\ell = 32 \text{ ft} \times 12 = 384 \text{ in.}$$
$$b_w = 18 \text{ in.}$$
$$h_f = 4 \text{ in.}$$
$$s = 10 \times 12 = 120 \text{ in.}$$

$$b_{\text{eff}} = \min\left\{ 18 + \frac{384}{12}, 18 + 6(4), \frac{(18 + 120)}{2} \right\}$$

$$b_{\text{eff}} = \min\{50 \text{ in.}, 42 \text{ in.}, 69 \text{ in.}\} = 42 \text{ in.}$$

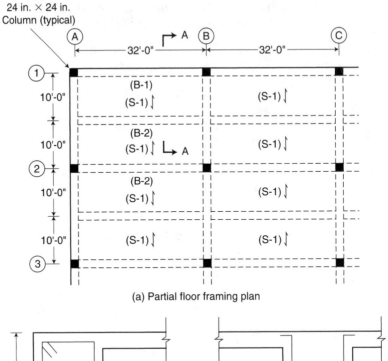

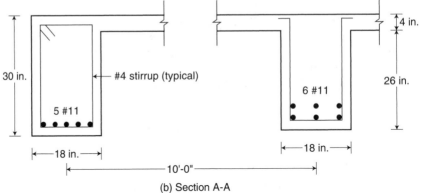

FIGURE 3–7   Floor framing plan and section for Example 3–1.

Step 2   From Table A2–4 ⟶ $\rho_{\min} = 0.0033$

$$d = h - \bar{y} = 30 \text{ in.} - \left(1.5 + \frac{4}{8} + \frac{1.41}{2} \right) = 27.3 \text{ in.}$$

$$A_{s,\min} = \rho_{\min} b_w d$$
$$A_{s,\min} = (0.0033)(18)(27.3)$$
$$A_{s,\min} = 1.62 \text{ in}^2$$

5 #11 ⟶ Table A2–9 ⟶ $A_s = 7.80 \text{ in}^2 > 1.62 \text{ in}^2$ ∴ ok

Step 3   Assuming that the steel yields at the nominal resisting moment ($f_s = f_y$), calculate the tensile force, T:

$$T = A_s f_y = 7.80 \times 60 = 468 \text{ kip}$$

Step 4   Determine the total compression force, C_f, assuming that the compression zone is within the flange:

$$C_f = 0.85 f_c' \, b_{\text{eff}} \, h_f$$
$$C_f = 0.85(3)(42)(4)$$
$$C_f = 428 \, k$$
$$T = 468 \, k > C_f = 428 \, \text{kip}$$

Because $T > C_f$, the assumption in step 4 was not correct, and the compression zone has to be larger in order for C_f to be equal to T. Thus, the compression zone extends below the flange.

Step 5   Determine the depth of the compression zone, a:

$$a = \frac{T - C_f}{0.85 f_c' b_w} + h_f$$

$$a = \frac{468 - 428}{0.85 \times 3 \times 18} + 4$$

$$a = 4.87 \, \text{in.}$$

Step 6   Calculate the location of the neutral axis, c:

$$c = \frac{a}{\beta_1} = \frac{4.87}{0.85}$$

$$c = 5.73 \, \text{in.}$$

Because there is only a single layer of reinforcement ($d_t = d = 27.3$ in.):

$$\frac{c}{d_t} = \frac{5.73}{27.3} = 0.210 < \frac{3}{7} = 0.429 \quad \therefore \text{ ok}$$

$$0.210 < \frac{3}{8} = 0.375 \quad \therefore \ \phi = 0.90$$

Therefore, the section is tension-controlled.

Step 6a   Locate the centroid of the compression zone (hatched area) in Figure 3–8. Divide the compression zone into two rectangular shapes and calculate $\bar{y}$ (measured from the top of the beam).

$$\bar{y} = \frac{\Sigma Ay}{\Sigma A}$$

$$\bar{y} = \frac{(42 \times 4)\left(\dfrac{4}{2}\right) + (18 \times 0.87)\left(4 + \dfrac{0.87}{2}\right)}{(42 \times 4) + (18 \times 0.87)}$$

$$\bar{y} = 2.21 \, \text{in.}$$

Calculate the moment arm, z:

$$z = d - \bar{y} = 27.3 - 2.21 = 25.09 \, \text{in.}$$

Step 7   The design resisting moment, M_R, is:

$$M_R = \phi Tz = 0.90 \times 468 \times 25.09 = \frac{10{,}568 \text{ kip-in.}}{12} = 881 \text{ ft-kip}$$

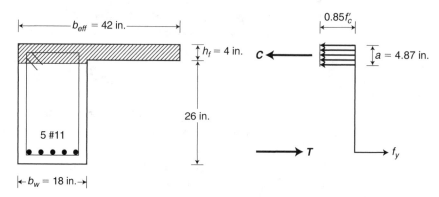

FIGURE 3–8 Forces acting on the beam section of Example 3–1a.

Step 8   To ensure that the beam can carry the loads, calculate the maximum factored moment after determining the loads.

$$\text{Weight of slab} = 150\left(\frac{4}{12}\right) = 50 \, \text{psf}$$

$$\text{Superimposed dead loads} = 24 \, \text{psf}$$
$$\text{Total dead load} = 74 \, \text{psf}$$

$$w_D = \frac{\left[74 \times 5.75 + 150\left(\frac{18}{12} \times \frac{26}{12}\right)\right]}{1000} = 0.913 \, \text{kip/ft}$$

$$w_L = \left[\frac{200 \times 5.75}{1000}\right] = 1.15 \, \text{kip/ft}$$

$$w_u = 1.2w_D + 1.6w_L = 1.2 \times 0.913 + 1.6 \times 1.15 = 2.94 \, \text{kip/ft}$$
$$\ell_n = 32 - 2 = 30 \, \text{ft}$$

From Table A2–1:

$$(M_u)^+ = \frac{w_u \ell_n^2}{14} = \frac{(2.94)(30)^2}{14} = 189 \, \text{ft-kip}$$

$$M_u = 189 \, \text{ft-kip} < M_R = 881 \, \text{ft-kip} \quad \therefore \text{ok}$$

Check the beam depth for deflection. See Figure 2–41. (B-1 is a one-end continuous beam):

$$h_{\min} = \frac{\ell}{18.5} = \frac{32 \times 12}{18.5}$$

$$h_{\min} = 21 \, \text{in.} < h = 30 \, \text{in.} \quad \therefore \text{ok}$$

Therefore, the deflection does not need to be checked.

B-1 is ok

(b) *Interior Beam (B-2)* B-2 is a T-beam for positive moment at midspan.

Step 1   Determine the effective flange width:

$$b_{\text{eff}} = \min\left\{\frac{\ell}{4}, b_w + 16h_f, s\right\}$$

where
$\ell = 32 \text{ ft} \times 12 = 384 \text{ in.}$
$b_w = 18 \text{ in.}$
$h_f = 4 \text{ in.}$
$s = 10 \times 12 = 120 \text{ in.}$

$$b_{\text{eff}} = \min\left\{\frac{384}{4}, 18 + 16(4), 120\right\}$$

$$b_{\text{eff}} = \min\{96 \text{ in.}, 82 \text{ in.}, 120 \text{ in.}\}$$

$$b_{\text{eff}} = 82 \text{ in.}$$

Step 2   From Table A2–4 $\rightarrow$ $\rho_{\min} = 0.0033$

$$d = h - \bar{y} = 30 - (1.5 + {}^4/_8 + 1.41 + {}^1/_2) = 26.1 \text{ in.}$$
$$A_{s,\min} = \rho_{\min} b_w d$$
$$A_{s,\min} = 0.0033(18)(26.1) = 1.55 \text{ in}^2$$

6 #11 $\longrightarrow$ Table A2–9 $\longrightarrow$ $A_s = 9.36 \text{ in}^2 > 1.55 \text{ in}^2$ $\therefore$ ok

Step 3   Assuming that the steel yields ($f_s = f_y$), calculate the tensile force, T:

$$T = A_s f_y = 9.36 \times 60 = 562 \text{ kip}$$

Step 4   Calculate the flange compression force, C_f:

$$C_f = 0.85 f_c' b_{\text{eff}} h_f = 0.85(3)(82)(4) = 836 \text{ kip}$$
$$T = 562 \text{ kip} < 836 \text{ kip}$$

Therefore, the compression zone is within the flange. In other words, $a < h_f$. Thus, the beam analysis is similar to that of a rectangular beam with a width of $b = b_{\text{eff}} = 82 \text{ in.}$

Step 5   Calculate the steel ratio, ρ:

$$\rho = \frac{A_s}{b_{\text{eff}} d} = \frac{9.36}{82 \times 26.1} = 0.0044$$

Step 6

$\rho = 0.0044$

$f_c' = 3 \text{ ksi} \longrightarrow$ Table A2–6a $\longrightarrow$ $\begin{array}{l} R = 225 \text{ psi} \\ \phi = 0.90 \end{array}$

$f_y = 60 \text{ ksi}$

Step 7   Calculate the design resisting moment, M_R:

$$M_R = \frac{b_{\text{eff}} d^2 R}{12,000}$$

$$M_R = \frac{(82)(26.1)^2(225)}{12,000}$$

$$M_R = 1,047 \text{ ft-kip}$$

Step 8   Determine maximum $(M_u)^+$:

From part a: Total dead load = 74 psf

Total live load = 200 psf

$$w_D = \frac{\left[74 \times 10 + 150\left(\dfrac{18}{12} \times \dfrac{26}{12} \right) \right]}{1,000} = 1.23 \text{ kip/ft}$$

$$w_L = \frac{[200 \times 10]}{1,000} = 2.0 \text{ kip/ft}$$

$$w_u = 1.2w_D + 1.6w_L$$

$$w_u = 1.2 \times 1.23 + 1.6 \times 2.0 = 4.68 \text{ kip/ft}$$

$$(M_u)^+ = \frac{w_u \ell_n^2}{14} = \frac{(4.68)(30)^2}{14} = 301 \text{ ft-kip}$$

$$M_u = 301 \text{ ft-kip} < M_R = 1,047 \text{ ft-kip} \quad \therefore \text{ ok}$$

Check the beam depth to determine whether deflection analysis is needed:

$$h_{\min} = 21 \text{ in. from case a}$$

$$h = 30 \text{ in.} > 21 \text{ in.} \quad \therefore \text{ ok}$$

$$B\text{-}2 \text{ is ok}$$

Design of T-beams

In theory, the design of T-beams involves finding the flange thickness, the width and depth of the web, and the amount of reinforcement required. In practice, however, the flange thickness is determined when designing the slab. The size of the web is selected to resist not only the moments at the supports (no T-beam action), but to provide adequate shear capacity, and to simplify formwork layout for ease of construction.

Hence, when designing a T-beam, the geometric dimensions of the beam typically are known. The only unknown is the amount of steel required to resist the loads. The T-beam design procedure, like beam analysis, depends on the required depth of the equivalent stress block. In most cases, the compression zone is within the flange area; so the design follows that of a simple rectangular beam, with a width equal to the effective width of the flange.

In some rare cases, however, the compression zone available within the depth of the flange, may not be adequate to develop the necessary factored moment. The difference then must be compensated by having an additional compression zone below the bottom of the flange (within the web).

The steps for the design of T- and L-beams follow. These are summarized in the flowchart of Figure 3–10.

Step 1   Calculate the maximum factored moment that the beam must carry (M_u).

Step 2   Determine the effective flange width (b_{eff}) based on the ACI requirements.

Step 3   Assume a single layer of reinforcement ($\bar{y} = 2.5$ in.) and the effective depth, $d = h - \bar{y}$. In addition, assume $\phi = 0.90$.

Step 4   Calculate M_R using Equation 3–8. M_R is the moment capacity when the compression zone is only within the flange.

$$M_{Rf} = \phi M_{nf} = \phi(0.85 f_c')b_{eff} h_f(d - {}^{h_f}/_2)$$ (3–8)

Step 5   Use *Case a.*   If $M_u \leq M_{Rf} \longrightarrow$ the compression zone is entirely within the flange.

Use *Case b.*   If $M_u > M_{Rf} \longrightarrow$ the flange area is not adequate to develop the required factored moment.

Case a: Compression zone is within the flange ($a \leq h_f$).

Step 6   Calculate the resistance coefficient, R:

$$R = \frac{12,000 M_u}{b_{eff} d^2}$$

Step 7   Use f_y, f_c', and R to obtain ρ and ϕ from Tables A2–5 through A2–7.

Step 8   Calculate the required area of steel, A_s:

$$A_s = \rho b_{eff} d$$

Check the result against the minimum reinforcement requirement $A_{s,min} = \rho_{min} b_w d$. ρ_{min} is given in Table A2–4. Select the size and number of the bars using Tables A2–8 and A2–9.

Step 9   Calculate the actual effective depth:

$$d = h - \bar{y}$$

If $d \geq d_{assumed}$, the design is a little conservative. Otherwise, you may revise the design by using this new value of effective depth.

Step 10 Compare the beam depth with the required minimum for deflection control.

Case b: Compression zone extends below the flange ($a > h_f$)—see Figure 3–9a.

Step 6   Calculate the area of steel required to balance the entire flange in compression. See Figure 3–9b. Assume $d_f = d = h - 2.5$ in., and the moment arm $z_f = d_f - {}^{h_f}/_2$.

Step 7   The area of steel necessary to develop the compression zone of the entire flange area (A_{sf}) is given in Equation 3–9.

$$A_{sf} = \frac{12 M_{Rf}}{\phi f_y z_f}$$ (3–9)

M_{Rf} is in kip-ft, f_y in ksi, z_f in inches, and A_{sf} in in^2.

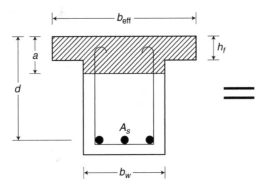

(a) Compression zone extends below the flange

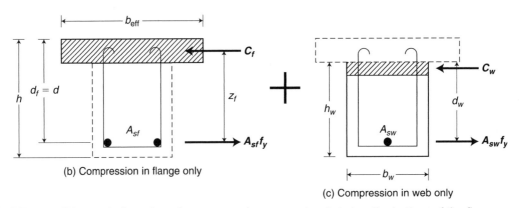

(b) Compression in flange only

(c) Compression in web only

FIGURE 3–9   T-beam design where the compression zone extends below the bottom of the flange.

Another way of calculating A_{sf} is to use equilibrium of forces in Figure 3–9b ($A_{sf} f_y = C_f$).

$$A_{sf}f_y = 0.85f_c' b_{eff} h_f$$

$$A_{sf} = \frac{0.85f_c' b_{eff} h_f}{f_y}$$

Step 8   In order to calculate the area of required steel for the part of the compression zone that is below the flange (A_{sw}), consider only the depth of the stem that is below the flange ($h_w = h - h_f$). Assume that the effective depth of the stem (d_w) is $d_w = h_w - 2.5$ in. See Figure 3–9c.

Then use Equation 3–10 to calculate the resistance coefficient for the required area of steel in the web (R_w).

$$R_w = \frac{12{,}000(M_u - M_{Rf})}{b_w d_w^2} \tag{3–10}$$

Step 9   Use f_y, f_c', and R_w to obtain ρ_w (steel ratio for the web) from Tables A2–5 through A2–7, and calculate the required area of steel in the web (A_{sw}) using Equation 3–11.

$$A_{sw} = \rho_w b_w d_w \tag{3–11}$$

Step 10 The total area of steel is:

$$A_s = A_{sf} + A_{sw} \qquad (3\text{--}12)$$

Step 11 Select the size and number of bars using Tables A2–8 and A2–9.

Step 12 Based on the size and number of the selected bars, compute the actual beam effective depth ($d = h - \bar{y}$). If this value is larger than what was assumed

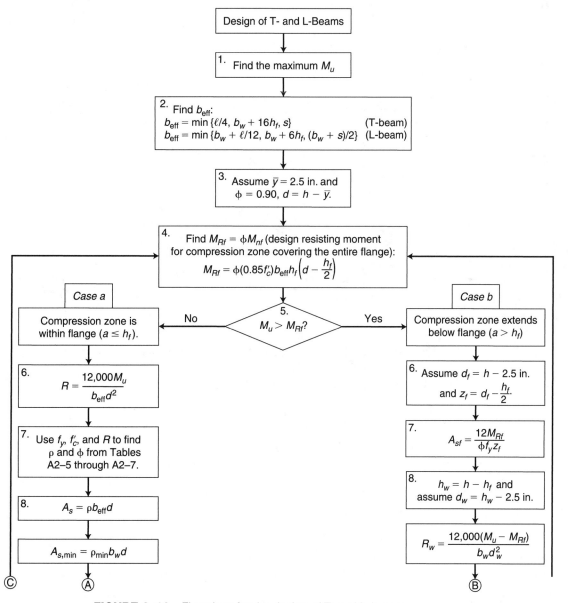

FIGURE 3–10   Flowchart for the design of T- and L-beams.

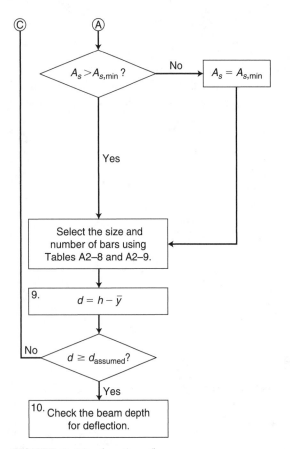

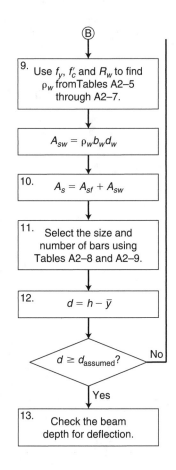

FIGURE 3–10 *(continued)*

in step 3, the design is a little conservative. Otherwise, revise the design as needed by using the new value of effective depth.

Step 13 Finally, check the beam depth (h) for deflection requirements.

EXAMPLE 3–2

Because beams B-1 and B-2 of Example 3–1 were overdesigned, redesign these L- and T-beams respectively for the maximum positive moments at midspan.

Solution *Use the flowchart of Figure 3–10.*

(a) *Edge Beams (B-1)*

Step 1 From Example 3–1a, step 8:

$$M_u = 189 \text{ ft-kip}$$

Step 2 From Example 3–1a, step 1, the effective flange width (b_{eff}) is:

$$b_{\text{eff}} = 42 \text{ in.}$$

Step 3   Assume $\bar{y} = 2.5$ in. and $\phi = 0.90$:

$$d = h - \bar{y} = 30 \text{ in.} - 2.5 \text{ in.} = 27.5 \text{ in.}$$

Step 4   Equation 3–8 gives the design resisting moment if the entire flange is in compression (M_{Rf}):

$$M_{Rf} = \phi M_{nf} = \phi\,(0.85 f_c')\,b_{\text{eff}} h_f\left(d - \frac{h_f}{2}\right)$$

$$M_{Rf} = 0.90\,(0.85 \times 3)\,(42)\,(4)\left(27.5 - \frac{4}{2}\right)$$

$$M_{Rf} = \frac{9{,}832 \text{ in-kip}}{12} = 819 \text{ ft-kip}$$

Step 5   Because $M_u = 189$ ft-kip $< M_{Rf} = 819$ ft-kip, the compression zone will be within the flange ($a < h_f$).

Step 6   Calculate the resistance coefficient, R:

$$R = \frac{12{,}000 M_u}{b_{\text{eff}} d^2}$$

$$R = \frac{12{,}000 \times 189}{42 \times 27.5^2} = 71 \text{ psi}$$

Step 7

$$f_c' = 3 \text{ ksi}$$
$$f_y = 60 \text{ ksi} \longrightarrow \text{Table A2–6a} \longrightarrow \rho = 0.0014$$
$$R = 71 \text{ psi}$$

(When obtaining ρ from Table A2–6a, because $R = 71$ psi is not in the table, we selected the value corresponding to $R = 74$ psi.)

Step 8

$$A_s = \rho b_{\text{eff}} d = (0.0014)(42)(27.5) = 1.62 \text{ in}^2$$

From Table A2–4, $\rho_{\min} = 0.0033$

Thus $A_{s,\min} = 0.0033 \times 18 \times 27.5 = 1.63 \text{ in}^2$.

Because A_s is less than $A_{s,\min}$ use $A_s = 1.63 \text{ in}^2$.

From Table A2–9, we select 3 #7 bars ($A_s = 1.8 \text{ in}^2$)

From Table A2–8 $\longrightarrow b_{\min} = 9$ in. < 18 in. $\therefore$ ok

From Table A2–8 $\longrightarrow b_{\max} = 24$ in. > 18 in. $\therefore$ ok

Step 9   Calculate the actual effective depth (d) of the beam:

$$\bar{y} = 1.5 \text{ in.} + \frac{4 \text{ in.}}{8} + \frac{0.875 \text{ in.}}{2} = 2.44 \text{ in.}$$

$$d = h - \bar{y} = 30 \text{ in.} - 2.44 \text{ in.} = 27.56 \text{ in.} \geq d_{\text{assumed}} = 27.5 \text{ in.}   \therefore \text{ ok}$$

Step 10 Check the beam depth for deflection:

$$h_{min} = \frac{\ell}{18.5} \text{ (one-end continuous beam)}$$

$$h_{min} = \frac{32 \times 12}{18.5} = 21 \text{ in.} < h = 30 \text{ in.} \quad \therefore \text{ ok}$$

Figure 3–11 shows a sketch of the beam.

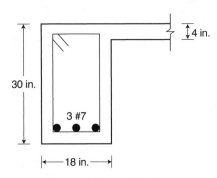

30 in.

4 in.

3 #7

|←—18 in.—→|

FIGURE 3–11   Sketch of beam B-1 for Example 3–2.

(b) *Interior Beam (B-2)*

Step 1   From Example 3-1b, step 8:

$$M_u = 301 \text{ ft-kip}$$

Step 2   From Example 3-1b, step 1, the effective flange width (b_{eff}) is:

$$b_{eff} = 82 \text{ in.}$$

Step 3   Assume $\phi = 0.90$ and $\bar{y} = 2.5$ in.:

$$d = h - \bar{y} = 30 \text{ in.} - 2.5 \text{ in.} = 27.5 \text{ in.}$$

Step 4   Equation 3–8 gives the design resisting moment for the beam with the entire flange in compression (M_{Rf}):

$$M_{Rf} = \phi M_{nf} = \phi \, (0.85 f'_c) b_{eff} hf \left(d - \frac{h_f}{2} \right)$$

$$M_{Rf} = 0.90(0.85 \times 3)(82)(4)(27.5 - {}^4/_2)$$

$$M_{Rf} = \frac{19,195 \text{ in.-kip}}{12} = 1,600 \text{ ft-kip}$$

Step 5   Because $M_u = 301$ ft-kip $< M_{Rf} = 1,600$ ft-kip, the compression zone will be within the flange.

Step 6   Calculate the resistance coefficient, R:

$$R = \frac{12{,}000 M_u}{b_{eff} d^2}$$

$$R = \frac{12{,}000 \times 301}{82(27.5)^2} = 58 \text{ psi}$$

Step 7

$$f'_c = 3 \text{ ksi}$$
$$f_y = 60 \text{ ksi} \longrightarrow \text{Table A2–6a} \longrightarrow \rho = 0.0011$$
$$R = 58 \text{ psi}$$

Step 8

$$A_s = \rho b_{eff} d = 0.0011 \times 82 \times 27.5$$
$$A_s = 2.48 \text{ in}^2$$

From Table A2–4 $\longrightarrow$ $\rho_{min} = 0.0033$

$$A_{s,min} = 0.0033 \times 18 \times 27.5 = 1.63 \text{ in}^2$$

From Table A2–9 $\longrightarrow$ 3 #9 bars ($A_s = 3 \text{ in}^2$)

From Table A2–8 $\longrightarrow$ $b_{min} = 10 \text{ in.} < 18 \text{ in.}$ $\therefore$ ok

From Table A2–8 $\longrightarrow$ $b_{max} = 24 \text{ in.} > 18 \text{ in.}$ $\therefore$ ok

Step 9   Calculate the actual effective depth (d):

$$\bar{y} = 1.5 + \frac{4}{8} + \frac{1.128}{2} = 2.56 \text{ in.}$$

$$d = h - \bar{y} = 30 \text{ in.} - 2.56 \text{ in.} = 27.44 \text{ in.} \approx d_{assumed} = 27.5 \text{ in.} \therefore \text{ ok}$$

Step 10   Check the beam depth for deflection:

$$h_{min} = \frac{\ell}{18.5} = \frac{32 \times 12}{18.5} = 21 \text{ in.} < h = 30 \text{ in.} \therefore \text{ ok}$$

Figure 3–12 shows a sketch of the beam.

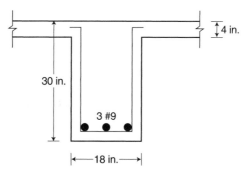

FIGURE 3–12   Sketch of beam B-2 for Example 3–2.

EXAMPLE 3-3

Design the T-beam shown in Figure 3–13. Assume that the effective flange width is 54 in. The T-beam is subjected to a total factored positive moment, $M_u = 950$ ft-kip. Use $f'_c = 3$ ksi, and $f_y = 60$ ksi. Assume #4 stirrups.

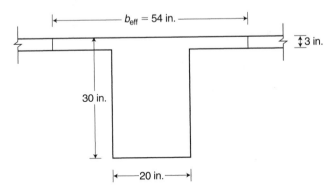

FIGURE 3-13   Sketch of T-beam for Example 3–3.

Solution   *Use the flowchart of Figure 3–10.*

Step 1

$$M_u = 950 \text{ ft-kip}   \text{(given)}$$

Step 2

$$b_{eff} = 54 \text{ in.}   \text{(given)}$$

Step 3   Assuming $\phi = 0.90$ and $\bar{y} = 2.5$ in.:

$$d = h - \bar{y} = 30 \text{ in.} - 2.5 \text{ in.} = 27.5 \text{ in.}$$

Step 4   From Equation 3–8:

$$M_{Rf} = \phi M_{nf} = \phi(0.85 f'_c) b_{eff} h_f \left(d - \frac{h_f}{2} \right)$$

$$M_{Rf} = (0.90)(0.85 \times 3)(54)(3)\left(27.5 - \frac{3}{2} \right)$$

$$M_{Rf} = \frac{9,667 \text{ in-kip}}{12} = 806 \text{ ft-kip}$$

Step 5   Because $M_u = 950$ ft-kip $> M_{Rf} = 806$ ft-kip, the compression zone will extend into the web area $(a > h_f)$. Use *Case b*.

Step 6   First calculate the amount of steel needed to work with the entire flange in compression (A_{sf}), and then the reinforcing needed to work with the part of the web that is in compression (A_{sw}). The total required area of steel (A_s) will then be:

$$A_s = A_{sf} + A_{sw}$$

To determine the area of steel required to work with the flange in compression (A_{sf}), assume d_f (effective beam depth for the entire flange in compression) as:

$$d_f = h - 2.5 \text{ in.} = 30 \text{ in.} - 2.5 \text{ in.} = 27.5 \text{ in.}$$

Then the moment arm of this internal couple is

$$z_f = d_f - \frac{h_f}{2} = 27.5 \text{ in.} - \frac{3 \text{ in.}}{2} = 26 \text{ in.}$$

Step 7   The area of the steel required to work with the flange (A_{sf}) is:

$$A_{sf} = \frac{12M_{Rf}}{\phi f_y z_f}$$

$$A_{sf} = \frac{12 \times 806}{0.9 \times 60 \times 26} = 6.89 \text{ in}^2$$

Step 8   The depth of the web (h_w) is:

$$h_w = h - h_f = 30 \text{ in.} - 3 \text{ in.} = 27 \text{ in.}$$

Then

$$d_w = h_w - 2.5 \text{ in.} = 27 \text{ in.} - 2.5 \text{ in.} = 24.5 \text{ in.}$$

The resistance coefficient for the area of steel required for the part of the compression in the web (R_w) is:

$$R_w = \frac{12{,}000(M_u - M_{Rf})}{b_w d_w^2}$$

$$R_w = \frac{12{,}000(950 - 806)}{20(24.5)^2} = 144 \text{ psi}$$

Step 9

$$f_c' = 3 \text{ ksi}$$
$$f_y = 60 \text{ ksi} \longrightarrow \text{Table A2–6a} \longrightarrow \rho = 0.0028$$
$$R_w = 144 \text{ psi}$$
$$A_{sw} = \rho_w b_w d_w = (0.0028)(20)(24.5)$$
$$A_{sw} = 1.37 \text{ in}^2$$

Step 10   The total required area of steel (A_s) is:

$$A_s = A_{sf} + A_{sw} = 6.89 + 1.37 = 8.26 \text{ in}^2$$

Step 11

From Table A2–9 $\longrightarrow$ Try 6 #11 bars ($A_s = 9.36 \text{ in}^2$)

From Table A2–8 $\longrightarrow$ $b_{min} = 19.5 \text{ in.} < 20 \text{ in.}$ $\therefore$ ok

From Table A2–8 $\longrightarrow$ $b_{max} = 54 \text{ in.} > 20 \text{ in.}$ $\therefore$ ok

Step 12   Calculate the actual effective depth (d):

$$\bar{y} = 1.5 + \frac{1}{2} + \frac{1.41}{2} = 2.71 \text{ in.}$$

$$d = h - \bar{y} = 30 \text{ in.} - 2.71 \text{ in.} = 27.3 \text{ in.} \approx d_{assumed} = 27.5 \quad \therefore \text{ ok}$$

The sketch of the final beam design is shown in Figure 3–14.

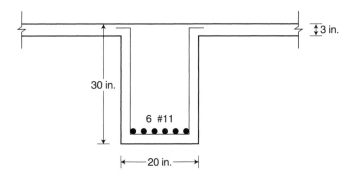

FIGURE 3–14 Sketch of final design of T-beam for Example 3–3.

3.2 DOUBLY-REINFORCED BEAMS

Introduction

To this point we have shown the use of steel reinforcement only for the tension part of a re-inforced concrete beam (tension steel). When reinforcement is also used in the compression zone of a reinforced concrete section (compression steel), the beam is referred to as a *doubly-reinforced* beam. Even though such a section in general is not economical, the use of compression steel has several advantages and applications, including the following:

1. It allows the use of a cross section smaller than that of a singly-reinforced beam. This is especially useful if the beam size is limited for architectural or aesthetic purposes.

2. It helps in reducing long-term deflections.

3. It can support stirrups or shear reinforcement by tying them to compression bars.

4. It adds significantly to the ductility of beams. Compression reinforcement enables the beam to withstand large levels of movement and deformation under extreme loading conditions that might occur during earthquakes.

5. It is frequently used where beams span more than two supports due to practical considerations. The ACI Code requires a percentage of the tensile steel at midspan to continue into the supports, and by a small extension this steel can easily be used as compression reinforcement at the face of the supporting column.

Analysis of Doubly-Reinforced Concrete Beams

The ACI Code (Section 10.3.5.1) allows the use of compression reinforcement, in conjunction with additional tensile reinforcement, to increase the strength of flexural members. To develop the internal couple in a reinforced concrete section, the total compression force, C, has to be equal to the total tensile force, T, which is provided by the steel. In a doubly-reinforced beam, however, the compression force is developed partly by the concrete, and partly by the compression steel.

Utilizing the principle of superposition, it is assumed that part of the steel in tension provides the tensile force to balance the compression force in the concrete ($C_1 = T_1$), and another part provides the tensile force that balances the force in the compression steel ($C_2 = T_2$). Figure 3–15 shows these forces.

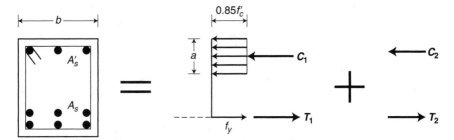

FIGURE 3–15   Compression and tensile forces in doubly-reinforced beams.

The following notations will be used in this section, and are shown in Figure 3–16.

A'_s = area of compression steel

d' = distance from the center of the compression steel to the compression edge of the beam

A_{s1} = area of tension steel for the concrete-steel couple

A_{s2} = area of tension steel required to work with the compression steel

A_s = total area of tension steel ($A_s = A_{s1} + A_{s2}$)

M_{n1} = nominal resisting moment of the concrete-steel couple

M_{n2} = nominal resisting moment of the steel-steel couple

d = effective depth of the section

d_t = effective depth of the extreme tension steel

ϵ_t = net tensile strain for extreme steel in tension

ϵ'_s = strain in the compression steel

f'_s = stress in the compression steel

E_s = modulus of elasticity of the steel

Figure 3–16a shows a doubly-reinforced beam represented by the superposition of two "beams": (1) a singly-reinforced beam with an area of steel A_{s1}, and (2) an imaginary tension-compression steel section, with A'_s as compression reinforcement and A_{s2} as tensile reinforcement. Therefore, the total area of tensile steel, A_s, is equal to the sum of A_{s1} and A_{s2} ($A_s = A_{s1} + A_{s2}$).

Figure 3–16b shows the distribution of strain in a doubly-reinforced beam at the ultimate moment. In order for the beam to remain tension-controlled, $\epsilon_t \geq 0.005$ when $\epsilon_c = 0.003$. If $\epsilon'_s \geq \epsilon_y$ then $f'_s = f_y$; however, when $\epsilon'_s < \epsilon_y$, $f'_s = E_s \epsilon_s$. Therefore, in order to determine the stress level in the compression steel, f'_s, it is always necessary to determine the strain, ϵ'_s, and check the above relationship.

Figures 3–16c and 3–16d show the forces generating the concrete–tensile steel and compression steel–tensile steel couples, respectively. Consider the compression steel–tensile steel couple. In order to form a couple, the compression force, C_2, must be equal to the tensile force, T_2, as shown in Equation 3–13.

$$C_2 = T_2 \qquad (3\text{–}13)$$
$$A'_s f'_s = A_{s2} f_y$$

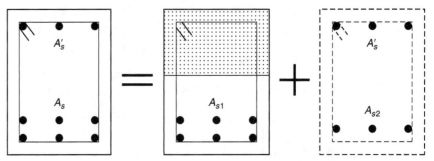

(a) Doubly-reinforced beam = singly-reinforced beam + tension-compression steel

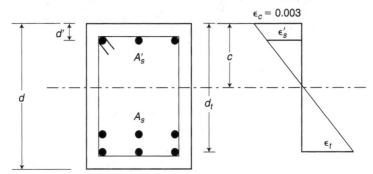

(b) Strain distribution in doubly-reinforced beam

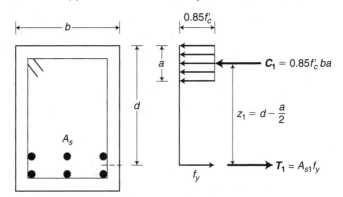

(c) Concrete-steel couple

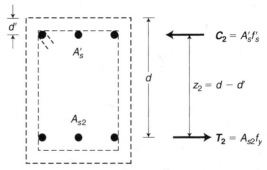

(d) Steel-steel couple

FIGURE 3–16 Analysis of doubly-reinforced beams.

The following steps present the analysis of a doubly-reinforced beam. Figure 3–17 summerizes these steps in a flowchart.

Step 1   Assume that the compression steel has yielded ($\epsilon_s' \geq \epsilon_y$) before the concrete in compression has reached its ultimate strain. Therefore, $f_s' = f_y$ and

$$A_{s2} f_y = A_s' f_y$$

from which

$$A_{s2} = A_s'$$

Because

$$A_s = A_{s1} + A_{s2}$$

A_{s1} can be calculated according to Equation 3–14.

$$A_{s1} = A_s - A_{s2} = A_s - A_s' \tag{3–14}$$

From Figure 3–16c, which is the part of the beam represented by the concrete–tensile steel couple, Equation 3–15 can be written.

$$C_1 = T_1 \tag{3–15}$$

$$0.85 f_c' ab = A_{s1} f_y$$

Step 2   Calculate the depth of the compression zone (a) using Equation 3–16.

$$a = \frac{A_{s1} f_y}{0.85 f_c' b} \tag{3–16}$$

Then determine the location of the neutral axis:

$$c = \frac{a}{\beta_1}$$

Step 3   Determine the strain levels for the tensile steel (ϵ_t) and the compression steel (ϵ_s') from the similarity of triangles (see Figure 3–16b).

$$\frac{\epsilon_t}{d_t - c} = \frac{0.003}{c}$$

$$\epsilon_t = \frac{0.003(d_t - c)}{c} \tag{3–17a}$$

Step 3a If $\epsilon_t < 0.005$ (transition-controlled section), calculate ϕ accordingly.

$\longrightarrow \phi = A_1 + B_1 \epsilon_t$ (A_1 and B_1 are obtained from Table A2–2a.)

Otherwise, $\phi = 0.90$.

Step 4   Determine the strain in the compression reinforcement using the similarity of triangles in Figure 3–16b:

$$\frac{\epsilon_s'}{c - d'} = \frac{0.003}{c}$$

$$\epsilon_s' = \frac{0.003(c - d')}{c} \tag{3–17b}$$

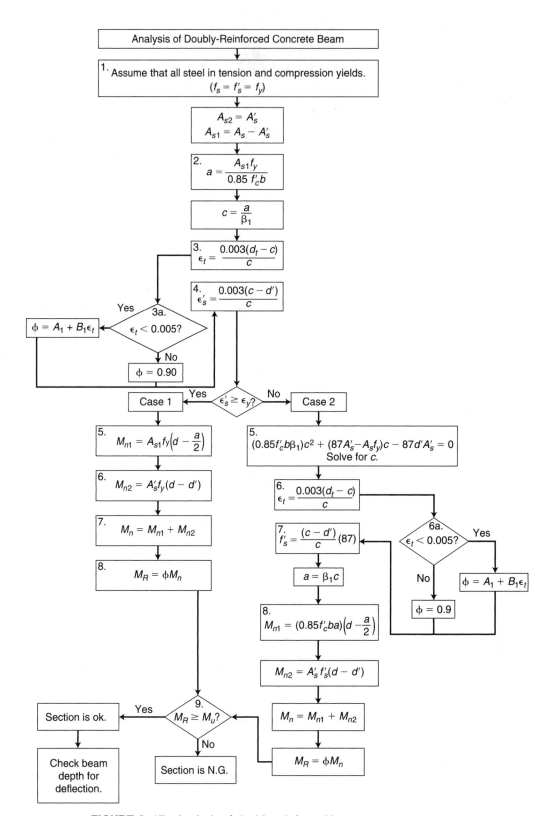

FIGURE 3–17 Analysis of doubly-reinforced beams.

Compare it to the yield strain of the compression reinforcement. There will be two possibilities: *Case 1* if $\epsilon'_s \geq \epsilon_y$, and *Case 2* if $\epsilon'_s < \epsilon_y$.

Case 1 – Compression reinforcement yields.

$$\epsilon'_s \geq \epsilon_y \tag{3–18}$$

This indicates that the compression steel yielded. In other words, $f'_s = f_y$, or the assumption made earlier in step 1 is correct. Hence, proceed directly to calculating the resisting moment of the section.

Step 5   Calculate the nominal resisting moment from the concrete–tensile steel couple according to Equation 3–19.

$$M_{n1} = A_{s1}f_y z_1 = (A_{s1}f_y)\left(d - \frac{a}{2}\right) \tag{3–19}$$

Step 6   Calculate the nominal resisting moment from the compression steel–tensile steel couple according to Equation 3–20.

$$M_{n2} = A'_s f'_s (d - d') = A'_s f_y (d - d') \tag{3–20}$$

Step 7   Calculate the nominal resisting moment for the doubly-reinforced beam according to Equation 3–21.

$$M_n = M_{n1} + M_{n2} \tag{3–21}$$

Step 8   Calculate the design resisting moment (M_R) using the strength reduction factor (ϕ) and Equation 3–22.

$$M_R = \phi M_n \tag{3–22}$$

Case 2 – Compression reinforcement does not yield.

$$\epsilon'_s < \epsilon_y \tag{3–23}$$

Step 5   Because $\epsilon'_s < \epsilon_y$, the compression steel did not yield when the strain at the extreme compression edge on the concrete section reached 0.003. From similar triangles (see Figure 3–16b) the strain in the compression steel can be calculated using Equation 3–24.

$$\epsilon'_s = \frac{0.003(c - d')}{c} \tag{3–24}$$

The stress in the compression steel (f'_s) can then be calculated using Equation 3–25.

$$f'_s = \epsilon'_s E_s = \left[\frac{0.003(c - d')}{c}\right] E_s \tag{3–25}$$

Thus, the assumption made in step 1 is not correct. The force provided by the compression steel is less than was assumed. Hence a smaller amount of tensile steel will work in the compression steel–tensile steel couple, and a new location has to be determined for the neutral axis.

Equilibrium requires that the total compression on the section be equal to the total tension, as expressed by Equations 3–26 and 3–27.

$$C_1 + C_2 = T_1 + T_2 \qquad (3\text{–}26)$$

$$0.85 f_c' ab + A_s' f_s' = A_{s1} f_y + A_{s2} f_y$$

$$0.85 f_c' ab + A_s' f_s' = (A_{s1} + A_{s2}) f_y \qquad (3\text{–}27)$$

Because $A_s = A_{s1} + A_{s2}$:

$$0.85 f_c' \, ab + A_s' f_s' = A_s f_y \qquad (3\text{–}28)$$

Substituting $a = \beta_1 c$ and f_s' from Equation 3–25 into the above equation:

$$0.85 f_c' \beta_1 cb + A_s' \left[\frac{0.003(c - d')}{c} \right] E_s = A_s f_y \qquad (3\text{–}29)$$

Multiplying the two sides of this equation by c:

$$(0.85 f_c' b \beta_1) c^2 + 0.003(c - d') E_s A_s' - A_s f_y c = 0$$

Rearranging the equation:

$$(0.85 f_c' b \beta_1) c^2 + (0.003 E_s A_s' - A_s f_y) c - 0.003 d' E_s A_s' = 0 \qquad (3\text{–}30)$$

Substituting $E_s = 29{,}000$ ksi, the location of the neutral axis (c) can be determined from the quadratic Equation 3–31.

$$(0.85 f_c' b \beta_1) c^2 + (87 A_s' - A_s f_y) c - 87 d' A_s' = 0 \qquad (3\text{–}31)$$

(Note that f_c' and f_y are in ksi.)

Step 6   Once c is known, determine the net tensile strain in the extreme layer of steel (ϵ_t) using Equation 3–32 (developed in step 3 above).

$$\epsilon_t = \frac{0.003(d_t - c)}{c} \qquad (3\text{–}32)$$

Step 6a   If $\epsilon_t < 0.005$ (transition-controlled section), calculate $\phi = A_1 + B_1 \epsilon_t$ (A_1 and B_1 are found from Table A2–2a). If $\epsilon_t \geq 0.005$ (tension-controlled section), set $\phi = 0.90$.

Step 7   Calculate the stress in the compression steel, f_s', using Equation 3–33, which is derived by substituting the value of E_s into Equation 3–25.

$$f_s' = \frac{0.003(c - d')}{c} E_s = \frac{(0.003)(c - d')(29{,}000)}{c}$$

$$f_s' = \frac{(c - d')}{c} (87) \qquad (3\text{–}33)$$

The depth of the equivalent stress block (a) is:

$$a = \beta_1 c$$

Calculate the component forces of the internal couples (see Figures 3–16c and 3–16d) and determine whether equilibrium is satisfied, as expressed in

Equation 3–35.

$$C_1 = 0.85 f'_c ba$$
$$C_2 = A'_s f'_s$$
$$T_1 + T_2 = A_s f_y \qquad (3\text{--}34)$$
$$T_1 + T_2 = C_1 + C_2 \qquad (3\text{--}35)$$

If Equation 3–35 is not satisfied, then most likely an error was made in the computation of c.

Step 8   Calculate the nominal resisting moment of the doubly-reinforced section by adding the concrete–tensile steel and compression steel–tensile steel couples as shown in Equations 3–36 through 3–38 (see Figures 3–16c and 3–16d).

$$M_{n1} = C_1 z_1 = C_1\left(d - \frac{a}{2}\right) = (0.85 f'_c ba)\left(d - \frac{a}{2}\right) \qquad (3\text{--}36)$$

$$M_{n2} = C_2 z_2 = C_2(d - d') = A'_s f'_s(d - d') \qquad (3\text{--}37)$$

$$M_n = M_{n1} + M_{n2} \qquad (3\text{--}38)$$

Calculate the design resisting moment, M_R, using Equation 3–39.

$$M_R = \phi M_n \qquad (3\text{--}39)$$

Step 9   Once M_R is calculated, determine whether the beam has enough capacity by comparing M_R with the maximum factored moment, M_u (i.e., the demand):

$$M_R \geq M_u \qquad \text{(Beam is adequate.)}$$
$$M_R < M_u \qquad \text{(Beam is not adequate.)}$$

Check if the beam depth is large enough so that deflection does not need to be computed.

EXAMPLE 3–4

Calculate the design resisting moment, M_R, of the doubly-reinforced beam shown in Figure 3–18. $f'_c = 4$ ksi, $f_y = 40$ ksi, $E_s = 29{,}000$ ksi. The stirrups are #4 bars. The beam is subjected to a positive bending moment.

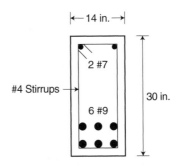

FIGURE 3–18   Beam section for Example 3–4.

Solution   *Use the flowchart of Figure 3–17.*

Step 1   Assume that the tension and compression steel yield. The validity of this assumption will be checked later in the analysis.

$$f_s = f_s' = f_y = 40 \, \text{ksi}$$

From Table A2–9 $\longrightarrow$ 6 #9 $\longrightarrow A_s = 6.0 \, \text{in}^2$

$\longrightarrow$ 2 #7 $\longrightarrow A_s' = 1.2 \, \text{in}^2$

$$A_{s1} = A_s - A_s' = 6 - 1.2 = 4.8 \, \text{in}^2$$

Step 2   Calculate the depth of the equivalent stress block (a) in the concrete for the section:

$$a = \frac{A_{s1}f_y}{0.85f_c'b} = \frac{4.8 \times 40}{0.85 \times 4 \times 14} = 4.03 \, \text{in.}$$

Therefore, the location of the neutral axis (c) is:

$$c = \frac{a}{\beta_1} = \frac{4.03 \, \text{in.}}{0.85} = 4.75 \, \text{in.}$$

Step 3   Calculate the strain in the tension and compression steel and check for the validity of the assumption made in step 1. Make a sketch of the strain distribution as shown in Figure 3–19 and calculate the strains from similar triangles.

$$d_t = 30 \, \text{in.} - \left(1.5 + \frac{4}{8} + \frac{1.128}{2}\right) = 27.44 \, \text{in.}$$

$$\epsilon_t = \frac{0.003(d_t - c)}{c}$$

$$\epsilon_t = \frac{0.003(27.44 - 4.75)}{4.75}$$

$$\epsilon_t = 0.0143$$

$$\epsilon_t = 0.0143 > \epsilon_y = \frac{f_y}{E_s} = \frac{40}{29,000} = 0.00138$$

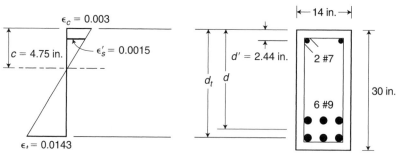

FIGURE 3–19   Strains for beam section of Example 2–10.

Step 3a

$$\epsilon_t = 0.0143 > 0.005 \quad \therefore \phi = 0.90$$

(The section is tension-controlled.)

Step 4   Calculate the strain level in the compression steel (ϵ_s'). Check to see if the steel yields when strain in the concrete reaches 0.003:

$$d' = 1.5 + \frac{4}{8} + \frac{0.875}{2} = 2.44 \text{ in.}$$

$$\epsilon_s' = \frac{0.003(c - d')}{c}$$

$$\epsilon_s' = \frac{0.003(4.75 - 2.44)}{4.75}$$

$$\epsilon_s' = 0.00146 > \epsilon_y = 0.00138$$

Therefore, the compression steel yields, and the assumption in step 1 was correct! Because the compression steel yields, follow the process under case 1:

Step 5   Calculate the effective depth (d).

$$d = 30 - \left(1.5 + \frac{4}{8} + 1.128 + \frac{1}{2}\right) = 26.37 \text{ in.}$$

Calculate the nominal resisting moment from the concrete–tensile steel couple, M_{n1}.

$$M_{n1} = A_{s1}f_y\left(d - \frac{a}{2}\right)$$

$$M_{n1} = \frac{(4.8)(40)\left(26.37 - \frac{4.03}{2}\right)}{12}$$

$$M_{n1} = 390 \text{ ft-kip}$$

Step 6   Calculate the nominal resisting moment for the compression steel–tensile steel couple, M_{n2}.

$$M_{n2} = A_s'f_y(d - d')$$

$$M_{n2} = \frac{(1.2)(40)(26.37 - 2.44)}{12}$$

$$M_{n2} = 95.7 \text{ ft-kip}$$

Step 7   Calculate the total nominal resisting moment, M_n, which is the sum of the concrete–tensile steel (M_{n1}) and compression steel–tensile steel (M_{n2}) couples:

$$M_n = M_{n1} + M_{n2}$$
$$M_n = 390 + 95.7 = 485.7 \text{ ft-kip}$$

Step 8   Calculate M_R.

$$M_R = \phi M_n = 0.90 \times 485.7 = 437 \text{ ft-kip}$$

EXAMPLE 3–5

Determine the design resisting moment, M_R, for the doubly-reinforced beam with 6 #8 bars used for steel in tension as shown in Figure 3–20. Use $f'_c = 4\,\text{ksi}$ and $f_y = 40\,\text{ksi}$, $E_s = 29{,}000\,\text{ksi}$. The stirrups are #4 bars. The beam is subject to a positive bending moment.

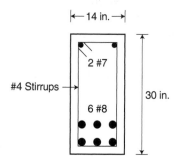

FIGURE 3–20 Beam section for Example 3–5.

Solution *Use the flowchart of Figure 3–17.*

Step 1 Assume $f_s = f'_s = f_y = 40\,\text{ksi}$

$$2\ \#7 \longrightarrow \text{Table A2–9} \longrightarrow A'_s = 1.20\,\text{in}^2$$

$$6\ \#8 \longrightarrow \text{Table A2–9} \longrightarrow A_s = 4.74\,\text{in}^2$$

$$A_{s1} = A_s - A'_s = 4.74 - 1.20 = 3.54\,\text{in}^2$$

Step 2 Calculate the depth of the equivalent stress block.

$$a = \frac{A_{s1}f_y}{0.85f'_c\,b} = \frac{3.54 \times 40}{0.85 \times 4 \times 14}$$

$$a = 2.97\,\text{in.}$$

The neutral axis is:

$$c = \frac{a}{\beta_1} = \frac{2.97}{0.85} = 3.5\,\text{in.}$$

Step 3 Calculate the strain in the tensile steel:

$$d_t = 30 - \left(1.5 + \frac{4}{8} + \frac{1}{2}\right) = 27.5\,\text{in.}$$

From Equation 3–17a:

$$\epsilon_t = \frac{0.003(d_t - c)}{c}$$

$$\epsilon_t = \frac{0.003(27.5\,\text{in.} - 3.5\,\text{in.})}{3.5\,\text{in.}}$$

$$\epsilon_t = 0.0206 > \epsilon_y = \frac{40}{29{,}000} = 0.00138$$

Step 3a

$$\epsilon_t = 0.0206 > 0.005 \quad \therefore \ \phi = 0.90$$

Step 4 Calculate the strain in the compression steel:

$$d' = 1.5 + \frac{4}{8} + \frac{0.875}{2} = 2.44 \text{ in.}$$

From Equation 3–17b:

$$\epsilon_s' = \frac{0.003(c - d')}{c}$$

$$\epsilon_s' = \frac{0.003(3.5 - 2.44)}{3.5}$$

$$\epsilon_s' = 0.00091 < \epsilon_y = 0.00138$$

The compression steel does not yield when the strain in the concrete reaches 0.003. Therefore, the assumption in step 1 was not correct. Hence, follow the procedure outlined in case 2.

Step 5 Determine the location of the neutral axis, c, using Equation 3–31.

$$(0.85 f_c' b \beta_1)c^2 + (87 A_s' - A_s f_y)c - 87 d' A_s' = 0$$
$$(0.85 \times 4 \times 14 \times 0.85)c^2 + (87 \times 1.2 - 4.74 \times 40)c - 87 \times 2.44 \times 1.2 = 0$$
$$40.46c^2 - 85.2c - 254.7 = 0$$

This is a second order equation in the form of:

$$Ax^2 + Bx + C = 0$$

The solutions for x are:

$$x = \frac{-B \pm \sqrt{B^2 - 4AC}}{2A}$$

where $A = 40.46$, $B = -85.20$, $C = -254.7$.

$$\text{Thus } c = \frac{85.20 \pm \sqrt{(-85.20)^2 - 4(40.46)(-254.7)}}{2(40.46)}$$

$$c = 3.77 \text{ in.}$$

(Quadratic equations have two roots. The one in the example has $c_1 = 3.77$ in. and $c_2 = -1.67$ in. c cannot be a negative value, so the second one obviously does not apply.)

Step 6 Determine the correct value of the net tensile strain at the extreme layer of the reinforcement using Equation 3–32.

$$\epsilon_t = \frac{0.003(d_t - c)}{c}$$

$$\epsilon_t = \frac{0.003(27.5 - 3.77)}{3.77}$$

Step 6a

$$\epsilon_t = 0.0189 > 0.005 \quad \therefore \; \phi = 0.90$$

Step 7 Calculate the stress in the compression steel (f_s') using Equation 3–33.

$$f_s' = \frac{(c - d')}{c} \quad (87)$$

$$f_s' = \frac{(3.77 - 2.44)}{3.77} \quad (87)$$

$$f_s' = 30.69 \, \text{ksi} < f_y = 40 \, \text{ksi}$$

The corrected depth of the compression zone (a) is:

$$a = \beta_1 c = 0.85 \times 3.77 = 3.20 \, \text{in.}$$

Step 8 Calculate M_{n1} and M_{n2}, the nominal resisting moments for the concrete–tensile steel couple and the compression steel–tensile steel couple, respectively.

$$d = 30 - \left(1.5 + \frac{4}{8} + 1 + \frac{1}{2}\right) = 26.5 \, \text{in.}$$

$$M_{n1} = (0.85 f_c' \, ba)\left(d - \frac{a}{2}\right)$$

$$M_{n1} = \frac{(0.85 \times 4 \times 14 \times 3.2)\left(26.5 - \dfrac{3.2}{2}\right)}{12}$$

$$M_{n1} = 316 \, \text{ft-kip}$$
$$M_{n2} = A_s' f_s'(d - d')$$

$$M_{n2} = \frac{1.2 \times 30.69(26.5 - 2.44)}{12}$$

$$M_{n2} = 74.0 \, \text{ft-kip}$$

Calculate the total nominal resisting moment, M_n.

$$M_n = M_{n1} + M_{n2}$$
$$M_n = 316 + 74.0 = 390 \, \text{ft-kip}$$

Calculate the design resisting moment, M_R.

$$M_R = \phi M_n = 0.90 \times 390$$
$$M_R = 351 \, \text{ft-kip}$$

Design of Doubly-Reinforced Concrete Beams

If a singly-reinforced section cannot develop the required factored moment and the beam size cannot be increased, a doubly-reinforced section may be appropriate. In the design of doubly-reinforced beams, the section sizes are known, so only the reinforcement needs to be determined.

The design of a doubly-reinforced section follows the same concept as that of the analysis: Calculate the amount of steel necessary for the concrete–tensile steel and compression steel–tensile steel couples and add the results. The step-by-step design procedure is outlined below and summarized in a flowchart in Figure 3–21.

Step 1   Calculate the maximum factored moment, M_u, from the loads acting at the section under consideration. Because the beam sizes (b and h) are known, estimate the effective depth (d) as

$$d = h - \bar{y} \quad \text{(assume } \bar{y} = 2.5 \text{ in.)}$$

Also, assume $d' = 2.5$ in. In the following two steps we determine whether a doubly-reinforced beam is required or a singly-reinforced beam will be adequate.

Step 2   In order to calculate the maximum moment capacity of a singly-reinforced tension-controlled section (ϕM_{n1}), obtain the maximum tension-controlled steel ratio (ρ_{tc}) permitted by the ACI code from Table A2–3 and the corresponding resistance coefficient (R) from Tables A2–5 through A2–7.

Step 3   Calculate the maximum ϕM_{n1} for a singly-reinforced beam:

$$\phi M_{n1} = Rbd^2/12{,}000 \qquad (3\text{–}40)$$

If $M_u > \phi M_{n1}$ a doubly-reinforced beam is required. If $M_u \leq \phi M_{n1}$ a singly-reinforced concrete beam will suffice. If a singly-reinforced concrete beam will suffice, design the beam accordingly using the flowchart of Figure 2–46. If a doubly-reinforced beam is needed (i.e., $M_u > \phi M_{n1}$), then proceed as follows:

Step 4   For the concrete–tensile steel couple:

$$A_{s1} = \rho_{tc}bd \qquad (3\text{–}41)$$

Calculate the difference between M_u and ϕM_{n1}. This difference is the moment that must be resisted by the compression steel–tensile steel couple.

$$\phi M_{n2} = M_u - \phi M_{n1}$$

Considering the compression steel–tensile steel couple, (see Figure 3–16d) calculate the compression force, C_2, from Equation 3–43.

$$\phi M_{n2} = \phi C_2 z_2 = \phi C_2(d - d') \qquad (3\text{–}42)$$

$$C_2 = A_s' f_s' = \frac{\phi M_{n2}}{\phi(d - d')} \qquad (3\text{–}43)$$

Step 5   To calculate A_s' we must first determine the value of f_s', as the compression strain in the reinforcing may be less than the yield strain. In order to do this, determine the location of the neutral axis and check the strain level in the compression steel (see Figure 3–16c):

$$C_1 = T_1$$

$$0.85f_c' ba = A_{s1}f_y$$

$$a = \frac{A_{s1}f_y}{0.85f_c' b}$$

$$c = \frac{a}{\beta_1}$$

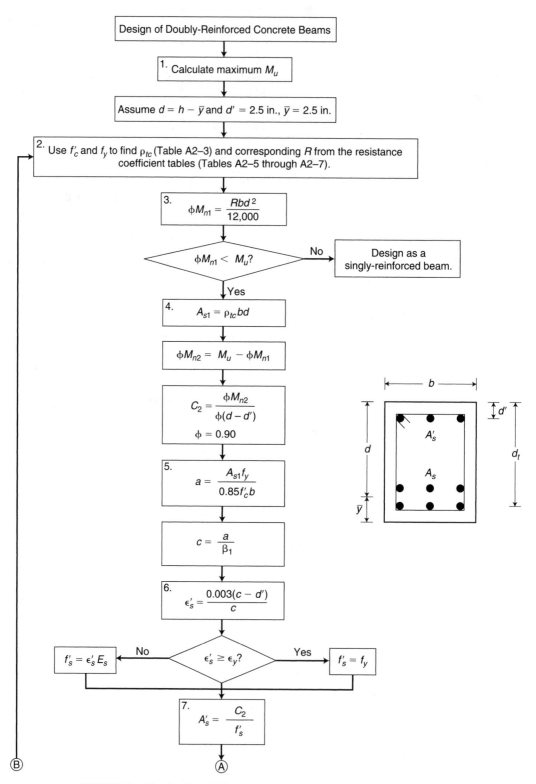

FIGURE 3–21 Design of doubly-reinforced beams.

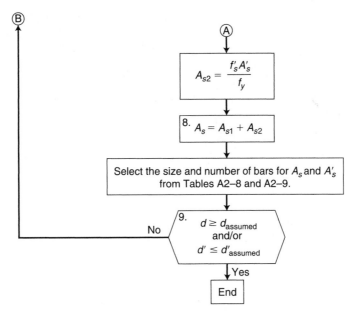

FIGURE 3–21   *(continued)*

Step 6   Calculate ϵ_s' (see Figure 3–16b):

$$\epsilon_s' = \frac{0.003(c - d')}{c}$$

If $\epsilon_s' \geq \epsilon_y \longrightarrow f_s' = f_y$ (i.e., the compression steel has yielded). If, however, $\epsilon_s' < \epsilon_y \longrightarrow f_s' = E_s\epsilon_s'$.

Step 7   Calculate A_s' from Equation 3–45.

$$C_2 = A_s'f_s' \tag{3–44}$$

$$A_s' = \frac{C_2}{f_s'} \tag{3–45}$$

Calculate A_{s2} using Equation 3–46 (see Figure 3–16d).

$$A_s'f_s' = A_{s2}f_y \tag{3–46}$$

$$A_{s2} = \frac{A_s'f_s'}{f_y}$$

Step 8   Calculate the total area of steel (A_s).

$$A_s = A_{s1} + A_{s2} \tag{3–47}$$

The result enables the selection of the size and number of bars for the compression steel (A_s') and the tension steel (A_s) using Table A2–9.

Step 9   After the selection of the tensile and compression steel, calculate the actual values of d and d', and compare to the assumed values in step 1. If

$d \geq d_{assumed}$ or $d' \leq d'_{assumed}$, the assumptions are conservative. However, if these relationships are violated by more than $^1/_2$ in., a recalculation of A_s and A'_s, is necessary using the adjusted d and d' values by repeating the process from step 2.

EXAMPLE 3–6

Figure 3–22 shows the floor framing plan and sections of a reinforced concrete building. The slab is 6 in. thick and there is a superimposed dead load of 25 psf. The floor live load is 125 psf. Assume that the beams are integral with the columns, $f'_c = 4$ ksi, $f_y = 60$ ksi,

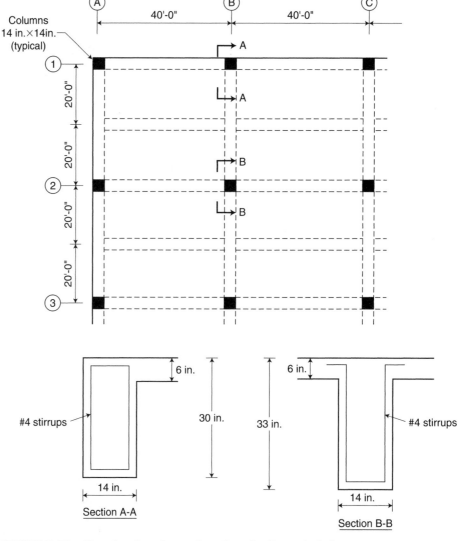

FIGURE 3–22 Floor framing plan and sections for Example 3–6.

and the unit weight of the concrete is 150 pcf. The stirrups are #4 bars. Design the reinforcements for the edge beam and the first interior beam along column lines 1 and 2 (as shown in the sections A-A and B-B of Figure 3–22) where the maximum negative moments occur. Consider doubly-reinforced beams if necessary. Use ACI coefficients for the calculation of moments.

Solution   *Use the flowchart of Figure 3–21.*

(a) *Edge Beam Along Line 1*

Step 1   Find the factored loads on the beam:

$$\text{Weight of slab} = 150\left(\frac{6}{12}\right) = 75 \text{ psf}$$

$$\underline{\text{Superimposed dead load} = 25 \text{ psf}}$$
$$\text{Total dead load} = 100 \text{ psf}$$
$$\text{Live load} = 125 \text{ psf}$$

The uniformly distributed dead and live loads on the beam are (tributary width = 10.58 ft):

$$w_D = \left[100 \times 10.58 + 150\left(\frac{14}{12} \times \frac{24}{12}\right)\right]\Big/1{,}000 = 1.41 \text{ kip/ft}$$

$$w_L = \frac{125 \times 10.58}{1{,}000} = 1.32 \text{ kip/ft}$$

(Note that live load reduction does not apply for the beams, because the unit live load is in excess of 100 psf.)

$$w_u = 1.2w_D + 1.6w_L = 1.2 \times 1.41 + 1.6 \times 1.32 = 3.8 \text{ kip/ft}$$

$$\text{Beam clear span} = 40 \text{ ft} - \left(\frac{7}{12} + \frac{7}{12}\right) = 38.8 \text{ ft}$$

The maximum factored bending moment is next to the first interior column (negative moment):

$$(M_u)^- = \frac{w_u \ell_n^2}{10} = \frac{3.8(38.8)^2}{10} = 572 \text{ ft-kip}$$

Assuming $\bar{y} = 2.5$ in., the effective depth, d, can be calculated.

$$d_{\text{assumed}} = h - \bar{y} = 30 \text{ in.} - 2.5 \text{ in.} = 27.5 \text{ in.}$$

Also,

$$d'_{\text{assumed}} = 2.5 \text{ in.}$$

Step 2   Use f'_c and f_y to obtain the maximum tension-controlled steel ratio (ρ_{tc}) from Table A2–3:

$$f'_c = 4 \text{ ksi} \longrightarrow \text{Table A2–3} \longrightarrow \rho_{tc} = 0.0180$$
$$f_y = 60 \text{ ksi}$$

The corresponding resistance coefficient, R, from Table A2–6b is:

$$\rho = 0.0180 \longrightarrow \text{Table A2–6b} \longrightarrow R = 818 \, \text{psi}$$

Step 3   Calculate the design resisting moment based on the limit of reinforcement for tension steel (ρ_{tc}):

$$\phi M_{n1} = \frac{Rbd^2}{12,000} = \frac{818(14)(27.5)^2}{12,000}$$

$$\phi M_{n1} = 722 \, \text{ft-kip} > 572 \, \text{ft-kip}$$

$$\therefore \phi M_{n1} > M_u \text{ Design as a singly-reinforced beam.}$$

Because the beam is to be designed as a singly-reinforced section, follow the flowchart of Figure 2–46. Continuing with step 3 of that flowchart:

$$R = \frac{12,000 M_u}{bd^2}$$

$$R = \frac{12,000 \times 572}{14(27.5)^2} = 648 \, \text{psi}$$

Step 4   From Table A2–6b, obtain the steel ratio (ρ) for this R value:

$$\rho = 0.0137 > \rho_{\min} = 0.0033 \, (\text{Table 2–4}) \quad \therefore \text{ ok}$$

Step 5   Calculate the required area of the steel (A_s):

$$A_s = \rho bd = (0.0137)(14)(27.5) = 5.27 \, \text{in}^2$$

From Table A2–9, use 6 #9 bars.

Step 6   Calculate the actual d.

$$\bar{y} = 1.5 \, \text{in.} + {}^4/_8 \, \text{in.} + 1.128/2 \, \text{in.} = 2.56 \, \text{in.}$$

$$\uparrow \qquad \uparrow \qquad \qquad \uparrow$$

Cover   #4 stirrup   #9 bar

$$d = 30 - 2.56 = 27.44 \, \text{in.} \approx d_{\text{assumed}} = 27.5 \, \text{in.} \quad \therefore \text{ ok}$$

Figure 3–23 shows the final cross section and reinforcement.

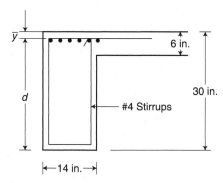

FIGURE 3–23   Final design for Example 3–6 (edge beam).

(b) *Interior Beam Along Line 2*

Step 1   Using the total dead and live loads calculated in part a and a tributary width of 20 ft, the uniformly distributed dead load (w_D) and live load (w_L) are:

$$w_D = \left[100 \times 20 + 150\left(\frac{14}{12} \times \frac{27}{12}\right) \right] \bigg/ 1{,}000 = 2.40 \text{ kip/ft}$$

$$w_L = \frac{125 \times 20}{1{,}000} = 2.5 \text{ kip/ft}$$

$$w_u = 1.2w_D + 1.6w_L = 1.2 \times 2.40 + 1.6 \times 2.5 = 6.88 \text{ kip/ft}$$

The maximum factored moment is:

$$(M_u)^- = \frac{w_u \ell_n^2}{10} = \frac{6.88(38.8)^2}{10} = 1{,}036 \text{ ft-kip}$$

$$d_{\text{assumed}} = h - \bar{y} = 33 \text{ in.} - 2.5 = 30.5 \text{ in.}$$

$$d'_{\text{assumed}} = 2.5 \text{ in.}$$

Step 2 $f'_c = 4 \text{ ksi} \longrightarrow \text{Table A2–3} \longrightarrow \rho_{tc} = 0.0180$

$f_y = 60 \text{ ksi}$

$\rho = 0.018 \longrightarrow \text{Table A2–6b} \longrightarrow R = 818 \text{ psi}$

Step 3   The limit of resisting moment for a singly-reinforced tension-controlled section is:

$$\phi M_{n1} = \frac{Rbd^2}{12{,}000} = \frac{818(14)(30.5)^2}{12{,}000}$$

$$\phi M_{n1} = 888 \text{ ft-kip} < M_u = 1{,}036 \text{ ft-kip}$$

Since $\phi M_{n1} < M_u$, the beam has to be designed as a doubly-reinforced section.

Step 4

$$A_{s1} = \rho_{tc} bd = (0.018)(14)(30.5)$$

$$A_{s1} = 7.69 \text{ in}^2$$

$$\phi M_{n2} = M_u - \phi M_{n1}$$

$$\phi M_{n2} = 1{,}036 - 888 = 148 \text{ ft-kip}$$

The compression force to be carried by the compression steel (C_2) is:

$$C_2 = \frac{\phi M_{n2}}{\phi(d - d')} = \frac{148 \times 12}{0.9(30.5 - 2.5)} = 70.5 \text{ kip}$$

Step 5   Determine the depth of the compression zone (a):

$$a = \frac{A_{s1} f_y}{0.85 f'_c b} = \frac{7.69 \times 60}{0.85 \times 4 \times 14}$$

$$a = 9.69 \text{ in.}$$

$$c = \frac{a}{\beta_1} = \frac{9.69}{0.85} = 11.4 \text{ in.}$$

Step 6 Calculate the strain in the compression steel (ϵ'_s):

$$\epsilon'_s = \frac{0.003(c - d')}{c}$$

$$\epsilon'_s = \frac{0.003(11.4 - 2.5)}{11.4} = 0.0023$$

$$\epsilon_y = \frac{f_y}{E_s} = \frac{60}{29,000} = 0.00207 < 0.0023$$

The compression steel will yield, thus

$$f'_s = f_y = 60\,\text{ksi}$$

Step 7 Calculate the required compression steel (A'_s) and the additional tension steel for the compression steel–tensile steel couple (A_{s2}).

$$A'_s = \frac{C_2}{f'_s} = \frac{70.5}{60} = 1.18\,\text{in}^2$$

$$A_{s2} = \frac{f'_s A'_s}{f_y} = \frac{60 \times 1.18}{60} = 1.18\,\text{in}^2$$

Step 8 Calculate the total required tensile steel (A_s) and select the bars.

$$A_s = A_{s1} + A_{s2}$$
$$A_s = 7.69 + 1.18 = 8.87\,\text{in}^2$$

For tensile steel, select 9 #9 bars ($A_{s,\text{provided}} = 9\,\text{in}^2$) and for compression reinforcement $A'_s = 1.18\,\text{in}^2$ select 2 #7 bars ($A'_{s,\text{provided}} = 1.20\,\text{in}^2$).

Step 9 Calculate the actual values of d and d', and compare to the assumed values.

$$d = h - \bar{y} = 33 - \left(1.5 + \frac{4}{8} + \frac{1.128}{2}\right)$$

$$= 30.44\,\text{in.} \simeq d_{\text{assumed}} = 30.5\,\text{in.}$$

$$d' = 1.5 + \frac{4}{8} + \frac{0.875}{2} = 2.44\,\text{in.} < d'_{\text{assumed}} = 2.5\,\text{in.} \quad \therefore \text{ ok}$$

The final design of the beam is shown in Figure 3–24.

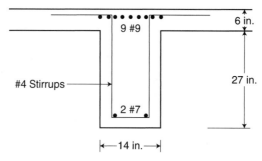

9 #9

6 in.

#4 Stirrups

27 in.

2 #7

|←—14 in.—→|

FIGURE 3–24 Final design for Example 3–6 (interior beam).

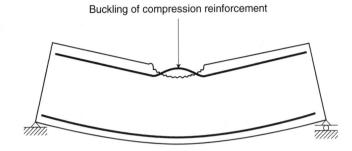

(a) Possible buckling of compression reinforcement without adequate stirrups

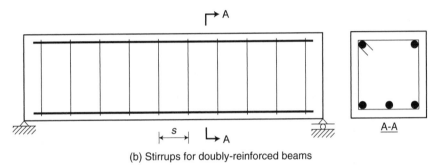

(b) Stirrups for doubly-reinforced beams

FIGURE 3–25   Lateral support for compression steel in doubly-reinforced beams.

Lateral Support for Compression Steel

Any slender compression member is susceptible to buckling. Compression steel is made up of slender reinforcing bars that can buckle and cause failure of the beam as shown in Figure 3–25a. To prevent such catastrophic failures, Section 7.11 of the ACI Code requires that compression reinforcement in beams be enclosed by ties or stirrups, as shown in Figure 3–25b. The size of the stirrups must be at least #3 for main bars that are #10 or smaller, and #4 for those that are #11 or larger, according to Section 7.10.5.1 of the ACI Code. The maximum spacing, s, of the stirrups for this purpose is given by ACI Code, Section 7.10.5.2 and is:

$$s_{max} = \min\{16d_b, 48d_t, b_{min}\}$$

where d_b is the diameter of the main bars, d_t is the diameter of the stirrups, and b_{min} is the smaller dimension of the beam section.

3.3   DEFLECTION OF REINFORCED CONCRETE BEAMS

Introduction

The analysis of deflections in reinforced concrete flexural members is a very complex and inexact process. The difficulty lies in three major uncertainties.

The first is the inelastic behavior of concrete. As previously discussed, concrete in compression does not follow Hooke's law: Stresses and strains are not linearly related, even at relatively low stress values. The inelastic behavior of concrete, however, may be the least of the difficulties, because the assumption of elastic response does not lead to very large errors up to working stress or service load stress levels.

The second uncertainty is much more difficult to get a handle on. In the formulae for deflection calculations, the product EI is in the denominator. For example, the elastic deflection formula for a simply-supported beam with uniformly distributed loads is:

$$\Delta = \frac{5}{384} \frac{w\ell^4}{EI}$$

The product $E \times I$ = modulus of elasticity × moment of inertia. As mentioned in Section 1.6 of Chapter 1, the Code gives the assumed modulus of elasticity of concrete as:

$$E = 33w_c^{1.5} \sqrt{f_c'}$$

where w_c is the weight of the concrete in pounds per cubic ft and f_c' is the 28-day cylinder strength of the concrete in psi. (Normal-weight concrete is about 145 pcf.)

The Code formula for the modulus of elasticity is accurate only within a range of about ±15%.

Calculating the moment of inertia is even more problematic. Concrete flexural members, as discussed earlier, develop cracks while subject to normal service load conditions. Between the cracked sections and the points where the moments are less than the cracking moment (M_{cr}), there is the full concrete section augmented by the reinforcing. At the cracked sections, however, only a much smaller moment of inertia is available. Correspondingly, the center region of a beam has considerably less rigidity as shown in Figure 3–26.

The third major uncertainty is due to the creep behavior of concrete in compression. The first two uncertainties influence the ambiguity of calculating the so-called *instantaneous deflections,* but creep influences long-term deformation (i.e., a gradually increasing deformation under sustained loads). Fortunately, the rate of increase of deformation dissipates with time, and it virtually stops after about 5 years.

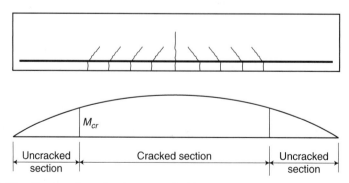

FIGURE 3–26   Regions in a simply-supported beam.

The Effective Moment of Inertia (I_e)

The ACI Code simplifies the complex problem posed by uncracked and cracked sections in different regions of beams by assuming that the effective moment of inertia (I_e) lies somewhere between the gross section's moment of inertia (I_g) and the cracked section's moment of inertia (I_{cr}). Equation 3–48 (ACI Equation 9–8) presents the ACI Code (Section 9.5.2.3) formula to calculate I_e.

$$I_e = \left(\frac{M_{cr}}{M_a}\right)^3 I_g + \left[1 - \left(\frac{M_{cr}}{M_a}\right)^3\right] I_{cr} \tag{3–48}$$

where

I_g is the moment of inertia of the gross concrete section about its centroidal axis, neglecting reinforcement

I_{cr} is the moment of inertia of the cracked concrete section

M_{cr} is the cracking moment

M_a is the actual (unfactored) maximum moment in the member

The ACI Code (Section 9.5.2.4) recommends using an average of values obtained from Equation 3–48 for the critical positive and negative moment sections in calculating I_e for continuous beams. This averaged value should be used in the appropriate deflection formulae for continuous beams.

a. Equation 3–49 gives the gross moment of inertia for a rectangular section.

$$I_g = \frac{bh^3}{12} \tag{3–49}$$

Equation 3–50 gives the cracking moment.

$$M_{cr} = \frac{f_r I_g}{y_t} \tag{3–50}$$

where f_r is the modulus of rupture given by Equation 3–51.

$$f_r = 7.5\sqrt{f_c'} \tag{3–51}$$

and y_t is the distance from the section's centroidal axis (neglecting reinforcement) to the extreme fiber in tension for a rectangular section, as shown in Figure 3–27:

$$y_t = \frac{h}{2}$$

b. For a typical T-beam section like the one shown in Figure 3–28, calculating y_t and the gross moment of inertia requires considerable computational effort. To ease the difficulty, Table 3–1 is provided, which gives coefficients (C_{yt}) as a function of the t/h and b_w/b ratios. Then we can calculate the distance from the section's centroidal axis to the bottom using Equation 3–52.

$$y_t = C_{yt}h \tag{3–52}$$

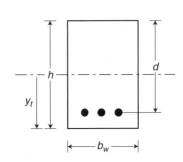

FIGURE 3–27 Rectangular section.

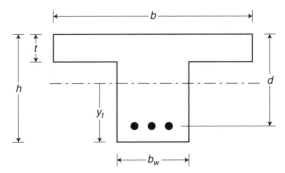

FIGURE 3–28 T-beam section.

The gross moment of inertia of T-beams about the centroidal axis can be determined with the help of Table 3–2, which gives coefficients (C_{Ig}) for different ratios of t/h and b_w/b. Then we can use Equation 3–53 to calculate the gross moment of inertia.

$$I_g = (C_{Ig})\, bh^3 \tag{3–53}$$

TABLE 3–1 Coefficients (C_{yt}) to Calculate y_t for T-Beams

b_w/b	t/h									
	0.05	0.1	0.15	0.2	0.25	0.3	0.35	0.4	0.45	0.5
0.1	0.647	0.713	0.744	0.757	0.760	0.755	0.747	0.735	0.721	0.705
0.2	0.579	0.629	0.659	0.678	0.688	0.691	0.690	0.685	0.677	0.667
0.3	0.550	0.585	0.610	0.627	0.638	0.644	0.646	0.645	0.641	0.635
0.4	0.533	0.559	0.578	0.592	0.602	0.609	0.612	0.613	0.611	0.607
0.5	0.523	0.541	0.555	0.567	0.575	0.581	0.584	0.586	0.585	0.583
0.6	0.515	0.528	0.539	0.547	0.554	0.558	0.561	0.563	0.563	0.563
0.7	0.510	0.518	0.526	0.532	0.536	0.540	0.542	0.544	0.544	0.544
0.8	0.506	0.511	0.515	0.519	0.522	0.524	0.526	0.527	0.528	0.528
0.9	0.503	0.505	0.507	0.509	0.510	0.511	0.512	0.513	0.513	0.513

TABLE 3–2 Coefficients (C_{Ig}) to Calculate I_g for T-Beams

b_w/b	t/h									
	0.05	0.1	0.15	0.2	0.25	0.3	0.35	0.4	0.45	0.5
0.1	0.01534	0.01800	0.01896	0.01922	0.01924	0.01930	0.01957	0.02018	0.02123	0.02282
0.2	0.02420	0.02830	0.03044	0.03142	0.03177	0.03183	0.03185	0.03201	0.03246	0.03333
0.3	0.03208	0.03655	0.03925	0.04074	0.04145	0.04171	0.04175	0.04177	0.04194	0.04239
0.4	0.03964	0.04395	0.04677	0.04850	0.04946	0.04989	0.05002	0.05003	0.05008	0.05030
0.5	0.04704	0.05091	0.05359	0.05533	0.05638	0.05693	0.05715	0.05719	0.05720	0.05729
0.6	0.05437	0.05763	0.05996	0.06156	0.06257	0.06315	0.06342	0.06350	0.06351	0.06354
0.7	0.06165	0.06418	0.06605	0.06738	0.06825	0.06878	0.06905	0.06915	0.06917	0.06918
0.8	0.06890	0.07063	0.07195	0.07290	0.07354	0.07395	0.07418	0.07428	0.07430	0.07431
0.9	0.07612	0.07701	0.07769	0.07820	0.07855	0.07878	0.07892	0.07898	0.07900	0.07900

EXAMPLE 3-7

Calculate y_t, I_g, f_r, and M_{cr} for a T-beam with the following data: $b = 60$ in., $b_w = 12$ in., $t = 4$ in., $h = 24$ in., $f'_c = 4,000$ psi, $\dfrac{b_w}{b} = \dfrac{12}{60} = 0.2$, $\dfrac{t}{h} = \dfrac{4}{24} = 0.167$

Solution

From Table 3–1 (interpolating) $\longrightarrow C_{yt} = 0.665 \longrightarrow y_t = C_{yt}h$
$$= 0.665 \times 24 = 15.96 \text{ in.}$$

From Table 3–2 (interpolating) $\longrightarrow C_{Ig} = 0.03077 \longrightarrow I_g = C_{Ig}bh^3$
$$= 0.03077 \times 60 \times 24^3 = 25,522 \text{ in}^4$$

The modulus of rupture is:
$$f_r = 7.5\sqrt{f'_c} = 7.5 \times \sqrt{4,000} = 474 \text{ psi}$$

The cracking moment in the positive moment regions (i.e., tension at the bottom) is:
$$M_{cr} = \frac{f_r I_g}{y_t} = \frac{474 \times 25,522}{15.96} = 757,984 \text{ in.-lb} = 63.2 \text{ ft-kip}$$

The cracking moment in the negative moment regions (i.e., tension at the top) is:
$$M_{cr} = \frac{f_r I_g}{h - y_t} = \frac{474 \times 25,522}{24 - 15.96} = 1,504,655 \text{ in.-lb} = 125.4 \text{ ft-kip}$$

Cracked Section Moment of Inertia (I_{cr})

Rectangular Section The calculation of the cracked section's properties is based on the transformed section concept. This is a useful tool borrowed from the theory of elasticity. Figure 3–29 shows the strains, the stresses, the internal couple, and the location of the cracked section's neutral axis for a rectangular section.

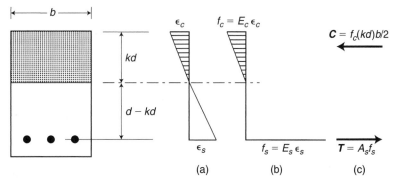

FIGURE 3-29   The cracked rectangular section: (a) strains, (b) stresses, and (c) the internal couple.

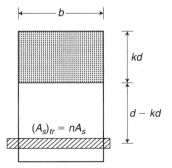

FIGURE 3–30   The transformed section.

The internal couple's components can be expressed as:

$$C = \frac{f_c(kd)b}{2} = E_c\epsilon_c\frac{(kd)b}{2} \quad \text{and} \quad T = A_sf_s = A_sE_s\epsilon_s$$

Substitute the area of steel with a "special kind" of material, $(A_s)_{tr}$, which can take tension and has an elastic response similar to that of concrete. The tension force then can be calculated as:

$$T = A_sE_s\epsilon_s = [(A_s)_{tr}]E_c\epsilon_s$$

The transformed steel area $(A_s)_{tr}$ is shown in Figure 3–30 and can be calculated using Equation 3–54.

$$(A_s)_{tr} = \frac{E_s}{E_c}A_s = nA_s \tag{3–54}$$

where n is the *modular ratio*, which is the ratio of the steel's and the concrete's modulus of elasticity. The value of $n = E_s/E_c$ may be rounded as shown in Table 3–3. The centroidal axis (measured as kd from the top) is located where the first moments of the areas above and below that axis balance each other.

$$b(kd)\frac{kd}{2} = (A_s)_{tr}(d - kd) = nA_s(d - kd)$$

By substituting $A_s = \rho bd$, the value for k can be calculated using Equation 3–55.

$$k = \sqrt{2n\rho + (n\rho)^2} - n\rho \tag{3–55}$$

The location of the neutral axis depends on only two parameters: the value of n that depends on the concrete's quality (because the modulus of elasticity of steel, E_s, is relatively constant and equal to 29,000 ksi), and the steel ratio, ρ, employed in the section.

TABLE 3–3   Values of n

	f_c'		
	3,000 psi	4,000 psi	5,000 psi
n	9	8	7

Then the moment of inertia about the centroidal axis can be expressed as:

$$I_{cr} = \frac{b(kd)^3}{3} + (n\rho bd)(d - kd)^2$$

After some mathematical manipulation, this equation can be written as shown in Equation 3–56.

$$I_{cr} = bd^3\left[\frac{k^3}{3} + n\rho(1 - k)^2\right] \qquad (3\text{–}56)$$

If the expression within the bracket is designated by C_r, the cracked moment of inertia can be calculated easily from Equation 3–57 using values obtained from Table 3–4.

$$I_{cr} = C_r bd^3 \qquad (3\text{–}57)$$

Table 3–4 lists k and C_r for different values of $n\rho$.

TABLE 3–4	Values of k and C_r for Rectangular Sections	
nρ	**k**	**C_r**
0.010	0.132	0.0083
0.020	0.181	0.0154
0.030	0.217	0.0218
0.040	0.246	0.0277
0.050	0.270	0.0332
0.060	0.292	0.0384
0.070	0.311	0.0433
0.080	0.328	0.0479
0.090	0.344	0.0523
0.100	0.358	0.0565
0.110	0.372	0.0605
0.120	0.384	0.0644
0.130	0.396	0.0681
0.140	0.407	0.0717
0.150	0.418	0.0752
0.160	0.428	0.0785
0.170	0.437	0.0817
0.180	0.446	0.0848
0.190	0.455	0.0878
0.200	0.463	0.0908
0.210	0.471	0.0936
0.220	0.479	0.0964
0.230	0.486	0.0990
0.240	0.493	0.1016
0.250	0.500	0.1042
0.260	0.507	0.1066
0.270	0.513	0.1090
0.280	0.519	0.1114
0.290	0.525	0.1137
0.300	0.531	0.1159

EXAMPLE 3–8

Given a rectangular section with $b = 12$ in., $h = 20$ in., A_s = three #8 bars, $f'_c = 4,000$ psi, and $f_y = 60,000$ psi; calculate the gross moment of inertia (I_g), the location of the neutral axis at service load conditions (kd), and the cracked section moment of inertia (I_{cr}). Assume $d = h - 2.5$ in. $= 17.5$ in.

Solution

The gross moment of inertia is:

$$I_g = \frac{bh^3}{12} = \frac{12 \times 20^3}{12} = 8,000 \text{ in}^4$$

The steel ratio is:

$$\rho = \frac{A_s}{bd} = \frac{2.37}{12 \times 17.5} = 0.0113$$

From Table 3–3:

$$f'_c = 4,000 \text{ psi} \longrightarrow n = 8$$

Then:

$$n\rho = 8 \times 0.0113 = 0.0904$$

From Table 3–4 (interpolating):

$$k \approx 0.344 \quad \text{and} \quad C_r \approx 0.0525$$

Hence, the location of the neutral axis from the top is:

$$kd = 0.344 \times 17.5 = 6.02 \text{ in.}$$

and the cracked section inertia is:

$$I_{cr} = C_r bd^3 = 0.0525 \times 12 \times 17.5^3 = 3,376 \text{ in}^4$$

T-Section Figure 3–31 shows a typical T-shaped concrete beam reinforced for positive moment. The expression for k and the cracked section moment of inertia are quite complicated for T-beams; however, there are easy solutions with certain simplifying assumptions. During service load conditions, especially with large amounts of reinforcing, the neutral axis may fall below the bottom of the flange (in other words $kd > t$). Figure 3–32 shows the neutral axis and stresses for the general case. If $kd \leq t$, or $n\rho$ is less than the value shown in Table 3–5, the neutral axis is within the flange and Equation 3–57 and Table 3–4 can be used to calculate I_{cr}.

In the introduction we stated that the calculation of deflections contains many uncertainties, so the errors introduced with simplifying assumptions are minimal and do not seriously influence the validity of the results. The main simplification for calculating

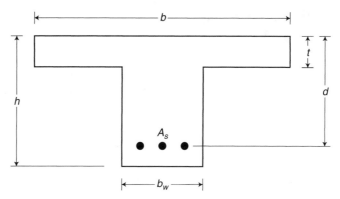

FIGURE 3–31   T-beam section.

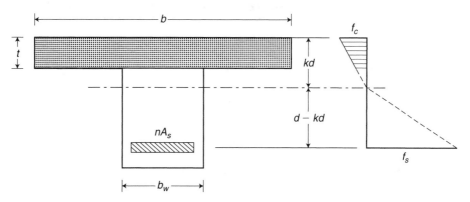

FIGURE 3–32   The transformed section of a T-beam.

deflection for T-beams is that when the neutral axis at service load conditions falls below the bottom of the flange, the portion of the compressive zone that is within the web is neglected. As shown in Figure 3–32, this is usually a small area combined with small stresses, and so the error is small. Thus, the neutral axis is located where the first moments of the transformed areas from above and from below are equal:

$$bt\left(kd - \frac{t}{2}\right) = nA_s(d - kd)$$

Introducing

$$A_s = \rho bd$$

TABLE 3–5	Values of $n\rho$ that Satisfy the Condition that $kd \leq t$				
t/d	0.1	0.2	0.3	0.4	0.5
$(n\rho)_{\text{limit}}$	0.0055	0.0250	0.0643	0.1333	0.2500

then solving for k gives:

$$k = \frac{n\rho + \frac{1}{2}\left(\dfrac{t}{d}\right)^2}{n\rho + \left(\dfrac{t}{d}\right)} \tag{3-58}$$

Table 3–6 provides k values for different $n\rho$ and t/d values. When $kd > t$, or $n\rho$ is greater than the value shown in Table 3–5, the neutral axis is below the flange and Table 3–6 should be used to find parameters required to calculate I_{cr}. The heavy horizontal line in each column of Table 3–6 represents the limiting values of Table 3–5.

TABLE 3–6 Values of k in T-Beams as a Function of $n\rho$ and t/d

			t/d		
$n\rho$	0.1	0.2	0.3	0.4	0.5
0.010	0.136				
0.020	0.208				
0.030	0.269	0.217			
0.040	0.321	0.250			
0.050	0.367	0.280			
0.060	0.406	0.308			
0.070	0.441	0.333	0.311		
0.080	0.472	0.357	0.329		
0.090	0.500	0.379	0.346		
0.100	0.525	0.400	0.363		
0.110	0.548	0.419	0.378		
0.120	0.568	0.438	0.393		
0.130	0.587	0.455	0.407		
0.140	0.604	0.471	0.420	0.407	
0.150	0.620	0.486	0.433	0.418	
0.160	0.635	0.500	0.446	0.429	
0.170	0.648	0.514	0.457	0.439	
0.180	0.661	0.526	0.469	0.448	
0.190	0.672	0.538	0.480	0.458	
0.200	0.683	0.550	0.490	0.467	
0.210	0.694	0.561	0.500	0.475	
0.220	0.703	0.571	0.510	0.484	
0.230	0.712	0.581	0.519	0.492	
0.240	0.721	0.591	0.528	0.500	
0.250	0.729	0.600	0.536	0.508	0.500
0.260	0.736	0.609	0.545	0.515	0.507
0.270	0.743	0.617	0.553	0.522	0.513
0.280	0.750	0.625	0.560	0.529	0.519
0.290	0.756	0.633	0.568	0.536	0.525
0.300	0.763	0.640	0.575	0.543	0.531

The cracked moment of inertia of a T-beam depends on many parameters. Thus, unlike a simple rectangular section, several tables would be required to give coefficients for the calculations. This text will rely on the reader to perform the necessary calculations.

By using the following parameters (Equation 3–59):

$$\rho = \frac{A_s}{bd} \qquad \alpha_\Delta = \frac{t}{d} \qquad \beta_\Delta = \frac{b_w}{b} \tag{3-59}$$

we can calculate the coefficient C_{rT} using Equation 3–60:

$$C_{rT} = \beta_\Delta \frac{k^3}{3} + n\rho(1-k)^2 + (1-\beta_\Delta)\left(\frac{\alpha_\Delta^3}{3} - \alpha_\Delta^2 k + \alpha_\Delta k^2\right) \tag{3-60}$$

The cracked section moment of inertia for T-beams, where the neutral axis is below the bottom of the flange, can finally be calculated using Equation 3–61.

$$I_{cr} = C_{rT} bd^3 \tag{3-61}$$

EXAMPLE 3–9

The cross section used in Example 3–7 has 6 #11 bars in two rows, $A_s = 6 \times 1.56 = 9.36 \text{ in}^2$. Calculate the cracked section moment of inertia of the T-beam.

Solution

Calculate the parameters required in Equation 3–60 for C_{rT}. Because there are two rows of reinforcing, use $d = h - 4$ in.; thus, $d = 24 - 4 = 20$ in.

$$f_c' = 4,000 \text{ psi} \longrightarrow n = 8 \quad \text{(from Table 3–3)}$$

$$\rho = \frac{A_s}{bd} = \frac{9.36}{60 \times 20} = 0.0078 \longrightarrow n\rho = 8 \times 0.0078 = 0.0624$$

$$\alpha_\Delta = \frac{t}{d} = \frac{4}{20} = 0.2 \qquad \beta_\Delta = \frac{b_w}{b} = \frac{12}{60} = 0.2$$

Solving from Equation 3–58 or using Table 3–6 (interpolating), $k \approx 0.314$.

$$kd = 0.314 \times 20 = 6.28 \text{ in.}$$

Thus, the neutral axis is below the bottom of the flange.

Hence, from Equation 3–60:

$$C_{rT} = 0.2 \times \frac{0.314^3}{3} + 0.0624 \times (1 - 0.314)^2 + (1 - 0.2)$$

$$\times \left(\frac{0.2^3}{3} - 0.2^2 \times 0.314 + 0.2 \times 0.314^2\right) = 0.0393$$

Then

$$I_{cr} = 0.0393 \times 60 \times 20^3 = 18,864 \text{ in}^4$$

Applications

As mentioned earlier, the main application of calculating I_{cr} is to find the effective moment of inertia (I_e) required for computing deflections. The following examples will demonstrate this.

EXAMPLE 3–10

The cross section of the simple span beam shown in Figure 3–33 is the same as that used in Example 3–8. Calculate the deflection due to the total service loads. Assume the beam is made of normal-weight concrete with a unit weight of 145 pcf.

Solution

From Example 3–8:

$$b = 12 \text{ in.}, h = 20 \text{ in.}, d = 17.5 \text{ in.}, A_s = 2.37 \text{ in}^2, I_g = 8{,}000 \text{ in}^4, I_{cr} = 3{,}376 \text{ in}^4$$
$$f_r = 7.5\sqrt{4{,}000} = 474 \text{ psi}$$

For the given loads, the service load moment is:

$$M_a = \frac{(0.7 + 0.4)(30)^2}{8} = 123.75 \text{ ft-kip}$$

The cracking moment is determined using Equation 3–50:

$$M_{cr} = \frac{f_r I_g}{y_t} = \frac{474 \times 8{,}000}{\dfrac{20}{2}} = 379{,}200 \text{ in-lb} = 31.6 \text{ ft-kip}$$

The effective moment of inertia from Equation 3–48 is:

$$I_e = \left(\frac{M_{cr}}{M_a}\right)^3 I_g + \left[1 - \left(\frac{M_{cr}}{M_a}\right)^3\right] I_{cr}$$
$$= \left(\frac{31.6}{123.75}\right)^3 \times 8{,}000 + \left[1 - \left(\frac{31.6}{123.75}\right)^3\right] \times 3{,}376 = 3{,}453 \text{ in}^4$$

This value is only about 2.3% higher than the cracked moment of inertia (I_{cr}). It hardly seems worth the trouble to go through the calculations.

The modulus of elasticity of the concrete is:

$$E_c = 33(145)^{1.5}\sqrt{4{,}000} = 3.64 \times 10^6 \text{ psi}$$

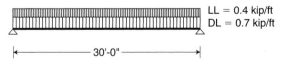

FIGURE 3–33 Loads for Example 3–10.

With these values the instantaneous deflection is:

$$\Delta_{DL+LL} = \frac{5\,w\ell^4}{384\,E_c I_e} = \frac{5 \times \dfrac{1{,}100}{12} \times (30 \times 12)^4}{384 \times 3.64 \times 10^6 \times 3{,}453} = 1.6 \text{ in.}$$

About 1 in. of this deflection is due to dead loads. The rest is due to live loads.

EXAMPLE 3–11

Given the T-beam used in Examples 3–7 and 3–9, calculate the instantaneous deflections due to the dead and live loads shown in Figure 3–34.

Solution

From dead loads:

$$M_{DL} = \frac{2.0 \times 40^2}{8} = 400.0 \text{ ft-kip}$$

From live loads:

$$M_{LL} = \frac{0.8 \times 40^2}{8} = 160 \text{ ft-kip}$$

From Example 3–7:

$$I_g = 25{,}522 \text{ in}^4 \quad \text{and} \quad M_{cr} = 63.2 \text{ ft-kip}$$

From Example 3–9:

$$I_{cr} = 18{,}864 \text{ in}^4$$

Substituting into Equation 3–48, the effective moment of inertia is:

$$I_e = \left(\frac{63.2}{400}\right)^3 \times 25{,}522 + \left[1 - \left(\frac{63.2}{400}\right)^3\right] \times 18{,}864 = 18{,}890 \text{ in}^4$$

Note again that the result is only slightly different from the value of I_{cr}. $f_c' = 4{,}000\,\text{psi}$, from Example 3–10, $E = 3.64 \times 10^6\,\text{psi}$, and therefore the instantaneous deflections are:

$$\Delta_{DL} = \frac{5}{384} \times \frac{\dfrac{2{,}000}{12}(40 \times 12)^4}{3.64 \times 10^6 \times 18{,}890} = 1.68 \text{ in.}$$

$$\Delta_{LL} = \frac{800}{2{,}000} \times 1.68 = 0.67 \text{ in.}$$

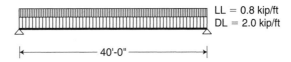

LL = 0.8 kip/ft
DL = 2.0 kip/ft

40'-0"

FIGURE 3–34   Loads for Example 3–11.

The calculation of the live load deflection (Δ_{LL}) is not exactly according to the ACI requirements, as this value has to be computed by subtracting the dead load deflection from the total dead and live load deflection. This total deflection is computed using the effective amount of inertia (I_e) based on the applied dead and live load amounts. The difference in the results, however, is negligible.

Comments on the Effective Moment of Inertia (I_e)

The values of the cracked section moment of inertia and the effective moment of inertia (as defined by the ACI Code) usually differ only slightly, as observed in Examples 3–10 and 3–11. To make it easy to understand the reason, we now rewrite Equation 3–48 as Equation 3–62.

$$I_e = \left(\frac{M_{cr}}{M_a}\right)^3 I_g + \left[1 - \left(\frac{M_{cr}}{M_a}\right)^3\right] I_{cr} = I_{cr} + \left(\frac{M_{cr}}{M_a}\right)^3 (I_g - I_{cr}) \qquad (3\text{--}62)$$

In other words, I_e is equal to I_{cr} plus a fraction of the difference between I_g and I_{cr}. Because the cracking moment (M_{cr}) is usually much smaller than the actual moment (M_a), their ratio, raised to the third power, is a small number. In building structures the actual moment is about 65%–75% of the ultimate moment; and in most members the ratio of the cracking moment to the actual moment is less than 0.3 where the required reinforcing is at least two or three times the minimum $A_{s,\min}$. Hence, the multiplier to the ($I_g - I_{cr}$) is only 0.027 or less. Thus

$$I_e \approx I_{cr} + 0.03(I_g - I_{cr}) \approx I_{cr}$$

Long-term Deflections

In addition to instantaneous (or elastic) deflections, designers must deal with deformations caused by shrinkage and creep. The ACI Code treats these as additional deformations obtained by using a few empirically obtained multipliers that represent "great national average" values.

Accordingly, Equation 3–63 gives us the *additional* long-term deflection multiplier (ACI Code, Section 9.5.2.5). The additional long-term deflection is computed by multiplying the immediate deflection by λ_Δ.

$$\lambda_\Delta = \frac{\xi}{1 + 50\rho'} \qquad (3\text{--}63)$$

where ρ' is the ratio of compressive reinforcing (if any) in positive moment regions of the beam $\left(\text{i.e., } \rho' = \dfrac{A_s'}{bd}\right)$; and the time-dependent factor ξ for sustained loads is equal to one of the following:

5 years or more	2.0
12 months	1.4
6 months	1.2
3 months	1.0

Tests have shown that the presence of compression reinforcing steel decreases the additional long-term deformation. If no reinforcing exists on the compression side, the deflection due to sustained loads may grow to three times the instantaneous deflection in 5 years or more. Fortunately the rate of growth dissipates and becomes very slow after about 3 years. The growth in deflections virtually disappears after about 5 years.

These additional deformations apply only to the part of the instantaneous deflections that the structure must sustain on a continuous basis. Thus they apply to the dead loads and the part of the live loads that is continuously present. For example, in a residential structure or an office structure, probably less than 15% of the design live loads are present continuously. In a library stack area or a storage facility, on the other hand, 75%–80% of the design live loads are present all the time; thus, the live loads in these facilities contribute a great deal to the long-term deformations as well.

Table 3–7 summarizes deflections that are permissible according to the ACI Code. Note that the main concern is damage to nonstructural elements that are supported by, or are attached to the concrete structure. These elements most frequently are walls, or, in some rare occasions, ceilings. So aside from the fact that some shallow elements with really long spans may also exhibit undesirable vibrations (very rare in concrete structures), the issue is not the magnitude of the deflection, but what it may cause. For example, when a beam or a slab deflects, a partition wall may unintentionally become a support to the beam or slab. If the partition wall cannot take that load without cracking or buckling, then that partition wall will fail, while nothing terrible happens to the beam or slab whose action caused the failure. So the designer's job is to evaluate the consequences arising from the inevitable deflections and take steps to avoid potential harm to neighboring elements. In the example cited above, the easy solution is to connect the partition wall at its top in such a way that it permits the deflection of the structure above it and, at the same time, provides lateral support to the wall.

TABLE 3–7 Maximum Permissible Computed Deflections [ACI Code Table 9–5(b)]

Type of Member	Deflection to be Considered	Deflection Limitation
Flat roofs not supporting or attached to nonstructural elements likely to be damaged by large deflections	Immediate deflection due to live load	$\ell/180$
Floors not supporting or attached to nonstructural elements likely to be damaged by large deflections	Immediate deflection due to live load	$\ell/360$
Roof or floor construction supporting or attached to nonstructural elements likely to be damaged by large deflections	That part of the total deflection occurring after attachment of nonstructural elements (sum of the long-term deflections due to all sustained loads and the immediate deflection due to any additional live load)	$\ell/480$
Roof or floor construction supporting or attached to nonstructural elements not likely to be damaged by large deflections		$\ell/240$

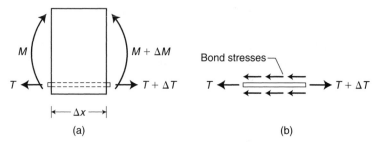

(a) (b)

FIGURE 3–35 (a) A Δx long portion of a beam; (b) isolated reinforcing as a free body.

3.4 REINFORCEMENT DEVELOPMENT AND SPLICES

Bond Stresses

The integrity of reinforced concrete requires that there be no slippage between the reinforcement and the surrounding concrete. The whole theory of design is based on that assumption.

Figure 3–35 shows a small piece of a beam with applied moments. As the moment changes along the length of a beam, so does the tension in the reinforcing steel.

When the reinforcing is isolated, as shown in Figure 3–35(b), the role of the bond stresses becomes quite clear. They transfer the difference in the tensile force, ΔT, from the steel to the concrete surface surrounding the bar, and vice versa.

The magnitude of the bond stresses varies along the length of the beam with the rate of change in the moments. Where the moments change rapidly, the bond stresses are high; and moments change rapidly where shears are high. Hence, where shears are high, the bond stresses also are high.

The use of deformed bars results in three distinct effects that resist relative slippage between the surface of the reinforcement and the concrete. The first is chemical adhesion between the two materials. The second is friction on the surface of the bar. (Reinforcing bars are not smooth; in fact, they have a rather rough surface.) The third comes from the concrete bearing on the ridges of the deformations. These effects are shown schematically in Figure 3–36.

Research has shown that the following sequence occurs at the bar/concrete interface. Initially, chemical adhesion bonds the two together. After the adhesion breaks down, friction and the reactions on the ribs become engaged. These reactions are at an angle to the axis of the bar, as seen in Figure 3–36b. The angle depends on the slope of the rib's surface

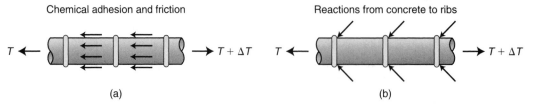

(a) (b)

FIGURE 3–36 (a) Adhesion and friction forces on the bar surface; (b) reactions on the deformation ribs.

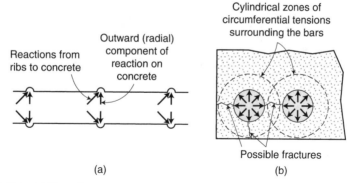

FIGURE 3–37   (a) Reactions on the surrounding concrete from the ribs; (b) outward-pointing radial pressures on the concrete.

and the rib configuration. For simplicity ribs are shown perpendicular to the bar, although they very often have different orientations.

Forces of the same magnitude but with opposite sense act on the surrounding concrete. The component of these forces that acts parallel with the axis of the bar counteracts ΔT. The component that is perpendicular to the bar axis, however, develops outward pressures from the bar to the concrete. Figure 3–37 shows these two components of the reaction. The perpendicular component, in turn, results in circumferential tensions in the concrete, similar to those in a pipe under pressure. The circumferential tensions affect a cylindrical portion of the concrete that surrounds the bar, as shown in Figure 3–37b. If the bar is too close to the outside of the concrete, the cylinder is too thin, and cracks may appear on the side or bottom of the beam, indicating a splitting failure. If bars are too close to each other, the two cylinders overlap, and a split may develop in a horizontal plane between the bars. Thus, the closer the bar is to the surface (small concrete cover), or the closer parallel bars are to each other, the greater is the likelihood of splitting failure due to bond stresses.

Bond stresses change along even a small length of the beam. Research has shown that bond stresses spike next to flexural cracks (there is no bond across the crack width), and also where a reinforcing bar terminates. These highly localized peak bond stresses do not significantly endanger the safety of the structure provided that an adequate length of bar extends beyond where the bar will be fully stressed to yield at ultimate strength. This extra bar length is called *embedment length* and is defined as the length necessary for the bar to develop its full capacity. Another, more common name for this length is *development length*.

Development Length for Bars in Tension

The ACI Code provides two ways to determine the required development length for deformed bars and deformed wires.

The first method (Section 12.2.2 of the ACI Code) is a simplified one, whereas the second method is more involved. In both of these methods, however, the formulae include all the important variables that influence the bond strength. The latter method, which is based on Equation 3–64 (ACI Code, Equation 12–1 of Section 12.2.3), is the general approach used to calculate the development length. Table A3–1 provides a description and values of the different factors.

$$l_d = \left[\frac{3}{40} \frac{f_y}{\sqrt{f_c'}} \frac{\psi_t \psi_e \psi_s \lambda}{\left(\dfrac{c_b + K_{tr}}{d_b} \right)} \right] d_b \geq 12 \text{ in.} \tag{3–64}$$

where

$$\frac{c_b + K_{tr}}{d_b} \leq 2.5$$

Table A3–2 summarizes the simplified equations allowed by the ACI for the calculation of the required development length for reinforcing bars in tension in lieu of using Equation 3–64. The required development length cannot be less than 12 in. ($\ell_d \geq$ 12 in.).

Table A3–3 shows the tensile bar development lengths (ℓ_d) for $f_y = 60$ ksi and $f_c' = 3$ ksi or 4 ksi. The reinforcing bars are assumed to be uncoated ($\psi_e = 1.0$) and not top bars ($\psi_t = 1.0$), and the concrete is normal weight ($\lambda = 1.0$).

EXAMPLE 3–12

Calculate the required development length, ℓ_d, for a #7 epoxy-coated bottom bar. Assume normal-weight concrete, $f_c' = 4,000$ psi, $f_y = 60,000$ psi, #3 stirrups, 1.5 in. concrete cover over the stirrups, and a 5 in. center-to-center spacing of bars.

Solution

Both methods will be used here:

(a) *Using Equation 3–64:*

Step 1   Obtain the factors' values from Table A3–1.

$$\psi_t = 1.0 \text{ (not a top bar)}$$
$$\psi_e = 1.5 \text{ (epoxy coated, cover is less than } 3d_b)$$
$$\psi_s = 1.0 \text{ (#7 bar)}$$
$$\lambda = 1.0 \text{ (normal-weight concrete)}$$
$$c_b = \min\left\{ 1.5 + 0.375 + 0.875/2 = 2.31 \text{ in.}, \frac{5 \text{ in.}}{2} \right\} = 2.31 \text{ in.}$$

Assume $K_{tr} = 0$ (conservative)

Step 2   Check the requirement for Equation 3–64.

$$\frac{c_b + K_{tr}}{d_b} = \frac{2.31 + 0}{0.875} = 2.64 > 2.5 \quad \therefore \text{ Use 2.5}$$

Step 3   Use values from steps 1 and 2 to calculate the required development length.

$$\ell_d = \left[\frac{3}{40} \frac{(60,000)\,1.0 \times 1.5 \times 1.0 \times 1.0}{\sqrt{4,000}\quad(2.5)} \right] \times 0.875 = 38 \text{ in.} > 12 \text{ in.} \quad \therefore \text{ ok}$$

(b) *Using the simplified expressions of Table A3–2:*

The bar diameter is 0.875 in. Because clear cover = 1.5 in. > 0.875 and clear spacing = 5 − 0.875 = 4.125 in. > 2(0.875), use condition A from Table A3–3:

$$\ell_d = \psi_e \ell_d = 1.5(42) = 63 \text{ in.} > 12 \text{ in.} \quad \therefore \text{ ok}$$

The simplified expression results in a more conservative development length.

Tension Bars Terminated in Hooks

When there is not enough "space" or length to transfer stresses using the entire development or embedment length (e.g., when a beam terminates into a column), the ACI Code permits the use of hooks or mechanical anchorage devices, or a combination of these. The bending radii and the extensions for hooks are standardized by the ACI Code and are shown in Figure 3–38. Equation 3–65 (ACI Code, Section 12.5.2) gives the required development length, ℓ_{dh}, for bars in tension when the end is terminated in a hook:

$$\ell_{dh} = \left(\frac{0.02\ \psi_e \lambda f_y}{\sqrt{f_c'}} \right) d_b \geq \min\{8d_b, 6 \text{ in.}\} \qquad (3\text{–}65)$$

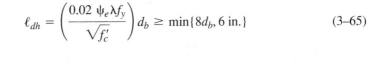

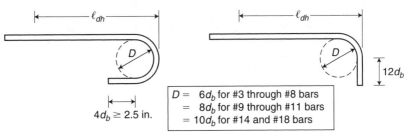

(a) For primary reinforcement

(b) For stirrups and ties

FIGURE 3–38   ACI Code standard hooks.

where

ψ_e = 1.0 (for uncoated bars)
   = 1.2 (for epoxy-coated bars)
λ = 1.0 (for normal weight concrete)
   = 1.3 (for lightweight aggregate concrete)

Bars that are developed by standard hooks at a discontinuous end of a member must be enclosed within ties or stirrups when both the top (bottom) cover and the side cover over the bar are less than 2.5 in. The stirrup or tie spacing may not exceed $3d_b$ along the development length ℓ_{dh}, and the first stirrup or tie must be within $2d_b$ of the outside of the bend, where d_b is the diameter of the hooked bar.

The development length resulting from Equation 3–65 may be multiplied by reduction factors, recommended by ACI Code, Section 12.5.3, as listed in Table A3–4.

EXAMPLE 3–13

Calculate the development length (ℓ_{dh}) for the bar in Example 3–12 if it is terminated in a standard hook.

Solution

$$\ell_{dh} = \frac{0.02 \times 1.2 \times 1.0 \times 60,000}{\sqrt{4,000}} \times 0.875 = 19.9 \text{ in.} > \min\{8(0.875), 6 \text{ in.}\} = 6 \text{ in.}$$

EXAMPLE 3–14

Figure 3–39 shows a beam/column connection. The beam clear span is 30.0 ft, w_u = 2.2 kip/ft, M_u = 194 kip-ft, and V_u = 33 kip at the face of the column. Calculate the cutoff points for the top bars. f_c' = 4,000 psi, f_y = 60,000 psi. Assume uncoated bars with normal-weight concrete.

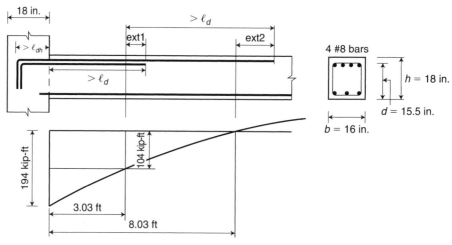

FIGURE 3–39   Sketch for Example 3–14.

Solution

Check to determine if there is any excess reinforcement:

$$\rho = \frac{A_s}{bd} = \frac{3.16}{16 \times 15.5} = 0.0127$$

$$\text{From Table A2–6b} \longrightarrow R = 609 \text{ psi}$$

$$M_R = bd^2R/12{,}000 = 16 \times 15.5^2 \times 609/12{,}000 = 195 \text{ kip-ft}$$

The reinforcement is just adequate; no excess is provided. Calculate the point of inflection, which is the theoretical point where the negative reinforcing is no longer needed; in other words, where $M = 0$. Cutting a section at the distance x from the face of column, the equation for moment, M_x, is:

$$M_x = -M_u + V_u x - w_u x^2/2 = 0$$

$$-194 + 33x - \frac{2.2x^2}{2} = 0 \longrightarrow x = 8.03 \text{ ft}$$

Calculate the theoretical point where two of the four bars may be terminated. With only 2 #8 bars, $\rho = 0.0064, R = 326$ psi, and $M_R = 104$ kip-ft. Then

$$-194 + 33x - \frac{2.2x^2}{2} = -104 \longrightarrow x = 3.03 \text{ ft}$$

Calculate ℓ_d using the general Equation 3–64. From Table A3–1 the necessary factors are:

$\psi_t = 1.3$ (top bars)
$\psi_e = 1.0$ (uncoated bars)
$\psi_s = 1.0$ (bars larger than #6)
$\lambda = 1.0$ (normal-weight concrete)
$c_b = 2.5$ in.
$K_{tr} = 0$ (transverse reinforcements present)

Then

$$\ell_d = \left[\frac{3}{40} \frac{60{,}000}{\sqrt{4{,}000}} \frac{1.3 \times 1.0 \times 1.0 \times 1.0}{\left(\dfrac{2.5 + 0}{1}\right)} \right] 1.0 = 37 \text{ in.} = 3.08 \text{ ft}$$

The theoretical cutoff point for two of the bars lies at 3.03 ft from the face of the column. In addition, the ACI Code requires that bars must extend beyond the theoretical cutoff point (see the dimension labelled as "ext1" in Figure 3–39) by the greater of d (the effective depth of the beam) or $12d_b$ (ACI Code, Section 12.10.3). Hence,

$$\text{ext1} \geq \max\{d, 12d_b\}$$

$$\geq \max\{15.5 \text{ in.}, 12 \times 1 \text{ in.}\} \qquad \text{Use } 15.5 \text{ in.} = 1.3 \text{ ft}$$

Thus, the two inner bars can be terminated at $3.03 + 1.3 = 4.33$ ft from the face of the column.

The other two bars, however, must be extended the dimension shown as "ext2" in Figure 3–39 according to ACI Code (Section 12.12.3).

$$\text{ext2} \geq \max\{d, 12d_b, \ell_n/16\},$$

where ℓ_n is the clear (net) span of the beam. The $\ell_n/16$ requirement covers the uncertainty of the true location of the point of inflection.

$$\text{ext2} \geq \max \{15.5 \text{ in.}, 12 \times 1 \text{ in.}, 30 \times 12/16\} \qquad \text{Use } 22.5 \text{ in.} = 1.88 \text{ ft}$$

Thus, the two outer bars may be cut off at $8.03 + 1.88 = 9.91$ ft $= 9'\text{-}11''$ from the face of the column.

These bars have to be long enough to develop their strength between their cut off point and the theoretical cutoff point for the first two bars. This length is $9.91 - 3.03 = 6.88$ ft, which is considerably greater than the required development length of 3.08 ft.

If the bars were extended straight into the column, they would not have the length needed to develop their strength because $\ell_d = 3.08$ ft, but the width of the column is only 18 in. These bars must be bent into the column with a 90-degree hook, as shown in Figure 3–39. Check for the adequacy of the available length to develop ℓ_{dh}.

$$\psi_e = 1.0 \text{ (uncoated bars)}$$
$$\lambda = 1.0 \text{ (normal-weight concrete)}$$
$$\ell_{dh} = \frac{0.02 \times 1.0 \times 1.0 \times 60,000}{\sqrt{4,000}} \times 1.0 = 19 \text{ in.} > \min\{8(1), 6 \text{ in.})\} = 6 \text{ in.}$$

With side cover ≥ 2.5 in. and with concrete cover of at least 2 in. beyond the 90-degree hook, a reduction factor of 0.7 is permitted according to Table A3–4. Hence:

$$\ell_{dh, \text{reduced}} = 0.7 \times 19 = 13.3 \text{ in.}$$

The anchorage into the column will be satisfactory.

Development Length for Bars in Compression

Bars in compression require considerably less development length than bars in tension, because there are no tensile cracks in the compression zone to weaken the bond. In addition, the bars transfer some of their forces to the concrete in end-bearing. Hooks are useless for bars in compression. Equation 3–66 is used to compute the compression development length, l_{dc} (ACI Code, Section 12.3.2):

$$\ell_{dc} = \max\left\{ \left(\frac{0.02 f_y}{\sqrt{f_c'}}\right) d_b; (0.0003 f_y) d_b \right\} \geq 8 \text{ in.} \qquad (3\text{-}66)$$

where the constant 0.0003 has the unit of in^2/lb. The calculated length (ℓ_{dc}) may be multiplied by the reduction factors given in Table A3–5.

Table A3–6 gives the compression bar development length (ℓ_{dc}) for $f_y = 60$ ksi, and $f_c' = 3$–5 ksi, or more. Reduction factors from Table A3–5 are not applied.

Splices of Reinforcement

Splices are often needed in construction, either because the required length of a bar cannot be supplied, or because of practical construction considerations, such as splicing column reinforcing just above the most recently cast floor.

Splices for #11 or smaller bars may be made either by simple overlap, by butt or lap welding them together, or by using a proprietary splicing device.

Tension Splices *Lap splices* The ACI Code (Section 12.15.1) gives the minimum length of lap for tensile reinforcing as

Class A splice...............................$(1.0 l_d)$
Class B splice...............................$(1.3 l_d)$

Generally speaking, Class B splice is required for most cases. Class A splice is permitted only when both of the following conditions are satisfied:

a. The area of reinforcement provided is at least twice that required by analysis everywhere along the length of the splice;
b. Only one-half or less of the total reinforcing is spliced within the required lap length.

Welded splices Splices may be butt welded or lap welded, as shown in Figure 3–40. The ACI Code (Section 12.14.3.4) requires that they be able to develop 125% of the yield strength of the reinforcing. Welding must conform to the *Structural Welding Code– Reinforcing Steel* (ANSI/AWS D1.4).

Butt-welded splices are preferred over welded lap splices. In the former the tensile force travels in a straight path. In the latter there is an eccentricity equal to the bar diameter. The resulting moment develops forces on the concrete perpendicular to the spliced bar. These forces may result in local cracking along the bars in the lap zone.

Welded splices are expensive, as they are very labor intensive. Butt-welded bars usually require extensive preparation of the ends, and lap-welded splices take more time to weld.

Proprietary mechanical splices Many patented devices are available for splicing reinforcing bars. Some are internally threaded sleeves into which the threaded bar ends are screwed from both ends. Others are sleeves that fit around the bar ends to be spliced; molten metallic filler is poured to provide the necessary interlock.

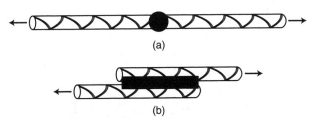

(a)

(b)

FIGURE 3–40   (a) Butt-welded bar, (b) lap-welded bar.

Compression Splices *Lap splices* The ACI Code (Section 12.16.1) defines compression lap splices as

$$(0.0005f_y d_b) \geq 12 \text{ in. for } f_y \leq 60,000 \text{ psi}$$

and

$$(0.0009f_y - 24)d_b \geq 12 \text{ in. for } f_y > 60,000 \text{ psi}$$

When $f_c' < 3,000$ psi, the calculated lap length must be increased by one-third.

Welded splices Welded compression splices are permitted. The rules for the welding are the same as those for welding tension splices.

PROBLEMS

3–1 Determine the nominal moment capacity, M_n, of the following T-beams. Use $f_c' = 3,000$ psi and $f_y = 60,000$ psi:

(a) $b_w = 12$ in., $b_{\text{eff}} = 30$ in., $h_f = 4$ in., $d = 21.5$ in., and three #10 bars
(b) $b_w = 14$ in., $b_{\text{eff}} = 36$ in., $h_f = 4$ in., $d = 27.5$ in., and four #10 bars
(c) $b_w = 16$ in., $b_{\text{eff}} = 36$ in., $h_f = 4$ in., $d = 33.5$ in., and five #10 bars

Assume single-layer reinforcement at the bottom of the beams.

3–2 Rework Problem 3–1 assuming the beam is rectangular (i.e., $b_{\text{eff}} = b_w$). How much does the T-beam nominal moment capacity increase (in percent) as compared to the rectangular beam assumption for each case?

3–3 Rework Problem 3–1 with $f_c' = 4,000$ psi. What is the percentage of increase in M_n for each case?

3–4 Calculate the positive moment capacity, M_R, of the T-beam shown below, which is part of a reinforced concrete floor system with beams spanning 25'-0" and spacing of 8'-0". Use $f_c' = 4,000$ psi, $f_y = 60,000$ psi, #4 stirrups and a cover of 1.5 in. Neglect the top bars in computing the moment capacity.

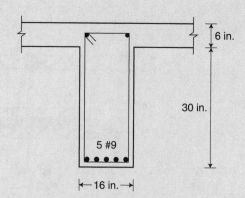

3–5   The Figure below shows a cross section of the interior bay of a floor system. The beam has a clear span of 24'-0". The superimposed dead load is 20 psf. What is the maximum allowable service live load on the floor in psf based on the moment capacity of the beam at the midspan? For simplicity assume the beam is simply-supported. Assume #4 stirrups and 1.5 in. cover. Use $f_c' = 4,000$ psi, $f_y = 60,000$ psi. Neglect the top bars. Do not consider live load reduction.

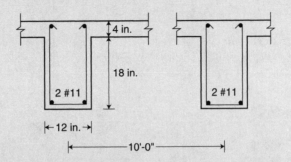

3–6   Consider the floor system of Problem 3–5 with $f_c' = 4,000$ psi, $f_y = 60,000$ psi, a superimposed dead load of 25 psf, and a live load of 60 psf. What is the maximum allowable clear span for the beam? For simplicity assume the beam is simply-supported. Do not consider live load reduction.

3–7   What are the required areas of reinforcement for the following T-beams? Use $f_c' = 4,000$ psi, and $f_y = 60,000$ psi.

(a) $b_{eff} = 66$ in., $b_w = 12$ in., $h_f = 4$ in., $h = 20$ in., $M_u = 200$ ft-kip
(b) $b_{eff} = 48$ in., $b_w = 12$ in., $h_f = 4$ in., $h = 18$ in., $M_u = 150$ ft-kip
(c) $b_{eff} = 32$ in., $b_w = 10$ in., $h_f = 3$ in., $h = 16$ in., $M_u = 100$ ft-kip

3–8   Select the reinforcement for the beam of Problem 3–5 if the superimposed dead load is 40 psf and live load is 60 psf. Assume the beam is singly reinforced with bottom bars (positive moment). Consider live load reduction.

3–9   What is the moment capacity, M_R, of the doubly-reinforced beam shown below? Use $f_y = 60,000$ psi, #4 stirrups, 1.5 in. cover, and $d' = 2.5$ in.

(a) $b = 12$ in., $h = 24$ in., $A_s = 3$ #9 and, $A_s' = 2$ #6; $f_c' = 4,000$ psi
(b) $b = 12$ in., $h = 30$ in., $A_s = 4$ #9 and, $A_s' = 2$ #7; $f_c' = 3,000$ psi
(c) $b = 16$ in., $h = 34$ in., $A_s = 5$ #10 and, $A_s' = 2$ #7; $f_c' = 3,000$ psi

Note: Reinforcements are in single layers.

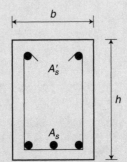

3–10 Calculate the moment capacity, M_R, of the rectangular beam shown below. How much will this capacity increase if 3 #9 bars are added as compression reinforcement? Assume $d' = 2.5$ in., $f'_c = 4,000$ psi, and $f_y = 60,000$ psi.

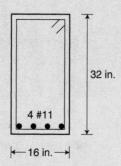

4 #11

32 in.

←— 16 in. —→

3–11 Design a rectangular reinforced concrete beam to resist service moments of 200 ft-kip from dead load (including the beam weight) and 150 ft-kip from live load. Architectural requirements limit the beam width to 14 in. and the total depth to 26 in. $f'_c = 3,000$ psi, and $f_y = 60,000$ psi. Assume #3 stirrups and 1.5 in. cover. Use compression reinforcements if needed.

3–12 Calculate the gross moment of inertia (I_g) and the cracked section moment of inertia (I_{cr}) for the following rectangular reinforced concrete beam. Use $f'_c = 4,000$ psi, and $f_y = 60,000$ psi.

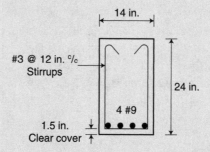

14 in.

#3 @ 12 in. $^c/_c$
Stirrups

24 in.

4 #9

1.5 in.
Clear cover

3–13 Calculate the cracked and effective moments of inertia for the beam of Problem 3–4 in the positive moment region (tension in the bottom). The actual service load moment is $M_a = 500$ kip-ft.

3–14 Calculate instantaneous deflections due to the dead and live loads for the T-beam of Problem 3–5. The floor live load is 100 psf.

3–15 The following rectangular reinforced concrete beam has a width $b_w = 12$ in., and a total depth $h = 24$ in. It is reinforced with 2 #9 bars and #3 @ 10 in. c/c stirrups. Use $f'_c = 3$ ksi, $f_y = 60$ ksi, and clear cover $= 1.5$ in. Assume normal-weight concrete. Answer the following questions:

(a) Use Table A3–3 to see whether sufficient development length is available for the 2 #9 bars.

(b) Check to see whether sufficient development length is available if 2 #7 and 2 #6 bars were used in lieu of the 2 #9 bars.

(c) Use the simplified formula given in Table A3–2 to see whether sufficient development length is available for the 2 #9 bars if $f'_c = 5$ ksi was used in the beam.

(d) Calculate the required development length for the 2 #9 bars when Equation 3–64 is used.

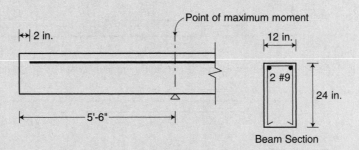

Beam Section

3–16 A 6-ft-wide wall footing supports a 12 in. thick concrete wall. $f'_c = 3$ ksi and $f_y = 60$ ksi. The maximum moment in the footing occurs at the face of the wall. Answer the following questions:

(a) Use the simplified expressions shown in Table A3–2 to determine whether #6 bars can develop their yield strength at the point of the maximum moment without hooks at the ends.

(b) Determine whether #6 bars could be used if they have standard hooks at their ends.

(c) What is the maximum bar size that could be used without hooks at the ends? Use Equation 3–64.

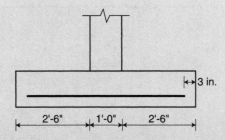

3–17 The reinforced concrete beam shown in the elevation below has a width of 15 in. and a total depth of 24 in. It is subjected to a factored moment of 260 kip-ft at the face of the column. Use $f_c' = 4$ ksi, $f_y = 60$ ksi, #3 stirrups, and 1.5 in. clear cover. Answer the following questions:

(a) Determine the required development length (ℓ_d) for the 4 #8 epoxy-coated top bars using the simplified formulae shown in Table A3–2. Calculate and use the permitted "excess reinforcement factor" (refer to Table A3–1).

(b) Recalculate the required development length by using Equation 3–64. Refer to Table A3–1 for definitions of the factors.

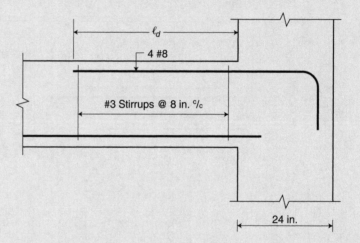

3–18 Calculate the location (measured from the face of the support) where two of the 4 #8 bars may be terminated. Use $f_c' = 4$ ksi, and $f_y = 60$ ksi.

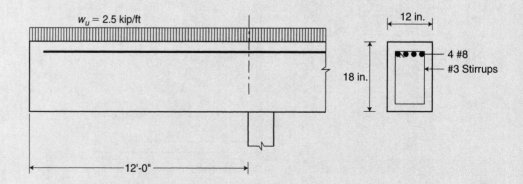

SELF-EXPERIMENTS

In the following self-experiments, you will learn the behavior of T-beams and doubly-reinforced beams. Include in the final report all the test details (sizes, time of day you cast the concrete, amounts of water/cement/aggregate, problems encountered, etc., with images showing steps of the tests).

Experiment 1

Cut several pieces of Styrofoam in the form of rectangular beams and a slab. Place the beams and slab on two supports as shown in Figure SE 3–1.

Apply a load (a few pounds) on top of the slab and observe how much the beams bend. Record the maximum deflection of the beams.

Next, glue the rectangular beams to the slab and repeat the test. How much do the beams deflect? Compare the maximum deflection between the two cases. Explain the difference, if any, in the results obtained. What is the importance of gluing the pieces together?

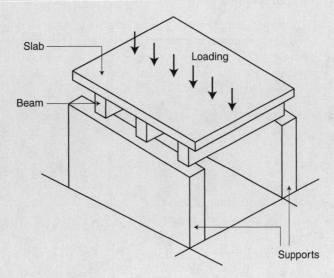

FIGURE SE 3–1   T-beam test.

Experiment 2

In this experiment, make the T-beams of Experiment 1 from concrete. First, build the forms for the beams and slab. Then, place wires at the bottom of the beams and slab (with about $1/4$ in. cover), and place concrete in the form. Make stirrups to hold the beam wires together as shown in Figure SE 3–2. Record all the different stages of casting the beams and slab and placing the wires. Record your observations and any problems encountered.

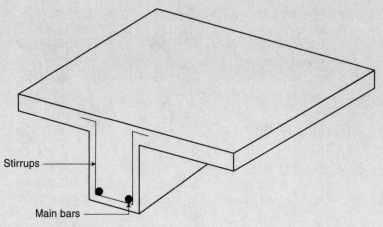

FIGURE SE 3–2   Reinforced concrete T-beam.

Experiment 3

This experiment demonstrates the behavior of doubly-reinforced beams. Get a piece of Styrofoam and cut it as a rectangular beam. Make two holes at the bottom of the beam using a heated wire. Then pass wires through these holes. Place the beams on two supports and apply a load at the center of beam. Record how much and the manner in which the beam bends under the load.

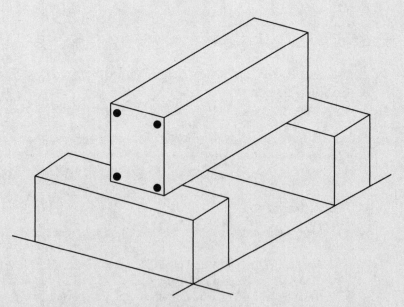

FIGURE SE 3–3   Doubly-reinforced Styrofoam beam.

Now, remove the wires and apply some glue to them. Again pass the wires through the holes and wait for the glue to harden. Load the beam as before and record now much and the manner in which the beam deflects under the load. Is there a change in the amount of deflection? Why?

Make two holes at the top of the beam, apply some glue to the two wires, pass the wires through the holes and wait for the glue to harden. Load the beam as before and record how much and the manner in which the beam deflects under the load. Do you notice any differences? Does the addition of top wires help the beam in resisting the load?

Experiment 4

In this experiment we construct a doubly-reinforced beam. Using wood and cardboard, make forms for the beam. Place two rows of wires at the top and bottom and tie them together with smaller-sized wire representing stirrups according to Figure SE 3–4.

Record all the different stages of making forms, placing bars, and casting concrete. Record your observations and any problems encountered.

FIGURE SE 3–4   Doubly-reinforced concrete beam.

4

SHEAR IN REINFORCED CONCRETE BEAMS

4.1 INTRODUCTION

In the classic two-dimensional structural studies of beams typically three separate internal forces are identified on any selected section. These are the *axial force, P,* (tension or compression) that acts along the axis of the member; the *shear force, V,* that acts in the plane of the section perpendicular to the axis of the member; and the *bending moment, M* (see Figure 4–1).

So far we have treated the bending moments as being the only force acting on a given section. It is true that the shear is zero where the bending moment is the largest, and thus it has no influence on the strength of the beam there. But elsewhere along the length of the element shear has a major effect on resulting tensile stresses. To fully understand the problem, we briefly review elementary strength of materials and the analysis of stresses and strains in homogeneous materials.

4.2 SHEAR IN BEAMS

Figure 4–2 shows a reinforced concrete beam and its shear and moment diagrams due to some applied load. A small length of the beam, *dx,* bounded by sections 1 and 2, is selected. The shear and the moment are different at the two respective sections (i.e., $V_1 > V_2$ and $M_1 < M_2$).

The change in the moment equals the area under the shear diagram, and the rate of change in the moment equals the magnitude of the shear. This is expressed mathematically as

$$\frac{dM}{dx} = V \quad \text{or} \quad \frac{M_2 - M_1}{dx} = V$$

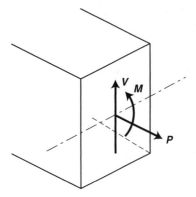

FIGURE 4-1 Internal forces on a section.

The internal couples are substituted for the moments (see Figure 4–3); (i.e., $M_1 = T_1 z = C_1 z$ and $M_2 = T_2 z = C_2 z$), therefore $T_1 < T_2$ because $M_1 < M_2$. When a small part of the beam that is below the horizontal section is isolated (see lower part of Figure 4–3), equilibrium requires that the horizontal force acting on that section balance the applied loads. The area of that horizontal section is bdx, and if the stress (i.e., the force per unit area) is designated by v, we can derive the following relationship:

$$T_2 - T_1 = \frac{M_2}{z} - \frac{M_1}{z} = \frac{dM}{z} \qquad (4-1)$$

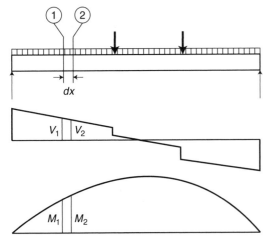

FIGURE 4-2 Shear and bending moment diagrams on a beam.

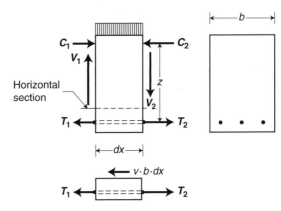

FIGURE 4-3 The internal couples on a short (dx) length of reinforced concrete beam.

From equilibrium:

$$T_2 - T_1 = vbdx \qquad (4-2)$$

Hence:

$$\frac{dM}{z} = vbdx \qquad (4-3)$$

Rearranging the terms of Equation 4–3:

$$V = \frac{dM}{dx} = vbz \qquad (4-4)$$

The horizontal shear stress value may then be calculated as

$$v = \frac{V}{bz} \qquad (4-5)$$

Figure 4–4 shows an isolated part of the beam in elevation. Inside this portion of the beam a small 1 in. × 1 in. × 1 in. cube is selected. The shear stresses are indicated on the elevation of this cube. Previously we showed what causes the horizontal shears. The horizontal shears form a couple (shown in Figure 4–4 as a counterclockwise couple). Because a couple can be kept in equilibrium only by another couple, a clockwise couple must be acting on this cube. The clockwise couple is furnished by equal-magnitude shears on the vertical sides of the cube. The existence of shears on both the horizontal and vertical sections of a beam is known as the "duality of shears." This means that shears of equal magnitude are always present on both the horizontal and the vertical surfaces of a small cube inside the beam.

Shears do not cause a problem for concrete. As a matter of fact, concrete is quite strong in shear. Figure 4–5 shows the cube with the components of the shear stresses parallel to the diagonals.

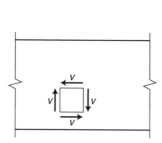

FIGURE 4–4   Shears on a
unit-size cube within the beam.

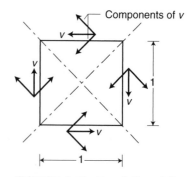

FIGURE 4–5   Resolution of the
shears into diagonal components.

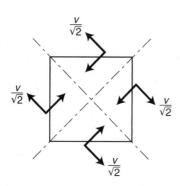

FIGURE 4–6 The shears substituted by their diagonal components.

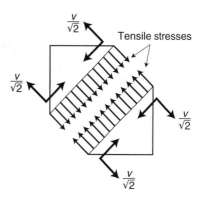

FIGURE 4–7 Equilibrium on one triangular wedge, resulting in tensile stresses.

In Figure 4–6 we substitute the shears with their components. Then we cut the cube into two triangular wedges along the diagonal planes. Figure 4–7 shows that tensile stresses are generated along one of the diagonal faces to maintain equilibrium. Figure 4–8 shows a section of the cube along the opposite diagonal. In this case, compression stresses will exist to maintain the equilibrium. So the conclusion here is that horizontal and vertical shears cause tension and compression in the diagonal directions.

By looking at one of the wedges in Figure 4–7 we can derive the magnitude of the *diagonal tension*. Because the area of the diagonal plane is $\sqrt{2}$ (the cube is 1 in. $\times$ 1 in.) the magnitude of the diagonal tension is $t\sqrt{2}$, where t is the diagonal tensile stress. The equilibrium equation for forces along the diagonal is:

$$t(\sqrt{2}) = 2\left(\frac{v}{\sqrt{2}}\right) \tag{4–6}$$

Thus

$$t = v \tag{4–7}$$

In other words, the magnitude of the *diagonal tensile stresses* equals that of the shear stresses.

We can perform a similar calculation for the wedges shown in Figure 4–8. The result would show that the magnitude of the compressive stresses on the diagonal face equals that of the shear stresses.

The above conclusion is valid only on a unit cube that is not subject to axial stresses, as these do not occur at the neutral axis. Flexural compressive stresses occur above the neutral axis while below the neutral axis we have flexural tensile stresses.

Figure 4–9 shows a cube that is above the neutral axis. This cube, therefore, is subject to axial compression as well as to shears. Imagine now a series of sections cut through this cube. These sections are rotated by an angle, ϕ, from the horizontal. A detailed mathematical analysis can show that among all the possible planes there exists a pair of planes, perpendicular to each other, where the resulting normal stresses are the

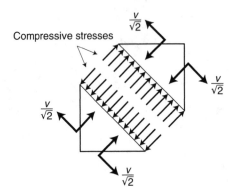

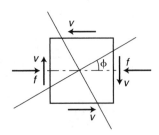

FIGURE 4–8   Equilibrium on the perpendicular triangular wedge, resulting in compressive stresses.

FIGURE 4–9   Stresses on a unit cube above the neutral axis.

largest compressions or tensions, respectively. These planes are called the *principal planes*. The stresses that act on the planes are the *principal stresses*. Figures 4–10 and 4–11 show the orientation of the principal compression and tensile stresses, respectively, on a section above the neutral axis.

The angle ϕ can be calculated as:

$$\tan 2\phi = \frac{2v}{f} \tag{4–8}$$

and the principal stresses as:

$$f_{1,2} = \tfrac{1}{2}(f \pm \sqrt{f^2 + 4v^2}) \tag{4–9}$$

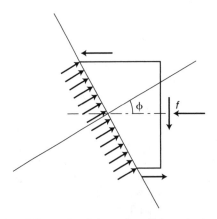

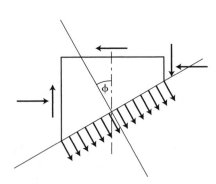

FIGURE 4–10   Orientation of the principal compressions above the neutral axis.

FIGURE 4–11   Orientation of the principal tensions above the neutral axis.

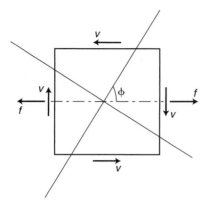

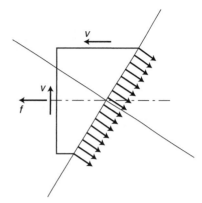

FIGURE 4–12 Stresses on a unit cube below the neutral axis.

FIGURE 4–13 Orientation of the principal tensions below the neutral axis.

The axial stress below the neutral axis is tension (see Figure 4–12). Thus, the orientations of the principal tensions will be similar to those shown in Figure 4–13.

Because the magnitudes of the flexural stress and the shearing stress vary along the beam as well as in relation to their distance from the neutral axis, the orientation and the magnitude of the principal stresses also vary accordingly. Of the two principal stresses, the tensile stress is the main concern here, as concrete is weak in tension. The diagonal tensions, therefore, may tear the beam apart. A potential crack starts out vertically at the bottom surface (because there is no shear at the outer edge), then changes orientation gradually as shear is introduced, causing a change in the principal stress direction. It crosses the neutral axis at 45°, because no axial forces exist at that location (pure shear), and then flattens out as it invades the zone of larger compressive stresses. Figure 4–14 shows two such cracks that follow the principal tensile stresses.

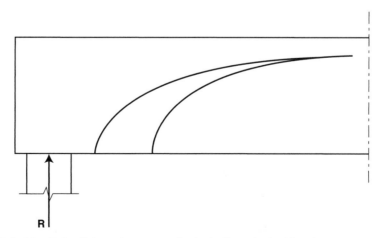

FIGURE 4–14 Potential cracks perpendicular to the principal tensions.

Shear (or, more precisely, diagonal tension) is a very complex problem. Thus a simplified approach is used in the analysis and design of beams for shear. Although simplified, the approach has been shown to provide safe and satisfactory design.

4.3 THE DESIGN OF SHEAR REINFORCEMENT

The basic concept of shear reinforcement is the same as that of flexural reinforcing. If cracks begin to open due to lack of tensile strength in the concrete, reinforcement is needed to transfer the tensile forces across the crack.

Vertical stirrups are used almost exclusively in modern concrete construction for shear reinforcement in beams (see Figure 4–15). The terminology *shear reinforcement* comes from the fact that shear is used as a measure of the diagonal tension. (To confuse the issue even further, shear reinforcement is often also referred to as *web reinforcement*.) Vertical stirrups typically are U-shaped #3 or #4 bars. They surround the tensile reinforcing on the bottom and are anchored into the compression zone by a hook at each end. (See the beam section in Figure 4–15.)

The relationship between the design resisting shear, V_R, and the nominal resisting shear is:

$$V_R = \phi V_n \tag{4–10a}$$

where ϕ is the strength reduction factor. As discussed in Chapter 2, the ACI Code uses this factor to account for possible understrength of the materials and construction inaccuracies. The ϕ factor for shear (ACI Code, Section 9.3.2.3) is:

$$\phi = 0.75 \tag{4–10b}$$

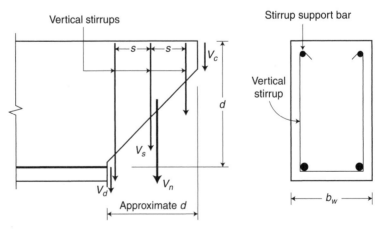

FIGURE 4–15   Model of shear resistance according to the ACI Code.

which is smaller than the value for bending ($\phi = 0.90$). The main reason for the difference is that reinforced concrete beams are less ductile in shear than in bending.

The design principle is to supply a greater strength than the required strength. Expressed mathematically (Equation 11–1 of the ACI Code):

$$V_R \geq V_u \qquad (4\text{–}11)$$

The left side of the equation, V_R, is the design shear *strength* of the section under investigation. The right side, V_u, is the *demand,* or the shear acting on the section.

To describe the V_n, or the nominal strength of the section against shear failure, Figure 4–15 shows a simple model that has been adapted in lieu of the very complex interaction of the concrete and the various reinforcements. Other theoretical models try to provide a practical solution to the problem. The empirical model adapted by the ACI Code and discussed here is an easy-to-follow representation of the different components of the available strength.

1. The first component of the model is the shear strength of the concrete section, V_c. The compression zone provides resistance due to friction and aggregate interlock. To make calculation easy, the ACI Code relates the value of V_c to an average shear over the whole working section of the concrete beam (ACI Code, Equation 11–3):

$$V_c = 2\sqrt{f_c'}\, b_w d \qquad (4\text{–}12)$$

In this expression f_c' must be entered in psi, and b_w and d are in inches. The resulting V_c is in pounds. Tables A4–1a, A4–1b, and A4–1c include V_c for different sizes of beam (b_w and h) and compressive strengths of concrete (f_c').

2. The second component of the model is the sum of the tensions developed by the vertical component of the diagonal tensions in the stirrup legs (V_s). All stirrup legs that cross a potential crack (Figure 4–15 shows three stirrups with two legs each) will provide this strength. Thus

$$V_s = n(A_v f_{yt}) \qquad (4\text{–}13)$$

where n is the number of stirrups crossing the potential 45° crack. Because the stirrups are placed at a spacing of s, and ns is approximately equal to d, we can calculate the strength, V_s, (ACI Equation 11–15) as:

$$V_s = \frac{A_v f_{yt} d}{s} \qquad (4\text{–}14)$$

where
- A_v = sum of the cross sectional areas of the stirrup legs in square inches
- s = spacing of the stirrups in inches
- b_w = width of the web of concrete beams in inches
- d = distance from the extreme compression fiber to the centroid of the tensile reinforcement in inches
- f_{yt} = the specified yield strength of the transverse reinforcing steel (i.e., stirrups) in ksi or psi (consistent units must be used)

Tables A4–2a and A4–2b show V_s for #3 and #4 stirrups with different s and h, and $f_{yt} = 60,000$ psi.

3. The third component of the model that provides strength against shear is the so-called "dowel action." This results from the vertical shear resistance of the horizontal reinforcing after the sides of the crack are vertically separated. The contribution from this source is neglected by the ACI Code.

The final design equation (based on ACI Code Equations 11–1 and 11–2) is:

$$V_R = \phi V_n = \phi(V_c + V_s) \geq V_u \qquad (4\text{--}15)$$

From what was mentioned above, if $V_u \leq \phi V_c$ (ϕV_c is the shear that concrete can carry), theoretically we do not need any stirrups. The ACI Code (Sections 11.5.6.1 and 11.5.6.3), however, requires a minimum amount of stirrups where $V_u > \phi V_c/2$. This minimum amount of stirrups (ACI Code Equation 11–13) is:

$$A_{v,min} = \max\left\{ 0.75\sqrt{f_c'}\ \frac{b_w s}{f_{yt}},\ \frac{50 b_w s}{f_{yt}} \right\} \qquad (4\text{--}16)$$

The use of minimum amount of stirrups required by the code prevents sudden shear failures when inclined cracking occurs. This rule has a few exceptions, such as for slabs, footings, and concrete joist construction; in these cases there is a possibility of load sharing between the weak and strong areas, so no shear reinforcements are needed when $V_u \leq \phi V_c$ (more on these topics in Chapters 6 and 7). In any case, the ACI Code requires no shear reinforcements where $V_u \leq \phi V_c/2$.

In general, therefore, a reinforced concrete beam has three possibilities (or zones) when designing for stirrups. Figure 4–16 shows these zones. Note that design of beams for shear involves finding the spacing of stirrups because almost all construction uses the same size stirrups for the entire beam. The spacing is changed based on the level of shear force the stirrups have to resist.

Zone 1 ($V_u \leq \phi V_c/2$)

No stirrups are needed where $V_u \leq \phi V_c/2$. For a symmetrically loaded beam this condition usually occurs in a region close to the center of the beam, as shown in Figure 4–16. Although the ACI Code does not require any stirrups in this zone, a few stirrups are used to hold the main reinforcements in place.

Zone 2 ($\phi V_c/2 < V_u \leq \phi V_c$)

This is a zone where theoretically no stirrups would be needed. The ACI Code, however, requires a minimum area of stirrups. Because our objective here is to determine the stirrup spacings, we rewrite Equation 4–16 to obtain the maximum allowable spacing (s_1) in

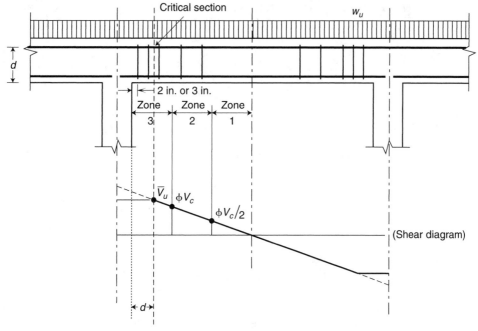

FIGURE 4–16   Different zones for stirrup spacing.

terms of the selected stirrup size, the width of the beam's web, and the materials used in the beam.

$$s_1 = \min\left\{\frac{A_v f_{yt}}{0.75\sqrt{f_c'}\,b_w}, \frac{A_v f_{yt}}{50 b_w}\right\} \tag{4-17}$$

The units used are as follows for each variable: A_v (in^2), f_{yt} (psi), f_c' (psi), b_w (in.), and s_1 (in.).

The ACI Code (Section 11.5.5.1) places a further restriction on the maximum allowable spacing in this zone ($s_{\max}$):

$$s_{\max} = \min\left\{s_1, \frac{d}{2}, 24 \text{ in.}\right\} \tag{4-18}$$

This requirement ensures that each 45° crack is intercepted by at least one stirrup (Figure 4–15). Therefore, Equation 4–18 determines the stirrup spacing in Zone 2 of the beam.

Zone 3 ($\phi V_c < V_u$)

This is the only part of the beam for which we need to design the stirrups (i.e., this zone may require closer stirrup spacing than the allowable maximum found in Equation 4–18). To determine the spacing of the stirrups in this zone, we need to calculate how much shear the stirrups must carry (V_s). This is accomplished by rearranging Equation 4–15:

$$V_u = \phi(V_c + V_s)$$

$$V_s = \frac{V_u}{\phi} - V_c \qquad (4\text{–}19)$$

The first term on the right side of this equation $\left(\dfrac{V_u}{\phi}\right)$ is the total factored shear on the beam at the section under consideration magnified by the strength reduction factor in the denominator and the second term (V_c) is the shear to be carried by the concrete. The remainder is to be resisted by the stirrups (V_s). Rearranging Equation 4–14 to find the stirrup spacing:

$$s = \frac{A_v f_{yt} d}{V_s} \qquad (4\text{–}20)$$

Usually A_v, f_{yt}, and d are the same for the entire beam. Therefore, the stirrup spacing (s) changes with the shear to be resisted by the stirrups (V_s). Clearly the stirrup spacing is smaller near the supports as V_s is larger. The calculated stirrup spacing increases continuously as we move toward the midspan and the shear diminishes. Although theoretically this is true, in reality only a few (two or three) different spacings are used. While we could save a few stirrups by continuously varying the stirrup spacings, constantly changing the spacing complicates construction, as locating and placing the stirrups become difficult. As Figure 4–16 shows, the first stirrup is usually placed 2 in. or 3 in. from the face of the support.

Because stirrups cannot resist shear unless they cross an inclined crack, the ACI Code (Sections 11.5.5.1 and 11.5.5.3) limits the maximum stirrup spacing. The maximum allowable stirrup spacing is:

$$\text{if } V_s \le 4\sqrt{f_c'}b_w d \text{ (or } V_s \le 2V_c) \longrightarrow s_{\max} = \min\left\{s_1, \frac{d}{2}, 24 \text{ in.}\right\}$$
$$(4\text{–}21)$$
$$\text{if } V_s > 4\sqrt{f_c'}b_w d \text{ (or } 2V_c < V_s \le 4V_c) \longrightarrow s_{\max} = \min\left\{s_1, \frac{d}{4}, 12 \text{ in.}\right\}$$

The first part of Equation 4–21 limits the stirrup spacing such that each potential 45° crack will be intercepted by at least one stirrup (Figure 4–15). Where the shears are so large that the stirrups are required to carry $V_s > 2V_c$, the maximum allowable spacing is limited to that shown in the second part of Equation 4–21. This is necessary to provide better control of the width of the potential inclined cracks.

EXAMPLE 4–1

Determine the total resisting shear, V_R, for the beam shown in Figure 4–17. The shear reinforcements provided are #3 stirrups @ 8 in. on center. Assume $f_c' = 4{,}000$ psi and $f_{yt} = 60{,}000$ psi.

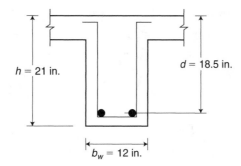

FIGURE 4–17   Section in Example 4–1.

Solution

From Equation 4–12

$$V_c = 2 \times \sqrt{4,000} \times 12 \times 18.5 = 28,081 \text{ lb} = 28.1 \text{ kip}$$

From Equation 4–14

$$V_s = \frac{(2 \times 0.11) \times 60,000 \times 18.5}{8} = 30,524 \text{ lb} = 30.5 \text{ kip}$$

From Equation 4–15

$$V_R = \phi V_n = 0.75(28,081 + 30,524) = 43,954 \text{ lb} = 43.95 \text{ kip}$$

Solution Using Tables

From Table A4–1b (interpolating for $h = 21$ in.)

$$V_c = 28.1 \text{ kip}$$

From Table A4–2a (interpolating for $h = 21$ in.)

$$V_s = 30.5 \text{ kip}$$

Thus

$$V_R = \phi V_n = 0.75(28.1 + 30.5) = 43.95 \text{ kip}$$

EXAMPLE 4–2

A reinforced concrete beam section with a width, $b_w = 15$ in., and, a total depth, $h = 24$ in., is subjected to a shear force, $V_u = 60$ kip. Find the spacing of #3 stirrups at the section. $f'_c = 3,000$ psi, and $f_{yt} = 60,000$ psi.

Solution

$$d_{est} = h - 2.5 = 24 - 2.5 = 21.5 \text{ in.}$$

From Equation 4–12

$$V_c = \frac{2 \times \sqrt{3{,}000} \times 15 \times 21.5}{1{,}000} = 35.3 \text{ kip}$$

From Equation 4–19

$$V_s = \frac{V_u}{\phi} - V_c = \frac{60}{0.75} - 35.3 = 44.7 \text{ kip}$$

From Equation 4–20

$$s = \frac{(2 \times 0.11) \times 60 \times 21.5}{44.7} = 6.35 \text{ in.}$$

Rounding down to the nearest ½ in., we use stirrups at 6 in. centers at this section.

Solution Using Tables

From Table A4–1a (interpolating for $b_w = 15$ in.)

$$V_c = 35.3 \text{ kip}$$

Then

$$V_s = \frac{60}{0.75} - 35.3 = 44.7 \text{ kip}$$

Entering into Table A4–2a with $h = 24$ in., #3 stirrups with two legs at 6 in. spacing will provide $V_s = 47.3$ kip, which is slightly more than we need.

4.4   ADDITIONAL REQUIREMENTS FOR THE DESIGN OF SHEAR REINFORCING

The following are additional ACI requirements:

a. The value of $\sqrt{f_c'}$ must be less than 100 psi (ACI Code, Section 11.1.2) unless minimum web reinforcement is used in the flexural member. This limitation has been placed because of the limited amount of experience with the use of concrete strength in excess of $f_c' > 10{,}000$ psi.

b. The design yield strength of the shear reinforcing bars is limited to 60,000 psi (ACI Code, Section 11.5.2). This requirement limits the crack width. The limit is 80,000 psi when welded wire reinforcement is used as shear reinforcing.

c. The value of V_s is limited to $8\sqrt{f_c'}b_w d$ (ACI Code, Section 11.5.7.9). This provision effectively limits the maximum value of V_n to $10\sqrt{f_c'}b_w d$. Stating it differently, V_s may not exceed $4V_c(V_s \leq 4V_c)$. Thus, if V_u is too large to satisfy this requirement,

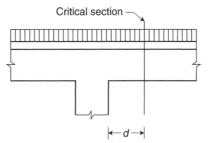

FIGURE 4-18 Location of the critical section.

the concrete section must be enlarged by making the beam wider or deeper. Note that $V_{s,\max}$ for a beam usually is at its critical section ($\overline{V}_s$).

d. The *critical section* for stirrup design (within zone 3) may be taken at distance d from the face of the support in beams and joists, when the loads are applied onto the top of the beam. In the zone between the *face of the support* and the *critical section* the support reaction introduces vertical compressions into the end zone of the member, which significantly increases the shear strength in that region. Sections located between the face of the support and the critical section may be designed for V_u at the critical section ($\overline{V}_u$). This means that for design purposes the shear force from the critical section to the face of the support is taken as $\overline{V}_u$, as shown in Figure 4–18. Note that we use the "bar" here to indicate the value at the critical section (i.e., $\overline{V}_u$, $\overline{V}_s$, and $\overline{s}$ represent the total shear at the critical section, shear to be resisted by the stirrups at the critical section, and the required stirrup spacing at the critical section, respectively).

e. Limit the stirrups' size to #3, #4, or #5, as these bar sizes are easier to bend. (This is only a recommendation, not an ACI requirement.) Also, the bend radii at the corners of the stirrups require a minimum beam width for each size of stirrup, as shown below.

Sometimes, to avoid very small (less than 3 in.) required spacing, the designer may employ four, six, or more legs for stirrups. This increases A_v in Equation 4–20, and consequently the calculated spacing, s. The use of multiple legs is also recommended in wide beams, as shown in Figure 4–19.

TABLE 4-1	Recommended Minimum Beam Width to Accommodate Different Stirrup Sizes
Stirrup Size	**Minimum Beam Width (b_w)**
#3	10 in.
#4	12 in.
#5	14 in.

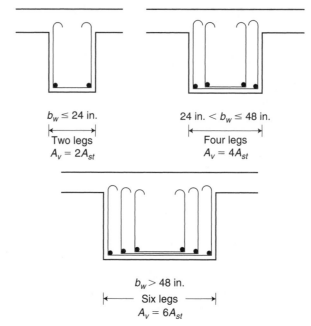

$b_w \leq 24$ in.

Two legs
$A_v = 2A_{st}$

24 in. $< b_w \leq 48$ in.

Four legs
$A_v = 4A_{st}$

$b_w > 48$ in.

Six legs
$A_v = 6A_{st}$

FIGURE 4–19 Recommended number of legs of stirrups based on beam width.
Note: $A_{st} =$ area of each leg of the stirrups.

4.5 STIRRUP DESIGN PROCEDURE

The steps for designing stirrups are summarized in Figure 4–20 and given below:

Step 1   Determine the distribution of shear along the beam to calculate the stirrup spacing. (You can do this by drawing the shear force diagram.) Use either the beam clear span or the center-to-center span. If you use the center-to-center span, include half the support width when locating the critical section.

Step 2   Determine the shear at the critical section ($\overline{V}_u$). As mentioned in Section 4.4, the critical section is at a distance d from the face of the support. $\overline{V}_u$ is the largest shear acting on the beam that needs to be considered.

Step 3   Calculate the shear capacity of concrete (V_c):

$$V_c = 2\sqrt{f_c'}\, b_w d$$

Step 4   Calculate the shear to be carried by the stirrups at the critical section ($\overline{V}_s$). If $\overline{V}_s > 4V_c$, we have to increase the beam section size, and repeat the procedure. Otherwise, the beam size is ok.

Step 5   If $\overline{V}_u \leq \dfrac{\phi V_c}{2}$, the beam does not require any stirrups to resist the shear force.

If $\overline{V}_u \geq \dfrac{\phi V_c}{2}$, we need at least a minimum area of stirrups.

Step 6   Determine the locations of $\phi V_c /2$ on the V_u diagram to identify the location where no stirrups are needed.

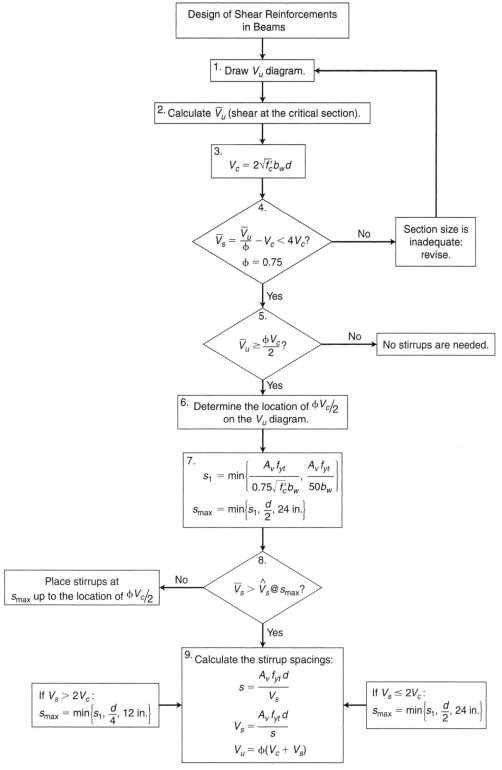

FIGURE 4–20 Flowchart for stirrup design.

Step 7   Determine the maximum spacing of the stirrups:

$$s_1 = \min\left\{ \frac{A_v f_{yt}}{0.75\sqrt{f_c'}\,b_w}, \frac{A_v f_{yt}}{50 b_w} \right\}$$

$$s_{\max} = \min\left\{ s_1, \frac{d}{2}, 24 \text{ in.} \right\}$$

Step 8   Calculate $\hat{V}_s = \dfrac{A_v f_{yt}\,d}{s_{\max}}$. If $\overline{V}_s \leq \hat{V}_s$, only minimum stirrups (at $s_{\max}$ spacing) are required. Place them up to the point of $\phi V_c/2$. If $\overline{V}_s > \hat{V}_s$, go to step 9.

Step 9   Calculate the stirrup spacing:

$$s = \frac{A_v f_{yt}\,d}{V_s}$$

Check for the maximum allowable spacing:

$$s_{\max} = \min\left\{ s_1, \frac{d}{2}, 24 \text{ in.} \right\} \qquad \text{if } V_s \leq 2V_c$$

$$s_{\max} = \min\left\{ s_1, \frac{d}{4}, 12 \text{ in.} \right\} \qquad \text{if } V_s > 2V_c$$

This calculation may need to be repeated at different locations. Use one or two different spacings, at most, in this zone.

EXAMPLE 4–3

Design the stirrups for the beam shown in Figure 4–21. The columns are 15 in. × 15 in. The load includes the beam's self-weight. Use $f_c' = 3,000$ psi and $f_{yt} = 60,000$ psi.

Solution

Step 1   Draw the V_u diagram:

$$R = \frac{w_u \ell}{2} = \frac{5(30)}{2} = 75 \text{ kip}$$

See the shear force diagram in Figure 4–21.

Step 2   Calculate $\overline{V}_u$:

The estimated effective depth (d) is:

$$d = h - 2.5 = 24 - 2.5 = 21.5 \text{ in.}$$

The critical section is at the distance d from the column face. Thus, from the column centerline this distance is:

$$x = \frac{15}{2} + 21.5 = 29 \text{ in.} = 2.42 \text{ ft}$$

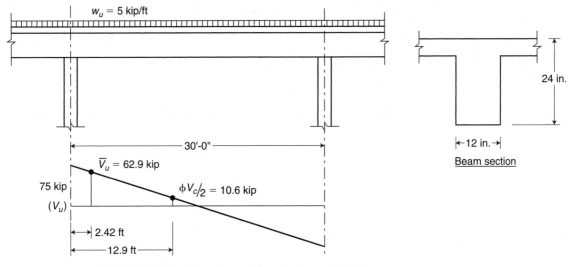

FIGURE 4–21   Elevation of beam in Example 4–3.

Because the factored shear decreases from the support to the midspan (Figure 4–21) at a rate of 5 kip/ft (the slope of the shear is the load), the shear at the critical section $(\overline{V}_u)$ is:

$$\overline{V}_u = 75 - 5(2.42) = 62.9 \text{ kip}$$

Step 3

$$V_c = 2\sqrt{f_c'}b_w d = 2\sqrt{3,000}(12)(21.5)/1,000 = 28.3 \text{ kip}$$

Step 4

$$\phi = 0.75$$

$$\overline{V}_s = \frac{\overline{V}_u}{\phi} - V_c = \frac{62.9}{0.75} - 28.3 = 55.6 \text{ kip}$$

$$4V_c = 4(28.3) = 113.2 \text{ kip} > 55.6 \text{ kip} \quad \therefore \text{ ok}$$

The beam size is adequate.

Step 5   Determine whether stirrups are required:

$$\frac{\phi V_c}{2} = \frac{0.75(28.3)}{2} = 10.6 \text{ kip} < \overline{V}_u = 62.9 \text{ kip} \quad \therefore \text{ Stirrups are needed!}$$

Step 6   Determine the location of $\phi V_c/2$ on the V_u diagram.
Write the equation for the shear force diagram:

$$V_u = 75 - 5x$$

Note that x in this equation is measured from the centerline of the column.

$$V_u = \frac{\phi V_c}{2}$$

$$75 - 5x = 10.6$$

$$x = 12.9 \text{ ft}$$

Therefore, from 12.9 ft to the center of the beam (i.e., $x = 15$ ft, no stirrups are needed because $V_u \le \dfrac{\phi V_c}{2}$ in this zone.

Step 7   Determine the maximum allowable stirrup spacing:
Using #3 stirrups with two legs ($b_w = 12$ in., see Figure 4–19) $\longrightarrow$
$A_v = 2(0.11) = 0.22$ in.^{2}

$$s_1 = \min\left\{\frac{A_v f_{yt}}{0.75\sqrt{f_c'}\,b_w}, \frac{A_v f_{yt}}{50 b_w}\right\}$$

$$s_1 = \min\left\{\frac{0.22(60{,}000)}{0.75\sqrt{3{,}000}(12)}, \frac{0.22(60{,}000)}{50(12)}\right\}$$

$$s_1 = \min\{26.8 \text{ in.}, 22 \text{ in.}\} = 22 \text{ in.}$$

$$s_{max} = \min\left\{s_1, \frac{d}{2}, 24 \text{ in.}\right\}$$

$$s_{max} = \min\left\{22 \text{ in.}, \frac{21.5 \text{ in.}}{2}, 24 \text{ in.}\right\}$$

$$s_{max} = 10.5 \text{ in. (rounded down to closest } \tfrac{1}{2} \text{ in.)}$$

Step 8   Determine whether the beam needs more than minimum stirrups. With #3 stirrups at $s_{max} = 10.5$ in.

$$\hat{V}_s = \frac{A_v f_{yt} d}{s_{max}} = \frac{(0.22)(60)(21.5)}{10.5} = 27.0 \text{ kip} < \overline{V}_s = 55.6 \text{ kip}$$

$\therefore$ Needs more than the minimum stirrups!

Step 9   Calculate the spacing of the stirrups:
The smallest required spacing of stirrups is at the critical section $(\overline{s})$, where $\overline{V}_s = 55.6$ kip (step 4):

$$\overline{s} = \frac{A_v f_{yt} d}{\overline{V}_s} = \frac{0.22(60)(21.5)}{55.6} = 5.1 \text{ in.}$$

We round the spacing down to the closest $\tfrac{1}{2}$ in., so $\overline{s} = 5$ in.
    Because $\overline{V}_s = 55.6$ kip $< 2V_c = 2(28.3) = 56.6$ kip, the maximum allowable stirrup spacing is:

$$s_{max} = \min\left\{s_1, \frac{d}{2}, 24 \text{ in.}\right\}$$

$$s_{max} = \min\left\{22, \frac{21.5}{2}, 24\right\} = 10.5 \text{ in.}$$

At this point we have calculated two stirrup spacings ($\bar{s} = 5$ in., and $s_{max} = 10.5$ in.). These spacings are acceptable, so we may just use them. But if we want to save a few stirrups, we could select another spacing between these two values (say, $s = 8$ in.).

Then we would have three different stirrup spacings, $s = 5$ in., 8 in., and 10.5 in. The question now is, where do these spacings start and where do they end? The first stirrup starts 2 in. from the face of the support, and then the stirrups are placed at $\bar{s} = 5$ in. This spacing ends where the 8 in. spacing starts. Use Equations 4–14 and 4–15 with $s = 8$ in.:

$$V_s = \frac{A_v f_{yt} d}{s} = \frac{0.22(60)(21.5)}{8} = 35.5 \text{ kip}$$

$$V_u = \phi(V_c + V_s) = 0.75(28.3 + 35.5) = 47.9 \text{ kip}$$

The $s = 8$ in. starts (or $s = 5$ in. ends) where $V_u = 47.9$ kip. Determine the location of this point on the shear diagram:

$$75 - 5x = 47.9$$

$$x = 5.42 \text{ ft}$$

from the center line of the column.

Therefore, the number of 5 in. spacings (N) is:

To column face   First spacing

$$N = \frac{5.42(12) - 7.5 - 2}{5} = 11.1$$

We conservatively round this spacing up to $N = 12$ or 12 @ 5 in.

Next we need to determine where $s = 8$ in. ends or the maximum allowable spacing ($s_{max} = 10.5$ in.) begins. To locate this point, use Equations 4–14 and 4–15 with $s = 10.5$ in.:

$$V_s = \frac{A_v f_{yt} d}{s} = \frac{0.22(60)(21.5)}{10.5} = 27 \text{ kip}$$

$$V_u = \phi(V_c + V_s) = 0.75(28.3 + 27) = 41.5 \text{ kip}$$

$$75 - 5x = 41.5$$

$$x = 6.7 \text{ ft}$$

This is the end point of $s = 8$ in.. The number of 8 in. spacings (N) is:

$$N = \frac{6.7(12) - (7.5 + 2 + 12 \times 5)}{8} = 1.4 \quad \therefore 2 @ 8 \text{ in.}$$

Finally, we will determine the portion of the beam in which a stirrup spacing of 10.5 in. can be used. Because 10.5 in. is the maximum spacing and no

Column center line

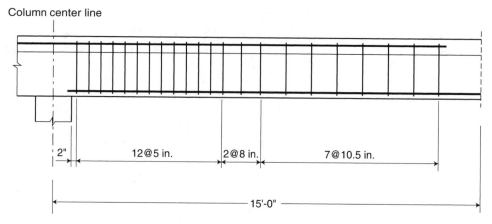

| 2" | 12@5 in. | 2@8 in. | 7@10.5 in. |

15'-0"

FIGURE 4–22   Resulting stirrup layout for Example 4–3.

stirrup is needed after that, the end point is $x = 12.9$ ft (found in step 6 for $V_u = \phi V_c/2$). The number of 10.5 in. spacings is:

$$N = \frac{12.9(12) - (7.5 + 2 + 12 \times 5 + 2 \times 8)}{10.5} = 6.6 \quad \therefore 7 @ 10.5 \text{ in.}$$

Therefore, the stirrup spacing from each end of the beam is 1 @ 2 in. + 12 @ 5 in. + 2 @ 8 in. + 7 @ 10.5 in. Figure 4–22 shows the resulting stirrup layout.

EXAMPLE 4–4

Design stirrups for the floor beam shown in Figure 4–23. The loads include the beam's self-weight. The columns are 15 in. × 15 in. Use $f'_c = 4{,}000$ psi, and $f_{yt} = 60{,}000$ psi. Assume that the minimum cover is 1.5 in.

Solution

Step 1   Draw the V_u diagram (see Figure 4–24):

$$\ell_n = 15 - {}^{15}/_{12} = 13.75 \text{ ft}$$

$$R = \frac{75 \times 2 + 1 \times 15}{2} = 82.5 \text{ kip}$$

Alternatively, we can use the clear span.
The reaction at the face of column is:

$$R = \frac{75 \times 2 + 1 \times 13.75}{2} = 81.88 \text{ kip}$$

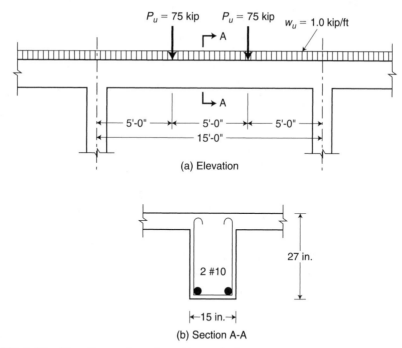

(a) Elevation

(b) Section A-A

FIGURE 4–23   Elevation and section of beam in Example 4–4.

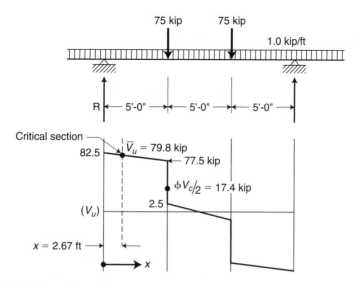

FIGURE 4–24   Shear force diagram for Example 4–4.

Step 2   Calculate $\overline{V}_u$:

$$d = h - (1.5 + \tfrac{3}{8} + 1.27/2)$$
$$= 27 - 2.5 = 24.5 \text{ in.}$$

The location of the critical section from the column center is:

$$x = 15/2 + 24.5 = 32 \text{ in.}/12 = 2.67 \text{ ft}$$
$$\overline{V}_u = 82.5 - 1.0(2.67) = 79.8 \text{ kip}$$

The shear at the critical section using the clear span is:

$$\overline{V}_u = 81.88 - 1.0(24.5/12) = 79.8 \text{ kip}$$

As expected, the shear at the critical section, $\overline{V}_u$, is the same when using the center-to-center span or clear span.

Step 3   Calculate the concrete resisting shear, V_c:

$$V_c = 2\sqrt{f_c'}b_w d = 2\sqrt{4{,}000}(15)(24.5)/1{,}000 = 46.5 \text{ kip}$$

Step 4   Calculate the shear to be resisted by the stirrups (determine whether the beam size is adequate):

$$\overline{V}_s = \frac{\overline{V}_u}{\phi} - V_c = \frac{79.8}{0.75} - 46.5 = 59.9 \text{ kip}$$

$$4V_c = 4(46.5) = 186 \text{ kip} > 59.9 \text{ kip} \quad \therefore \text{ ok}$$

The beam size is adequate.

Step 5   Determine whether stirrups are required.

$$\frac{\phi V_c}{2} = \frac{0.75(46.5)}{2} = 17.4 \text{ kip} < \overline{V}_u = 79.8 \text{ kip} \quad \therefore \text{ Stirrups are required.}$$

Step 6   Locate $\phi V_c/2$ (17.4 kip) on the V_u diagram. This point lies on the vertical part of the shear force diagram (Figure 4–24). Determine the stirrup spacing for the first 5 feet of beam from each end; the 5 feet at the center do not require any stirrups for shear.

Step 7   Calculate the maximum stirrup spacing:
Use #3 stirrups with two legs ($b_w = 15$ in., see Figure 4–19), $A_v = 2 \times 0.11 = 0.22 \text{ in.}^2$

$$s_1 = \min\left\{\frac{A_v f_{yt}}{0.75\sqrt{f_c'}b_w}, \frac{A_v f_{yt}}{50 b_w}\right\}$$

$$s_1 = \min\left\{\frac{0.22(60{,}000)}{0.75\sqrt{4{,}000}(15)}, \frac{0.22(60{,}000)}{50(15)}\right\}$$

$$s_1 = \min\{18.6, 17.6\} = 17.6 \text{ in.}$$

The maximum allowable spacing is:

$$s_{max} = \min\left\{ s_1, \frac{d}{2}, 24 \text{ in.} \right\}$$

$$s_{max} = \min\left\{ 17.6 \text{ in.}, \frac{24.5 \text{ in.}}{2}, 24 \text{ in.} \right\}$$

$$s_{max} = 12 \text{ in.}$$

Step 8   Determine whether more than the minimum amount of stirrups (or maximum stirrup spacing) are required.

$$\hat{V}_s = \frac{A_v f_y d}{s_{max}}$$

$$= \frac{0.22 \times 60 \times 24.5}{12}$$

$$= 27.0 \text{ kip} < \overline{V}_s = 59.9 \text{ kip}$$

∴ Needs more than the minimum amount of stirrups. (or stirrups to be closer than the maximum allowable spacing).

Step 9   Calculate the stirrup spacing.
The stirrup spacing at the critical section ($\bar{s}$) is:

$$\bar{s} = \frac{A_v f_{yt} d}{\overline{V}_s} = \frac{0.22(60)(24.5)}{59.9} = 5.4 \text{ in.} \quad \therefore \bar{s} = 5 \text{ in.}$$

Find the stirrup spacing at the end of this zone, where $V_u = 77.5$ kip:

$$V_s = \frac{V_u}{\phi} - V_c = \frac{77.5}{0.75} - 46.5 = 56.8 \text{ kip}$$

$$s = \frac{A_v f_{yt} d}{\overline{V}_s} = \frac{0.22(60)(24.5)}{56.8} = 5.7 \text{ in.}$$

Because $\overline{V}_s = 59.9$ kip $< 2V_c = 2(46.5) = 93$ kip the maximum stirrup spacing is the same value found in step 7.

$$s_{max} = 12.0 \text{ in.} > \bar{s} = 5 \text{ in.} \quad \therefore \text{ ok}$$

Because the portion of the beam to have $\bar{s} = 5$ in. is a short distance (5 feet), we use 5 in. spacing for the entire 5 feet. The number of spacings (N) is:

$$N = \frac{5(12) - (7.5 + 2)}{5} = 10.1 \quad \therefore \text{ Use 11 @ 5 in.}$$

Figure 4–25 shows the stirrup layout. As a practical matter, we usually place a few stirrups where stirrups are not required to hold the main reinforcements in place.

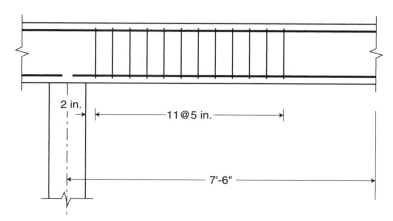

FIGURE 4–25 Stirrup layout for Example 4–4.

4.6 ADDITIONAL FORMULAS TO CALCULATE THE SHEAR STRENGTH OF A BEAM SECTION

Beams Subject to Flexure and Shear Only

Equation 4–12 ($V_c = 2\sqrt{f_c'}b_w d$) is the simplest expression that the ACI Code permits in calculating V_c. This equation neglects the influence of the longitudinal reinforcing, $\rho_w = \dfrac{A_s}{b_w d}$ and the ratio $\dfrac{V_u d}{M_u}$ both of which affect the shear strength. If the designer wishes to take the contribution of these parameters into account as well, then the following equation (ACI Code, Equation 11–5) may be used (ACI Code, Section 11.3.2.1):

$$V_c = \left(1.9\sqrt{f_c'} + 2{,}500\rho_w\frac{V_u d}{M_u}\right)b_w d \le 3.5\sqrt{f_c'}b_w d \tag{4–22}$$

where

$$\frac{V_u d}{M_u} \le 1.0$$

Members Subject to Axial Compression

The presence of significant axial compression (in addition to flexure and shear) increases the shear strength of a section. This is because the compressive loads can prevent cracks from developing. The ACI Code provides the following equation (ACI Code, Equation 11–4) to account for the contribution of axial compression:

$$V_c = 2\left(1 + \frac{N_u}{2{,}000\,A_g}\right)\sqrt{f_c'}b_w d \tag{4–23}$$

where N_u is the axial compression calculated from factored loads and A_g is the gross cross-sectional area of the concrete section. In the formula N_u/A_g must be expressed in psi.

Members Subject to Significant Axial Tension

The presence of significant axial tension (in addition to flexure and shear) decreases the shear strength of the section. The ACI Code mandates the use of the following equation (ACI Code, Equation 11–8) to account for the presence of axial tension:

$$V_c = 2\left(1 + \frac{N_u}{500A_g}\right)\sqrt{f_c'}\, b_w d \qquad (4\text{--}24)$$

where N_u is negative. N_u/A_g must be expressed in psi.

EXAMPLE 4–5

Calculate the nominal shear capacity, V_c, of the section shown below for the following cases ($f_c' = 3{,}000$ psi):

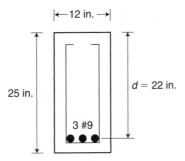

a. Without any axial load or consideration of flexure.
b. Considering the effects of flexure where $V_u = 20$ kip and $M_u = 100$ kip-ft.
c. The section is subjected to an axial compressive force, $N_u = 100$ kip.
d. The section is subjected to an axial tensile force, $N_u = -100$ kip.

Solution

a. The shear capacity of the concrete section according to Equation 4–12 is:

$$V_c = 2\sqrt{f_c'}\, b_w d = 2\sqrt{3{,}000}(12)(22)/1{,}000 \quad \therefore\ V_c = 28.9\ \text{kip}$$

b. Using Equation 4–22:

$$V_c = \left(1.9\sqrt{f_c'} + 2{,}500\ \rho_w \frac{V_u d}{M_u}\right) b_w d \le 3.5\sqrt{f_c'}\,b_w d$$

$$\frac{V_u d}{M_u} \le 1.0$$

$$\rho_w = \frac{A_s}{b_w d} = \frac{3.0}{(12)(22)} = 0.0114$$

$$\frac{V_u d}{M_u} = \frac{20(22)}{100(12)} = 0.37 < 1.0 \quad \therefore \text{ ok}$$

$$V_c = [1.9\sqrt{3{,}000} + 2{,}500(0.0114)(0.37)](12)(22)/1{,}000$$
$$\le 3.5\sqrt{3{,}000}(12)(22)/1{,}000$$

$$V_c = 30.3\ \text{kip} \le 50.6\ \text{kip}$$

$$\therefore V_c = 30.3\ \text{kip}$$

c. Using Equation 4–23:

$$V_c = 2\left(1 + \frac{N_u}{2{,}000\ A_g}\right)\sqrt{f_c'}\,b_w d$$

$$V_c = 2\left(1 + \frac{100(1{,}000)}{2{,}000(12)(25)}\right)\sqrt{3{,}000}(12)(22)/1{,}000$$

$$\therefore V_c = 33.7\ \text{kip}$$

d. Using Equation 4–24:

$$V_c = 2\left(1 + \frac{N_u}{500\ A_g}\right)\sqrt{f_c'}\,b_w d$$

$$V_c = 2\left(1 + \frac{(-100)(1{,}000)}{(500)(12)(25)}\right)\sqrt{3{,}000}(12)(22)/1{,}000$$

$$\therefore V_c = 9.6\ \text{kip}$$

4.7   CORBELS AND BRACKETS

The ACI Code has special provisions for brackets and corbels. Figure 4–26 shows a typical corbel. These are special elements on the side of a column or at the end of a wall. In Figure 4–26, V_u is the factored vertical load from some building element, which may be a precast or prestressed building girder, or a crane girder. N_{uc} is the factored tension force on the corbel acting simultaneously with V_u. This horizontal tension force results from any restraint against free relative horizontal movement between the bracket and the supported element. Most often N_{uc} comes from frictional restraint that occurs in the presence of volumetric changes in the supported girder. The use of special bearing pads helps to minimize the magnitude of N_{uc}.

The ACI Code's design methodology (Section 11.9) is based on the satisfying of the equilibrium of four forces (assumed to be concurrent). The method is applicable when the

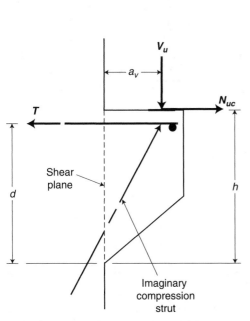

FIGURE 4–26 Corbel (or bracket).

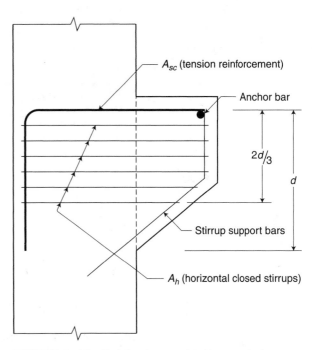

FIGURE 4–27 Reinforcing required in a corbel.

following conditions are satisfied: (1) $a_v/d < 1.0$, (2) $N_{uc} < V_u$, and (3) the depth of the bracket at the front is not less than $d/2$.

From Figure 4–26 it is clear that the shear plane at the level of the primary tension reinforcement is subject to a moment:

$$M_u = V_u a_v + N_{uc}(h - d) \tag{4–25}$$

It is also subject to the tensile force, N_{uc}, and the shear force, V_u.

Figure 4–27 shows the typical reinforcement of a corbel. The required amount of primary reinforcement, A_{sc}, is determined from two parts. The first part, A_f, resists the moment in Equation 4–25. Its design follows the procedure of the flexural design of rectangular sections. The second part, A_n, resists the tensile force, N_{uc}.

Hence:

$$A_{sc} = A_f + A_n \tag{4–26}$$

where

$$A_n = \frac{N_{uc}}{\phi f_y}, \; \phi = 0.90 \tag{4–27}$$

The design of A_h in the form of closed stirrups is based on the shear-friction concept. If a crack forms at a shear plane, reinforcing is needed to prevent a relative displacement (slippage) between the surfaces. This type of reinforcement is shown in Figure 4–28. The

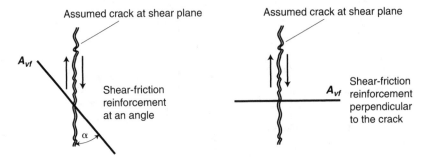

FIGURE 4–28   Reinforcing at an assumed crack.

reinforcement ties together the two halves and ensures that the friction resistance parallel to the crack is maintained.

When the shear-friction reinforcing is perpendicular to the shear plane, as is the case for the corbel shown in Figure 4–27, the shear strength can be calculated as:

$$\phi V_n = \phi(A_{vf}f_y\mu), \ \phi = 0.75 \tag{4–28}$$

where μ is the coefficient of friction defined by the ACI Code for different types of concrete and different pouring sequence scenarios. (Refer to Section 11.7.4.3 of the ACI Code.) For a corbel cast monolithically with the column (always the case), $\mu = 1.4$ for normal-weight concrete.

The ACI Code imposes the following limitations to ensure that the corbel will act in concurrence with the proposed design model:

1. The depth of the corbel at the outside edge of the bearing area shall be not less than $d/2$.
2. The corbel must be deep enough so that V_n for normal-weight concrete may not exceed the smaller of $0.2 f'_c b_w d$ and $800 \ b_w d$.
3. The corbel's minimum primary reinforcement must be the greater of the following:

$$(A_f + A_n) \quad \text{or} \quad (2A_{vf}/3 + A_n)$$

Assuming $V_u = \phi V_n$ and rearranging Equation 4–28, the expression for A_{vf} is obtained as

$$A_{vf} = \frac{V_u}{\phi f_y \mu} = \frac{V_u}{0.75 f_y 1.4} = \frac{V_u}{1.05 f_y} \tag{4–29}$$

for corbels cast monolithically with normal-weight concrete.

A few words must also be said about the anchor bar. The tensile reinforcement (A_{sc}) must develop its strength between the outer edge of the corbel and the face of the column. (For a discussion of development of tensile reinforcing, see Chapter 3.) This length is not adequate in most cases, hence some device is needed to add mechanical anchorage. One such device is a large-diameter bar (#9 or larger) to which the reinforcing bars representing A_{sc} are welded (see Figure 4–27). Another way to provide mechanical anchorage is to weld the bars to an edge angle, as shown in Figure 4–29.

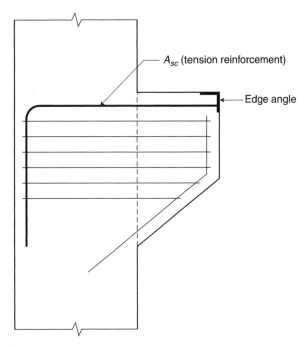

FIGURE 4–29 Tension bars anchored to an edge angle.

EXAMPLE 4-6

Design the required reinforcement for the corbel shown in Figure 4–30.
Assume the following data: $V_u = 62$ kip, $N_{uc} = 6$ kip, $a_v = 8$ in., $b = 16$ in., $f'_c = 5{,}000$ psi, and $f_y = 60{,}000$ psi.

Solution

Step 1 Assume 1.5 in. cover and #6 bars for the primary reinforcement.

$$d = 18 - 1.5 - \frac{0.75}{2} = 16.12 \text{ in.}$$

From Equation 4–25:

$$M_u = 62 \times 8 + 6(18 - 16.12) = 507 \text{ kip-in.}$$

$$R = \frac{M_u}{bd^2} = \frac{507 \times 1{,}000}{16 \times 16.12^2} = 122 \text{ psi}$$

From Table A2–6c

$$\rho = 0.0023$$

$$(A_f)_{rqd} = \rho bd = 0.0023 \times 16 \times 16.12 = 0.59 \text{ in.}^2$$

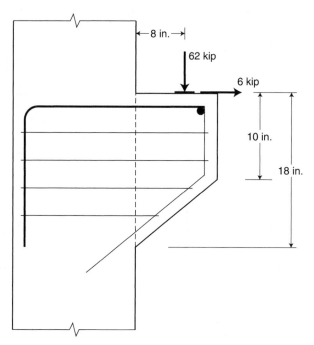

FIGURE 4–30   Sketch of the corbel in Example 4–6.

From Equation 4–27

$$(A_n)_{\text{rqd}} = \frac{N_{uc}}{\phi f_y} = \frac{6}{0.9 \times 60} = 0.11 \text{ in.}^2$$

Hence

$$(A_{sc})_{\text{rqd}} = 0.59 + 0.11 = 0.7 \text{ in.}^2$$

Select 2 #6 bars Table A2–9 $\longrightarrow$ 0.88 in.^{2}

Step 2   Design the required shear reinforcement:
Check for $V_{u,\text{max}}$:

$$\phi(800 b_w d) = 0.75 \times 800 \times 16 \times 16.12/1{,}000$$
$$= 154.8 \text{ kip} > 62 \text{ kip} \quad \therefore \text{ ok}$$
$$\phi(0.2 f'_c b_w d) = 0.75 \times 0.2 \times 5{,}000 \times 16 \times 16.12/1{,}000$$
$$= 193.4 \text{ kip} > 62 \text{ kip} \quad \therefore \text{ ok}$$

From Equation 4–29

$$(A_{vf})_{\text{rqd}} = \frac{V_u}{1.05 f_y} = \frac{62}{1.05 \times 60} = 0.98 \text{ in.}^2$$

Using #4 stirrups $\longrightarrow$ two legs provide $2 \times 0.2 = 0.4$ in.^{2} Thus, the required number of horizontal stirrups is

$$n = 0.98/0.40 = 2.45 \longrightarrow \text{Use a minimum of three stirrups.}$$

Because $(2/3)d = (2/3) \times 16.12 = 10.7$ in., place stirrups at 3.5 in. center-to-center to have the three stirrups within the ACI Code-required distance $2/3d$. Then use additional stirrups at the same spacing to the bottom of the corbel.

Step 3   Check for $A_{sc,min}$:

$$A_{sc,min} = {}^2/_3 A_{vf} + A_n = {}^2/_3 \times 0.98 + 0.11 = 0.76 \text{ in.}^2 < (2 \# 6 \text{ bars})$$

$$= 0.88 \text{ in.}^2 \quad \therefore \text{ ok}$$

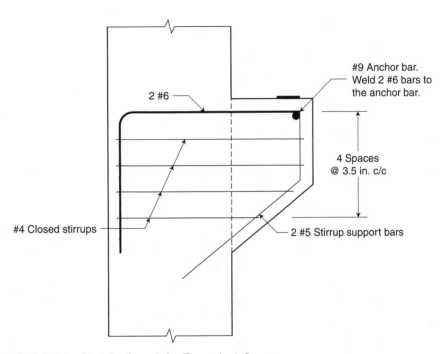

FIGURE 4–31   Sketch of result for Example 4–6.

PROBLEMS

4–1   A rectangular reinforced concrete beam has been designed for moment *only*, without any stirrups for shear. It is, however, subjected to a shear at the critical section, $\overline{V}_u = 10$ kip. The beam width, b, is 12 in., and the effective depth, d, is 26 in. Use $f'_c = 4,000$ psi. Determine whether this beam is adequate.

4–2   A beam is subjected to a uniformly distributed load and has a maximum shear of 60 kip at the face of its supports. The beam clear span is 30'-0 in., $b = 12$ in., $d = 24$ in., $f'_c = 4,000$ psi, and $f_{yt} = 60,000$ psi. What is the shear at the critical section? What is the required spacing for #3 stirrups at the critical section?

4–3 Design stirrups for the beam shown below. The dead load includes the beam's self-weight. Use $f'_c = 4{,}000$ psi, $f_{yt} = 60{,}000$ psi, and 1½ in. cover.

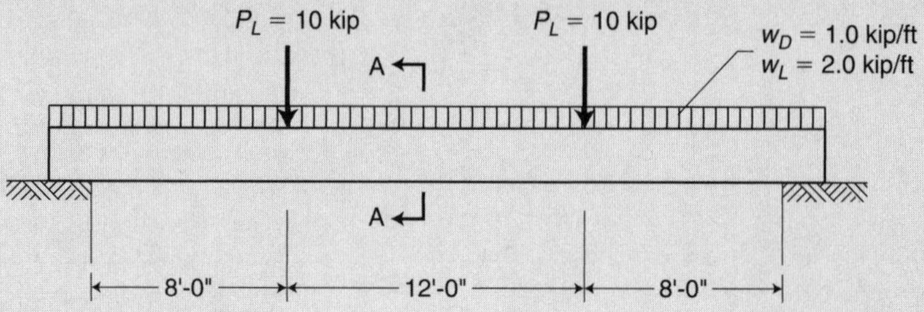

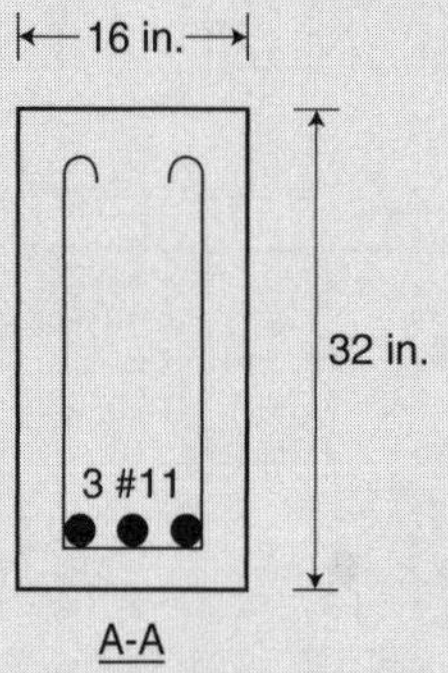

4–4 Rework Problem 4–3 for a beam subjected only to the concentrated load on the left in addition to the distributed loads.

4–5 The shear force at the critical section, $\overline{V}_u$, of a reinforced concrete beam is 60 kip. If the beam has $b_w = 14$ in., $f'_c = 3{,}000$ psi, and $f_{yt} = 60{,}000$ psi, what is the required effective depth, d, so that the minimum spacing of #3 stirrups is 9 in.?

4–6 Design stirrups for the beam shown below. The dead load is 0.70 kip/ft (beam weight not included), and the live load is 1.5 kip/ft. Use $f'_c = 4{,}000$ psi, $f_{yt} = 60{,}000$ psi, and $1\frac{1}{2}$ in. cover. The unit weight of the concrete is 150 lb/ft^3.

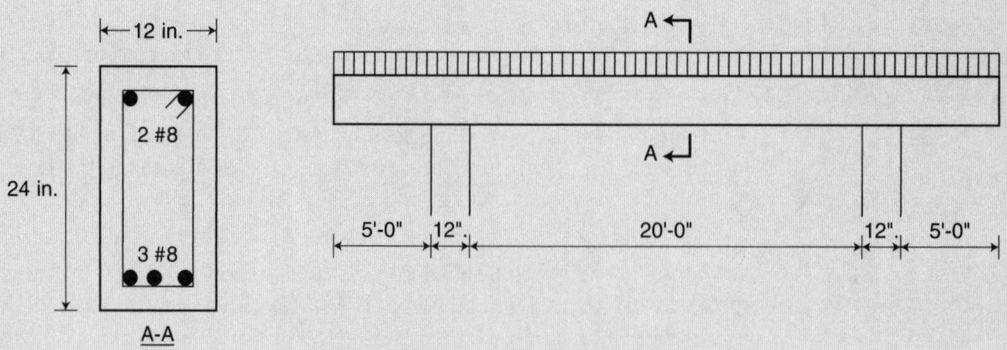

4–7 A 6 in. thick one-way reinforced concrete slab has #6 @ 8 in. main reinforcement. The cover is $^3/_4$ in. and f'_c = 3,000 psi. The unit weight of the concrete is 150 pcf. Answer the following questions:

(a) What is the maximum shear (V_u) the slab can carry?

(b) What is the maximum live load the slab can support based on shear requirements? Assume that the slab is simply-supported and has a clear span of 10 ft-0 in.

4–8 Design stirrups for the interior beam (B-1) shown below. The mechanical/electrical systems weigh 5 psf, the partitions are 20 psf, and the ceiling, carpeting, and so on, weigh 5 psf. The floor live load is 80 psf. Consider live load reduction, if applicable. Use f'_c = 4,000 psi, f_{yt} = 60,000 psi, and $1^1/_2$ in. for cover. Use ACI Code coefficients to determine the beam shear force. The unit weight of the concrete is 150 pcf.

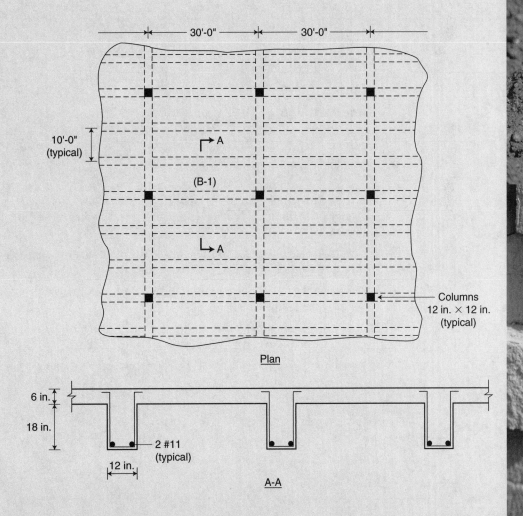

Plan

A-A

4–9 Design stirrups for the beam shown below. The applied loads do not include the beam's self-weight. Use $f_c' = 4,000$ psi, $f_{yt} = 60,000$ psi, and $1\frac{1}{2}$ in. for cover. The unit weight of concrete is 110 pcf.

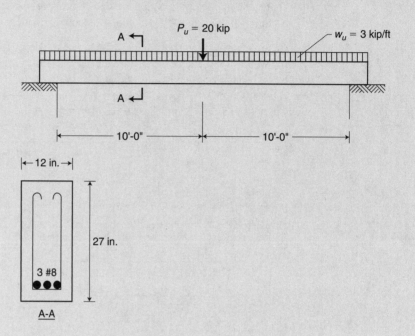

4–10 Rework Problem 4–3, considering the effects of moment on the shear strength of concrete. Use the moment and the shear at the critical section for purposes of simplification.

4–11 Rework Problem 4–3 for a beam subjected to an axial compressive live load of 150 kip. Compare the results with Problem 4–3.

4–12 Rework Problem 4–3 for a beam subjected to an axial tensile live load of 50 kip. Compare the results with Problem 4–3.

4–13 Design the required reinforcement for the corbel shown below. Use $f_c' = 4,000$ psi, $f_y = f_{yt} = 60,000$ psi, $b = 18$ in., and $1\frac{1}{2}$ in. for cover.

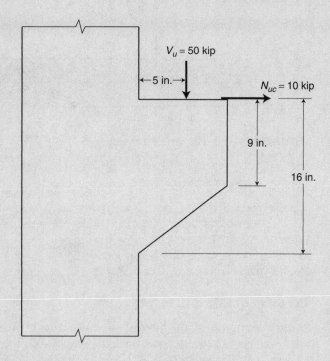

SELF-EXPERIMENTS

In these tests you learn about shear in beams. You will use both Styrofoam and reinforced concrete models. Remember to include in your report all the details of your tests (sizes, time of day you poured, amount of water/cement/aggregate, problems that you encountered, etc.) together with images showing the steps of the tests.

Experiment 1

In this experiment we use Styrofoam models to learn about the vertical and horizontal components of shear in beams.

Test 1: Horizontal Shear

Stack several layers of Styrofoam, one on top of the other, and place them on two supports, as shown in Figure SE 4–1(a). Apply a load, P, at the center of the beams.

Measure how much the beam deflects at the center under the load. Now glue the layers together and repeat the test. Compare the measured deflection for the two cases, and discuss your observation.

FIGURE SE 4–1a   Horizontal shear test.

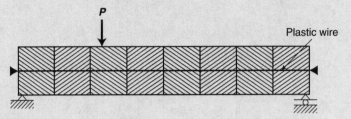

FIGURE SE 4–1b   Vertical shear test.

Test 2: Vertical Shear

Place layers of Styrofoam next to each other and run a plastic wire through them. Anchor the wire at both ends. Place the beam on two supports and apply a load on the beam, as shown in Figure SE 4–1(b).

Observe how the different pieces of Styrofoam move with respect to each other. Record your observations.

Experiment 2

In this experiment we use different sizes of wire as main and shear reinforcements for a reinforced concrete beam, as shown in Figure SE 4–2. The sizes of beam, reinforcement, and span length are your choices. However, they have to be in reasonable proportions for further testing. Cast the reinforced concrete beam. Describe all the different stages of casting the beam and placing the bars. Also, include a drawing and show the sizes and dimensions you used. After the beam cures, apply a load at the center of the beam. Increase the load until you notice the concrete cracks. Record your observations and any problems encountered.

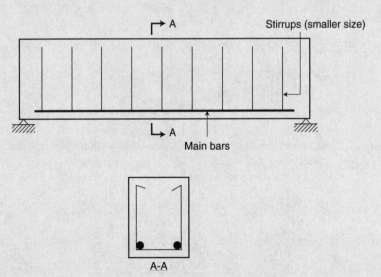

FIGURE SE 4–2   Reinforced concrete beam with main and shear reinforcements.

5

COLUMNS

5.1 INTRODUCTION

Columns are the main supporting elements of a building structure. If we compare a building to a tree, we can think of columns as the trunk of the tree. Any damage to columns may result in catastrophic failure of at least part of the building. Columns mainly carry loads in compression, although they also may be subjected to bending moments transferred by the beams and girders connected to them.

Aside from walls, the compression members of reinforced concrete structures are divided into two groups: pedestals and columns. Section 2.2 of the ACI Code indicates that an upright compression member is considered to be a pedestal if its height is less than three times its least lateral dimension. Pedestals may be designed with plain or reinforced concrete. Figure 5–1a shows a pedestal.

Columns, on the other hand, are compression members whose height is more than 3 times their least lateral dimension. Figure 5–1b shows a typical column. The ACI Code requires all structural columns to be reinforced in order to prevent unexpected brittle failure.

5.2 TYPES OF COLUMNS

Figure 5–2 shows the various classifications of reinforced concrete columns. Columns can be classified by the type of their reinforcement (main and lateral), by their shape, by the type of loads that they will resist, by the type of structural system of which they are part, and by their length. We now will study each class of columns.

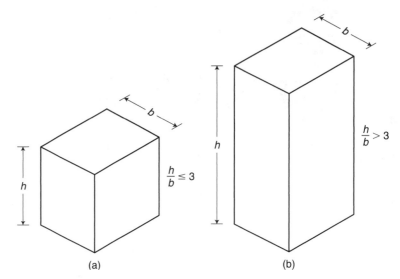

$$\frac{h}{b} \leq 3$$

$$\frac{h}{b} > 3$$

(a)   (b)

FIGURE 5–1   (a) Pedestal, (b) column.

Based on Reinforcement

Three main types of columns fall in this category: tied columns, spiral columns, and composite columns.

Tied Columns Because columns are subjected mainly to axial loads, they are reinforced by longitudinal bars along their length. Because these bars are very slender, they need to be laterally supported to keep them in place during concrete placement, and they need lateral support when subjected to loads. Small-diameter (#3 or #4) bars, referred to as *ties,* are used to fulfill these requirements. Columns that use ties for lateral reinforcement are called *tied columns*. Figure 5–3a shows a square tied column. The ties are wired to the longitudinal bars to make a cage, which then is placed into the form and properly positioned before casting the concrete. The cage of bars and ties keeps the longitudinal bars straight and the ties provides resistance against buckling. Ties generally follow the perimeter of the column's cross section (rectangular in rectangular columns and circular in circular columns).

Tied columns are the most common because their construction costs are lower than those for spiral and composite columns. In fact, over 95% of all columns in concrete buildings located outside earthquake-prone regions are tied columns. The area inside the ties is called *core,* and the area outside them is the *shell* of the column (see Figure 5–3a).

Spirally Reinforced Columns Spirals are used in *spirally reinforced columns* to provide lateral support to the main reinforcements. Spirals are helical-shape wires, which are placed around the main reinforcements as shown in Figure 5–3b. Because most spiral columns are circular in shape, a spirally reinforced core sometimes may be placed inside a square cross section. Spiral reinforcing is a more expensive construction (about twice as much) than using ties. Spiral columns provide larger capacity than do tied columns, but their main advantage is their ductility and toughness when large overloads, such as loads

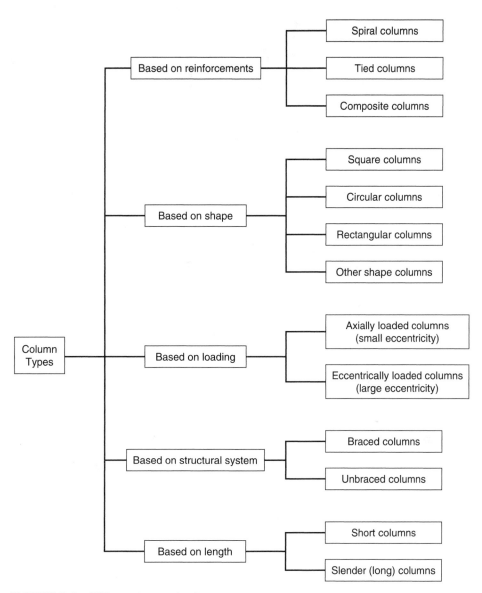

FIGURE 5–2   Different types of columns.

occur in earthquakes, are expected. Similar to tied columns, the area confined by the spirals is the core, and the area outside them is the shell (see Figure 5–3b).

Composite Columns *Composite columns* are constructed by placing a steel shape, such as a pipe or I-section, inside the form and casting concrete around it. These columns may have additional reinforcing bars around the steel shape, as shown in Figure 5–3c. Composite columns are often used in multistory buildings to increase the capacity of the steel sections. The surrounding concrete also provides fireproofing to the steel core.

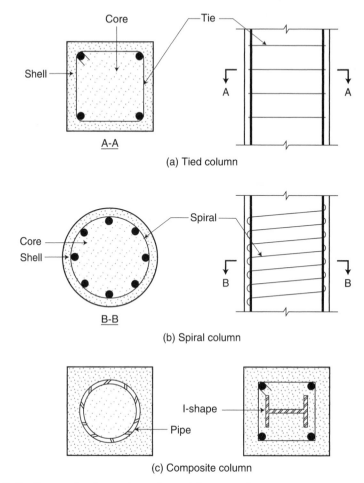

Core

Tie

Shell

A

A

A-A

(a) Tied column

Spiral

Core

Shell

B

B

B-B

(b) Spiral column

I-shape

Pipe

(c) Composite column

FIGURE 5–3   Types of columns based on reinforcements.

Based on Shape

Selecting a column shape is generally an architectural and structural decision and depends on the framing system, costs, reinforcement arrangement, and aesthetics. Square and rectangular shapes are the most common, as they are the simplest to form and construct. Circular columns may be formed by using cardboard or plastic tubes, or by using hinged steel forms, which can be removed easily. Other column shapes besides circular and rectangular are also used. Figure 5–4 shows a few of them.

Based on Loading

Columns primarily carry loads in compression. But they can also be subjected to moments, depending on the building's geometry and loading. Therefore, columns are grouped into two classes: *axially loaded columns,* and *eccentrically loaded columns.* (Sometimes these are referred to as *columns with small eccentricity,* and *columns with large eccentricity,* respectively.)

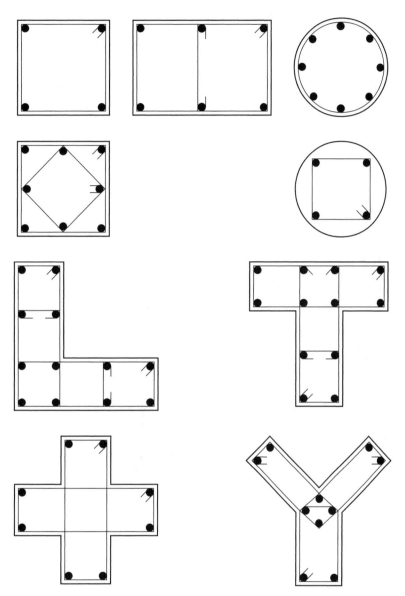

FIGURE 5–4 Different column shapes.

A concentric axial load and a moment can be combined into an eccentric load. The term *eccentricity* refers to the distance between the point of load application and the center of the section. To better understand the consequence of an eccentricity on the behavior of columns, consider Figure 5–5a, which shows a column subjected to a compression force, P, acting at the center of the section (point 1). Because the force acts at the center of the section, the internal compression stresses are distributed uniformly on the section.

If we move P to a new location (point 2) at a distance e from the column center, as shown in Figure 5–5b, the load generates bending stresses in addition to axial compressive

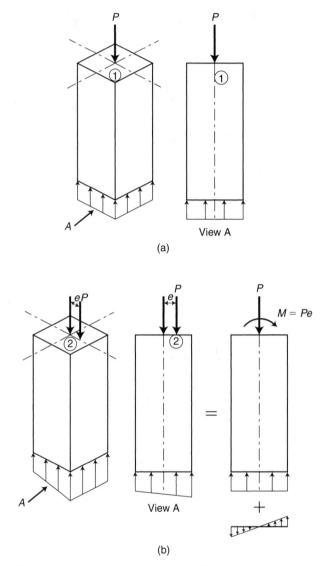

FIGURE 5–5 (a) Axially loaded column (column with small or no eccentricity); (b) eccentrically loaded column.

stresses. The bending stress is the result of the moment caused by the off-center load ($M = Pe$). The action of P at the eccentricity, e, is equivalent to the load P acting at the center and an additional moment, $M = Pe$.

If the moment acting on the column is negligible compared to its axial load, we consider the column to be an axially loaded column, or a column with small eccentricity. If the applied moment is large, the column is an eccentrically loaded column, or a column with large eccentricity. In former ACI Codes, tied and spiral columns were considered to be axially loaded columns when the eccentricity was less than $0.1h$ and $0.05h$, respectively (h = the cross-sectional dimension in the direction of the eccentricity).

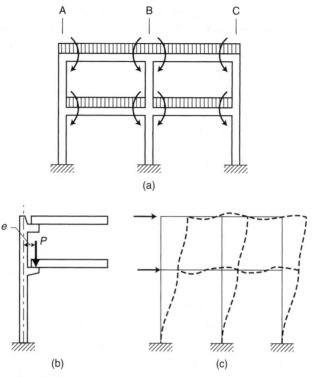

FIGURE 5–6   (a) and (b) Gravity loading, (c) lateral loading.

Now that we know that the effects of eccentric loads are the same as adding moments on columns and vice versa, let us review the sources of moments or eccentricities. Figure 5–6a shows a reinforced concrete building frame under gravity loads (Refer to Figures 2–10 and 2–11). The column on line B is subjected to moments from the adjoining beams. If the beam spans and loads are equal, the applied moments have the same magnitude but opposite directions, thus canceling each other. As a result, the column on line B is subjected only to an axial load. Even though this is theoretically correct, in reality there is always some moment on the column because the loads on the neighboring beams are never the same and the column is not perfectly straight.

The columns on lines A and C, on the other hand, are subjected to moments from the beams on one side in addition to axial loads. Therefore, these columns are subjected to large moments or have large eccentricities. Also, the column between the two bays (column B) will be subjected to moment in addition to axial loads if the live load is larger on one span than on its adjacent span.

Another example of a column with large eccentricity is shown in Figure 5–6b, in which the column is part of a precast concrete structure. The beams and columns are cast off the construction site, and then transported to the site for assembly. In precast construction, beams are often placed on brackets and connected together through steel plates embedded in both the beam and the bracket. As a result, there is always an eccentricity between where the beam is supported and the column centerline. This eccentricity generates a moment on the column, which needs to be considered in the analysis and design of the column.

Lateral loads, such as high winds and earthquakes, can generate large moments on the columns of monolithic concrete structures. Such columns usually have large eccentricity. Figure 5–6c shows how the columns of a two-story building undergo large bending moments when subjected to lateral loads. Columns subjected to loads with large eccentricity will be studied in greater detail later in this chapter.

Based on Structural System

Column and beam assemblies can be divided into two categories, depending on the building structural framing systems used: *braced frames* and *unbraced frames*. The columns within such systems are called *braced columns* and *unbraced columns,* respectively.

In a braced frame, lateral loads are resisted by shear walls, elevator or stairwell shafts, diagonal braces, or a combination thereof. The large stiffness of these elements prevents the columns of such a frame from undergoing large lateral motion or sidesway, and from experiencing significant moments due to lateral loads. In an unbraced frame, on the other hand, the columns (unbraced columns) are subject to large bending moments due to the lateral loads and have to withstand large lateral motions. These columns generally have large eccentricities. Figure 5–7 shows braced and unbraced columns in two different structural framing systems.

Based on Length

Columns may be divided into two groups based on their length, or more accurately their slenderness ratio. *Slenderness ratio* ($k\ell/r$) is the ratio of the column's effective length ($k\ell$) to the least radius of gyration (r) of the section. *Short columns* are columns whose slenderness ratio is low enough that their failure occurs from excessive stress levels rather than by buckling. *Slender columns,* on the other hand, may buckle when subjected to large axial loads.

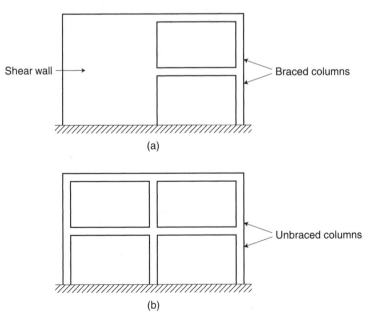

FIGURE 5–7   (a) Braced columns, (b) unbraced columns.

Most reinforced concrete columns in normal building structures are short columns. In fact, the results of a study conducted by the ACI (*Notes on ACI 318-71, Building Code with Design Applications,* p. 10-2) indicate that 90% of braced columns and about 40% of unbraced columns could be considered to be short columns. As a result, the emphasis in this book will be on short columns.

5.3 BEHAVIOR OF SHORT COLUMNS WITH SMALL ECCENTRICITY UNDER LOAD

Figure 5–8 shows the failure mechanisms of an axially loaded square tied column and spiral column. When a short tied column is subjected to increasing axial loads, the column fails suddenly. First the longitudinal reinforcing reaches yield, and then the concrete fails when the ultimate strain is reached. The failure is usually accompanied by plastic buckling of the longitudinal bars. Figure 5–9 shows a typical load-deformation relationship for tied and spiral columns. The tied column reaches the maximum capacity at point A, and fails soon thereafter at point B.

A spiral column, on the other hand, does not fail suddenly because the closely spaced spirals keep the core confined while the column shell spalls (Figure 5–8b). This confinement does increase the column's deformability significantly. The outer shell is not confined, thus it falls away readily. The inner core, however, is still able to carry loads, even after the concrete has been crushed by large compressive stresses. The column behaves like a bag of flour: As long as the paper sack does not burst, the flour column will support loads. For this reason, in this type of column, the ACI Code requires a minimum spiral reinforcement that will prevent the column from bursting until well after the concrete has reached its assumed ultimate strain of 0.003. Thus, a typical spiral column will have a second maximum point in its load-deformation diagram (point C in Figure 5–9). The yielding of the spirals makes the column failure ductile, which makes a spiral column ideal for unexpected large overloads such as seismic loads.

5.4 GENERAL ACI CODE REQUIREMENTS FOR COLUMNS

The ACI Code has several requirements for the design of columns:

1. Limits on the amount of longitudinal reinforcements. Column steel ratio, ρ_g, is defined as the ratio of the area of the longitudinal reinforcement (A_{st}) to the gross area of the column (A_g):

$$\rho_g = \frac{A_{st}}{A_g} \qquad (5–1)$$

The ACI 318-05 no longer uses the column steel ratio notation (ρ_g); however, the authors have kept it for purposes of clarity.

The ACI Code (Section 10.9.1) limits the area of the longitudinal reinforcement, A_{st}, in columns between $0.01\,A_g$ to $0.08\,A_g$. This means that the steel ratio, ρ_g, can average as shown in Equation 5–2:

$$0.01 \leq \rho_g \leq 0.08 \qquad (5–2)$$

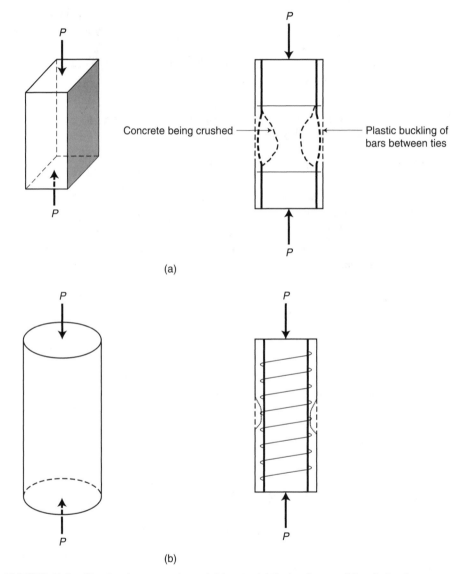

(a)

(b)

FIGURE 5–8 Short columns under axial loads: (a) tied columns; (b) spiral columns.

The minimum steel ratio of 0.01 provides resistance to bending, which may exist whether or not calculations show the column is subjected to bending moments. In addition, a minimum amount of steel reduces creep and shrinkage of the concrete under sustained compression loads. It is common practice to use a minimum bar size of #5 for the longitudinal reinforcement.

Although the maximum steel ratio is 0.08, in practice it is very difficult to use such a high amount of steel in the column, especially where the bars are spliced above a floor level. Such congestion may be avoided by using #14 or #18 bars. (A #14 bar has a cross-sectional area of 2.25 in^2, a #18 bar has one of 4.0 in^2.) These bars are not used in beams, but are very useful in columns.

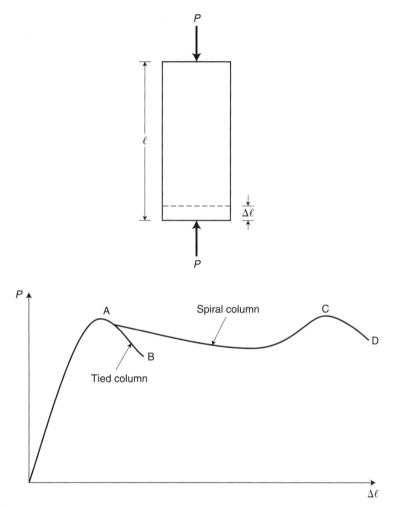

FIGURE 5–9 Load-deformation relationship of axially loaded columns.

Bundled bars may be used if the column is subjected to a large load and a large number of bars is needed. Bundles consist of three or four bars (a maximum of four bars according to the ACI Code, Section 7.6.6.1) tied together in direct contact, and are usually placed at the corners of the column, as seen in Figure 5–10. Each bundle of bars is treated as if it were a single round bar of area equal to the sum of the areas of the bundled bars. The main drawback of bundled bars is that they cannot be lap-spliced.

2. *Limit on the number of bars.* According to the ACI Code (Section 10.9.2), the minimum number of main longitudinal bars is four for rectangular or circular tied columns, three for triangular tied columns, and six for spiral columns (see Figure 5–11). This requirement is based mainly on construction needs.

3. *Limit on the clear cover.* According to the ACI Code (Section 7.7.1), the minimum clear cover for columns is 1.5 in., measured from the edge of the column to the transverse reinforcement. This cover is for interior columns that are not exposed to weather or in contact with the ground. The clear cover is 2 in. for formed surfaces exposed to weather

FIGURE 5–10   Column reinforced with bundled bars.

or in contact with the ground. If the concrete is cast directly against the earth without form-ing (as in drilled piles), the cover must be increased to 3 in.

   4. Limit on tie spacing. In general, there are four main reasons for having ties in a column:

> A. They hold the longitudinal reinforcement in place during construction.
> B. They provide a confined core and, as a result, increase the column's strength and ductility.
> C. They act as shear reinforcement.
> D. They provide lateral support for the longitudinal bars and prevent them from elastic buckling. Columns need sufficiently large tie sizes that are well con-nected to the longitudinal bars at sufficiently close vertical spacings (*s*). To satisfy the above requirements, the ACI Code (Section 7.10.5.1) requires that at least #3 ties be used for #10 or smaller longitudinal bars, and #4 ties be used for #11, #14, and #18 and bundled bars.

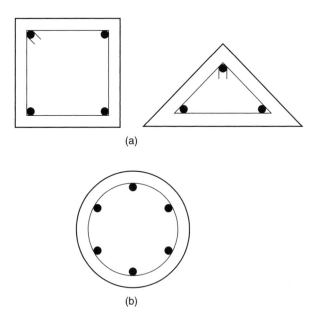

(a)

(b)

FIGURE 5–11   Minimum number of bars: (a) tied columns, (b) spiral columns.

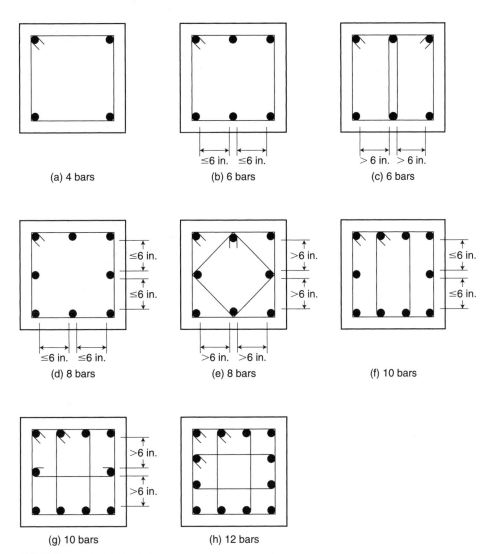

FIGURE 5–12 Typical tie arrangements.

In addition, the ACI Code (Section 7.10.5.2) requires that the vertical spacing of ties (s) be limited to:

$$s \leq \min\{16d_b, 48d_t, b_{\min}\}$$

where d_b is the diameter of the longitudinal bars, d_t is the diameter of the ties, and $b_{\min}$ is the minimum dimension of the column. According to the ACI Code (Section 7.10.5.3), a bar is adequately supported laterally if it is located at a corner of a tie, with an enclosed angle not exceeding 135°, or if it is located between laterally supported bars with a clear spacing of 6 in. or less. Figure 5–12 shows typical tie arrangements that satisfy this requirement.

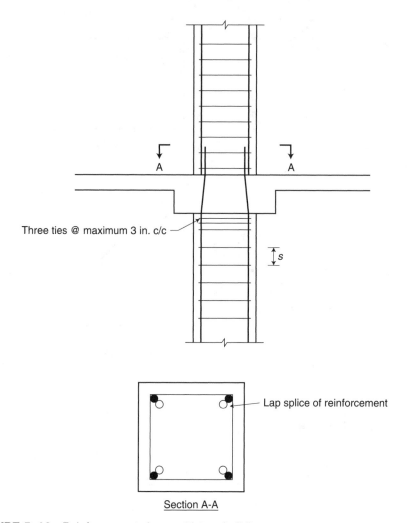

Three ties @ maximum 3 in. c/c

s

Lap splice of reinforcement

Section A-A

FIGURE 5–13 Reinforcements in a multistory building.

   5. *Limit on longitudinal bar spacing.* Section 7.6.3 of the ACI Code requires that the clear distance between longitudinal bars be at least 1.5 times the bar diameter or 1.5 in. to allow concrete to flow between the reinforcements. Reinforced concrete columns in multistory buildings are generally cast one level at a time. Therefore, the longitudinal reinforcements in columns typically are spliced above every floor. An exception is columns in seismic zones where splicing is usually near midheight between floors. There are different methods of splicing bars in columns. Figure 5–13 shows one common method of splicing reinforcements in a multistory building. Table A5–1 lists the maximum number of bars that can be placed in a square or circular column based on minimum bar spacing requirements and the splicing method shown in Figure 5–13. The spiral and tie sizes are assumed to be #4 with 1.5 in. cover.

   6. *Limit on spacing and amount of spiral reinforcement.* Spirals are often made of smooth bars rather than deformed bars; and the spacing and amount of spirals need to be such that they confine the column core. For these reasons, the ACI Code (Section 7.10.4.3)

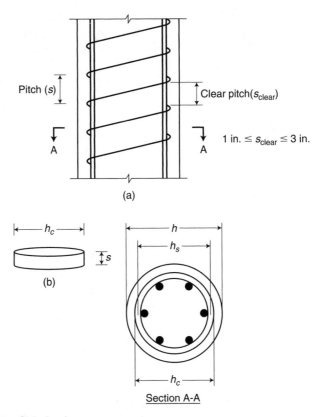

FIGURE 5–14 Spiral columns.

requires that clear spacing between spirals (s_{clear}) be between 1 and 3 in. (see Figure 5–14a). In cast-in-place construction spirals must be at least $^3/_8$ in. in diameter.

Spiral steel ratio is defined as:

$$\rho_s = \frac{\text{Volume of spiral steel in one turn, } s}{\text{Volume of column core in height, } s} \tag{5–3}$$

If the diameter of the spiral steel is d_{sp} and the area of the spiral steel is A_{sp}, the volume of column core in height s (see Figure 5–14b) is:

$$\frac{\pi h_c^2}{4} \times s$$

The volume of spiral steel in one turn, s, is:

$$\pi h_s \times A_{sp}$$

Substituting the above into Equation 5–3, we get:

$$\rho_s = \frac{\pi h_s \times A_{sp}}{\pi \dfrac{h_c^2}{4} \times s} = \frac{4A_{sp}h_s}{h_c^2 s}$$

Because $h_s = h_c - d_{sp}$, and d_{sp} is negligible compared to h_s, we can assume that h_s and h_c are approximately equal ($h_s \cong h_c$). Substituting h_c for h_s into the above equation allows us to calculate the spiral steel ratio for columns, ρ_s, using the simplified Equation 5–4.

$$\rho_s = \frac{4A_{sp}}{h_c s} \tag{5–4}$$

The ACI Code requires a minimum spiral steel ratio to ensure ductility and toughness. According to Equation 10–5 of Section 10.9.3 of the ACI Code, the minimum spiral steel ratio ($\rho_{s,\min}$) is:

$$\rho_{s,\min} = 0.45 \left(\frac{A_g}{A_{ch}} - 1 \right) \frac{f_c'}{f_{yt}} \tag{5–5}$$

In this equation, A_g is the gross area of the column:

$$A_g = \frac{\pi h^2}{4}$$

A_{ch} is the area of the core (measured from outside-to-outside edge of spiral reinforcement). See Figure 5–14:

$$A_{ch} = \frac{\pi h_c^2}{4}$$

and f_{yt} is the specified yield strength of the transverse (spiral) reinforcement. The required spiral pitch, s, can then be calculated from Equation 5–4 based on an assumed spiral size, which must be at least $^3/_8$ in. in diameter.

5.5   SOME CONSIDERATIONS ON THE DESIGN OF REINFORCED CONCRETE COLUMNS

When designing reinforced concrete columns, we must consider a number of factors in order to minimize the overall cost of construction.

Column Size

In general, columns in multistory buildings are designed based on their floor-to-floor height. In order to simplify the formwork, the size of columns in a multistory building is usually kept the same throughout the height of the structure whereas the amount of reinforcement, and perhaps the concrete's compressive strength, are increased for the lower stories. Smaller size columns are easier to conceal in walls and less intrusive architecturally, which results in larger rentable floor spaces for building owners. Therefore, the structural designer tends to select as small a column size as possible. The ACI Code does not require a minimum column size, but in practice rectangular columns are at least 10 in.

wide and round columns have a minimum diameter of 12 in. Smaller columns are very difficult to construct properly.

High-Strength Material Use

Because most columns are in compression, it is more economical to use high-strength concrete. High-strength concrete with compressive strength exceeding 16,000 psi has been used for the columns supporting lower stories in large, tall buildings. The reliable production of such ultra-high-strength concrete requires very special technology, so it is less commonly used; but 8,000 to 10,000 psi concrete is commonly available.

In most cases, however, the compressive strength of concrete in columns in low or mid-rise buildings is in the 5,000–6,000 psi range. Although the cost of concrete increases as compression strength increases, the strength increases at a greater rate than the cost. Grade 60 rebars are used in most concrete structures. Grade 75 bars may provide better economy for columns in high-rise structures, especially when they are used in conjunction with high-strength concrete.

5.6   ANALYSIS OF SHORT COLUMNS WITH SMALL ECCENTRICITY

Most reinforced concrete columns are categorized as short columns. This means that they will fail in compression under large loads rather than undergo elastic buckling. In this section we will study the load carrying capacity and the design of short columns with small eccentricity.

We first find the axial load strength of a column. Figure 5–15a shows a typical column subject to a concentrated load at its center. From equilibrium of forces in the vertical direction (along the column axis), the axial load capacity of a column (P_o) is equal to the sum of the volume of stresses in the concrete and the steel (see Figure 5–15b). Based on the

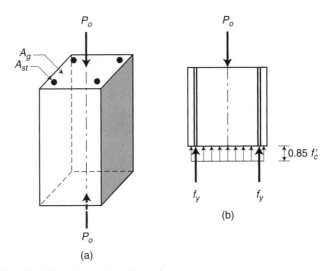

(b)

(a)

FIGURE 5–15   Axial load capacity of a column.

results of tests carried out at the University of Illinois and Lehigh University from 1927 to 1933, the ACI Code uses $0.85f'_c$ for the ultimate concrete compression stress. At ultimate load the stress in the steel is equal to the yield stress (f_y). Therefore,

$$P_o = \underbrace{0.85f'_c\,(A_g - A_{st})}_{\text{Concrete contribution}} + \underbrace{f_y A_{st}}_{\text{Steel contribution}} \qquad (5\text{--}6)$$

The stress in the concrete $(0.85f'_c)$ is applied on the net column area $(A_g - A_{st})$, which is only the area of the concrete. In reality, however, the loads acting on columns always have an eccentricity (e.g., due to vertical misalignment of the form). The ACI Code accounts for "accidental eccentricity" by requiring that the theoretical capacity be reduced by 20% for tied columns and 15% for spirally reinforced columns. Then the *nominal load capacity* of columns is:

$$P_n = 0.8P_o = 0.8[0.85f'_c\,(A_g - A_{st}) + f_y A_{st}] \quad \text{(tied columns)} \qquad (5\text{--}7)$$

$$P_n = 0.85P_o = 0.85[0.85f'_c\,(A_g - A_{st}) + f_y A_{st}] \quad \text{(spiral columns)} \qquad (5\text{--}8)$$

To find the design resisting load, P_R, we must reduce P_n by the strength reduction factor, ϕ:

$$P_R = \phi P_n \qquad (5\text{--}9)$$

According to the ACI Code (Section 9.3.2.2), columns with small eccentricity have the following strength reduction factors:

$$\begin{aligned} \phi &= 0.65 \quad \text{(tied columns)} \\ \phi &= 0.70 \quad \text{(spiral columns)} \end{aligned} \qquad (5\text{--}10)$$

The values of ϕ for pure compression are less than those used for beams in bending (0.90) and in shear (0.75). The main reasons that the strength reduction factors are considerably lower are the following:

1. A column failure is a much more severe event than the local failure of a beam, because a column supports larger areas of a building.

2. The quality of concrete used in columns is less reliable than that used in beams and slabs. The difficulty of consolidating the concrete in narrow column forms and between the longitudinal and lateral reinforcements often leads to honeycombs that are difficult to repair (even when visible).

3. The strength of the concrete has a much greater role in the ultimate strength of a column than it does in beams and slabs, where the reinforcing has the most influence on the ultimate strength.

After introducing the ϕ factor, we calculate the strength of an axially loaded column (ACI Code Equations 10–2 and 10–1, respectively) as follows:

$$P_R = \phi P_n = 0.8\phi[0.85f'_c\,(A_g - A_{st}) + f_y A_{st}] \quad \text{(tied columns)} \qquad (5\text{--}11)$$

$$P_R = \phi P_n = 0.85\phi[0.85f'_c\,(A_g - A_{st}) + f_y A_{st}] \quad \text{(spiral columns)} \qquad (5\text{--}12)$$

The steps of the analysis of short columns with small eccentricity are shown in Figure 5–16 and are as follows:

Step 1   Check the steel ratio. When we analyze a column, we know its dimensions and the size and number of its reinforcements. Therefore, we have the gross

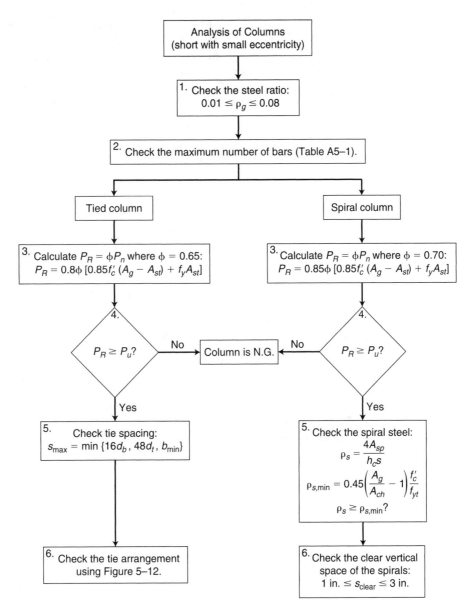

FIGURE 5–16 Flowchart for analysis of reinforced concrete columns.

area of concrete (A_g) and the total area of steel (A_{st}), from which we can determine the column steel ratio (ρ_g):

$$\rho_g = \frac{A_{st}}{A_g}$$

The steel ratio is limited by:

$$0.01 \leq \rho_g \leq 0.08$$

If the steel ratio does not fall within these limits, the column does not conform to the current ACI Code requirements.

Step 2   Determine whether the spacing between the longitudinal bars meets the ACI Code requirements by obtaining the maximum number of bars that can be placed in the column according to Table A5–1. The remaining steps differ depending on whether the column is tied or spiral:

(a)   *Tied Columns*

Step 3   Calculate the column capacity, P_R:

$$\phi = 0.65$$
$$P_R = \phi P_n = 0.8\phi[0.85f_c' (A_g - A_{st}) + f_y A_{st}]$$

Step 4   Calculate the factored loads, P_u, and determine whether the column can resist the applied loads $(P_R \geq P_u)$.

Step 5   Check the tie spacing:

$$s_{max} = \min\{16d_b, 48d_t, b_{min}\}$$

Step 6   Check the arrangement of the ties using Figure 5–12.

(b)   *Spiral Columns*

Step 3   Calculate the spiral column capacity, P_R:

$$\phi = 0.70$$
$$P_R = \phi P_n = 0.85\phi[0.85f_c' (A_g - A_{st}) + f_y A_{st}]$$

Step 4   Calculate the factored loads, P_u, and determine whether the column can resist the applied loads $(P_R \geq P_u)$.

Step 5   Check the spiral steel. Calculate the spiral steel ratio (ρ_s) and compare it with the minimum amount required by the ACI Code:

$$\rho_s = \frac{4A_{sp}}{h_c s}$$

$$\rho_{s,min} = 0.45\left(\frac{A_g}{A_{ch}} - 1\right)\frac{f_c'}{f_{yt}}$$

$$\rho_s \geq \rho_{s,min}$$

Step 6   Check the clear space (s_{clear}) between each turn of the spirals:

$$1 \text{ in.} \leq s_{clear} = s - d_{sp} \leq 3 \text{ in.}$$

EXAMPLE 5–1

Determine the maximum factored axial load that a short tied column with the cross section shown below can resist. There is no moment on the column. Determine whether the ties are appropriate. The compressive strength of the concrete is 4,000 psi, and the reinforcement is A615 grade 60 steel.

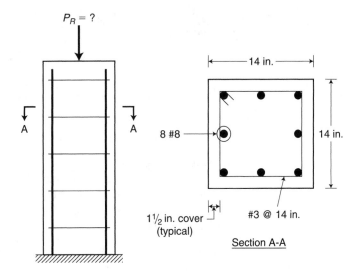

Solution

Step 1   Determine and check the steel ratio, ρ_g:

$$8 \text{ #8} \longrightarrow \text{Table A2–9} \longrightarrow A_{st} = 6.32 \text{ in}^2$$
$$14 \text{ in.} \times 14 \text{ in. column} \longrightarrow A_g = 14 \times 14 = 196 \text{ in}^2$$

$$\rho_g = \frac{A_{st}}{A_g} = \frac{6.32}{196} = 0.032$$

$$0.01 < 0.032 < 0.08 \quad \therefore \text{ ok}$$

Step 2   Check the spacing of the longitudinal bars by obtaining the maximum number of #8 bars that can be placed into the column from Table A5–1:

$$h = 14 \text{ in.} \longrightarrow \text{Maximum of 12 #8 bars}$$

Step 3   Calculate the column load capacity, P_R:

$$P_R = \phi P_n = 0.8\phi[0.85f_c' (A_g - A_{st}) + f_y A_{st}]$$
$$P_R = 0.8 \times 0.65[0.85 \times 4(196 - 6.32) + 60 \times 6.32]$$
$$P_R = 533 \text{ kip}$$

Therefore, the maximum design (factored) load for this column is 533 kip.

Step 4   We skip this step because we need only the load capacity.

Step 5   Check the adequacy of the ties. The maximum spacing of the ties (s_{max}) is:

$$s_{max} = \min\{16d_b, 48d_t, b_{min}\}$$
$$s_{max} = \min\{16 \times 1, 48 \times \tfrac{3}{8}, 14 \text{ in.}\}$$
$$s_{max} = \min\{16 \text{ in.}, 18 \text{ in.}, 14 \text{ in.}\} = 14 \text{ in.}$$

$\therefore$ Therefore, #3 @ 14 in. for the ties is adequate.

Step 6 Check the tie arrangement, according to Figure 5–12. Determine the clear space between the bars:

$$\text{Clear space} = \frac{\overset{\text{Cover}}{\downarrow} \quad \overset{\text{Tie}}{\downarrow} \quad \overset{\#8}{\downarrow}}{14 - 2(1.5) - 2(\tfrac{3}{8}) - 3(1.0)}{2}$$

Clear space = 3.6 in. < 6 in.

Because the clear space between the bars is less than 6 in., no additional ties are necessary on the non-corner longitudinal reinforcing. Therefore, the tie arrangement meets the ACI Code requirements.

EXAMPLE 5–2

The circular spiral column shown below is subjected to a dead load of 200 kip and a live load of 225 kip. The eccentricity of the loads is small. The compressive strength of the concrete is 4,000 psi, and the reinforcement is A615 grade 60 steel. Check the adequacy of the column including the spirals.

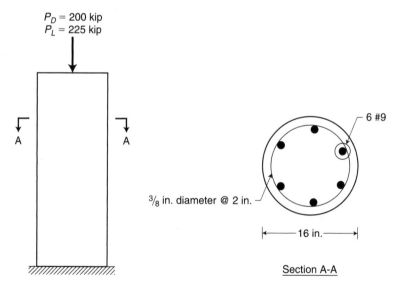

Solution

Step 1 Determine and check the steel ratio:

$$6 \ \#9 \longrightarrow \text{Table A2–9} \longrightarrow A_{st} = 6.0 \ \text{in}^2$$

$$16 \ \text{in. diameter column} \longrightarrow A_g = \frac{\pi(16)^2}{4} = 201.1 \ \text{in}^2$$

$$\rho_g = \frac{6.0}{201.1} = 0.03$$

$$0.01 < 0.03 < 0.08 \quad \therefore \ \text{ok}$$

Step 2   Obtain the maximum number of #9 bars for a 16 in. diameter column from Table A5–1. The answer is nine bars. Therefore, 6 #9 bars can easily fit into the column.

Step 3   Calculate the design resisting load, P_R, using $\phi = 0.70$ for spiral columns:

$$P_R = \phi P_n = 0.85\phi[0.85f'_c\,(A_g - A_{st}) + f_y A_{st}]$$
$$P_R = 0.85 \times 0.70[0.85 \times 4(201.1 - 6.0) + 60 \times 6.0]$$
$$P_R = 609\,\text{kip}$$

Step 4   Determine the total factored load on the column:

$$P_u = 1.2P_D + 1.6P_L = (1.2)(200) + (1.6)(225)$$
$$P_u = 600\,\text{kip}$$

Because $P_R = 609\,\text{kip} > P_u = 600\,\text{kip}$, the column has enough strength to carry the load.

Step 5   Check the amount of spiral steel:

$$A_{sp} = 0.11\,\text{in}^2 \quad (\tfrac{3}{8}\,\text{in. dia. spiral})$$
$$h_c = h - 2(1.5) = 16 - 3 = 13\,\text{in.}$$
$$s = 2.0\,\text{in.} \quad (\text{from figure})$$

Calculate the spiral steel ratio, ρ_s:

$$\rho_s = \frac{4A_{sp}}{h_c s} = \frac{4(0.11)}{(13)(2.0)} = 0.0169$$

The minimum spiral steel ratio, $\rho_{s,\text{min}}$ is:

$$\rho_{s,\text{min}} = 0.45\left(\frac{A_g}{A_{ch}} - 1\right)\frac{f'_c}{f_{yt}}$$
$$= 0.45\left(\frac{201.1}{\dfrac{\pi(13)^2}{4}} - 1\right)\frac{4}{60}$$
$$= 0.0155 < 0.0169 \quad \therefore\ \text{ok}$$

Therefore, enough spiral steel is provided.

Step 6   Check the clear space between each turn of spiral:

$$s_{\text{clear}} = s - d_{sp} = 2 - \tfrac{3}{8} = 1.625\,\text{in.}$$
$$1\,\text{in.} < 1.625\,\text{in.} < 3\,\text{in.} \quad \therefore\ \text{ok}$$

Therefore, the column is adequate for the given loading condition.

5.7   DESIGN OF SHORT COLUMNS WITH SMALL ECCENTRICITY

Design of reinforced concrete columns is a task that requires the involvement of both the architect and the structural engineer. The shapes and sizes are usually based on architectural requirements such as aesthetics and space needs. The construction costs also play an important role. These costs can be reduced by doing the following:

1. Make the forms reusable by making the column shapes and sizes as uniform as possible.

2. Typically it is cost effective to use the fewest longitudinal reinforcements (or largest bar size) possible. This also reduces the cost of ties, as fewer ties will be required. In addition, difficulties in placement of the concrete will be reduced.

Often the column size (A_g) is preselected, or decided by factors other than strictly structural considerations. In these cases, the structural designer needs only to find the required amount of steel (A_{st}) in addition to designing the ties or spirals. In other cases, however, the structural designer may want to determine the minimum size of a "workable" column.

In the following we consider two cases: A_g = known, A_{st} = unknown; and A_g and A_{st} = unknown.

A_g = Known, A_{st} = Unknown

A safe column requires that:

$$PR = \phi P_n \geq P_u$$

The load capacity of a tied column according to Equation 5–11 is:

$$P_R = 0.8\phi[0.85f'_c (A_g - A_{st}) + f_y A_{st}]$$

The useful capacity of the column (P_R) must be at least equal to the factored load on the column (P_u). Thus,

$$P_u = 0.8\phi[0.85f'_c (A_g - A_{st}) + f_y A_{st}]$$

Because the area of concrete (A_g) is known we can solve the above equation for the area of steel (A_{st}):

$$A_{st}[0.8\phi(f_y - 0.85f'_c)] = P_u - 0.8\phi(0.85f'_c A_g)$$

$$\phi = 0.65$$

$$A_{st} = \frac{P_u - 0.8\phi(0.85f'_c A_g)}{0.8\phi(f_y - 0.85f'_c)} \tag{5–13}$$

Similarly, the required area of steel, A_{st}, for spiral columns is:

$$\phi = 0.70$$

$$A_{st} = \frac{P_u - 0.85\phi(0.85f'_c A_g)}{0.85\phi(f_y - 0.85f'_c)} \tag{5–14}$$

Note that if the numerator in Equations 5–13 or 5–14 results in a negative value, the column requires only 1% longitudinal reinforcement (ρ_{min}). The design steps are shown in Figure 5–17 and are as follows:

Step 1   Determine the factored axial load on the column, P_u:

$$P_u = 1.2P_D + 1.6P_L$$

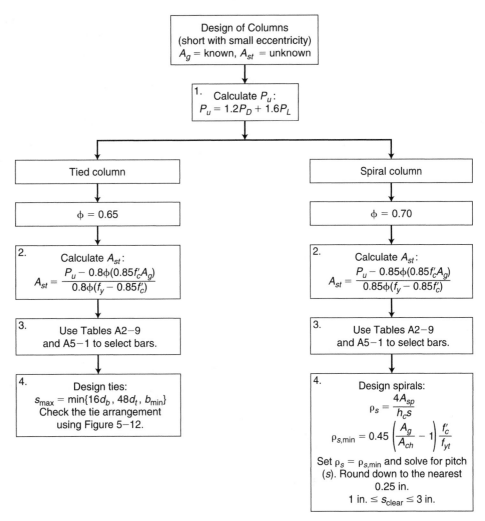

FIGURE 5–17 Flowchart for the design of a short column with small eccentricity (A_g = known, A_{st} = unknown).

Which steps you perform next depends on whether the column is to be tied or spirally reinforced:

Tied Columns

Step 2   Calculate the required area of steel using Equation 5–13. Use $\phi = 0.65$.

Step 3   Use Tables A2–9 and A5–1 to select bars. The minimum number of bars for tied square columns is four. Determine whether $0.01 \leq \rho_g \leq 0.08$. If $\rho_g < 0.01$, use $\rho_g = 0.01$. Also, if $\rho_g > 0.08$, or you cannot find any arrangements of bars to fit inside the column, the column dimensions are not enough and its cross-sectional area (A_g) must be increased.

Step 4   Design the ties. Use #3 ties for #10 and smaller longitudinal bars. Other-wise, use #4 ties. The tie spacing, s_{max}, is:

$$s_{max} = \min\{16d_b, 48d_t, b_{min}\}$$

Round down s_{max} to the nearest 0.5 in. Check the tie arrangement using Figure 5–12.

Spiral Columns

Step 2   Calculate the required area of steel using Equation 5–14. Use $\phi = 0.70$.

Step 3   Use Tables A2–9 and A5–1 to select bars. The minimum number of bars for spiral columns is six. Determine whether $0.01 \leq \rho_g \leq 0.08$. Similar to the tied columns, if $\rho_g < 0.01$, use $\rho_g = 0.01$; and if $\rho_g > 0.08$ or bars do not fit inside the column, increase the column cross-sectional sizes.

Step 4   Design the spiral steel by equating the spiral steel ratio (ρ_s) to $\rho_{s,min}$ (use a minimum spiral diameter of $3/8$ in.):

$$\rho_{s,min} = 0.45\left(\frac{A_g}{A_{ch}} - 1\right)\frac{f_c'}{f_{yt}}$$

$$\rho_s = \frac{4A_{sp}}{h_c s}$$

$$\rho_s = \rho_{s,min}$$

Solve for s (spiral pitch) and round down to the nearest $1/4$ in. Check the clear pitch where $s_{clear} = s - d_{sp}$, which must be between 1 and 3 in. If s_{clear} is less than 1 in., increase the spiral size; if s_{clear} is more than 3 in., use 3 in.

EXAMPLE 5–3

Figure 5–18a shows the typical partial floor plan and sections of a three-story reinforced concrete office building. The mechanical and electrical systems for the floor and the roofing and insulation weigh 5 psf. The weight of the partitions is 15 psf. The floor live load is 50 psf and the roof snow load is 30 psf. Assume $f_c' = 4{,}000$ psi and $f_y = 60{,}000$ psi. Design the square 16 in. × 16 in. tied interior columns between the ground and second levels. Moments acting on the columns are not significant, and you should not consider live load reduction in load calculations. Assume that the unit weight of the concrete is 150 pcf. Neglect the self-weight of the columns.

Solution

Step 1   Determine the loads acting on the columns:

Floor Loads

Weight of concrete slab	$= 150(6/12) = 75$ psf
Mechanical and electrical	$= 5$ psf
Partitions	$= 15$ psf
Floor dead load	$= 95$ psf

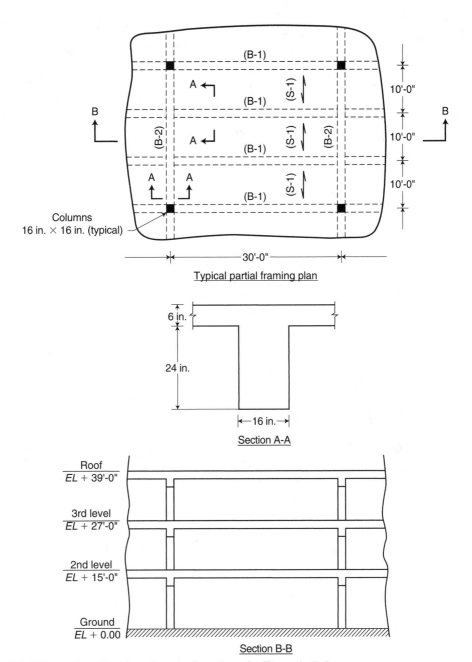

FIGURE 5–18a   Framing plans and sections for Example 5–3.

The tributary area for the columns is 30 ft $\times$ 30 ft = 900 ft^2. In addition to supporting the slab, the columns also support beams B-1 and B-2.

$$P_{D,\text{floor}} = \overbrace{\frac{95}{1000}(900)}^{\text{Slab}} + \overbrace{\frac{150}{1000}\left(\frac{24}{12} \times \frac{16}{12} \times 3 \times 30\right)}^{\text{B-1}} + \overbrace{\frac{150}{1000}\left(\frac{24}{12} \times \frac{16}{12} \times 30\right)}^{\text{B-2}}$$

$$P_{D,\text{floor}} = 85.5 + 36 + 12.0 = 133.5 \text{ kip}$$

$$P_{L,\text{floor}} = \frac{50}{1000}(900) = 45 \text{ kip}$$

Roof Loads

$$\begin{array}{ll}
\text{Weight of concrete slab} & = 150(^6/_{12}) = 75 \text{ psf} \\
\text{Roofing and insulation} & = 5 \text{ psf} \\
\hline
\text{Roof dead load} & = 80 \text{ psf}
\end{array}$$

The tributary area of the columns at roof level is the same as that of a floor (900 ft^2). Therefore

$$P_{D,\text{roof}} = \overbrace{\frac{80}{1000}(900)}^{\text{Slab}} + \overbrace{\frac{150}{1000}\left(\frac{24}{12} \times \frac{16}{12} \times 3 \times 30\right)}^{\text{B-1}} + \overbrace{\frac{150}{1000}\left(\frac{24}{12} \times \frac{16}{12} \times 30\right)}^{\text{B-2}}$$

$$P_{D,\text{roof}} = 72.0 + 36.0 + 12.0 = 120.0 \text{ kip}$$

$$P_{L,\text{roof}} = \text{snow load} = \frac{30}{1000}(900) = 27.0 \text{ kip}$$

Because the column self-weight is small compared to the applied loads, we neglect the column weight. The columns between the ground and second levels carry two floor loads and one roof load:

$$P_{D,\text{total}} = 2P_{D,\text{floor}} + P_{D,\text{roof}}$$
$$P_{D,\text{total}} = 2(133.5) + 120.0 = 387.0 \text{ kip}$$
$$P_{L,\text{total}} = 2P_{L,\text{floor}} + P_{L,\text{roof}}$$
$$P_{L,\text{total}} = 2(45) + 27.0$$
$$P_{L,\text{total}} = 117.0 \text{ kip}$$
$$P_u = 1.2P_D + 1.6P_L$$
$$P_u = 1.2(387) + 1.6(117.0)$$
$$P_u = 651.6 \text{ kip}$$

Step 2   Determine the required area of steel, A_{st}:

$$A_{st} = \frac{P_u - 0.8\phi[0.85f'_c A_g]}{0.8\phi[f_y - 0.85f'_c]}$$

$$A_{st} = \frac{651.6 - 0.8(0.65)[0.85(4.0)(16 \times 16)]}{0.8(0.65)[60 - 0.85](4.0)]}$$

$$A_{st} = 6.76 \text{ in}^2$$

Step 3   Using Table A2–9, select 8 #9 bars ($A_s = 8.0\,\text{in}^2$). Based on Table A5–1, the maximum number of #9 bars for a 16 in. $\times$ 16 in. column is 12. Therefore 8 #9 bars are ok. The provided column steel ratio, ρ_g, is:

$$\rho_g = \frac{A_{st}}{A_g} = \frac{8.0}{16 \times 16} = 0.0313$$

$$0.01 < \rho_g = 0.0313 < 0.08 \quad \therefore \text{ ok}$$

Step 4   Design ties:
Use #3 ties for #9 longitudinal bars. The maximum tie spacing, s_{max}, is:

$$s_{max} = \min\{16d_b, 48d_t, b_{min}\}$$
$$s_{max} = \min\{16(1.128), 48(^3/_8), 16\}$$
$$s_{max} = \{18.0, 18.0, 16.0\}$$
$$s_{max} = 16.0\,\text{in.}$$

Check the tie arrangement based on Figure 5–12:

$$\text{Clear space} = \frac{16 - 2(1.5) - 2(^3/_8) - 3(1.128)}{2}$$

$$\text{Clear space} = 4.4\,\text{in.} < 6\,\text{in.} \quad \therefore \text{ One tie per set}$$

Figure 5–18b shows the cross section of the designed column.

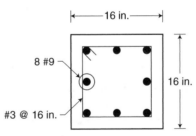

FIGURE 5–18b   Final design of Example 5–3.

A_g and A_{st} = Unknown

Because we have to determine both the size of the column and the required area of steel, and only one equation defines the column load capacity, we must assume one unknown. According to the ACI Code, ρ_g can vary between 0.01 and 0.08. If $\rho_g = 0.01$, the column size may be excessively large. On the other hand, $\rho_g = 0.08$ is not practical as the reinforcement will be very congested. Exceeding $\rho_g = 0.04$ is not recommended, so for this process we use $\rho_g = 0.03$.

A safe column must satisfy the following relationship:

$$P_R = \phi P_n \geq P_u$$

From Equation 5–11 the load capacity of a tied column is:

$$P_R = 0.8\phi[0.85f_c' (A_g - A_{st}) + f_y A_{st}] \geq P_u$$

For design, we consider $P_R = P_u$. The column steel ratio, ρ_g, is defined as:

$$\rho_g = \frac{A_{st}}{A_g} \longrightarrow A_{st} = \rho_g A_g$$

Substituting A_{st} in the equation for P_u:

$$P_u = 0.8\phi[0.85f_c' (A_g - \rho_g A_g) + f_y \rho_g A_g]$$

and simplifying:

$$P_u = 0.8\phi A_g[0.85f_c'(1 - \rho_g) + f_y \rho_g]$$

Solving A_g for tied columns:

$$A_g = \frac{P_u}{0.8\phi[0.85f_c' (1 - \rho_g) + f_y \rho_g]} \qquad \phi = 0.65 \qquad (5\text{–}15)$$

Similarly, for spiral columns, the required column area, A_g, is:

$$A_g = \frac{P_u}{0.85\phi[0.85f_c' (1 - \rho_g) + f_y \rho_g]} \qquad \phi = 0.70 \qquad (5\text{–}16)$$

Now that we have determined the column area, we can calculate the column dimensions, h and b, as follows:

$$A_g = h^2 \longrightarrow h = \sqrt{A_g} \qquad \text{(square column)}$$
$$A_g = h \times b \qquad \text{(rectangular column)}$$
$$A_g = \frac{\pi h^2}{4} \longrightarrow h = 2\sqrt{\frac{A_g}{\pi}} \quad \text{(round column)}$$

As we mentioned before, the minimum practical size for a rectangular or square column is 10 in. and for a round column is a diameter of 12 in.

Use the column area, A_g, to calculate the required area of steel, A_{st} using Equation 5–13 or 5–14:

$$A_{st} = \frac{P_u - 0.8\phi(0.85f_c' A_g)}{0.8\phi(f_y - 0.85f_c')} \qquad \text{(tied column)}$$

$$A_{st} = \frac{P_u - 0.85\phi(0.85f_c' A_g)}{0.85\phi(f_y - 0.85f_c')} \qquad \text{(spiral column)} \qquad (5\text{–}17)$$

The steps for the design of short columns with small eccentricity are shown in Figure 5–19 and are as follows:

Step 1   Determine the factored axial load on column, P_u:

$$P_u = 1.2P_D + 1.6P_L$$

and assume a column steel ratio $\rho_g = 0.03$.

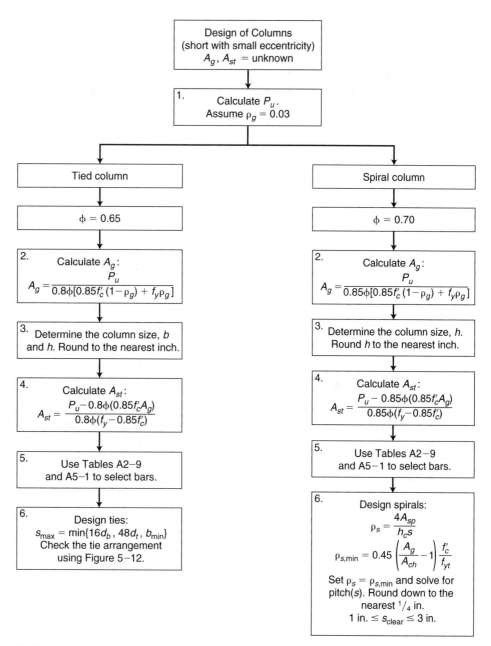

FIGURE 5–19   Flowchart for the design of a short column with small eccentricity.

Based on the type of column (i.e., tied or spiral), follow the appropriate subsequent steps:

Tied Columns

Step 2   Calculate the required gross area of the column, A_g:

$$A_g = \frac{P_u}{0.8\phi[0.85f_c'\ (1 - \rho_g) + f_y\rho_g]} \qquad \phi = 0.65$$

Step 3   Determine the column size:

$$h = \sqrt{A_g} \qquad \text{(square column)}$$
$$h \times b = A_g \qquad \text{(rectangular column)}$$
$$h = 2\sqrt{\frac{A_g}{\pi}} \quad \text{(round column)}$$

Round h or b to the nearest full or even inch.

Step 4   Calculate the required area of steel, A_{st}:

$$A_{st} = \frac{P_u - 0.8\phi(0.85f_c'\ A_g)}{0.8\phi(f_y - 0.85f_c')}$$

Step 5   Use Tables A2–9 and A5–1 to select the size and number of longitudinal bars. Remember that the minimum number of bars for square tied columns is four.

Step 6   Design the ties. Use #3 ties for #10 and smaller longitudinal bars. Otherwise, use #4 ties. The tie spacing, s_{max}, is:

$$s_{max} = \min\{16d_b, 48d_t, b_{min}\}$$

Round down s_{max} to the nearest $\frac{1}{2}$ in. Check the tie arrangement using Figure 5–12.

Spiral Columns

Step 2   Calculate the required gross area of the column, A_g:

$$A_g = \frac{P_u}{0.85\phi[0.85f_c'(1 - \rho_g) + f_y\rho_g]} \qquad \phi = 0.70$$

Step 3   Calculate the column size:

$$h = \sqrt{A_g} \qquad \text{(square column)}$$
$$h \times b = A_g \qquad \text{(rectangular column)}$$
$$h = 2\sqrt{\frac{A_g}{\pi}} \quad \text{(round column)}$$

Round h to the nearest full or even inch.

Step 4   Determine the required area of steel, A_{st}:

$$A_{st} = \frac{P_u - 0.85\phi(0.85f_c'A_g)}{0.85\phi(f_y - 0.85f_c')}$$

Step 5   Use Tables A2–9 and A5–1 to select the size and number of longitudinal bars. Remember that the minimum number of bars for spiral columns is six.

Step 6   Design the spiral steel by equating the spiral steel ratio, ρ_s, to $\rho_{s,min}$. Use a minimum spiral diameter of $^3/_8$ in.:

$$\rho_{s,min} = 0.45\left(\frac{A_g}{A_{ch}} - 1\right)\frac{f_c'}{f_{yt}}$$

$$\rho_s = \frac{4A_{sp}}{h_c s}$$

$$\rho_s = \rho_{s,min}$$

Solve for s (spiral pitch) and round down to the nearest $^1/_4$ in. Check the clear pitch, $s_{clear} = s - d_{sp}$, which must be between 1 and 3 in. If s_{clear} is less than 1 in., increase the spiral size; if s_{clear} is more than 3 in., use 3 in.

EXAMPLE 5–4

Design a short square tied column to carry an axial dead load of 300 kip and a live load of 200 kip. Assume that the applied moments on the column are negligible. Use $f_c' = 4,000$ psi and $f_y = 60,000$ psi.

Solution

Step 1   The factored load, P_u, is:

$$P_u = 1.2P_D + 1.6P_L$$
$$P_u = 1.2(300) + 1.6(200)$$
$$P_u = 680\ kip$$

Assume $\rho_g = 0.03$.

Step 2   The required area of the column, A_g, is:

$$A_g = \frac{P_u}{0.8\phi[0.85f_c'(1 - \rho_g) + f_y\rho_g]}$$

$$A_g = \frac{680}{0.80\,(0.65)[0.85\,(4)(1 - 0.03) + 60(0.03)]}$$

$$A_g = 257\ in^2$$

Step 3   For a square column, the size, h, is:

$$h = \sqrt{A_g} = \sqrt{257}$$
$$\therefore h = 16.0\ in.$$

Try a 16 in. × 16 in. column:

$$A_g = (16)(16) = 256\ in^2$$

Step 4   The required amount of steel, A_{st}, is:

$$A_{st} = \frac{P_u - 0.8\phi(0.85f'_c A_g)}{0.8\phi(f_y - 0.85f'_c)}$$

$$A_{st} = \frac{680 - 0.8 \times 0.65(0.85 \times 4 \times 256)}{0.8 \times 0.65(60 - 0.85 \times 4)} = 7.73 \text{ in}^2$$

Step 5   Select the size and number of bars. For a square column with bars uni-formly distributed along the edges, we keep the number of bars as multi-ples of four. Using Table A2–9, 8 #9 bars ($A_s = 8 \text{ in}^2$) are selected.

From Table A5–1 $\longrightarrow$ Maximum of 12 #9 bars $\therefore$ ok

Step 6   Because the longitudinal bars are #9, select #3 bars for the ties. The maxi-mum spacing of the ties (s_{max}) is:

$$s_{max} = \min\{16d_b, 48d_t, b_{min}\}$$
$$s_{max} = \min\{16(1.128), 48(^3/_8), 16\}$$
$$s_{max} = \min\{18.0, 18.0, 16.0\}$$
$$\therefore s_{max} = 16 \text{ in.}$$

The selected ties are #3 @ 16 in.

To check the tie arrangement, use Figure 5–12. To check the number of ties per set, calculate the clear space between the longitudinal bars:

$$\text{Clear space} = \frac{\overset{\text{Cover}}{16} - \overset{\text{#3 Ties}}{2(1.5)} - 2(^3/_8) - \overset{\text{#9 Bars}}{3(1.128)}}{2}$$

$$\text{Clear space} = 4.4 \text{ in.} < 6.0 \text{ in.}$$

Therefore, one tie per set is enough, as shown below:

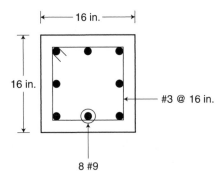

EXAMPLE 5–5

Solve Example 5–4 for a circular spiral column. $f_{yt} = 60,000 \, psi$.

Solution

Step 1   The factored load was determined in Example 5–4: $P_u = 680 \, kip$. Assume $\rho_g = 0.03$.

Step 2   The required gross area of column, A_g, is:

$$A_g = \frac{P_u}{0.85\phi[0.85f_c'(1 - \rho_g) + f_y\rho_g]}$$

$$A_g = \frac{680}{0.85(0.70)[0.85(4)(1 - 0.03) + 60(0.03)]}$$

$$A_g = 224 \, in^2$$

Step 3   The column size, h, is:

$$h = 2\sqrt{\frac{A_g}{\pi}} = 2\sqrt{\frac{224}{3.14}} = 16.9 \, in.$$

Because 17 in. diameter is an odd size, we round up to 18 in.

$$\therefore h = 18 \, in.$$

The provided gross area of the columns, A_g, is:

$$A_g = \frac{\pi h^2}{4} = \frac{3.14(18)^2}{4} = 254 \, in^2$$

Step 4   The area of steel required, A_{st}, is:

$$A_{st} = \frac{P_u - 0.85\phi(0.85f_c'A_g)}{0.85\phi(f_y - 0.85f_c')}$$

$$A_{st} = \frac{680 - 0.85 \times 0.70(0.85 \times 4 \times 254)}{0.85 \times 0.70(60 - 0.85 \times 4)}$$

$$A_{st} = 4.93 \, in^2$$

Step 5   Using Table A2–9, select 7 #8 bars. The provided area of steel is 5.53 in^2.

Table A5–1 $\longrightarrow$ Maximum of 13 #8 bars $\therefore$ ok

Step 6   Design the required spiral:
Because the longitudinal bars are #8 bars, try $\frac{3}{8}$ in. diameter spirals. The cross-sectional area of the spiral, A_{sp}, is 0.11 in^2. The column core size, h_c, is:

$$h_c = h - 2(1.5) = 18 - 2(1.5) = 15 \, in.$$

Therefore, the spiral steel ratio, ρ_s, is:

$$\rho_s = \frac{4A_{sp}}{h_c s} = \frac{4(0.11)}{15s}$$

In the above equation, the pitch of spiral, s, is the unknown. The minimum required spiral steel ratio, $\rho_{s,min}$, is:

$$\rho_{s,min} = 0.45\left(\frac{A_g}{A_{ch}} - 1\right)\frac{f'_c}{f_{yt}}$$

where A_{ch} is the area of core.

$$A_{ch} = \frac{\pi h_c^2}{4} = \frac{3.14(15)^2}{4} = 176.6 \text{ in}^2$$

Substituting the above into the equation for $\rho_{s,min}$:

$$\rho_{s,min} = 0.45\left(\frac{254}{176.6} - 1\right)\frac{4}{60}$$

$$\rho_{s,min} = 0.0131$$

Calculate the spiral maximum pitch, s_{max}:

$$\rho_s = \rho_{s,min}$$

$$\frac{4(0.11)}{15 s_{max}} = 0.0131$$

$$s_{max} = \frac{4(0.11)}{15(0.0131)} = 2.23 \text{ in.}$$

$$s_{max} = 2.0 \text{ in.}$$

In addition, the spiral clear pitch, s_{clear}, should be between 1 and 3 in.:

$$s_{clear} = s - d_{sp} = 2 - \tfrac{3}{8} = 1.625 \text{ in.}$$
$$1 \text{ in.} < 1.625 \text{ in.} < 3 \text{ in.} \quad \therefore \text{ ok}$$

Therefore, the spiral to be used for this column is $\tfrac{3}{8}$ in. diameter at 2 in. The following figure shows the final design of the column.

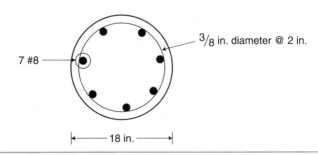

7 #8

$\tfrac{3}{8}$ in. diameter @ 2 in.

18 in.

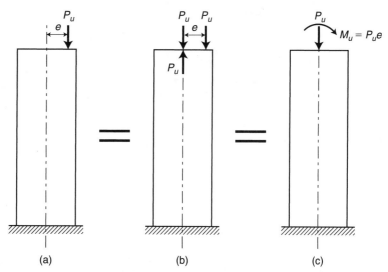

FIGURE 5–20 Eccentrically loaded column.

5.8 BEHAVIOR OF SHORT COLUMNS UNDER ECCENTRIC LOADS

There are two types of columns, based on the applied loads: *axially loaded* and *eccentrically loaded*. In monolithic concrete construction, most columns are eccentrically loaded, which means that the applied load is not acting at the center of the column. In other words, the column is subjected to a moment in addition to the axial load. In the following, we explore the behavior of such columns in more detail.

Figure 5–20a shows a column subject to a load, P_u, at a distance e from the center of the column. By adding two equal and opposite forces of magnitude P_u at the center of the column, as shown in Figure 5–20b, we nullify the net effect because these forces cancel each other out.

The two equal and opposite forces at a distance e form a couple or moment with magnitude $M_u = P_u e$, and a concurrent axial load of P_u applied concentrically as shown in Figure 5–20c. Hence, we conclude that a column subjected to a load P_u at a distance e from its center is equivalent to a concentric load, P_u, and a moment, $M_u = P_u e$. Similarly, a concentric load, P_u, and a moment, M_u, may be represented by an eccentric load, P_u, at an eccentricity e equal to M_u/P_u from the centroid.

The concentric load, P_u, creates a uniform compression stress while the applied moment, M_u, adds bending stresses, as shown in Figure 5–5. Suppose a column has a nominal axial load strength of P_n. If the load is applied at an eccentricity, e, the column axial load capacity, P_n, will be reduced because it is subjected to a moment in addition to the load. The moment, as shown on Figure 5–20c, adds compressive stresses to the already compressed column. Thus, as the eccentricity of the load increases, the applied moment increases, and the axial load capacity of the column decreases.

Figure 5–21 shows the deformations and strains of a typical column subjected to an axial load, P_u, and a bending moment, M_u. When the load is concentric, the deformations across the section are uniform (i.e., the section shortens uniformly). Because strain (ϵ) is

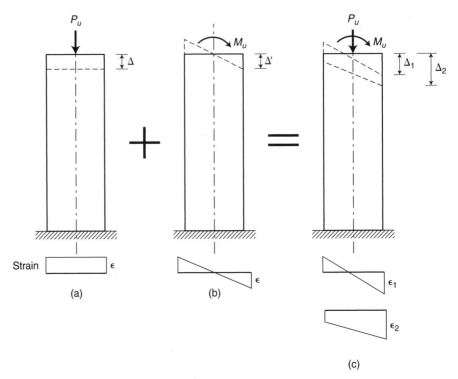

FIGURE 5–21 Deformations and strains from axial load and moment.

defined as the ratio of the change in length to the original length, the distribution of strain across such a section is uniform, as shown in Figure 5–21a. Assume that the same column section now is subjected to only a moment, M_u, which causes one side of the section to be in tension and the other side in compression. The result is a linear distribution of deformation (Δ') or strain (ϵ), as shown in Figure 5–21b.

When a column is subjected to the combined action of an axial load, P_u, and a moment, M_u, the compressive deformations and strains due to P_u and those due to M_u add up, while the compressive deformations and strains due to P_u and the tensile deformations and strains due to M_u reduce each other. Depending on how large M_u is in comparison to P_u, a part of the section may be in tension (large M_u), or the entire section may be in compression (small M_u). In Figure 5–21c, ϵ_1 is the shape of the strain distribution for a large M_u/P_u ratio, and ϵ_2 is the strain distribution for a small M_u/P_u ratio. Therefore, a column with a given amount of reinforcing may fail due either to excessive compression, where the effects of the load and the moment are added up, or to excessive tension, where tension from a large moment overcomes the compression from the axial force.

The ultimate useful strain in the concrete is assumed to be 0.003. Any reinforced concrete column with a given amount of reinforcing has a combination of P_u and M_u that causes the compressive strain in the concrete to reach 0.003 while tensile strain in the steel at the opposite side of the section reaches the yield strain. This state is called a *balanced failure condition,* which is somewhat similar to that defined for reinforced concrete beams (see Chapter 2).

EXAMPLE 5–6

Determine the nominal axial load strength, P_n, and the nominal moment, M_n, for the short tied column shown in Figure 5–22a for the following cases: (1) axial load (i.e., $e = 0.0$); (2) $e = 5$ in.; (3) balanced condition; (4) no load but moment (i.e., $e = \infty$); and (5) axial tensile load. Assume $f'_c = 4,000$ psi, $f_y = 60,000$ psi, and bending about the x-x axis. Do not consider reduction in P_n due to accidental eccentricity.

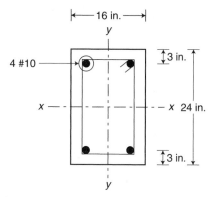

FIGURE 5–22a   Column of Example 5–6.

Solution

1. Assuming the load is concentric, the nominal axial load capacity of the column is the sum of the compressive strengths of the concrete and the steel:

$$P_n = P_o = 0.85f'_c(A_g - A_{st}) + f_y A_{st}$$
$$A_g = (16)(24) = 384 \text{ in}^2$$
$$4 \text{ #10} \longrightarrow \text{Table A2–9} \longrightarrow 5.08 \text{ in}^2$$
$$P_n = 0.85(4)(384 - 5.08) + 60(5.08)$$
$$P_n = 1,593 \text{ kip}$$

Thus, for case 1, $P_n = 1,593$ kip, and $M_n = 0$.

2. Figure 5–22b shows case 2, which is $e = 5$ in. about the x-x axis. In order to determine the value of P_n we must determine the stress in the steel and the distribution of stress in the concrete at the time of failure. The stress and strain in the steel are proportional up to the yield point. Because $e = 5$ in. is small compared to the column depth (i.e., $h = 24$ in., and $e/h = \frac{5}{24} = 0.21$), *assume* that the tensile steel has not reached yield ($\epsilon_t < \epsilon_y$) when the concrete reaches the compressive strain of 0.003. Also, because the yield strain for grade 60 steel is $\epsilon_y = 0.00207$ and the compression steel is close to the compression edge of the column, we can *assume* that the strain in the compression steel is more than the yield ($\epsilon'_s > \epsilon_y$). Therefore,

$$\epsilon'_s > \epsilon_y \longrightarrow f'_s = f_y$$
$$\epsilon_t < \epsilon_y \longrightarrow f_s < f_y$$

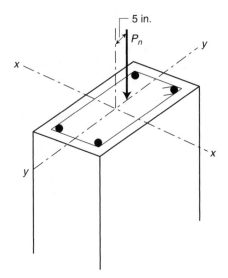

FIGURE 5–22b   Isometric view of column.

Figure 5–23 shows the assumed distribution of strain and stress at failure for this section. The strains in the tension and the compression steel (ϵ_t and ϵ_s') depend on the location of the neutral axis (c). From similarity of the triangles of Figure 5–23b, determine the relationship between ϵ_t and c:

$$\frac{\epsilon_t}{0.003} = \frac{d - c}{c}$$

$$\epsilon_t = 0.003 \left(\frac{d - c}{c} \right)$$

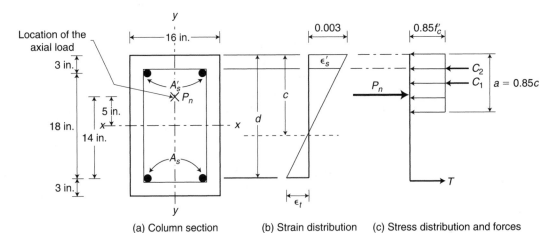

(a) Column section   (b) Strain distribution   (c) Stress distribution and forces

FIGURE 5–23   Assumed strain and stress distribution for $e = 5$ in. for Example 5–6.

The strain in the tensile steel is:

$$f_s = E_s \epsilon_t = 29{,}000(0.003)\left(\frac{d - c}{c}\right)$$

$$f_s = 87\left(\frac{d - c}{c}\right)$$

Use the volumes under the stresses shown in Figure 5–23c to calculate the compression and tensile forces acting on the section:

$$C_1 = 0.85f_c' \, ab = 0.85(4)(0.85c)(16) = 46.24c \text{ kip}$$
$$C_2 = A_s' (f_y - 0.85f_c')$$
$$C_2 = 2.54[60 - 0.85(4)] = 143.8 \text{ kip}$$

The tensile force, T, is:

$$T = f_s A_s$$

Substituting f_s:

$$T = 87\left(\frac{d - c}{c}\right)A_s$$

$$T = 87\left(\frac{21 - c}{c}\right)(2.54) = 221\left(\frac{21 - c}{c}\right)\text{kip}$$

Equilibrium requires that the sum of forces be equal to zero.

$$P_n - C_1 - C_2 + T = 0$$

or

$$P_n = C_1 + C_2 - T$$

$$P_n = 46.24c + 143.8 - 221\left(\frac{21 - c}{c}\right)\text{kip}$$

In addition, the section needs to satisfy the second equilibrium equation (i.e., the sum of moments must equal zero). Taking the moments about the location of tensile steel (A_s) for simplicity:

$$P_n(14) - C_1\left(d - \frac{a}{2}\right) - C_2(18) = 0$$

or

$$P_n(14) = C_1\left(d - \frac{a}{2}\right) + C_2(18)$$

$$P_n = \frac{1}{14}\left[(46.24c)\left(21 - \frac{0.85c}{2}\right) + (143.8)(18)\right]$$

$$P_n = -1.40c^2 + 69.36c + 184.89 \text{ kip}$$

Equating the two expressions for P_n:

$$46.24c + 143.8 - 221\left(\frac{21 - c}{c}\right) = -1.40c^2 + 69.36c + 184.89$$

After some simplifications, the following third order equation results:

$$1.40c^3 - 23.12c^2 + 179.91c - 4{,}641 = 0$$

Solving for c by trial and error:

$$c = 18.93 \text{ in.}$$

Substituting c into either equation for P_n:

$$P_n = 995 \text{ kip}$$

Having determined c and P_n, we now check the correctness of our assumptions. First, calculate the strain in the compression steel (ϵ'_s). From similarity of the triangles of Figure 5–23b:

$$\frac{\epsilon'_s}{0.003} = \frac{c - 3}{c}$$

$$\epsilon'_s = 0.003\left(\frac{c - 3}{c}\right)$$

$$\epsilon'_s = 0.003\left(\frac{18.93 - 3}{18.93}\right)$$

$$\epsilon'_s = 0.00252 > \epsilon_y = 0.00207$$

Strain in the compression steel is more than the yield strain; therefore, the stress is equal to the yield stress ($f'_s = f_y$). Thus, the assumption that the compression reinforcement had yielded is correct. Now we need to determine the level of strain in the tensile steel:

$$\epsilon_t = 0.003\left(\frac{d - c}{c}\right)$$

$$\epsilon_t = 0.003\left(\frac{21 - 18.93}{18.93}\right)$$

$$\epsilon_t = 0.00033 < \epsilon_y = 0.00207$$

Hence, the second assumption (i.e., the tensile steel has not yielded) was also correct.

Calculate the nominal moment:

$$M_n = P_n e = 995\left(\frac{5}{12}\right)$$

$$M_n = 415 \text{ ft-kip}$$

Thus for case 2, $P_n = 995$ kip and $M_n = 415$ ft-kip.

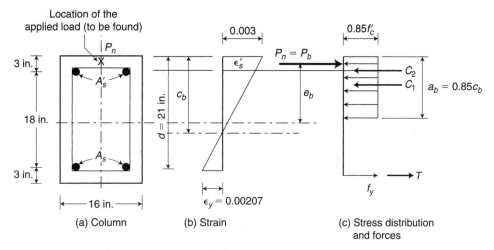

FIGURE 5–24   Strain, stress, and force distributions at balanced condition.

3. Case 3 is the balanced condition. The balanced failure condition occurs when the extreme concrete compression strain is 0.003 and the steel tensile strain is equal to the yield strain, ϵ_y. In this case, the strain distribution across the section is defined, so there is no need to make any assumptions. Figure 5–24 shows the strain and stress at balance condition.

Using similarity of the triangles of Figure 5–24b, locate the neutral axis:

$$\frac{0.003}{\epsilon_y} = \frac{c_b}{d - c_b}$$

$$0.003(d - c_b) = \epsilon_y c_b$$

$$c_b = \frac{0.003d}{0.003 + \epsilon_y}$$

$$c_b = \frac{0.003(21)}{0.003 + 0.00207}$$

$$c_b = 12.43 \text{ in.}$$

Also from similarity of the triangles, calculate the strain (and stress) in the compression steel (ϵ_s' and f_s'):

$$\frac{\epsilon_s'}{0.003} = \frac{c_b - 3}{c_b}$$

$$\epsilon_s' = \frac{0.003(c_b - 3)}{c_b}$$

$$\epsilon_s' = \frac{0.003\,(12.43 - 3)}{12.43}$$

$$\epsilon_s' = 0.00228$$

Because $\epsilon'_s = 0.00228 > \epsilon_y = 0.00207$, $f'_s = f_y = 60$ ksi. The sum of the forces acting on the section is P_n or P_b (see Figure 5–24c):

$$C_1 = 0.85f'_c \, ab = 0.85(4.0)(0.85 \times 12.43)(16) = 574.6 \text{ kip}$$
$$C_2 = A'_s(f_y - 0.85f'_c) = 2.54[60 - 0.85(4)] = 143.8 \text{ kip}$$
$$T = A_s f_y = 2.54(60) = 152.4 \text{ kip}$$
$$P_n = P_b = C_1 + C_2 - T = 574.6 + 143.8 - 152.4$$
$$P_n = 566 \text{ kip}$$

From the sum of moments about the tensile steel, determine the balanced eccentricity, or e_b:

$$P_n(e_b + 9) = C_1\left(d - \frac{0.85c_b}{2}\right) + C_2(18)$$

$$566(e_b + 9) = 574.6\left(21 - \frac{0.85 \times 12.43}{2}\right) + 143.8(18)$$

$$e_b + 9 = 20.53$$
$$e_b = 11.53 \text{ in.}$$

and the nominal moment at the balanced condition, $M_n = M_b$, is:

$$M_n = P_n e = 566\left(\frac{11.53}{12}\right) = 544 \text{ ft-kip}$$

Thus, for case 3, $P_n = 566$ kip and $M_n = 544$ ft-kip. Note that as the moment increases, the axial load decreases.

4. In case 4 the column is subjected only to moment. This is obviously only a theoretical case, as columns always have an axial load. Because the eccentricity is the ratio of moment to applied load, this condition represents a very large (infinite) eccentricity. Essentially, columns subjected to pure moments behave like doubly-reinforced beams. Assuming steel in tension yields ($f_s = f_y$) before concrete crushes in compression, the stress in the compression steel has to be less than the yield stress ($f'_s < f_y$) to make the sum of the compression forces equal to the tensile force. This is because the areas of the tension steel and the compression steel are equal ($A_s = A'_s = 2.54$ in^2), and if both of them yield, the force in the concrete would have to be zero, which definitely can not be true.

Figure 5–25 shows the strain and stress distributions of the section for this condition. Because we do not know the exact level of strain in the tension steel, we cannot determine the location of the neutral axis, c, by using similarity of the triangles of Figure 5–25b. Therefore, determine c through the use of equilibrium equations. First calculate the stress in the compression steel as a function of c. From similarity of the triangles in Figure 5–25b:

$$\frac{\epsilon'_s}{0.003} = \frac{c - 3}{c}$$

$$\epsilon'_s = \frac{0.003(c - 3)}{c}$$

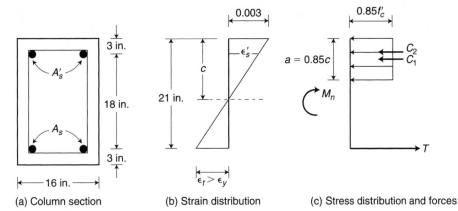

(a) Column section   (b) Strain distribution   (c) Stress distribution and forces

FIGURE 5–25   Strain, stress, and force distribution of column subjected only to moment.

Because $\epsilon'_s < \epsilon_y \longrightarrow f'_s = E_s \epsilon'_s$

$$f'_s = 29{,}000 \times \frac{0.003(c - 3)}{c}$$

$$f'_s = 87 \frac{(c - 3)}{c}$$

The forces acting on the section are:

$$C_1 = 0.85 f'_c \, ab$$
$$C_1 = 0.85(4)(0.85c)(16) = 46.24c \text{ kip}$$
$$C_2 = A'_s [f'_s - 0.85 f'_c]$$
$$C_2 = 2.54[f'_s - 0.85(4)] = 2.54 \left[87 \frac{(c - 3)}{c} - 3.4 \right]$$
$$C_2 = 221 \frac{(c - 3)}{c} - 8.64 \text{ kip}$$
$$T = A_s f_y = 2.54(60) = 152.4 \text{ kip}$$

Equilibrium of the forces acting on the section (Figure 5–25c) requires that:

$$C_1 + C_2 = T$$

$$46.24c + \frac{221(c - 3)}{c} - 8.64 = 152.4$$

Simplifying the above equation, we get:

$$46.24c^2 + 60c - 663 = 0$$

which is a second order equation. Solving for c:

$$c = \frac{-60 + \sqrt{(60)^2 + 4(46.24)(663)}}{2(46.24)}$$

$$c = 3.2 \text{ in.}$$

The stress in the compression steel (f_s') is:

$$f_s' = 87\frac{(c - 3)}{c} = \frac{87(3.2 - 3)}{3.2}$$

$$f_s' = 5.44\,\text{ksi}$$

and the magnitude of the forces acting on the section is:

$$C_1 = 46.24c = 46.24\,(3.2) = 148.0\,\text{kip}$$

$$C_2 = \frac{221\,(c - 3)}{c} - 8.64 = \frac{221(3.2 - 3)}{3.2} - 8.64 = 5.2\,\text{kip}$$

$$T = A_s f_y = 2.54(60) = 152.4\,\text{kip}$$

$$C_1 + C_2 = 148.0 + 5.2 = 153.2\,\text{kip}$$

The small difference between $C_1 + C_2$ and T is due to round-off errors and is negligible. To calculate the nominal moment capacity, M_n, we determine the moment of these forces about the tensile steel:

$$M_n = C_1\left(d - \frac{0.85c}{2}\right) + C_2(d - 3)$$

$$M_n = 148\left(21 - \frac{0.85 \times 3.2}{2}\right) + 5.2(21 - 3)$$

$$M_n = 2{,}907 + 93 = 3{,}000\,\text{in.-kip}/12 = 250\,\text{ft-kip}$$

Thus, for case 4, $P_n = 0$ and $M_n = 250$ ft-kip.

5. Normally, concrete columns are not subjected to pure tension. To obtain a complete picture of the effects of loads and moments, however, we consider the case of a tensile member. Concrete cracks in tension and does not provide any strength. Therefore, the column's tensile strength is provided only by the steel (4 #10 bars):

$$P_n = -A_s f_y = -5.08(60) = -305\,\text{kip}$$

Thus, for case 5, $P_n = -305$ kip and $M_n = 0$. The negative sign means that the column is in tension.

The following table shows the calculated values of P_n, M_n, and corresponding e.

M_n (ft-kip)	P_n (kip)	e (in.)
0	1,593	0
415	995	5
544	566	11.53
250	0	∞
0	−305	0

Figure 5–26 shows the plot of these values to better visualize the results obtained. The horizontal axis in the graph is $M_n = P_n e$ and the vertical axis is P_n.

The graph shows the combinations of moment, M_n, and load, P_n, at which the column may fail. This graph is called a *column interaction diagram*. The interaction

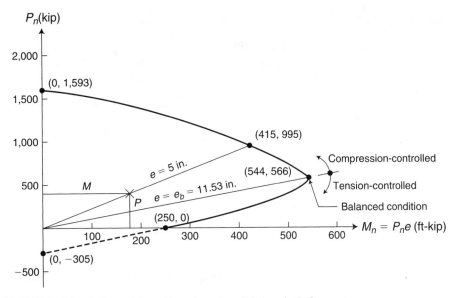

FIGURE 5–26   Column interaction diagram of Example 5–6.

diagram in Figure 5–26 is a unique property of a specific column with given dimensions, materials, and amount of reinforcing. When the load is applied with no eccentricity or moment, the column has a nominal axial load capacity of $P_n = 1,593$ kip. As the eccentricity, e, increases (or the moment on the column increases), the axial load capacity of the column decreases until it reaches the balanced failure condition, which is when the failure of concrete in compression and the yielding of steel in tension occur simultaneously. Values of P_n and M_n above the balanced condition cause the concrete to crush in compression ($\epsilon_c = 0.003$) before the steel yields in tension. Therefore, the column section in this region is *compression controlled*. If the eccentricity increases from the balanced condition (e_b), the failure of the section occurs at decreasing values of P_n and M_n. This may seem odd; however, when the eccentricity increases from a balanced condition ($e > e_b$), the steel in tension yields before concrete in compression crushes (i.e., the element, in fact, may act as a flexural or bending member). In this region the failure of the section is *tension controlled*. Because the column fails in tension in this region, an increasing compression force, P_n, keeps the section from failing and results in an increase in its moment capacity, M_n, as well.

The interaction diagram also shows that, if we draw a line connecting a point on the diagram to the origin ($P_n = M_n = 0$), any point on this line that represents a P and M combination has the same eccentricity (e), because the ratios of P and M are constant.

Here we repeat Example 5–6 to gain a better understanding of the interaction diagram and its relationship to the distribution of strain across the section. This time, however, we examine P_n and M_n values for different levels of strain in the tensile steel (ϵ_t).

EXAMPLE 5–7

Determine M_n and P_n from the interaction diagram of Example 5–6 for the following different levels of strain in the tensile steel: (1) $\epsilon_t = 0.0$; (2) $\epsilon_t = 0.25\epsilon_y$; (3) $\epsilon_t = 0.50\epsilon_y$; (4) $\epsilon_t = 0.75\epsilon_y$; (5) $\epsilon_t = \epsilon_y$; (6) $\epsilon_t = 0.0035$; (7) $\epsilon_t = 0.0040$; and (8) $\epsilon_t = 0.0050$.

Solution

For $f_y = 60,000\,\text{psi}$, the yield strain, ϵ_y, is:

$$\epsilon_y = \frac{f_y}{E_s} = \frac{60,000}{29,000,000} = 0.00207$$

The stress in the steel depends on its strain level:

$$\begin{array}{lll} \text{if} & \epsilon_t < \epsilon_y \longrightarrow f_s = E_s\epsilon_t \\ \text{if} & \epsilon_t \geq \epsilon_y \longrightarrow f_s = f_y \end{array} \tag{1}$$

Now, we consider the distribution of strain and stress on the column section, as shown in Figure 5–27.

In order to determine P_n and M_n for different values of ϵ_t, we must calculate the stress levels of the steel in tension (f_s) and in compression (f_s'). The tensile stress of steel can be determined from the strains (ϵ_t) given in the problem statement. To calculate f_s', however, we must determine the location of the neutral axis (c) for each strain case.

From similarity of the triangles of Figure 5–27b:

$$\frac{c}{0.003} = \frac{d - c}{\epsilon_t}$$

$$c = \frac{0.003d}{0.003 + \epsilon_t} = \frac{0.003(21)}{0.003 + \epsilon_t} = \frac{0.063}{0.003 + \epsilon_t} \tag{2}$$

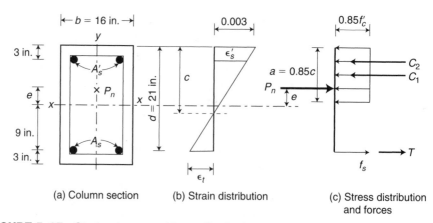

(a) Column section   (b) Strain distribution   (c) Stress distribution and forces

FIGURE 5–27   Strain, stress, and force distribution of column of Example 5–7.

The strain in the compression steel (ϵ_s') is:

$$\frac{c}{0.003} = \frac{c - 3}{\epsilon_s'}$$

$$\epsilon_s' = \frac{0.003(c - 3)}{c}$$

$$\begin{aligned} \text{if} \quad & \epsilon_s' < \epsilon_y \longrightarrow f_s' = E_s \epsilon_s' \\ \text{if} \quad & \epsilon_s' \geq \epsilon_y \longrightarrow f_s' = f_y \end{aligned} \tag{3}$$

The forces acting on the section shown in Figure 5–27c are:

$$C_1 = 0.85 f_c' ab = 0.85(4)(0.85c)(16) = 46.24c \text{ kip}$$
$$C_2 = f_s' A_s' - 0.85 f_c' A_s' = f_s'(2.54) - 0.85(4)(2.54) = 2.54 f_s' - 8.64 \text{ kip}$$
$$T = A_s f_s = 2.54 f_s \text{ kip}$$

Equilibrium requires that:

$$P_n = C_1 + C_2 - T$$
$$P_n = 46.24c + 2.54 f_s' - 8.64 - 2.54 f_s \text{ kip} \tag{4}$$

Taking moments about the location of the tensile force:

$$P_n(e + 9) = C_1\left(d - \frac{a}{2}\right) + C_2(d - 3)$$

$$P_n(e + 9) = 46.24c\left(21 - \frac{0.85c}{2}\right) + (2.54 f_s' - 8.64)(21 - 3)$$

$$e = \frac{46.24c(21 - 0.425c) + 18(2.54 f_s' - 8.64)}{46.24c + 2.54 f_s' - 8.64 - 2.54 f_s} - 9 \tag{5}$$

$$M_n = P_n \frac{e}{12} \tag{6}$$

For each strain case we can use equations 1, 2, and 3 to determine the location of the neutral axis (c) and the stress in the tensile and the compression steels (f_s and f_s'). Having c, f_s, and f_s', we can calculate P_n, e, and M_n from equations 4 to 6. The table below shows the results for each case:

Case	ϵ_t	ϵ_s'	c (in.)	M_n (ft-kip)	P_n (kip)	e (in.)
1	0.00	0.0026	21.0	357	1,115	3.84
2	0.25(0.00207)	0.0025	17.91	439	934	5.64
3	0.5(0.00207)	0.0024	15.62	488	790	7.41
4	0.75(0.00207)	0.0023	13.84	520	670	9.32
5	1.0(0.00207)	0.0023	12.43	544	566	11.53
6	0.0035	0.0021	9.69	517	440	14.1
7	0.0040	0.0020	9.0	502	402	14.96
8	0.0050	0.00186	7.875	473	340	16.7

Figure 5–28 shows the interaction diagram generated from the results of Examples 5–6 and 5–7. This diagram is the same as the one shown in Figure 5–26. The levels of the tensile

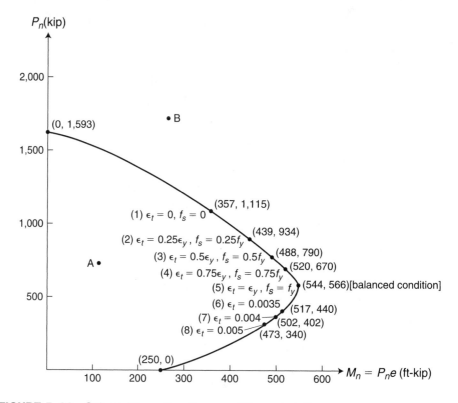

FIGURE 5–28 Column interaction diagram of Example 5–7.

steel strain (ϵ_t) and stress (f_s) along the curve are also shown. This interaction diagram shows the maximum nominal capacity of the column. Any combination of P_u and M_u that lies inside the curve (e.g., point A) is safe for the column; however, any combination of P_u and M_u that lies outside the interaction diagram (e.g., point B) will cause the column to fail.

5.9 ACI COLUMN INTERACTION DIAGRAMS

In Examples 5–6 and 5–7, the rectangular column had 4 #10 bars. But if we increase the area of reinforcements (e.g., to 8 or 12 #10 bars), the shape of the interaction diagram would remain approximately the same but would have larger values of P_n and M_n, as shown in Figure 5–29.

Column interaction diagrams exist for rectangular and round columns with different reinforcement arrangements, concrete compression strength, and steel tensile strength. These curves are similar to the one shown in Figure 5–29. To make these curves more versatile, however, the P_n values are substituted by $K_n = \dfrac{P_n}{f_c' A_g}$, which is a nondimensional parameter as long as the values are substituted with consistent units (e.g., kip and inches). Also,

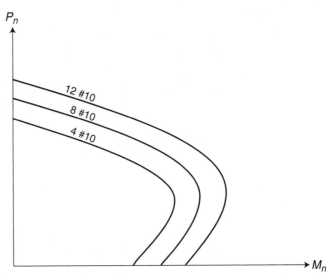

FIGURE 5–29   Column interaction diagrams for different areas of steel reinforcements.

$R_n = \dfrac{M_n}{f_c' A_g h}$ is used for the horizontal axis instead of M_n,, which is a nondimensional value. Each set of curves is made for a specific arrangement of reinforcement, compressive strength of concrete (f_c'), yield strength of steel (f_y), steel ratio (ρ_g), and parameter γ, which represents the spread of reinforcements in the column:

$$\gamma = \frac{h'}{h} \tag{5–18}$$

where h' and h are the distance between the center to center of the extreme steel in the column, and the total depth of the column, respectively, as shown in Figure 5–30. The dimensions h' and h are measured perpendicular to the bending axis of the column.

Figure 5–31 shows the interaction diagram for a rectangular column with steel reinforcement uniformly distributed around the column. It is from the ACI Design Handbook, SP–17(97). In Figure 5–31, $f_c' = 4$ ksi, $f_y = 60$ ksi, and $\gamma = 0.6$. The interaction diagrams

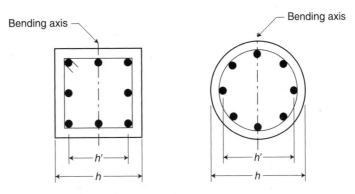

FIGURE 5–30   Definition of parameter γ ($\gamma = h'/h$).

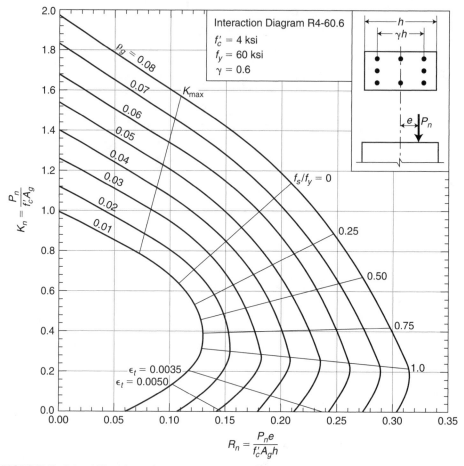

FIGURE 5–31 ACI column interaction diagram [SP-17(97)].

are for $\rho_g = 0.01$ to 0.08. Figure 5–32 is a similar interaction diagram for a circular column with $f_c' = 4$ ksi, $f_y = 60$ ksi, and $\gamma = 0.8$. The diagrams show the levels of stress in the tension steel (f_s) as a fraction of the steel yield strength (f_y). In addition, they indicate tensile strains of $\epsilon_t = 0.0035$ and 0.0050. Another value that is given is K_{max}, which is the maximum useable nominal axial load capacity for a tied column:

$$K_{max} = 0.80[0.85f_c'(A_g - A_{st}) + A_{st}f_y] \qquad (5-19)$$

In essence K_{max} represents a cut-off level for K_n (and consequently for P_n). The value of K_{max} in the diagrams is defined for *tied* columns, as just mentioned. For a spiral column, the value from the diagrams has to be multiplied by $\dfrac{0.85}{0.80} = 1.0625$ (i.e., the ratio of the limiting coefficients that account for accidental moments). These computer-generated interaction diagrams assume the reinforcement to be a thin rectangular tube for rectangular cross sections that have longitudinal reinforcements distributed along all four faces, and a thin circular tube for patterns of longitudinal steel bars arranged in a circle.

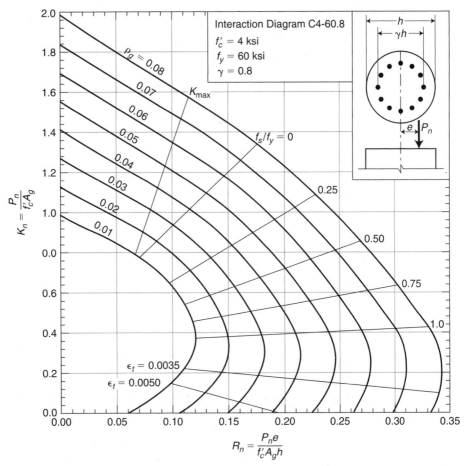

FIGURE 5–32 ACI column interaction diagram [SP-17(97)].

5.10 DESIGN AXIAL LOAD STRENGTH (ϕP_n), AND MOMENT CAPACITY (ϕM_n)

The latest ACI column interaction diagrams [ACI Design Handbook, SP-17(97)] take no consideration of the strength reduction factor, ϕ, which is a significant change from previous versions. This change is intended mainly to make the diagrams as universal as possible.

As in the case of beams, the design resisting moment of columns (M_R) and their design axial load strength (P_R) are:

$$M_R = \phi M_n$$
$$P_R = \phi P_n$$

where ϕ is the column strength reduction factor, which depends on the level of strain in the tension steel. Figure 5–33 shows the column strength reduction factor as a function of net tensile steel strain (ε_t) for tied and spiral columns. Most building columns are compression

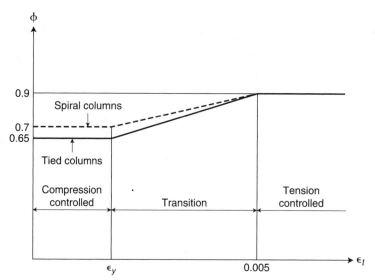

FIGURE 5–33   Variation of strength reduction factor (ϕ) with the net tensile strain in steel (ϵ_t).

controlled, that is, the concrete reaches the ultimate useable compressive strain of 0.003 before the strain in the tensile steel reaches the yield strain (ϵ_y). The ϕ factor is constant for compression-controlled sections (0.70 for spiral columns and 0.65 for tied columns). If the moment on the column is relatively large compared to the axial load, the column section may be in the transition zone. Then the ϕ factor varies between 0.70 for spiral columns, or 0.65 for tied columns, and 0.90 as ϵ_t varies between ϵ_y and 0.005. If ϵ_t is more than 0.005, the section is tension controlled, and ϕ is constant and equal to 0.9. Such a section acts like a flexure member (beam) rather than a compression member (column).

    If the column section is compression controlled, we can easily calculate the design resisting axial load, P_R, and the design resisting moment, M_R; however, when the column is in the transition zone, we must calculate ϕ from ϵ_t as given below:

$$\phi = \frac{0.0035 - 0.9\epsilon_y}{0.005 - \epsilon_y} + \frac{0.20}{0.005 - \epsilon_y}\epsilon_t \quad \text{(spiral column)} \qquad (5\text{--}20)$$

$$\phi = \frac{0.00325 - 0.9\epsilon_y}{0.005 - \epsilon_y} + \frac{0.25}{0.005 - \epsilon_y}\epsilon_t \quad \text{(tied column)} \qquad (5\text{--}21)$$

These values have significance in only a small area of the interaction diagrams (i.e., the zone between $\frac{f_s}{f_y} = 1.0$ and $\epsilon_t = 0.005$).

    Graphs that relate ϕ and K_n have been developed for each interaction diagram to simplify the computation of ϕ. Figures 5–34 and 5–35 are examples of such graphs and are to be used in conjunction with the interaction diagrams of Figures 5–31 and 5–32, respectively. Appendix A contains additional interaction diagrams and their corresponding K_n versus ϕ graphs.

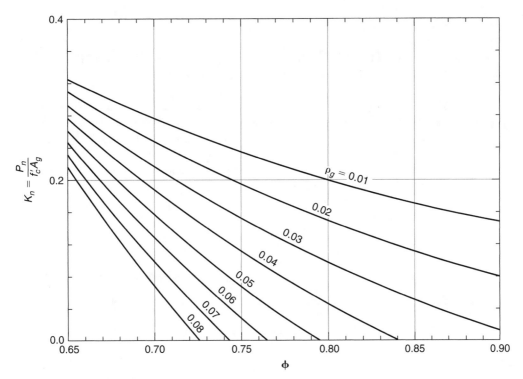

FIGURE 5–34 K_n vs. ϕ diagram for the interaction diagram of Figure 5–31.

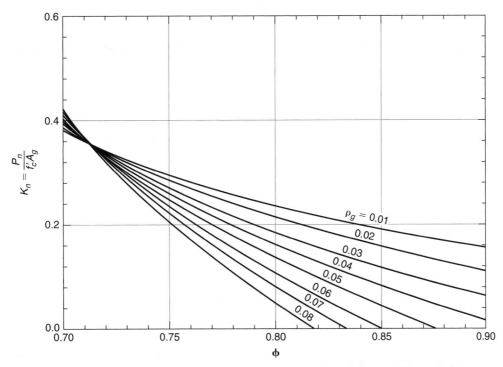

FIGURE 5–35 K_n vs. ϕ diagram for the interaction diagram of Figure 5–32.

5.11   ANALYSIS OF SHORT COLUMNS WITH LARGE ECCENTRICITY USING INTERACTION DIAGRAMS

The analysis of columns with large eccentricities can be approached in many different ways. A possible approach is to ask the question, "Is a particular column safe or not for a given set of P_u and M_u?" Another approach is to determine the largest factored axial load that the column may take with a given moment. Yet another approach is to ask the question, "What is the largest eccentricity (e) that a given factored load may safely have?" Regardless of the approach, one can always take advantage of the interaction diagrams.

The value of ϕ is constant if the section is compression controlled. For columns in the transition zone, however, ϕ must be adjusted accordingly. Therefore, the procedures for the analysis of columns in compression-controlled and non–compression-controlled zones are somewhat different.

Analysis of Columns with Compression-Controlled Behavior

The following are steps for the analysis of compression-controlled members which are summarized in Figure 5–36:

Step 1   Calculate and check the column steel ratio, ρ_g:

$$0.01 \leq \rho_g = \frac{A_{st}}{A_g} \leq 0.08$$

The strength reduction factor, ϕ, is equal to 0.65 for tied, and 0.70 for spiral columns.

Step 2   Calculate $\gamma = \dfrac{h'}{h}$ (see Figure 5–30) and the nondimensional factors K_n or R_n:

$$K_n = \frac{P_u}{\phi f_c' A_g}$$

$$R_n = \frac{M_u}{\phi f_c' A_g h}$$

If we know both P_u and M_u, then we may take these calculated K_n and R_n values as the "demand" on the section. We now enter into the appropriate interaction diagram (based on f_c', f_y, and γ) and locate the point defined by the calculated K_n and R_n. If this point falls within the curve defined by ρ_g (calculated in step 1), the P_u and M_u combination is safe for this column.

Another approach is to calculate only K_n (or R_n) and, from the appropriate interaction diagram, obtain the corresponding R_n (or K_n) using the calculated ρ_g.

We need to consider two important points here:

1. If the point defined by the calculated K_n or R_n on the interaction diagram falls between $K_{\max}$ and $\dfrac{f_s}{f_y} = 1.0$ (balanced condition), the column section is compression controlled, and the assumed ϕ is correct. Therefore, proceed to step 3.

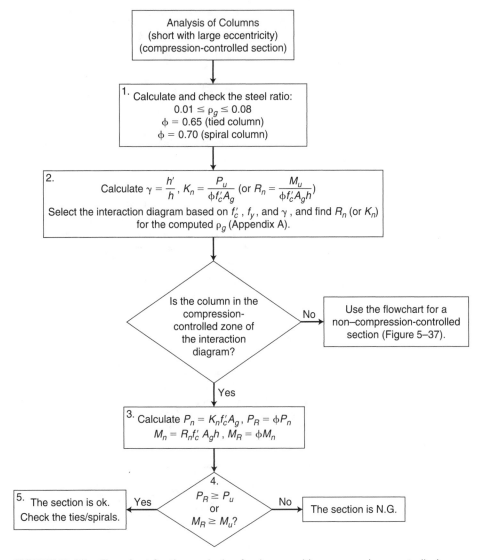

FIGURE 5–36 Flowchart for the analysis of columns with compression-controlled section.

2. If the point defined by the calculated K_n or R_n falls below $\dfrac{f_s}{f_y} = 1.0$, the column section is either in the transitional or the tension-controlled zone. This means that the ϕ value used in step 1 is not correct and must be adjusted. We need to proceed with the steps for the analysis of non–compression-controlled columns, as discussed in step 2a and summarized in Figure 5–37.

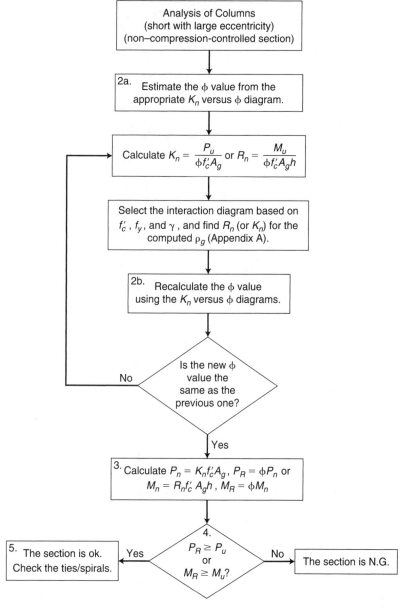

FIGURE 5–37   Flowchart for the analysis of columns with non–compression-controlled section.

Step 3   Calculate the axial load (or moment) capacities:

$$P_n = K_n f'_c A_g, \; P_R = \phi P_n$$
$$M_n = R_n f'_c A_g h, \; M_R = \phi M_n$$

$\phi = 0.65$ for the tied columns and 0.70 for the spiral columns.

Step 4   Check the column capacity. For the column to be adequate, its axial load capacity, P_R, has to be greater than the applied load, P_u or the column moment capacity, M_R, has to be larger than the applied moment, M_u:

$$P_R \geq P_u \quad \text{or} \quad M_R \geq M_u$$

Step 5   Check the ties or spirals. If the section is ok, we can check the ties or spirals as we did for the columns with small eccentricity (step 5 of Figure 5–16).

Analysis of Non–Compression-Controlled Columns

The following are steps for the analysis of non–compression-controlled columns, which are summarized in Figure 5–37.

Step 1   Same as that for compression-controlled columns.

Step 2a   Estimate the ϕ value. If the column is not compression controlled, the assumption made for the ϕ factor is not correct. A larger ϕ value, which will increase the P_R and M_R capacities, can be used in the non–compression-controlled region of the interaction diagrams. The ϕ factor, however, varies with the tensile steel strain (ϵ_t), as shown in Figure 5–33. Because the value of ϵ_t at this stage is unknown, we estimate a new ϕ value and recalculate K_n or R_n if only one of them is given.

$$K_n = \frac{P_u}{\phi f'_c A_g}$$

$$R_n = \frac{M_u}{\phi f'_c A_g h}$$

Use f'_c, f_y, and γ to select the appropriate interaction diagram. Having K_n or R_n and ρ_g, we can either locate the corresponding point on the diagram, when both K_n and R_n are known, or obtain the corresponding R_n (or K_n), when only one of them is known.

Step 2b   Recalculate the ϕ value. At this point, recalculate the ϕ value using the K_n value obtained in step 2a. We can determine the corrected ϕ factor by using K_n versus ϕ diagrams such as those in Figures 5–34 and 5–35. If the new ϕ factor is close to the previous estimate, proceed to step 3. Otherwise, move back to step 2a and use the new ϕ factor to revise K_n or R_n, then repeat the process.

Step 3   Calculate the column's resisting load and moment:

$$P_n = K_n f'_c A_g \qquad P_R = \phi P_n$$
$$M_n = R_n f'_c A_g h \qquad M_R = \phi M_n$$

Step 4   Check the adequacy of the column. The following relationships must be satisfied for the column to be adequate:

$$P_R \geq P_u \quad \text{or} \quad M_R \geq M_u$$

Step 5   Check the ties or spirals. If the column section is adequate, we can check the ties or spirals (step 5 of Figure 5–16).

EXAMPLE 5-8

Determine the maximum axial load that can be applied on the short tied column section shown below (this is the column used in Example 5–1). The applied dead and live load moments are $M_D = M_L = 35$ ft-kip. Use $f'_c = 4,000$ psi and $f_y = 60,000$ psi.

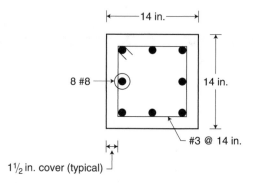

Solution

Step 1   Check the column steel ratio, ρ_g:

$$\rho_g = \frac{A_{st}}{A_g} = \frac{6.32}{196} = 0.032$$

$$0.01 < \rho_g = 0.032 < 0.08 \quad \therefore \text{ ok}$$

$$h = 14 \text{ in.} \longrightarrow \text{Table A5–1} \longrightarrow \text{Maximum of 12 \#8 bars} \quad \therefore \text{ ok}$$

This is a tied column; therefore, $\phi = 0.65$.

Step 2   Select the interaction diagram to be used.

$$h' = 14 - 2(1.5) - 2(\tfrac{3}{8}) - 2(\tfrac{1}{2}) = 9.25 \text{ in.}$$
$$\gamma = h'/h = 9.25/14 = 0.66$$
$$M_u = 1.2M_D + 1.6M_L = 1.2 \times 35 + 1.6 \times 35 = 98 \text{ ft-kip}$$
$$R_n = \frac{M_u}{\phi f'_c A_g h} = \frac{98 \times 12}{0.65(4)(196)14} = 0.165$$

$f'_c = 4$ ksi, $f_y = 60$ ksi, and $\gamma = 0.66$; therefore, we use the interaction diagrams of Figures A5–1a and A5–2a and interpolate:

Figure A5–1a ($\gamma = 0.60$) $\longrightarrow$ $\rho_g = 0.032$, $\quad R_n = 0.165 \longrightarrow K_n = 0.64$
Figure A5–2a ($\gamma = 0.70$) $\longrightarrow$ $\rho_g = 0.032$, $\quad R_n = 0.165 \longrightarrow K_n = 0.74$

Interpolating between the above K_n values for $\gamma = 0.66$:

$$K_n = 0.64 + \frac{(0.74 - 0.64)(0.66 - 0.60)}{0.1} = 0.70$$

The point $\rho_g = 0.032$ and $R_n = 0.165$ on both the interaction diagrams is in the compression-controlled zone; therefore, proceed with step 3.

Step 3 & 4   The nominal axial load capacity, P_n, is:

$$P_n = K_n f'_c A_g$$
$$P_n = 0.70(4)(196) = 549 \text{ kip}$$
$$P_R = \phi P_n = 0.65(549) = 357 \text{ kip}$$

If we compare $P_R = 357$ kip with the result of Example 5–1 for the same column but with small eccentricity, $P_R = 533$ kip, the effect of added moment on the reduction of the column axial load capacity is evident.

Step 5   The procedure of checking the ties is the same as that shown in Example 5–1.

EXAMPLE 5–9

Check the adequacy of the short spiral column shown below (the same as the one used in Example 5–2) if it is subjected to $M_D = 36$ ft-kip and $M_L = 43$ ft-kip. Use $P_D = 200$ kip, $P_L = 225$ kip, $f'_c = 4,000$ psi, and $f_y = 60,000$ psi.

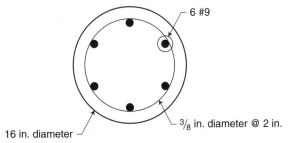

6 #9

3/8 in. diameter @ 2 in.

16 in. diameter

Solution

Step 1   Check the column steel ratio, ρ_g:

$$\rho_g = \frac{A_{st}}{A_g} = \frac{6.0}{201.1} = 0.03$$

$$0.01 < \rho_g = 0.03 < 0.08 \quad \therefore \text{ ok}$$

If we assume that the column is compression controlled, $\phi = 0.70$.

Step 2, 3 & 4   To show the different ways of solving this problem, steps 2, 3, and 4 steps are combined. However, one can choose one of these methods to solve the problem. Because we have both the axial loads and the moments, we proceed as follows:

$$h' = 16 - 2(1.5) - 2(\tfrac{3}{8}) - 2(1.128/2) = 11.12 \text{ in.}$$

$$\gamma = h'/h = 11.12/16 = 0.70$$

$$M_u = 1.2M_D + 1.6M_L = 1.2 \times 36 + 1.6 \times 43 = 112 \text{ ft-kip}$$

$$P_u = 1.2P_D + 1.6P_L = 1.2 \times 200 + 1.6 \times 225 = 600 \text{ kip}$$

$$R_n = \frac{M_u}{\phi f'_c A_g h} = \frac{112 \times 12}{0.7(4)(201.1)16} = 0.15$$

$$K_n = \frac{P_u}{\phi f'_c A_g} = \frac{600}{0.7 \times 4 \times 201.1} = 1.07$$

If we enter these values into the interaction diagram shown in Figure A5–10a, the point falls in the compression-controlled region. The point representing $K_n = 1.07$ and $R_n = 0.15$ requires a ρ_g value of about 0.058; the provided value, however, is 0.03. Thus, the column is *not* adequate.

Alternatively, to calculate the maximum factored load that may be used in conjunction with the given moments, proceed as follows.

For $R_n = 0.15$ and $\rho_g = 0.03$, the corresponding K_n using Figure A5–10a is

$$K_n = 0.56$$

Thus

$$P_n = K_n f_c' A_g = 0.56 \times 4 \times 201.1 = 450\,\text{kip}$$

and

$$P_R = \phi P_n = 0.70 \times 450 = 315\,\text{kip}$$

Therefore, another easy way of solving the problem is just to compare P_u with P_R:

$$P_R = 315\,\text{kip} < P_u = 600\,\text{kip} \quad \therefore\ N.G.$$

In Example 5–2, the axial load capacity of this column, P_R, was 609 kip, which was satisfactory. Comparison of the axial load capacities again shows the significant decrease due to the applied moment.

Alternatively, the following question could be asked: "How large an eccentricity may the given loads safely have?" Then we proceed as follows.

For $K_n = 1.07$ and $\rho_g = 0.03$, the corresponding R_n from the interaction diagram is

$$R_n = 0.061$$

Thus

$$M_n = P_n e = R_n f_c' A_g h = 0.061 \times 4 \times 201.1 \times 16$$
$$= 785\,\text{kip-in} = 65.4\,\text{ft-kip}$$

and

$$M_R = \phi M_n = 0.70 \times 65.4 = 45.8\,\text{ft-kip}$$

$P_u = 600\,\text{kip}$ (see above calculation), so

$$e = \frac{M_R}{P_u} = \frac{45.8 \times 12}{600} = 0.92\,\text{in.}$$

EXAMPLE 5–10

Determine whether the tied column shown below is adequate. The applied factored axial load and bending moment are $P_u = 70$ kip, and $M_u = 60$ ft-kip, respectively. Use $f_c' = 4$ ksi, and $f_y = 60$ ksi. Also, check the adequacy of the ties. The typical clear cover is 1.5 in.

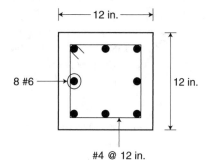

8 #6

12 in.

12 in.

#4 @ 12 in.

Solution

Step 1   Check the steel ratio, ρ_g:

$$8 \text{ #6 bars} \longrightarrow \text{Table A2–9} \longrightarrow A_{st} = 3.52 \text{ in}^2$$

$$\rho_g = \frac{A_{st}}{A_g} = \frac{3.52}{12 \times 12} = 0.024$$

$$0.01 < 0.024 < 0.08 \quad \therefore \text{ ok}$$

We assume that the column is compression controlled; therefore, $\phi = 0.65$.

Step 2

$$h' = 12 - 2(1.5) - 2(\tfrac{1}{2}) - 2(0.75/2) = 7.25 \text{ in.}$$

$$\gamma = \frac{7.25}{12} = 0.60$$

$$K_n = \frac{P_u}{\phi f_c' A_g}$$

$$K_n = \frac{70}{0.65(4)(12 \times 12)} = 0.19$$

Figure A5–1a is the interaction diagram for $f_c' = 4 \text{ ksi}, f_y = 60 \text{ ksi}$, and $\gamma = 0.60$ with uniformly distributed reinforcements for a rectangular column. The point for $\rho_g = 0.024$ and $K_n = 0.19$ falls in the transition zone (i.e., the region between $\dfrac{f_s}{f_y} = 1.0$ and $\epsilon_t = 0.005$). Therefore, continue with the analysis for columns with non–compression-controlled sections using the flowchart of Figure 5–37.

Step 2a   Use Figure A5–1b with $\rho_g = 0.024$ and $K_n = 0.19$ to obtain the strength reduction factor, ϕ:

$$\phi = 0.735$$

Therefore, the new value of K_n is:

$$K_n = \frac{P_u}{\phi f_c' A_g} = \frac{70}{0.735(4.0)(12 \times 12)} = 0.165$$

Using $\rho_g = 0.024$ and $K_n = 0.165$, from Figure A5–1a, we find that the section is in the transition zone and $R_n = 0.155$.

Step 2b   We use the K_n versus ϕ graph of Figure A5–1b to obtain the new ϕ value with the new values of K_n and ρ_g. The new strength reduction factor, ϕ, is 0.76. Because this new ϕ factor is different from the one obtained in step 2a, repeat the process:

$$K_n = \frac{P_u}{\phi f_c' A_g} = \frac{70}{0.76(4.0)(12 \times 12)} = 0.160$$

Using $\rho_g = 0.024$ and $K_n = 0.160$, from Figure A5–1a, we find that the section is in the transition zone, $R_n = 0.152$. From Figure A5–1b, $\phi = 0.77$. Repeat the process as this new ϕ factor is different from the previous value:

$$K_n = \frac{P_u}{\phi f_c' A_g} = \frac{70}{0.77(4.0)(12 \times 12)} = 0.158$$

From Figure A5–1a, we can conclude that the section is in the transition zone and obtain $R_n = 0.152$. Use Figure A5–1b to obtain $\phi = 0.77$. Because the new ϕ value is about the same as the one obtained in the previous iteration, we proceed with step 3.

Step 3

$$M_n = R_n f_c' A_g h$$

$$M_n = \frac{0.152(4)(12 \times 12)(12)}{12}$$

$$M_n = 87.6 \text{ ft-kip}$$

$$M_R = \phi M_n = 0.77(87.6) = 67.5 \text{ ft-kip}$$

Step 4

$$M_R = 67.5 \text{ ft-kip} > M_u = 60 \text{ ft-kip} \quad \therefore \text{ ok}$$

Therefore, the section is adequate.

Step 5   Check the adequacy of the ties:

$$s_{max} = \min\{16d_b, 48d_t, b_{min}\}$$

$$s_{max} = \min\{16(^6\!/_8), 48(^1\!/_2), 12 \text{ in.}\}$$

$$s_{max} = \min\{12 \text{ in.}, 24 \text{ in.}, 12 \text{ in.}\}$$

Therefore, #4 @ 12 in. is ok.

Step 6   Check the tie arrangement using Figure 5–12:

$$\text{Clear space} = \frac{12 - 2(1.5) - 2(0.5) - 3(0.75)}{2} = 2.9 \text{ in.}$$

$$2.9 \text{ in.} < 6 \text{ in.} \quad \therefore \text{ One set of ties is required.}$$

Therefore, the tie arrangement is ok.

5.12   DESIGN OF SHORT COLUMNS WITH LARGE ECCENTRICITY

The design of columns with large eccentricity, similar to the analysis, depends on the column behavior under the load and moment. If the column is compression controlled, the ϕ factor is constant, and the design process is straightforward. If the column is not compression controlled, however, the ϕ factor varies with the tensile strain in the steel (ϵ_t). Thus, using the ϕ value as if the column were compression controlled would lead to a conservative design. An iterative approach is necessary to calculate the correct ϕ value.

Design of Columns with Compression-Controlled Behavior

The following are steps for the design of compression-controlled columns, which are also summarized in the flowchart of Figure 5–38:

Step 1   Determine the factored loads and moments acting on the column. The loads on the column come from the beams and slabs connected to it. The moments acting on the column are due to either gravity loads or lateral loads such as wind or earthquake loads. If gravity loads are considered:

$$P_u = 1.2P_D + 1.6P_L$$
$$M_u = 1.2M_D + 1.6M_L$$

Step 2   Estimate the column size. In most cases, the column size is preselected based on architectural considerations or ease of construction. If it is necessary to estimate the column size, however, we must make reasonable assumptions because both the area of the column (A_g) and the area of steel (A_{st}) are unknown. Different simplifying assumptions may be used to obtain a preliminary size. The most common assumptions are that the column's capacity is the same as that of an axially loaded column and that the area of steel, A_{st}, is neglected. The latter is made in an attempt to account for the effects of the moments. The preliminary design equations are given below:

$$P_R = 0.8\phi[0.85f_c'(A_g - A_{st}) + f_y A_{st}] \quad \text{(tied column)}$$
$$P_R = 0.85\phi[0.85f_c'(A_g - A_{st}) + f_y A_{st}] \quad \text{(spiral column)}$$

Assuming $A_{st} = 0$, these equations become:
Thus:

$$P_R = 0.8\phi(0.85f_c' A_g) \quad \text{(tied column)}$$
$$P_R = 0.85\phi(0.85f_c' A_g) \quad \text{(spiral column)}$$

When designing columns, $P_R = P_u$. Therefore, if this substitution is made, we can calculate the column area, A_g, as follows:

$$A_g = \frac{P_u}{0.8\phi(0.85f_c')} \quad \text{(tied column)}$$

$$A_g = \frac{P_u}{0.85\phi(0.85f_c')} \quad \text{(spiral column)}$$

After calculating the gross area of the column, A_g, we can select the column size. Round the column size to the nearest inch.

Step 3 & 4   Calculate γ, K_n, and R_n. In order to utilize interaction diagrams it is necessary to calculate $\gamma = \dfrac{h'}{h}$. We assume that the centerline of the longitudinal reinforcing is 2.5 in. from the face of the column.

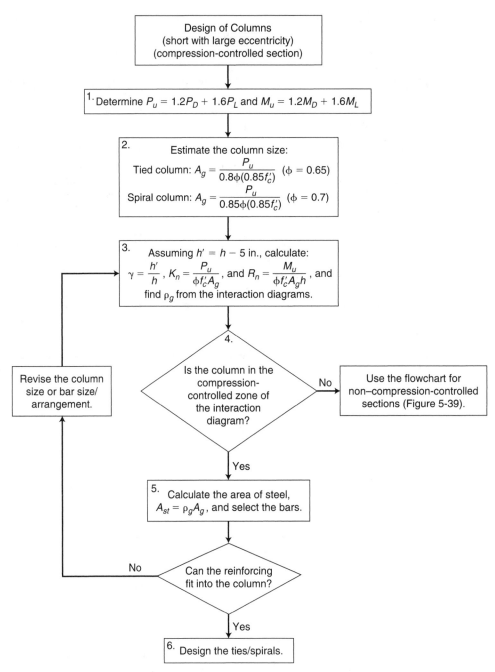

FIGURE 5–38   Flowchart for the design of compression-controlled columns.

Thus

$$h' = h - 2(2.5) = h - 5 \text{ in.}$$

Using f'_c, f_y, and γ, we select the appropriate interaction diagram, and then compute K_n and R_n:

$$K_n = \frac{P_u}{\phi f'_c A_g} \qquad R_n = \frac{M_u}{\phi f'_c A_g h}$$

Entering the interaction diagram with the known values of K_n and R_n, we determine whether the column is compression controlled. If the column is not compression controlled or if R_n is out of the range of the interaction diagram, we proceed to the flowchart for the design of non–compression-controlled sections (Figure 5–39). Otherwise, use K_n and R_n to obtain ρ_g and move to step 5.

Step 5   Calculate the required area of steel, A_{st}:

$$A_{st} = \rho_g A_g$$

Select bars using Tables A2–9 and A5–1. If the selected bars can fit into the column based on Table A5–1, we proceed to step 6. Otherwise, increase the column size, which will increase the space between the bars, and repeat the process starting back in step 3.

Step 6   Design the ties or spirals. This step is the same as that for columns with small eccentricity (as shown in Figure 5–17).

Design of Non–Compression-Controlled Columns

The following are steps for the design of non–compression-controlled columns, which are also summarized in the flowchart of Figure 5–39.

Steps 1   through 3 are the same as those for compression-controlled columns.

Step 4a   Assume a ϕ between 0.65 and 0.90 (say $\phi = 0.80$), because the column is not compression controlled.

Step 4b   Check the column size. If the column is in the non–compression-controlled region or R_n is completely outside the range of the interaction diagram (due to very large moment), the ϕ factor used in step 2 of the design of compression-controlled columns is overly conservative. If R_n is out of the range of the interaction diagram, we need to increase the column size, as the moment on the column is too large for the column dimensions. If this is the case, we proceed to step 4c. Otherwise, the column dimensions are ok and we go directly to step 4e.

Step 4c   If the column size is not enough for the applied moment, we need to resize it. In order to do so, assume an R_n value in the transition zone for a steel ratio of ρ_g about 0.02. Recall that an estimated $\rho_g = 0.03$ was used for the design of columns with small eccentricity. We try to work with a smaller steel ratio for columns in the non–compression-controlled zone, however, as they behave more like flexural members.

Step 4d   Resize the column. The resisting moment of the column, M_R, is:

$$M_R = \phi R_n f'_c A_g h$$

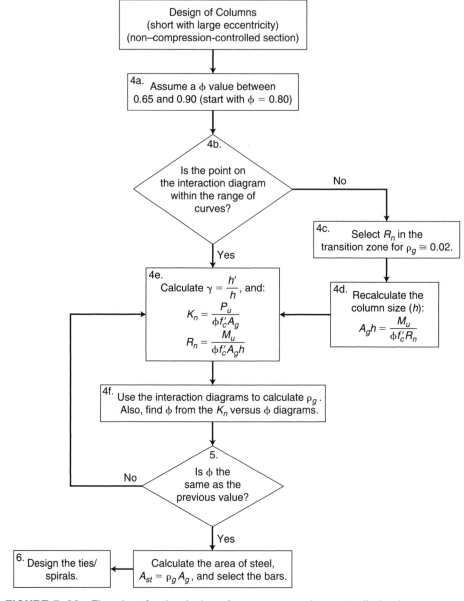

FIGURE 5–39 Flowchart for the design of non–compression-controlled columns.

To resize the column, make $M_R = M_u$ and solve for $A_g h$ as shown below:

$$M_u = \phi R_n f_c' A_g h$$

$$A_g h = \frac{M_u}{\phi f_c' R_n}$$

We select the dimensions for a square column. For a rectangular column se-lect one side of the cross section, either the width (b) or depth (h), and solve for the other dimension. Note that selecting a larger h value (elongating the section in the direction of bending) is more beneficial. Round the column sizes to the nearest inch.

Step 4e   Determine γ, K_n, and R_n. Once you know the new size of the column, de-termine $\gamma = \dfrac{h'}{h}$. (As in step 3, assume $h' = h - 2(2.5) = h - 5$ in.) Also, calculate K_n and R_n as

$$K_n = \frac{P_u}{\phi f_c' A_g} \qquad R_n = \frac{M_u}{\phi f_c' A_g h}$$

Step 4f   Obtain ρ_g from the appropriate interaction diagram, and update the ϕ value. To do so, use γ, f_c', and f_y, and the assumed ϕ value to select the appropri-ate interaction diagram. Then use the values of R_n and K_n (determined in the previous step) to obtain ρ_g from the interaction diagram. Also, determine ϕ from the K_n versus ϕ diagram.

Step 5   Determine whether convergence has been achieved. If the new ϕ factor is not approximately the same as the one obtained in the previous cycle, repeat the process with this new ϕ value starting in step 4e. Repeat the procedure cycle until the ϕ value converges. After achieving convergence, compute the required area of steel by using the steel ratio, ρ_g, determined in the last cycle:

$$A_{st} = \rho_g A_g$$

Select the reinforcing bars using Table A2–9 and check the layout using Table A5–1.

Step 6   Design the ties or spirals. This step is similar to that of the design of compression-controlled columns.

EXAMPLE 5–11

Design a short tied square column to carry $P_D = 300$ kip, $P_L = 200$ kip, $M_D = 150$ ft-kip, and $M_L = 100$ ft-kip. Assume $f_c' = 4,000$ psi, $f_y = 60,000$ psi, and that the main rein-forcements are distributed uniformly around the column edges.

Solution

Step 1   Calculate the factored load, P_u, and the factored moment, M_u:

$$P_u = 1.2P_D + 1.6P_L$$
$$P_u = 1.2(300) + 1.6(200) = 680 \text{ kip}$$
$$M_u = 1.2M_D + 1.6M_L$$
$$M_u = 1.2(150) + 1.6(100) = 340 \text{ ft-kip}$$

Step 2   Estimate the column size:

$$A_g = \frac{P_u}{0.8\phi(0.85f_c')}$$

$$A_g = \frac{680}{0.8(0.65)(0.85)(4.0)} = 385 \text{ in}^2$$

$$h = \sqrt{A_g} = \sqrt{385} = 19.6 \text{ in.}$$

$$\therefore \text{ Try a 20 in.} \times 20 \text{ in. column}$$

Step 3

$$h' = h - 2(2.5) = 20 - 2(2.5) = 15 \text{ in.}$$

$$\gamma = \frac{h'}{h} = \frac{15}{20} = 0.75$$

Calculate K_n and R_n:

$$K_n = \frac{P_u}{\phi f_c' A_g} = \frac{680}{0.65(4)(20 \times 20)} = 0.65$$

$$R_n = \frac{M_u}{\phi f_c' A_g h} = \frac{340(12)}{0.65(4)(20 \times 20)(20)} = 0.20$$

Because $f_c' = 4$ ksi, $f_y = 60$ ksi, and $\gamma = 0.75$, interpolate between Figures A5–2a and A5–3a. Enter these diagrams with the values of K_n and R_n to obtain the corresponding steel ratio, ρ_g:

From Figure A5–2a $\gamma = 0.70 \longrightarrow \rho_g = 0.04$

From Figure A5–3a $\gamma = 0.80 \longrightarrow \rho_g = 0.034$

Interpolation between the values for $\gamma = 0.75$ gives us:

$$\rho_g = \frac{0.034 + 0.04}{2} = 0.037$$

Step 4   We now check our assumption for column behavior. Because the column is in the compression-controlled region of the interaction diagram, our assumption for the ϕ factor is correct. Go to step 5.

Step 5   The required area of steel, A_{st}, is:

$$A_{st} = \rho_g A_g = 0.037(20 \times 20) = 14.8 \text{ in}^2$$

Using Tables A2–9 and A5–1 $\longrightarrow$ $\therefore$ Use 12 #10 bars ($A_{st} = 15.24 \text{ in}^2$)

Step 6   Design of the ties: Using #3 ties, the maximum spacing of the ties is:

$$s_{max} = \min\{16d_b, 48d_t, b_{min}\}$$

$$s_{max} = \min\{16(1.27), 48(\tfrac{3}{8}), 20\}$$

$$s_{max} = \min\{20.3 \text{ in.}, 18 \text{ in.}, 20 \text{ in.}\} = 18 \text{ in.}   \therefore \text{ #3 @ 18 in. ties}$$

Based on Figure 5–12, this column requires three sets of ties, as shown in the figure below.

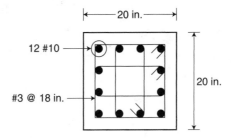

EXAMPLE 5–12

Solve Example 5–11 for a round spiral column. $f_{yt} = 60,000$ psi.

Solution

Step 1   From Example 5–11, $P_u = 680$ kip, and $M_u = 340$ ft-kip.

Step 2   Estimating the column size:

$$A_g = \frac{P_u}{0.85\phi(0.85f_c')}$$

$$A_g = \frac{680}{0.85(0.70)(0.85)(4.0)} = 336 \text{ in}^2$$

$$h = 2\sqrt{\frac{A_g}{\pi}} = 2\sqrt{\frac{336}{3.14}} = 20.7 \text{ in.}$$

$$\therefore \text{ Assume } h = 20 \text{ in} \longrightarrow A_g = \frac{\pi(20)^2}{4} = 314 \text{ in}^2$$

Step 3 & 4

$$h' = h - 2(2.5) = 20 - 2(2.5) = 15 \text{ in.}$$

$$\gamma = \frac{h'}{h} = \frac{15}{20} = 0.75$$

$$K_n = \frac{P_u}{\phi f_c' A_g} = \frac{680}{0.70(4)(314)} = 0.77$$

$$R_n = \frac{M_u}{\phi f_c' A_g h} = \frac{340(12)}{0.70(4)(314)(20)} = 0.23$$

Because $\gamma = 0.75$, we interpolate between the interaction diagrams for $\gamma = 0.70$ and $\gamma = 0.80$. From Figures A5–10a and A5–11a $\longrightarrow \rho_g \cong 0.070$ (compression-controlled zone). Although ρ_g is within the allowable range of 0.01–0.08, in practice a steel ratio of more than 0.04 will cause congestion of reinforcements and is not acceptable. To make this point clear, continue to step 5.

Step 5

$$A_{st} = \rho_g A_g = 0.070(314) = 22.0 \text{ in}^2$$

We cannot fit this much steel reinforcement into a 20 in. diameter column, based on Tables A2–9 and A5–1 (unless we use six #18 bars, which are special-ordered). Therefore, we should increase the column dimension by 2 in. to $h = 22$ in. and repeat steps 3 and 4:

Step 3 & 4

$$h' = 22 - 2(2.5) = 17 \text{ in.}$$

$$\gamma = \frac{h'}{h} = \frac{17}{22} = 0.77$$

$$A_g = \frac{\pi(22)^2}{4} = 380 \text{ in}^2$$

$$K_n = \frac{P_u}{\phi f_c' A_g} = \frac{680}{0.70(4)(380)} = 0.64$$

$$R_n = \frac{M_u}{\phi f_c' A_g h} = \frac{340(12)}{0.70(4)(380)(22)} = 0.174$$

Using the interaction diagrams:

Figure A5–10a $\gamma = 0.70 \longrightarrow \rho_g = 0.043$ (compression controlled)

Figure A5–11a $\gamma = 0.80 \longrightarrow \rho_g = 0.036$ (compression controlled)

Interpolating between the above values:

$$\rho_g = 0.036 + \frac{(0.043 - 0.036)(0.8 - 0.77)}{0.1} = 0.038$$

Step 5

$$A_{st} = \rho_g A_g = 0.038(380) = 14.4 \text{ in}^2$$
$$\therefore \text{ Use 12 #10 bars.}\quad (A_{st} = 15.24 \text{ in}^2)$$

Step 6   Design the spirals. Assume $\frac{3}{8}$ in. diameter spirals:

$$h_c = 22 - 2(1.5) = 19 \text{ in.}$$

$$A_{ch} = \frac{\pi(19)^2}{4} = 284 \text{ in}^2$$

$$\rho_{s,min} = 0.45 \left(\frac{A_g}{A_{ch}} - 1 \right) \frac{f_c'}{f_{yt}}$$

$$\rho_{s,min} = 0.45 \left(\frac{380}{284} - 1 \right) \frac{4.0}{60} = 0.0101$$

$$\rho_s = \frac{4 A_{sp}}{h_c s} = \frac{4(0.11)}{19(s)}$$

$$\rho_{s,min} = \rho_s$$

$$0.0101 = \frac{4(0.11)}{19(s)}$$

$$s = 2.29 \text{ in.} \longrightarrow s = 2\,\tfrac{1}{4} \text{ in.}$$

$$s_{clear} = 2\tfrac{1}{4} \text{ in.} - \tfrac{3}{8} \text{ in.} = 1\tfrac{7}{8} < 3 \text{ in.}\quad \therefore \text{ ok}$$

Therefore, the spiral pitch (s) is:

$$s = 2\tfrac{1}{4} \text{ in.}$$

The following is a sketch of the final design:

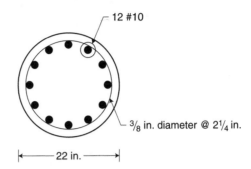

12 #10

$\tfrac{3}{8}$ in. diameter @ $2\tfrac{1}{4}$ in.

22 in.

EXAMPLE 5–13

Design a circular spiral column to resist $P_u = 300$ kip, and $M_u = 400$ ft-kip. Use $f'_c = 4$ ksi and $f_y = 60$ ksi. Use a maximum steel ratio of 0.02. The design of the spirals is not required.

Solution

Step 1 $P_u = 300$ kip, and $M_u = 400$ ft-kip.

Step 2   Estimate the column size (h):

$$A_g = \frac{P_u}{0.85\phi(0.85 f'_c)} = \frac{300}{0.85(0.70)(0.85 \times 4)} = 148.3 \text{ in}^2$$

$$h = 2\sqrt{\frac{A_g}{\pi}} = 2\sqrt{\frac{148.3}{3.14}} = 13.7 \text{ in.}$$

$$\therefore \text{ Try } h = 14 \text{ in.}$$

$$A_g = \frac{\pi(14)^2}{4} = 154 \text{ in}^2$$

Step 3

$$h' = h - 2(2.5) = 14 - 2(2.5) = 9 \text{ in.}$$

$$\gamma = \frac{h'}{h} = \frac{9}{14} = 0.64$$

$$K_n = \frac{P_u}{\phi f'_c A_g} = \frac{300}{0.70(4)(154)} = 0.70$$

$$R_n = \frac{M_u}{\phi f'_c A_g h} = \frac{400(12)}{0.70(4)(154)(14)} = 0.80 > 0.30 \quad \therefore \text{ Out of range.}$$

Because R_n is out of the range of the interaction diagram, use the flowchart for non–compression-controlled columns (Figure 5–39).

Step 4a   Assume $\phi = 0.80$.

Step 4b As noted in step 3, the value of R_n is not within the diagram's range; therefore, proceed to step 4c.

Step 4c In order to resize the column, we must have a reasonable value of R_n. Using the interaction diagram of Figure A5–10a ($f_c' = 4\,\text{ksi}, f_y = 60\,\text{ksi}$, and $\gamma = 0.70$) for a steel ratio, $\rho_g \cong 0.02$, we obtain the maximum value of R_n. This value is about 0.14. We use this value to select a reasonable size for the column.

Step 4d

$$A_g h = \frac{M_u}{\phi f_c' R_n}$$

$$A_g h = \frac{400(12)}{0.80(4)(0.14)} = 10{,}714\,\text{in}^3$$

$$\left(\frac{\pi h^2}{4}\right) h = 10{,}714$$

$$\frac{\pi h^3}{4} = 10{,}714 \longrightarrow h^3 = 13{,}641\,\text{in}^3$$

$$h = 23.9\,\text{in.} \quad \therefore \text{Try } h = 24\,\text{in.}$$

Step 4e

$$h' = 24 - 2(2.5) = 19\,\text{in.}$$

$$\gamma = \frac{h'}{h} = \frac{19}{24} = 0.79 \quad \therefore \gamma \approx 0.80$$

Calculate the nondimensional parameters K_n and R_n:

$$A_g = \frac{\pi(24)^2}{4} = 452\,\text{in}^2$$

$$K_n = \frac{P_u}{\phi f_c' A_g} = \frac{300}{0.80(4)(452)} = 0.207$$

$$R_n = \frac{M_u}{\phi f_c' A_g h} = \frac{400(12)}{0.80(4)(452)(24)} = 0.138$$

Step 4f Because $f_c' = 4\,\text{ksi}, f_y = 60\,\text{ksi}$, and $\gamma = 0.80$, and the column is a round section; therefore, use the interaction diagram of Figure A5–11a. Using K_n and R_n, we obtain a steel ratio, $\rho_g = 0.019$. In addition, using the K_n versus ϕ diagram of Figure A5–11b and $K_n = 0.207$, we obtain a value of $\phi = 0.808$.

Step 5 Because the new ϕ factor (0.808) is slightly higher than that of the previous cycle (0.80), repeat step 4e with the new ϕ.

Step 4e

$$K_n = \frac{P_u}{\phi f_c' A_g} = \frac{300}{0.808(4)(452)} = 0.205$$

$$R_n = \frac{M_u}{\phi f_c' A_g h} = \frac{400(12)}{0.808(4)(452)(24)} = 0.137$$

Step 4f   From Figure A5–11a, $\rho_g = 0.019$. From Figure A5–11b, $\phi = 0.810$.

Step 5   Because the new value of ϕ is about the same as that of the previous step (0.810 vs. 0.808), accept this value and calculate the area of steel (A_{st}):

$$A_{st} = \rho_g A_g = 0.019(452) = 8.59 \text{ in}^2$$

Using Tables A2–9 and A5–1, select 7 #10 bars $(A_s = 8.89 \text{ in}^2)$. The following shows a sketch of the final design.

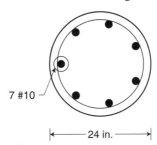

7 #10

|← 24 in. →|

Note: The spiral was not designed.

5.13 SLENDER COLUMNS

Column Buckling and Slenderness Ratio

Columns are divided into two classes based on their slenderness: short columns and slender columns. Short columns crush under large axial force, whereas columns with great slenderness may buckle before they fail in crushing.

Figure 5–40 shows a slender column subjected to an increasing axial compression force, P. As P increases the column may buckle (i.e., suddenly show large lateral movement (see Figure 5–40). The stress at which the column starts buckling is called the *Euler buckling stress* (f_E):

$$f_E = \frac{\pi^2 E}{\left(\dfrac{k\ell_u}{r}\right)^2} \tag{5–22}$$

where
   E = modulus of elasticity of the concrete
   k = effective length factor for the column
   ℓ_u = the unsupported length of the column, which is the clear distance between the floor slabs, beams, or other members that provide lateral support for the column
   r = radius of gyration of column section, equal to $\sqrt{\dfrac{I}{A}}$, in which I is the moment of inertia and A is the cross-sectional area of the column. The ACI Code (Section 10.11.2) suggests a rounded value of r of $0.3h$ for rectangular columns, and $0.25h$ for circular columns, where h is the dimension of the column in the direction perpendicular to the axis of bending.

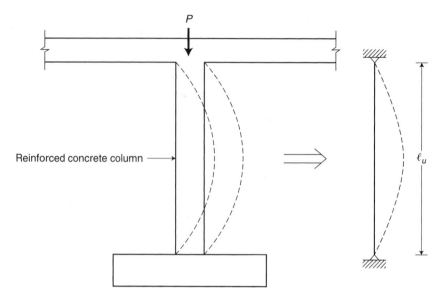

FIGURE 5–40 Buckling of a reinforced concrete slender column.

(Note: Reinforced concrete columns cannot usually be assumed to be pin-ended due to the continuity of these members. Here we show a pin-supported column as a theoretical example.)

The term $\dfrac{k\ell_u}{r}$ is called the *slenderness ratio*. From the expression of the Euler buckling stress we can see that as the slenderness ratio increases, the stress at which the column buckles decreases. In other words, the column buckles at a lower stress when it is more slender. The numerator of the slenderness ratio ($k\ell_u$) is called the *effective length,* which depends on the unsupported column length (ℓ_u), the type of end supports (i.e., pinned, partially fixed, or fully fixed), and whether or not the column is allowed to move laterally (i.e., unbraced or braced). The effective length is the length of that portion of the column that lies between two points of inflection of the buckled shape.

Column ends are connected to a foundation (at the base of the building), or to slabs, beams and girders, or to both. Thus, there is no such thing as a true pin or a fully fixed support for columns. The amount of the column's fixity depends on the relative stiffness of the build-ing elements at the ends of the column. For example, if the size of the beams and girders at the ends of column is small compared to the column size, the end supports can be assumed to be pinned because the beams and girders at the ends will provide little restraint to the free rotation of the column ends when buckling. This condition is depicted in Figure 5–41a. But if the size of beams and girders is large compared to the column, the end supports may be treated as fixed because the beams and girders will prevent the column ends from rotating, as shown in Figure 5–41b. When the column's ends are "pinned," the entire column buckles; as a result, the col-umn's effective length, $k\ell_u$, is almost the same as the column length, ℓ_u, in other words $k = 1.0$. When the column ends are "fixed," its buckled shape has two inflection points; thus, $k\ell_u$ is smaller than ℓ_u. In the theoretical case of perfect fixity, $k\ell_u = 0.5\ell_u$ (i.e., $k = 0.5$).

Another major factor influencing the effective length of a column is the type of structural system. If a column is part of the lateral load carrying system, it is subject to

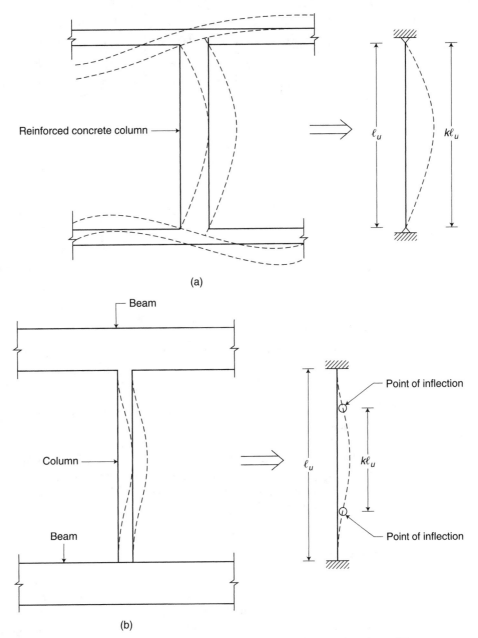

(a)

(b)

FIGURE 5–41   Relationship between relative member sizes and end conditions of columns.

sidesway, which means it can have significant lateral motion and is called an *unbraced column*. When other elements such as shear walls, however, are used as the lateral load carrying system, the column will not have significant lateral motion and is called a *braced column*. The ends of an unbraced column have significant relative horizontal motion or sidesway. This relative movement is small when the column is braced. A

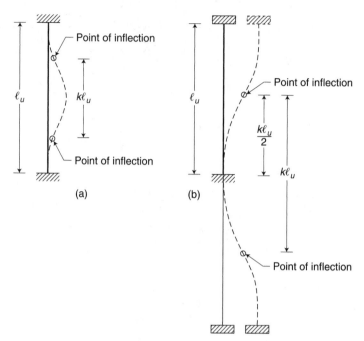

FIGURE 5–42 Effective lengths for (a) column without sidesway, and (b) column with sidesway.

braced column is referred to as one *without sidesway,* and an unbraced column as one *with sidesway.*

Sidesway affects the column's effective lengths. Figure 5–42 shows two columns with fixed-end supports. The column in Figure 5–42a is without sidesway and theoretically its effective length is $k\ell_u = 0.5\ell_u$. If the same column is subjected to sidesway (unbraced), the effective column length, which is the theoretical length between two points of inflection, is $k\ell_u = \ell_u$, as shown in Figure 5–42b.

P–Δ Effects

There is no such thing as a perfectly straight and vertical column. Applied moments at the ends also bend the columns into a curvilinear shape. When a slender column bends into a curve while subjected to an axial load, P, added moments, M, are generated on the column. These moments have a magnitude of P multiplied by the lateral deformation, Δ. This is called the *P–Δ effect.*

Figures 5–43 and 5–44 show *P–Δ* effects on columns. The *P–Δ* effects on columns with sidesway are more severe than those on columns without sidesway.

The ACI Code requires a magnification of moments on slender columns due to *P–Δ* effects. ACI Code (Section 10.12.2) allows *P–Δ* effects to be ignored for columns without sidesway if the column's slenderness ratio $\left(\dfrac{k\ell_u}{r}\right)$ satisfies the following relationships

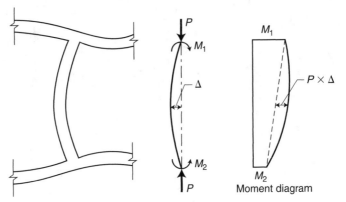

(a) Column bent into a single curve by the end moments

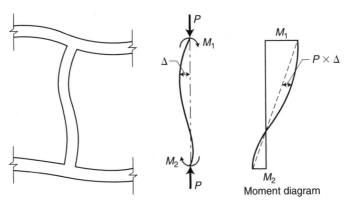

(b) Column bent into a double curve by the end moments

FIGURE 5–43 P–Δ effect on columns without sidesway superimposed over the end moments.

(ACI Equation 10–7):

$$\frac{k\ell_u}{r} \le 34 - 12\frac{M_1}{M_2} \qquad (5\text{--}23)$$

where M_1 is the smaller end moment. $\dfrac{M_1}{M_2}$ is positive if M_1 and M_2 are acting in the same direction (column is bent into a double curve; see Figure 5–43b), and is negative if M_1 and M_2 act in opposite directions (column is bent into a single curve; see Figure 5–43a). The right side of Equation 5–23 is limited to a maximum value of 40.

    The ACI Code (Section 10.13.2) permits the effects of slenderness to be ignored for columns with sidesway when $\dfrac{k\ell_u}{r}$ is less than 22. Figure 5–44 shows the P–Δ effects on columns with sidesway.

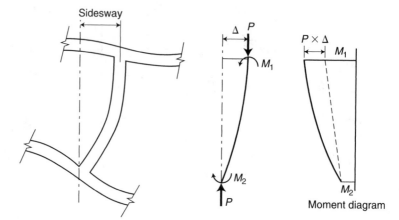

(a) Column with sidesway bent into a single curve by the end moments

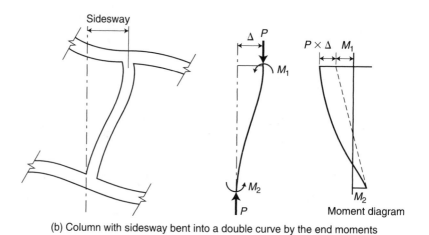

(b) Column with sidesway bent into a double curve by the end moments

FIGURE 5–44 P–Δ effect on columns with sidesway superimposed over the end moments.

Because of the complexity of the design of slender columns (as well as for visual reasons), designers prefer to work with column dimensions that are not slender. In general, we can ignore the slenderness effect of braced frames (columns without sidesway) if $\dfrac{\ell_u}{h}$ is equal to 10 or less on lower floors, and 12 or less on upper floors. For unbraced columns $\dfrac{\ell_u}{h}$ must be smaller than 6 to have negligible slenderness effects.

Designing slender reinforced concrete columns is a complex procedure. Computer software is available to help the structural designer analyze and design slender columns. The detailed discussion of this subject is beyond the scope of this text.

PROBLEMS

5–1 Calculate the axial load capacity, P_R, of the following columns with small eccentricity. Use $f'_c = 3,000$ psi and $f_y = f_{yt} = 60,000$ psi. Determine whether the ties/spirals are adequate based on ACI requirements. The clear cover is 1.5 in.

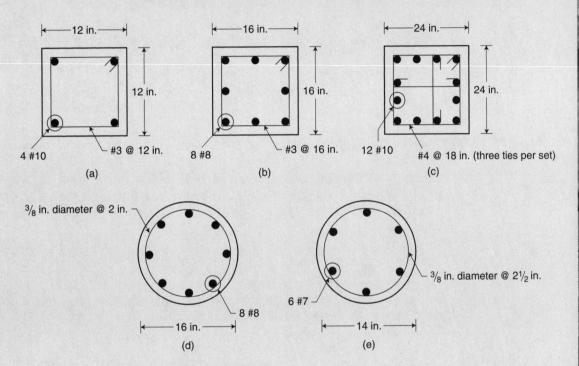

4 #10   #3 @ 12 in.   8 #8   #3 @ 16 in.   12 #10   #4 @ 18 in. (three ties per set)

(a)   (b)   (c)

$\frac{3}{8}$ in. diameter @ 2 in.

8 #8

$\frac{3}{8}$ in. diameter @ 2½ in.

6 #7

(d)   (e)

5–2 Rework Problem 5–1 for $f'_c = 4,000$ psi and $f'_c = 5,000$ psi. What is the percentage of change in the axial load capacity for each case? Do not check the ties/spirals.

5–3 Repeat Problem 5–1 for $f_y = 40,000$ psi and $f_y = 75,000$ psi. What is the percentage of change in the axial load capacity for each case? Do not check the ties/spirals.

5–4 The square reinforced concrete tied column shown below is subjected to a dead load of 200 kip and a live load of 220 kip. Determine whether the column is adequate. The clear cover is 1.5 in. The load's eccentricity is negligible. Use $f'_c = 4,000$ psi and $f_y = 60,000$ psi. Do not check the ties.

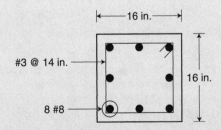

#3 @ 14 in.

8 #8

5–5   Rework Problem 5–4 for the following circular spiral column. Do not check spirals.

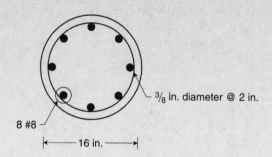

3/8 in. diameter @ 2 in.

8 #8

16 in.

Compare the P_R determined in this problem with that of the square column of Problem 5–4.

5–6   Determine the required reinforcements for a 12 in. $\times$ 12 in. tied reinforced concrete column subjected to a dead load of 20 kip and a live load of 30 kip. Assume that the loads have small eccentricity. Use $f_c' = 5,000$ psi, $f_y = 60,000$ psi, and 2 in. clear cover.

5–7   Redesign the column of Problem 5–6 for a dead load of 125 kip and a live load of 200 kip.

5–8   Design a square tied reinforced concrete column subjected to a dead load of 250 kip and a live load of 300 kip. The moments due to the loads are negligible. Use $f_c' = 4,000$ psi, $f_y = 60,000$ psi, and 1.5 in. clear cover.

5–9   Redesign the column of Problem 5–8 as a circular spiral reinforced concrete section. Assume $f_{yt} = 60,000$ psi.

5–10 A 12 in. $\times$ 12 in. column reinforced with 4 #9 bars is subjected to an axial load with small eccentricity. If the ratio $P_{dead}/P_{live} = 1.5$, determine the maximum compressive axial service loads that the column can carry. Use $f_c' = 4,000$ psi and $f_y = 60,000$ psi. Use #3 ties, and 1.5 in. clear cover.

5–11 The following figures show the typical framing plan, elevation, and beam section of a three-story reinforced concrete building. The floor live load is 50 psf and the roof live load is 15 psf. Assume 5 psf for mechanical/electrical systems and 20 psf for partitions. Determine the required reinforcements for a typical interior tied column between the ground and second levels. Use $f_c' = 4,000$ psi, $f_y = 60,000$ psi, and 1.5 in. for cover. Do not reduce live loads. Assume small eccentricity for the loads. The unit weight of the concrete is 150 pcf. Neglect the self-weight of the column.

5–12 Rework Problem 5–4. Assume the column is subjected to a dead load moment, $M_D = 20$ ft-kip, and a live load moment, $M_L = 30$ ft-kip.

5–13 Determine the maximum factored moment, M_R, that a 24 in. $\times$ 24 in. column with 12 #10 bars distributed uniformly around the column can carry when subjected to a factored axial load, $P_u = 750$ kip. Use $f_c' = 4,000$ psi, $f_y = 60,000$ psi, #3 ties, and $1\frac{1}{2}$ in. for cover.

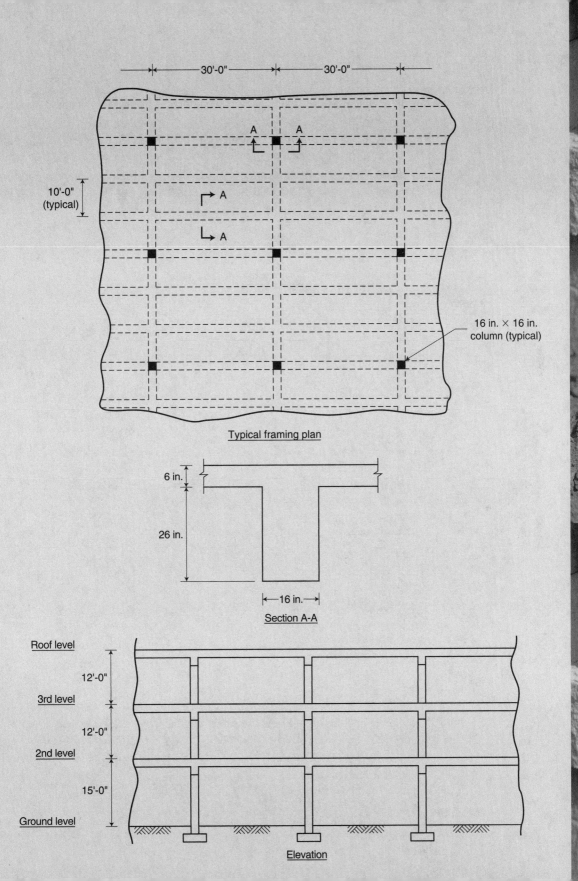

30'-0"   30'-0"

A   A

10'-0"
(typical)

A

A

16 in. × 16 in.
column (typical)

Typical framing plan

6 in.

26 in.

16 in.

Section A-A

Roof level

12'-0"

3rd level

12'-0"

2nd level

15'-0"

Ground level

Elevation

5–14 Rework Problem 5–10. Assume that the column is subjected to a factored moment, $M_u = 50$ ft-kip.

5–15 Rework Problem 5–11. Assume that the column is subjected to a factored moment, $M_u = 110$ ft-kip. Place the longitudinal reinforcing uniformly around the four faces.

5–16 Determine P_R values for the columns shown below subjected to the factored moments indicated. Use $f'_c = 4,000$ psi, $f_y = 60,000$ psi, and 2.0 in. cover. Do not check ties or spirals.

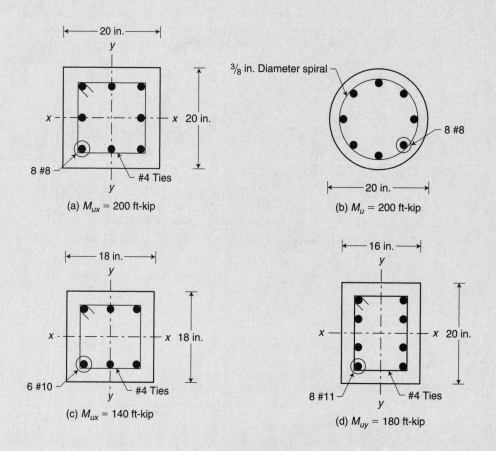

(a) $M_{ux} = 200$ ft-kip

(b) $M_u = 200$ ft-kip

(c) $M_{ux} = 140$ ft-kip

(d) $M_{uy} = 180$ ft-kip

5–17 A square tied column is subjected to a factored load, $P_u = 250$ kip, and a factored moment, $M_u = 50$ ft-kip. Design this column. Use $f'_c = 4,000$ psi, $f_y = 60,000$ psi, and 1.5 in. cover. Do not design the ties. Place the longitudinal reinforcing uniformly in the four faces.

5–18 A circular spiral column is subjected to a factored load, $P_u = 400$ kip, and a factored moment, $M_u = 150$ ft-kip. Design this column. Use $f'_c = 4,000$ *psi*, $f_y = 60,000$ psi, and 1.5 in. for cover. Do not design the spirals.

SELF-EXPERIMENTS

These self-experiments focus on the behavior of columns subjected to axial load and moment. Include all the details of the tests in your report along with images showing the different steps.

Experiment 1

In this experiment, we look at the beam-column action. Glue together two pieces of Styrofoam to make a beam-column assembly. Glue the base of the column to a rigid surface. Apply a concentrated load to the beam and move the load along the beam, as shown in Figure SE 5–1. How does the load affect the column? How does the location of the load affect the behavior of the beam and column? Discuss any other observations.

Experiment 2

In this experiment, you will cast the square tied column shown in Figure SE 5–2. Two sets of wires are needed, larger wires for the main reinforcement and smaller wires for the ties. The column dimensions and reinforcement sizes are optional. Discuss the different stages of construction. Estimate how much load the column can carry. Discuss any other observations.

Experiment 3

Repeat Experiment 1 by casting the column and beam using reinforced concrete. Make sure that the beam reinforcements are bent into the column to provide a moment connection. How does the reinforced concrete beam-column behave differently from that of Experiment 1? Discuss any other observations.

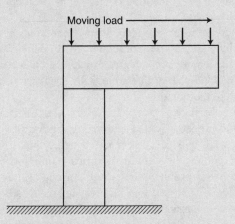

FIGURE SE 5–1   Beam-column assembly model.

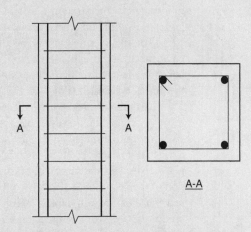

FIGURE SE 5–2   Tied column model.

6

FLOOR SYSTEMS

6.1 INTRODUCTION

The appropriate selection of a floor system heavily influences the overall cost of a building. A designer has to take many factors into account when making such a selection; and, unfortunately, the structural scheme cheapest to construct may not be the best bargain in terms of overall construction cost of the building.

The first element that forms the floor surface is the slab. Beams or columns, in mathematical abstraction can be described by a single line. These are called *linear elements,* because one of their three dimensions, the length, is much greater than the other two (i.e., the dimensions of the cross-section). The load-path of linear elements is easy to describe: They carry their loads along their length to the supports.

A single line, however, cannot describe slabs or plates. As discussed in Chapter Two, slabs have two dimensions that are significantly larger than the third one, the thickness. They are usually described mathematically as thin plates. The "exact" bending theory of thin plates, based on the theory of elasticity, requires the solution of a partial differential equation of the fourth order. This is completely impossible for any practicing engineer or architect. Furthermore, so-called "exact solutions" fail to deal with everyday realities, such as reinforced concrete that does not follow strict elastic behavior, or load distributions that are not nice and uniform, and so on.

Fortunately, we can understand how a slab behaves by carefully considering how it deforms under loads. Slabs bend in two directions, so a single line cannot describe the bent shape. A way of describing a bent surface is therefore necessary for understanding slab behavior. The *load-path* (i.e., the way a slab transfers any load to the supports) depends on the way the slab bends between those supports, which in turn depends on the way the designer chooses to support the slab.

The types of supports for slabs are divided into three groups.

1. *Point supports* These consist of columns, posts, suspension points, and so on. Slabs supported by supports of this type are referred to as *flat slabs* or *flat plates*.

2. *Line supports* Examples of line supports are beams, girders, and walls. Slabs supported by supports of this type are referred to as *one-* or *two-way slabs*.

3. *Continuous media* (Slabs on grade supported by soil).

Admittedly the classifications of the first two types of support are somewhat arbitrary. Often a designer may employ linear support elements (beams and walls) in conjunction with point support elements, which makes the referencing more difficult.

The overall cost of a monolithic concrete floor system depends on several factors. First and foremost among these is the cost of shoring and forming. A strong shoring system must be constructed. It should safely support the weight of the freshly poured concrete and the associated construction loads (i.e., the people and the equipment necessary for placing and finishing the wet concrete). The forms that will serve as the mold must also be built.

The costs in a reinforced concrete floor system usually break down as follows:

Formwork (and shoring)	50%–60%
Concrete, including placing and finishing	25%–30%
Reinforcement, including placement	15%–20%

These figures are based on current material and labor costs in the United States and may not necessarily be the same elsewhere in the world. They clearly show, however, that the cost of labor in the United States usually is greater than the raw cost of materials. Thus, the selection of the right floor system for a building is not an easy task.

6.2 FLAT SLABS AND PLATES

Figure 6–1 shows a typical flat plate floor system. *Flat plate* is a slab of uniform thickness resting on column supports. It is the most economical system to form, as it requires only a wood deck on adequate shoring. It also provides for the least structural depth and thus for

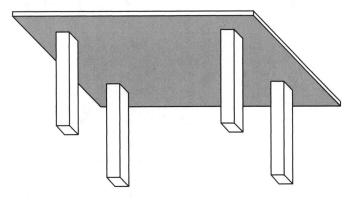

FIGURE 6–1   Flat plate floor system.

TABLE 6–1 Minimum Slab Thicknesses Recommended by the ACI Code (Without Interior Beams)

Yield Strength, f_y (psi)	Without Drop Panels			With Drop Panels		
	Exterior Panels		**Interior Panels**	**Exterior Panels**		**Interior Panels**
	Without Edge Beams	**With Edge Beams**		**Without Edge Beams**	**With Edge Beams**	
40,000	$\dfrac{\ell_n}{33}$	$\dfrac{\ell_n}{36}$	$\dfrac{\ell_n}{36}$	$\dfrac{\ell_n}{36}$	$\dfrac{\ell_n}{40}$	$\dfrac{\ell_n}{40}$
60,000	$\dfrac{\ell_n}{30}$	$\dfrac{\ell_n}{33}$	$\dfrac{\ell_n}{33}$	$\dfrac{\ell_n}{33}$	$\dfrac{\ell_n}{36}$	$\dfrac{\ell_n}{36}$
75,000	$\dfrac{\ell_n}{28}$	$\dfrac{\ell_n}{31}$	$\dfrac{\ell_n}{31}$	$\dfrac{\ell_n}{31}$	$\dfrac{\ell_n}{34}$	$\dfrac{\ell_n}{34}$

minimal floor-to-floor height, which is a very important cost consideration in the overall economy of a multistory building.

Table 6–1, which is Table 9.5(c) of the ACI Code, provides guidelines for the selection of minimum thickness of slabs without interior beams or flat plates as a function of the clear span. The guidelines enable the designer to select a slab thick enough to prevent excessive short- and long-term deflections in the system. The authors believe, on the basis of decades of experience in designing and observing similar structures, that the ACI-recommended minimum thicknesses are somewhat small and probably will result in excessive deflections, especially in exterior or corner panels. Thus, we recommend selecting slabs about 7%–10% thicker than what the Code requires. In Table 6–1, ℓ_n is the clear span from the face of one column to the face of the next. Use the longer clear span when selecting a slab thickness for rectangular bays.

Flat plates are the most economical for square (or nearly so) bays, and for spans of about 26 ft or less. Beyond that span length the slab becomes too thick, with corresponding increase of self-weight. Flat slabs (i.e., plates strengthened around the columns by additional depth provided by drop panels, as shown in Figure 6–2, and with

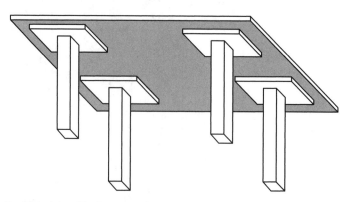

FIGURE 6–2 Flat slab with drop panels.

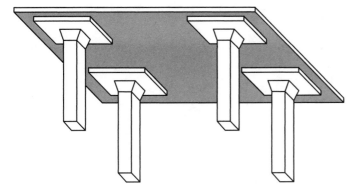

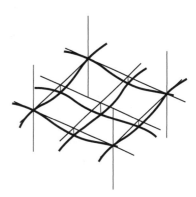

FIGURE 6–3 Flat slab with drop panels and column capitals.

FIGURE 6–4 Schematic deflection of an interior bay under load.

column capitals, as shown in Figure 6–3) are an economical choice for bays up to about 35 ft.

Figure 6–4 shows the schematic deformation diagram of a flat plate under load. The largest deflections are in the center of the bay, and the most highly stressed zones occur around the supports.

6.3 SHEARS IN FLAT SLABS AND PLATES

Because loads must travel toward the columns, the available zone through which shear forces must travel becomes smaller and smaller; thus, the shear stress increases, reaching a maximum at or near the interface of the column and the slab, as shown in Figure 6–5. The large shears also indicate a sharp change in the moments that occur around the columns. Shears cause diagonal tensions in concrete structures that are subject to flexure, and because concrete is quite weak in resisting tension, failure can result. The failure surface may be envisioned as a truncated pyramid similar to the one shown in Figure 6–6. This phenomenon is known as *punching shear:* The column "punches" through the slab, or, more precisely, the slab fails and falls down around the column.

This type of failure (i.e., punching) can be quite catastrophic. Many spectacular failures in the history of construction have happened due to punching shear. The intensity of the shear stress and the resulting diagonal tension depends on the cross-sectional area through which the shear forces must travel toward the column. This cross-sectional area, or shear surface area, depends on two parameters: the thickness of the slab around the column, and the cross-sectional size of the column. So the selection of these dimensions plays a very important role in the preliminary design of the system. The column size is also influenced by the loads and moments that the column must resist at the floor level under consideration. In this discussion, however, the column size is considered only from the point of view of the punching shear in the slab.

Figure 6–7 shows a plan view of a typical interior column. The ACI Code approach is based on a simple analytical model. It assumes that the critical shear surface lies at a $d/2$ distance from the face of the column, where d is the effective depth of the slab. The shear

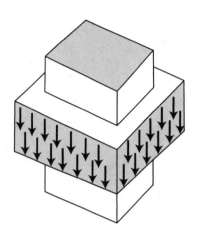

FIGURE 6–5 Gravity load shear transfer from slab to column.

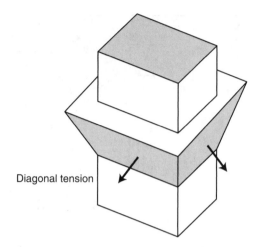

FIGURE 6–6 Diagonal tensions around the columns.

surface area then is the length of the critical periphery (or perimeter of the critical section) multiplied by d.

The ACI Code gives the maximum factored shear to be transferred by stresses on the concrete from the slab to the column as the smallest of Equations 6–1, 6–2, and 6–3 (ACI Code Equations 11–33, 11–34, and 11–35, respectively).

$$\phi V_c = \phi \left(2 + \frac{4}{\beta} \right) \sqrt{f_c'} \, b_o d \qquad (6\text{–}1)$$

where β is the ratio of the long side to the short side of the column's cross section, and b_o is the shear periphery (perimeter of the critical section for shear). Equation 6–1 is the governing

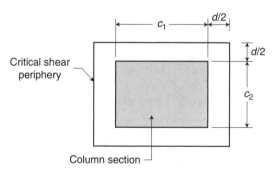

FIGURE 6–7 Definition of the critical shear periphery.

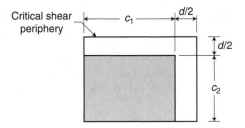

FIGURE 6–8 Definition of the shear periphery at a corner column, $b_o = c_1 + c_2 + 2\dfrac{d}{2}$.

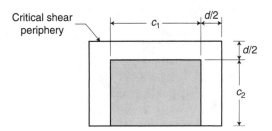

FIGURE 6–9 Definition of the shear periphery at an edge column, $b_o = c_1 + 2c_2 + 4\dfrac{d}{2}$.

formula when the column's cross section is an elongated rectangle, with the ratio of longer side to shorter side greater than 2.

$$\phi V_c = \phi\left(\frac{\alpha_s d}{b_o} + 2\right)\sqrt{f_c'}\,b_o d \tag{6–2}$$

where α_s is 40 for interior columns, 30 for edge columns, and 20 for corner columns.

$$\phi V_c = \phi(4\sqrt{f_c'}\,b_o d) \tag{6–3}$$

Figures 6–8 and 6–9 show the shear periphery for a corner column and for an edge column, respectively.

Experience has shown that vertical chases, ducts, pipes, and so on are somehow "attracted" to columns. Although structures can tolerate openings near columns, openings reduce the available shear periphery. Figure 6–10 shows some examples of openings near a column and the resulting reduction in the effectiveness of the shear transfer.

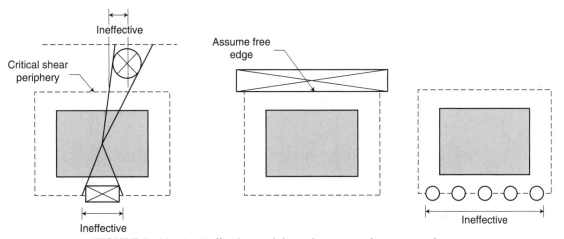

FIGURE 6–10 Lost effective periphery due to openings near columns.

EXAMPLE 6–1

For a floor structure with typical 24 ft by 26 ft bay sizes, the superimposed dead loads are 20 psf and the superimposed live loads are 100 psf. As a preliminary design, select an appropriate flat slab thickness (without drop panels) and a column size as governed by punching shear. Assume $f_c' = 4,000$ psi and $f_y = 60,000$ psi. The unit weight of the concrete is 150 pcf.

Solution

The slab thickness will be governed by the longer span. Because we do not yet know the column size, assume 18 in. × 18 in. square. Even if we under- or overestimate the column size, the error will have little effect on the slab thickness selection. Thus, the larger net (clear) span is:

$$26 \times 12 - 18 = 294 \text{ in.}$$

From Table 6–1 (for the exterior panel):

$$h_{min} = \ell_n/30 = 294/30 = 9.8 \text{ in.}$$

Select 10 in. as a practical dimension.
Load analysis:

$$\text{Self weight of 10 in. slab} = 150\left(\frac{10}{12}\right) = 125 \text{ psf}$$

Superimposed dead loads	20 psf
Live loads	100 psf

The factored load per square foot is:

$$q_u = 1.2(125 + 20) + 1.6(100) = 334 \text{ psf}$$

The factored load on a typical interior column is:

$$P_u = q_u A = 24 \text{ ft} \times 26 \text{ ft} \times 334 = 208,416 \text{ lb}$$

This value is larger than the actual shear that must be transmitted through the critical shear periphery because the loads within the periphery do not contribute to it. We will use this approximate value, however, for the factored shear as well as to estimate the size of the required column.
Thus:

$$V_u = 208,416 \text{ lb}$$

$$d = 10 - 0.75 - 0.75 = 8.5 \text{ in. (average } d \text{ assuming \#6 bars in both directions}$$
$$\text{with } ^3\!/_4 \text{ in. cover)}$$

From Equation 6–3:

$$\phi V_c = 0.75 \times 4 \times \sqrt{4,000} \times b_o \times 8.5 = 1,613 b_o \geq 208,416 \text{ lb}$$

Hence:

$$b_o \geq 129 \text{ in.}$$

Because b_o = column periphery + 8 × $d/2$ = column periphery + 34 in., the column periphery for a typical interior column must be equal to or greater than $129 - 34 = 95$ in.

Thus, several possibilities exist. We can use 24 in. × 24 in. square column, 20 in. × 28 in. rectangular column, or 18 in. × 30 in. rectangular column, and so on, as long as the aspect ratio of the longer side to the shorter side remains less than 2. Thus, if the column size is 14 in. × 34 in., the column periphery will satisfy the minimum 95 in. requirement; but the aspect ratio $\beta = 34/14 = 2.43$ is greater than 2, which would require using Equation 6–1 as the governing equation.

$$\phi V_c = 0.75 \times \left(2 + \frac{4}{2.43} \right) \times \sqrt{4{,}000} \times (2 \times 14 + 2 \times 34 + 4 \times 8.5) \times 8.5$$

$$= 191{,}109 \text{ lb}$$

which is less than the required $V_u = 208{,}416$ lb.

Moment transfer between the slab and the columns increases shear stresses around the columns. How this moment transfer occurs precisely is still a subject of discussion and research. The ACI Code (Sections 11.12.6.1 and 13.5.3.2) rather arbitrarily assumes that 60% of the moment transfer for square columns occurs via flexure at the column's face, and 40% is assigned to a shear distribution model over the critical periphery. Figure 6–11 shows the assumed model. These shears then must be combined with those from the gravity loads that were discussed above.

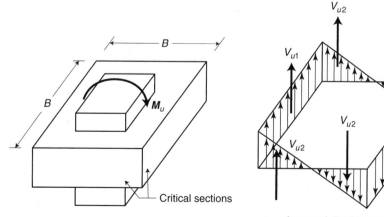

Assumed distribution of shears
providing for moment transfer

FIGURE 6–11 Model of moment transfer from slab to column via shears.

EXAMPLE 6–2

Assume that it is necessary to transfer a factored moment of $M_u = 120$ kip-ft between a flat slab and an interior column. The moment acts clockwise, as shown on Figure 6–11.

The column is 20 in. $\times$ 20 in. and $d = 8.5$ in. for the slab. Calculate the factored shear stress due to the moment.

Solution

Per the ACI Code, the shears will transfer 40% of the moment. Thus, the shears will be responsible for an $M_u = 0.4 \times 120 = 48$ kip-ft.

Designating the maximum shear as V_{uM}, then:

$$B = 8.5 + 20 = 28.5 \text{ in.}$$

$$V_{u1} = v_{uM}Bd = v_{uM} \times 28.5 \times 8.5 = 242.25 v_{uM}$$

$$V_{u2} = \frac{1}{2} v_{uM} \frac{Bd}{2} = \frac{1}{2} v_{uM} \frac{28.5 \times 8.5}{2} = 60.56 v_{uM}$$

$$M_u = 2\left(\frac{V_{u1} \times B}{2}\right) + 4\left(V_{u2} \times \frac{2}{3} \times \frac{B}{2}\right)$$

$$48 \times 12{,}000 = 2\left(242.25 v_{uM} \times \frac{28.5}{2}\right) + 4\left[60.56 v_{uM}\left(\frac{2}{3}\right)\left(\frac{28.5}{2}\right)\right]$$

$$576{,}000 = 9{,}205 v_{uM} \longrightarrow v_{uM} = 62.6 \text{ psi}$$

If the shear stresses exceed what the ACI Code allows, we can reinforce the column/slab interface. The reinforcing may be a shear head manufactured from crossing steel shapes, or sets of closed stirrups. The last two decades have witnessed the development of proprietary premanufactured shear reinforcement.

6.4   FLEXURE IN FLAT SLABS AND PLATES

Flexure in flat slabs and plates is a very complex problem. The simple representation of the deformation shown in Figure 6–4 does not truly describe the deflections, which have a rather intricate topography. But the magnitude of bending moments in any direction is related to the slope of the deflection curve, so the largest moments occur where the curvature of the deflection surface is greatest. Because the surface curves in all directions, bending moments will occur in all directions at any location on the slab. Moments in any direction can be represented by their component moments in a preselected coordinate system, so the ACI Code uses a design method based on a simple model that is easy to visualize and proven to be safe. (Refer to Figure 6–12 in the following discussion of this model.)

Each slab bay is divided into strips in both directions, as shown in Figure 6–12. By studying the deflection pattern shown in Figure 6–4, we easily understand how flexures occur in the structure. In the zone where two middle strips cross (Zone A), the slab bends downward in both directions; thus, there will be tensions in the bottom in both directions (positive moment regions). Where two column strips cross (Zone B), the slab bends upward

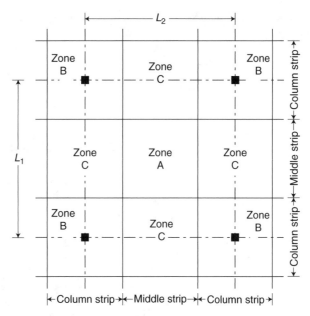

The width of the column strip within a panel is defined
as $0.25L_1$ or $0.25L_2$ (whichever is less) on each
side of the column centerline.

FIGURE 6–12   Definition of column strips and middle strips in flat plates and slabs.

in both directions, generating tensions at the top in both directions (negative moment regions). Where a middle strip crosses a column strip (Zone C), the slab bends downward in the direction of the column strip, but bends upward in the direction of the middle strip; thus, there will be positive moments in the column strip's direction and negative moments in the middle strip's direction.

The ACI Code suggests two methods for the flexural analysis of flat slabs and plates. The first (and simpler) is called the *direct design method;* the second is called the *equivalent frame method*. These methods are not exclusively for flat slabs or flat plates. They may also be used when beams exist on the column lines, which are commonly known as two-way slabs on beams. In this chapter we discuss only the direct design method. (Discussion of the equivalent frame method is beyond the scope of this book. The interested reader is referred to Section 13.7 of the ACI Code.)

The direct design method may be used only when the plan geometry conforms to the following set of limitations (ACI Code, Section 13.6.1):

1. There are at least three consecutive spans in each direction.

2. The panels are rectangular and the ratio of the longer span to the shorter span is not greater than 2.

3. The neighboring span lengths differ by no more than one-third of the longer span.

4. Columns can be offset by a maximum of 10% of the span (in direction of offset) from either axis between centerlines of successive columns.

The flexure analysis of slab then proceeds as follows:

Step 1   Calculate the absolute sum of the average positive and negative moments in each direction on a panel (ACI Code, Equation 13–4):

$$M_o = \frac{q_u \ell_2 \ell_n^2}{8}$$

where
q_u = the factored load on a unit area (psf);
ℓ_n = the clear (net) span length in the direction for which moments are being determined;
ℓ_2 = the length of span (center to center) transverse to ℓ_n;

Step 2   Divide M_o into positive and negative moments.

The value of M_o is divided between total factored positive and negative moments in the span under consideration. Figure 6–13 shows a schematic moment diagram for a slab and the value of M_o. The values assigned are not the result of theoretical studies, but rather observations from testing. They appear to be safe and reasonable values given the highly indeterminate nature of the problem. (A few percentage points of difference one way or the other does not change the overall ultimate strength of the system.)

In an *interior* span (ACI Code, Section 13.6.3.2), the negative factored moment is $0.65\,M_o$, and the positive factored moment is $0.35\,M_o$. In an *exterior* span, the ratios of the total negative moments and the positive moments are strongly dependent on the presence of beams between columns. Edge beams on the exterior perimeter help support the exterior wall system and better control deflections around the periphery, where deflections may be harmful to the wall system. Table 6–2 (from the ACI Code, Section 13.6.3.3) shows the proportions of M_o to be used in *exterior* spans according to the edge support condition.

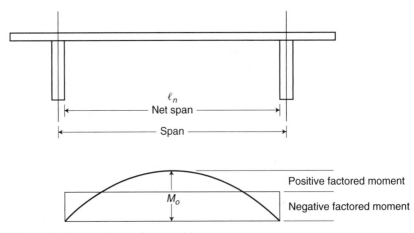

FIGURE 6–13   Moments in a flat panel bay.

TABLE 6-2 Percent Distribution of Moments into Positive Moments and Negative Moments in an End Bay

	(1)	(2)	(3)	(4)	(5)
			Slabs without Beams between Interior Supports		
	Exterior Edge Unrestrained	Slab with Beams Between All Supports	Without Edge Beam	With Edge Beam	Exterior Edge Fully Restrained
Interior Negative Factored Moment	0.75	0.70	0.70	0.70	0.65
Positive Factored Moment	0.63	0.57	0.52	0.50	0.35
Exterior Negative Factored Moment	0	0.16	0.26	0.30	0.65

Step 3 Divide positive and negative moments between column strips and middle strips.

Now that we have determined the values of the positive and negative moments across the full width (ℓ_2) of the panel, we divide these moments to the appropriate column strips and middle strips.

The calculations are slightly more involved when beams are incorporated into the floor system.

The ACI Code defines a coefficient, α_f, which is the ratio of flexural stiffness of a beam section to the flexural stiffness of a width of slab bounded laterally by the centerlines of adjacent panels on each side of the beam. The coefficient α_f is calculated using Equation 6–4 (ACI Equation 13–3).

$$\alpha_f = \frac{E_{cb}I_b}{E_{cs}I_s} \qquad (6\text{–}4)$$

where E_{cb} and E_{cs} are the modulus of elasticity of the concrete in the beam and slab respectively (these two values are usually the same in cast-in-place concrete construction); and I_b and I_s are the moment of inertia of the gross concrete section of the beam and slab, respectively. For flat plates and slabs, $\alpha_f = 1.0$.

Table 6–3 summarizes the percentages of the negative moment that is assigned to the column strips at an interior support (ACI Code, Section

TABLE 6-3 Percent of Interior Negative Moments Assigned to Column Strips

ℓ_2/ℓ_1	0.5	1.0	2.0
$(\alpha_{f1}\ell_2/\ell_1) = 0$	75	75	75
$(\alpha_{f1}\ell_2/\ell_1) \geq 1.0$	90	75	45

TABLE 6–4 Percent of Positive Moments Assigned to Column Strips

ℓ_2/ℓ_1	0.5	1.0	2.0
$(\alpha_{f1}\ell_2/\ell_1) = 0$	60	60	60
$(\alpha_{f1}\ell_2/\ell_1) \geq 1.0$	90	75	45

13.6.4.1). The remainder of the moment is assigned to the middle strip. Linear interpolation is permitted between the values shown.

Table 6–4 summarizes the percentages of positive moment assigned to the column strips (ACI Code, Section 13.6.4.4). The remainder of the moment is assigned to the middle strip. Linear interpolation is permitted between the values shown.

Column strips at an exterior support are assigned percentages of the negative moments according to Table 6–5 (ACI Code, Section 13.6.4.2). Again, linear interpolation is permitted between the values shown. In Table 6–5, the main distinction is whether or not an edge beam connects into the column. A large edge beam with a significant torsional stiffness attracts negative moments away from the column strip, or in other words, it provides some fixity for the middle strip as well. In the case of free slab edge (no edge beam), we assign the total exterior negative moment calculated in step 2 to the column strip.

In Table 6–5, β_t is the ratio of the torsional stiffness of the beam to the flexural stiffness of a width of slab equal to the span length of the beam. The term "beam" here refers to a T-section attached to a certain amount of the slab that helps to increase the beam's torsional stiffness. See Figure 6–14 for an illustration of the T-section.

The cross-sectional constant of the combined stem and attached slabs may be evaluated from rectangular parts as given by Equation 6–5 (ACI Code Equation 13–6).

$$C = \sum \left(1 - 0.63\frac{x}{y}\right)\frac{x^3 y}{3} \tag{6–5}$$

TABLE 6–5 Percent Distribution of Negative Moment at an Exterior Column into the Column Strip

ℓ_2/ℓ_1		0.5	1.0	2.0
$(\alpha_{f1}\ell_2/\ell_1) = 0$	$\beta_t = 0$	100	100	100
	$\beta_t \geq 2.5$	75	75	75
$(\alpha_{f1}\ell_2/\ell_1) \geq 1.0$	$\beta_t = 0$	100	100	100
	$\beta_t \geq 2.5$	90	75	45

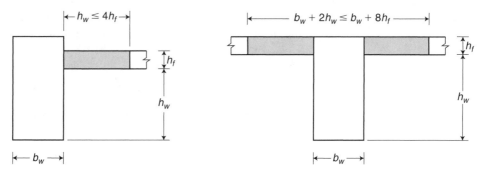

FIGURE 6–14 Slab width increasing the beam's torsional stiffness.

where x and y are the shorter and longer overall dimensions, respectively, of the rectangular part of the cross section. The value for β_t can then be calculated using Equation 6–6 (ACI Equation 13–5).

$$\beta_t = \frac{E_{cb}C}{2E_{cs}I_s} \tag{6–6}$$

In typical cast-in-place concrete construction, $E_{cb} = E_{cs}$.

Step 4   Determine reinforcement.

From the moments calculated in step 3, determine the reinforcement required in the column strips and middle strips using the flexural design methods discussed in Chapter 2.

EXAMPLE 6–3

Design the reinforcement for a typical interior bay of the flat plate floor system of Example 6–1. Assume that the columns are 20 in. × 20 in. and that the slab's thickness is 10 in. Use $f_c' = 4{,}000$ psi and $f_y = 60{,}000$ psi.

The solution shows detailed calculations for the longer span (26 ft) direction only.

Solution

Step 1   Calculate M_o.

$$\ell_1 = 26.0 \text{ ft} \qquad \ell_2 = 24.0 \text{ ft} \qquad q_u = 334 \text{ psf} \quad \text{(from Example 6–1)}$$

$$\ell_n = 26 - 20/12 = 24.33 \text{ ft}$$

$$M_o = \frac{0.334 \times 24 \times 24.33^2}{8} = 593.3 \text{ kip-ft}$$

Step 2   Distribute M_o between the negative and positive moments.

$$M_{\text{neg}} = 0.65 \times 593.3 = 385.6 \text{ kip-ft}$$

$$M_{\text{pos}} = 0.35 \times 593.3 = 207.7 \text{ kip-ft}$$

Step 3   Distribute M_{neg} and M_{pos} between the column strip and the middle strip. Because no beams are incorporated into the system, $\alpha_{f1} = 0$ and $\alpha_{f2} = 0$. Thus, using the coefficient from Table 6–3, the factored negative moment assigned to the column strip at an interior support line is:

$$-M_{\text{column strip}} = 0.75 \times 385.6 = 289.2 \text{ kip-ft}$$

The remainder ($100 - 75 = 25\%$) is assigned to the middle strip.

$$-M_{\text{middle strip}} = 0.25 \times 385.6 = 96.4 \text{ kip-ft}$$

Using the coefficient from Table 6–4, the factored positive moment assigned to the column strip is:

$$+M_{\text{column strip}} = 0.60 \times 207.7 = 124.6 \text{ kip-ft}$$

The remainder ($100 - 60 = 40\%$) is assigned to the middle strip.

$$+M_{\text{middle strip}} = 0.40 \times 207.7 = 83.1 \text{ kip-ft}$$

Step 4   Determine the required reinforcement.

In the 26-ft long span direction, the width of a column strip or a middle strip is one-half the perpendicular span. Thus, use the following cross-sectional data to calculate the required reinforcing:

$b = 24 \times 12/2 = 144$ in.

$d = 10 - 0.75 - .375 = 8.87$ in.   (assuming #6 bars in the outer layer and $\frac{3}{4}$ in. concrete cover)

Then the calculated column strip negative reinforcing is:

$$R = \frac{12{,}000 M_u}{bd^2} = \frac{12{,}000 \times 289.2}{144 \times 8.87^2} = 306 \text{ psi} \longrightarrow$$
$$\text{from Table A2–6b} \longrightarrow \rho = 0.0060$$
$$then \, A_S = \rho bd = 0.0060 \times 144 \times 8.87 = 7.66 \text{ in}^2$$

From Table A2–9, select 18 #6 bars ($A_s = 7.92 \text{ in}^2$).

The middle strip negative reinforcing is:

$$R = \frac{12{,}000 \times 96.4}{144 \times 8.87^2} = 102 \text{ psi} \longrightarrow \text{from Table A2–6b} \longrightarrow \rho = 0.0019$$
$$then \, A_S = 0.0019 \times 144 \times 8.87 = 2.43 \text{ in}^2$$

From Table A2–9, select 13 #4 bars ($A_s = 2.60 \text{ in}^2$).

The column strip positive reinforcing is:

$$R = \frac{12{,}000 \times 124.6}{144 \times 8.87^2} = 132 \text{ psi} \longrightarrow \text{from Table A2–6b} \longrightarrow \rho = 0.0025$$
$$then \, A_S = 0.0025 \times 144 \times 8.87 = 3.19 \text{ in}^2$$

From Table A2–9, select 16 #4 bars ($A_s = 3.20 \text{ in}^2$).

The middle strip positive reinforcing is:

$$R = \frac{12{,}000 \times 83.1}{144 \times 8.87^2} = 88 \text{ psi} \longrightarrow \text{from Table A2--6b} \longrightarrow \rho = 0.0017$$

then $A_s = 0.0017 \times 144 \times 8.87 = 2.17 \text{ in}^2$

From Table A2–9, select 11 #4 bars ($A_s = 2.20 \text{ in}^2$). Figure 6–15 shows the selected reinforcing in the different strip zones. Final results include the required reinforcing in the 24-ft span direction as well. The resulting moments in the short span strips are somewhat less (due to the shorter span), but the reinforcing required is almost identical to that of the long span strip. This reinforcement will be placed in a second layer, with the working depth d estimated as only 8.12 in. The student is encouraged to verify these reinforcing requirements. The placement order (i.e., which layer of reinforcement must be laid first and which layer onto the second layer) must be clearly noted by the designer on the structural plans.

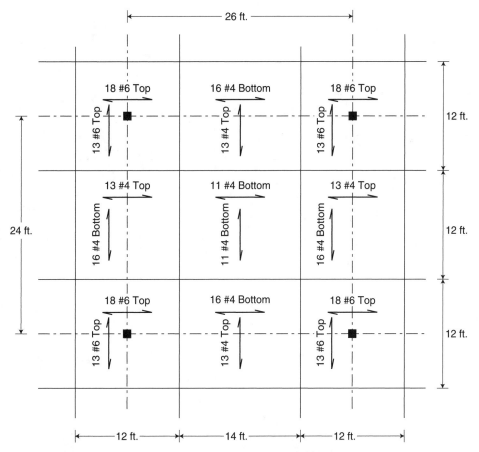

FIGURE 6–15 The calculated reinforcing from Example 6–3.

EXAMPLE 6–4

Calculate the column and middle strip moments for an end bay of the floor system in Example 6–3 with the addition of a 12 in. wide by 20 in. deep edge beam. Figure 6–16 shows the slab divided into column and middle strips. As in Example 6–3 only the 26 ft span direction calculations are shown.

Solution

Step 1 Calculate M_o: $\ell_1 = 26.0\,\text{ft},$ $\ell_2 = 24.0\,\text{ft},$ $q_u = 334\,\text{psf},$

$$\ell_n = 26 - 20/12 = 24.33\,\text{ft}$$

$$M_o = \frac{0.334 \times 24 \times 24.33^2}{8} = 593.3\,\text{kip-ft}$$

Step 2 In this problem, the values listed in the fourth column of Table 6–2—slabs without beams between interior supports, but with edge beam—will apply.

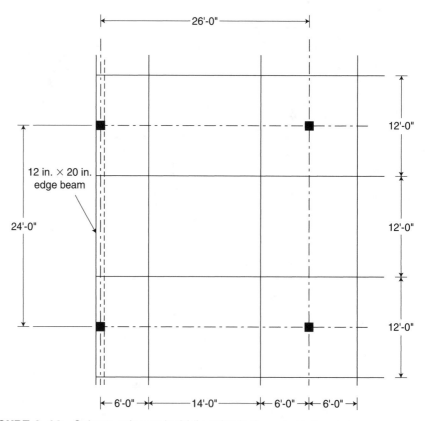

FIGURE 6–16 Column strips and middle strips in the exterior bay.

Thus:

The total factored negative moment at the first interior support is:

$$0.70 \times 593.3 = 415.3 \text{ kip-ft}$$

The total factored positive moment in the first span is:

$$0.50 \times 593.3 = 296.7 \text{ kip-ft}$$

The total factored negative moment at the exterior support is:

$$0.30 \times 593.3 = 178.0 \text{ kip-ft}$$

Step 3   Distribute the moments to the column strips and the middle strips.

a. Negative moments at the first interior support:
   Because $\alpha_{f1} = 0$ (no beams in the span direction), from Table 6–3:

$$-M_{\text{col. strip}} = 0.75 \times 415.3 = 311.5 \text{ kip-ft}$$
$$-M_{\text{mid. strip}} = 0.25 \times 415.3 = 103.8 \text{ kip-ft}$$

b. Positive moments in the first span, from Table 6–4:

$$+M_{\text{col. strip}} = 0.60 \times 296.7 = 178.0 \text{ kip-ft}$$
$$+M_{\text{mid. strip}} = 0.40 \times 296.7 = 118.7 \text{ kip-ft}$$

c. Negative moments at the exterior support:
   In order to use Table 6–5, calculate the value of β_t.
   Figure 6–17 shows the definition of the edge beam per Figure 6–14. Calculating the required parameters:

$$y_t = \frac{20 \times 12 \times 10 + 10 \times 10 \times 5}{20 \times 12 + 10 \times 10} = 8.53 \text{ in.}$$

$$I_b = \frac{20^3 \times 12}{3} + \frac{10^3 \times 10}{3} - 340 \times 8.53^2 = 10{,}595 \text{ in}^4$$

$$I_S = \frac{(24 \times 12) \times 10^3}{12} = 24{,}000 \text{ in}^4$$

$$C = \left(1 - 0.63 \times \frac{12}{20}\right)\frac{12^3 \times 20}{3} + \left(1 - 0.63 \times \frac{10}{10}\right)\frac{10^3 \times 10}{3} = 8{,}399 \text{ in}^4$$

$$\beta_t = \frac{8{,}399}{2 \times 24{,}000} = 0.175$$

By interpolating between the values listed in Table 6–5, we obtain the percentage needed to calculate the negative moment at the exterior support:

$$-M_{\text{col. strip}} = 0.982 \times 178.0 = 174.8 \text{ kip-ft}$$

The remainder, which is assigned to the middle strip, is:

$$-M_{\text{mid. strip}} = 178.0 - 174.8 = 3.2 \text{ kip-ft}$$

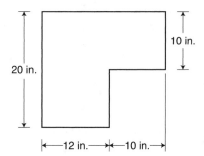

FIGURE 6–17 The edge beam in Example 6–4.

This value is very small due to the relatively small torsional stiffness of the edge beam.

6.5 FLAT SLABS AND THE USE OF DROP PANELS

Flat plates are usually the most economical choice when spans are about 26 ft or less. Beyond this length the slab thickness required to control deflections becomes too large, thus making the slab too heavy. Moments and shears are highest in the areas around the columns. Hence, it makes eminent sense to increase the depth of the plate in these critical areas, as shown in Figures 6–2 and 6–3. Although the use of drop panels may be attractive from the standpoint of structural behavior, the associated forming costs are considerable. On the other hand, drop panels allow thinner slabs to be used in most of the areas. This results in weight and concrete savings that offset some of the excess forming costs.

According to the ACI Code (Section 13.2.5), drop panels must extend at least one-sixth of the span length in each direction from the column center line, and their thickness must be at least 25% of the slab thickness beyond them.

EXAMPLE 6–5

Calculate the appropriate size of a flat slab system for the floor in Example 6–3. The columns are 20 in. × 20 in. Figure 6–18 shows the plan layout indicating the outlines of the drop panels.

Solution

The minimum plan dimension of the drop panel is

$$\left[26 \times \frac{1}{6}\right](2) = 8\text{'-}8\text{''} \text{ by} \left[24 \times \frac{1}{6}\right](2) = 8\text{'-}0\text{''}$$

From Table 6–1 the recommended minimum slab thickness is:

$$h_{\min} = \frac{\ell_n}{36} = \frac{(26 \times 12 - 20)}{36} = 8.11 \text{ in.} \quad \text{select} \quad h = 8.5 \text{ in.}$$

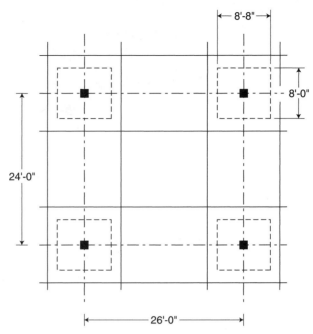

FIGURE 6–18 Plan of flat slab.

The minimum drop panel thickness below the slab is:

$$0.25 \text{ in.} \times 8.5 \text{ in.} = 2.13 \text{ in.} \qquad \text{Use } 2.5 \text{ in.}$$

Thus, the total thickness within the drop panel will be $8.5 + 2.5 = 11$ in.

The total volume of concrete within one typical bay is:

$$24 \times 26 \times \frac{8.5}{12} + 8.67 \times 8 \times \frac{2.5}{12} = 456.5 \text{ ft}^3$$

or an equivalent uniform thickness of 8.78 in., as opposed to the 10 in. thickness used in the flat plate structure. This represents a 12.2% saving in concrete use and, correspondingly, in the self-weight of the structural slab. After the superimposed dead and live loads are added, however, the savings in the average factored loads diminish to about 5.4%. The reduced working depth in the zones outside the drop panels results in increased reinforcing as well, further diminishing the economical advantages gained from the use of less concrete.

6.6 WAFFLE SLAB STRUCTURES

Figure 6–19 shows a typical *waffle slab* or *two-way joist floor* structure. Waffle slab floor structures are thick flat plate structures, with the concrete removed from zones where it is not required by strength considerations. Waffle slab provides economical structures for spans up to 60 ft, in square bays, loaded with light and moderate loads. The voids are formed by steel (or fiberglass) "domes" that are reusable, thus very economical. These forms are

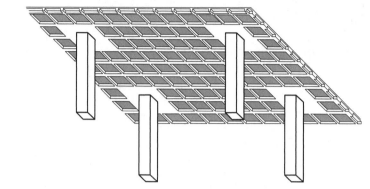

FIGURE 6–19   Underside view of a waffle slab.

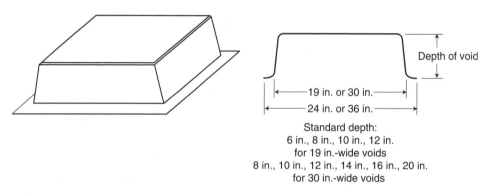

Depth of void

19 in. or 30 in.

24 in. or 36 in.

Standard depth:
6 in., 8 in., 10 in., 12 in.
for 19 in.-wide voids
8 in., 10 in., 12 in., 14 in., 16 in., 20 in.
for 30 in.-wide voids

FIGURE 6–20   Standard forming pans (domes) for waffle slabs.

available in standard sizes, as shown in Figure 6–20, although wider or odd-shaped domes are also used to satisfy some design objectives. The sides of the domes are tapered (usually 1 in 12) to permit easy removal after the concrete has cured. The two-way joists, when carefully finished after the removal of the forms, provide a pleasing appearance as well.

The lips on the domes, when laid out side by side, form 5 in.-wide joists for the 19 in.-wide voids and 6 in.-wide joists for the 30 in.-wide voids. But the designer does not have to work with 24 in. or 36 in. planning modules. Because the domes are always laid out on a flat plywood deck, the spacing between the domes can be easily adjusted to make the joists wider than standard at the base. This accommodates virtually any column spacing while maintaining a uniform appearance. Leaving out the domes around the columns forms a shear head that provides increased shear strength as well as concrete in the bottom for the high negative moments. The slab over the domes is typically 3 to 4.5 in. thick, unless large concentrated loads, increased fire rating requirements, or embedded electrical boxes and conduits warrant the use of a thicker slab. The 3 in. minimum is quite adequate for roofs. The slab is reinforced with a light welded wire reinforcement to prevent shrinkage and temperature cracks.

For overall depth selection, the span/depth ratios given in Table 6–1 under the heading "Without Drop Panels" will result in a very serviceable structure. The solid concrete area around the column ideally should approach the size required for drop panels; in other words, it should extend about one-sixth of the span length measured from the column centerline.

EXAMPLE 6–6

Select an appropriate waffle slab floor structure for 36 ft × 36 ft bays. Columns are 20 in. × 20 in.

Solution

Select a 4.5 in.-thick slab, anticipating electrical conduits or junction boxes in the slab. Also, use the same depth structure in the end bays without edge beams.

From Table 6–1:

$$h \geq \frac{\ell_n}{30} = \frac{36 \times 12 - 20}{30} = 13.73 \text{ in.}$$

Select 10 in.-deep pans and a 4.5 in. slab for a total structural depth of 14.5 in. Figure 6–21 shows the resulting layout.

Each 30 in. × 30 in. × 10 in. dome displaces 4.92 ft³ of concrete (*CRSI Design Handbook,* p. 11-1). A typical bay contains 128 domes. Thus, the total volume of concrete in the bay is:

$$\text{Volume} = 36 \times 36 \times \frac{14.5}{12} - 128 \times 4.92 = 936.2 \text{ ft}^3$$

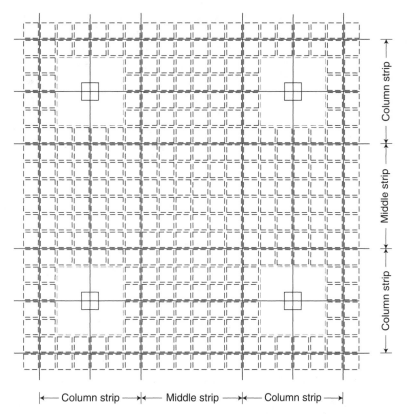

FIGURE 6–21 Plan view of a waffle slab showing the defined column and middle strips.

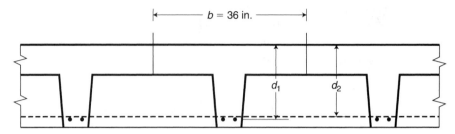

FIGURE 6–22   Reinforcing in positive moment zones.

The average concrete thickness is only:

$$t_{avg} = \frac{936.2}{36 \times 36} (12) = 8.67 \text{ in.}$$

The analysis of the system is very similar to that of flat plates. When finding the reinforcement take into account that, with the exception of the filled areas around the columns, the slab is no longer solid, but rather a set of joists. Thus:

a. In positive moment areas, divide the strip moment by the number of joists and design the joists for that fraction. Figure 6–22 shows that joists in these areas are like T-beams, and the slab on top provides a wide compression flange. The joists will have two layers of reinforcement, one for each joist direction; thus, the working depth will be slightly less in one direction.

b. In negative moment regions (where middle strips bend upwards over column strips), the bottom width of the stem is in compression and the slab on the top is in tension. Thus, distributed reinforcing is used over the width of the flange, as shown in Figure 6–23.

c. Shear and bending must be determined in the joists around the solid section surrounding the column. On rare occasions, shear reinforcing is necessary in the joists along a short distance.

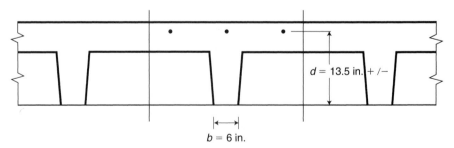

FIGURE 6–23   Reinforcing in negative moment zones.

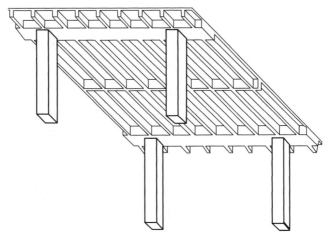

FIGURE 6–24   One-way joists and beams.

6.7   ONE-WAY JOISTS

Figure 6–24 shows a typical one-way joist system. One-way joists spanning between beams are essentially closely spaced beam elements. The clear space between them must not exceed 30 in. in order to qualify for the *joist* designation used by the ACI Code. The forms are made of various materials, such as steel, fiberglass, fiber board, and corrugated cardboard, and are made with or without the edge lip, as shown in Figure 6–25. Forms without the edge lip, however, tend to bulge sideways during construction under the lateral pressure of the freshly placed concrete, and the resulting joist widths are uneven. Forms with square or tapered ends are also available, as shown in Figure 6–26. The tapered ends provide increased shear capacity as well as increased moment capacity at the negative moment regions.

One-way joist systems are often used when the bays are elongated (i.e., the column spacing in one direction exceeds the spacing in the other direction by about 40% or more). At such span ratios the advantage of two-way behavior is greatly reduced, and it is more economical to use one-way systems in which beams span between columns, and joists span between the beams. It is also more economical to orient the beams in the shorter spans and the joists in the longer span. For ease of forming, the selected depth of the beams is often equal to

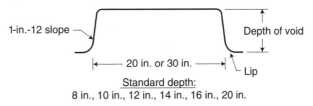

FIGURE 6–25   Standard one-way joist pans.

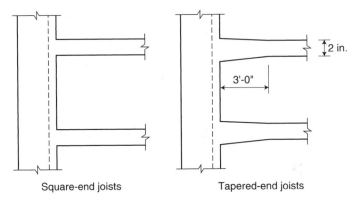

FIGURE 6-26 Square- and tapered-end pan layouts.

that of the joists. In order to provide for the necessary shear and moment capacity, the beams are made considerably wider than the faces of the columns into which they frame. These wide and shallow beams are not as efficient as deeper beams, but the savings achieved by the reduced forming cost more than make up for that. Beams deeper than the joists, such as those shown in Figure 6–27, occupy additional ceiling space and require additional forming cost.

The slab over the voids is typically 3 in. thick, unless large concentrated loads or increased fire rating requirements warrant the use of a thicker slab. The slab is reinforced with a light welded wire reinforcement to prevent shrinkage and temperature cracks. The overall depth of the joist (including the slab's thickness) should be selected in accordance with Table 9.5(a) of the ACI Code, which is shown in Table 6–6. (This table was discussed in Chapter 2, but it is repeated here for the readers' convenience.) The ratios listed therein give satisfactory performance for most structural elements. The designer should be aware, however, that these are minimum depth values, and should be used for members not supporting or attached to partitions or other construction likely to be damaged by large deflections. Thus, special attention should be paid when attaching walls to the underside of concrete structural elements to ensure that such elements do not bear on the walls when deflecting. Furthermore, the deflection of the supporting beams should not exceed span/600 to ensure crack-free masonry walls.

If the depth must be minimized beyond the values listed in Table 6–6, the designer may use Grade 40 (f_y = 40,000 psi) reinforcement. This will result in about 50% more

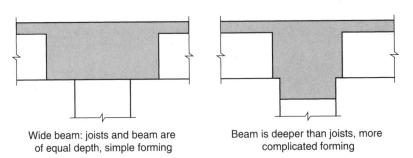

Wide beam: joists and beam are
of equal depth, simple forming

Beam is deeper than joists, more
complicated forming

FIGURE 6-27 Beam sections with joists.

TABLE 6–6 Recommended Minimum Span/Depth Ratios [ACI Code, Table 9.5(a)]. (Minimum Thickness of Non-Prestressed Beams or One-Way Slabs Unless Deflections are Computed.)

Member	Simply Supported	One End Continuous	Both Ends Continuous	Cantilever
Solid One-way Slabs	span/20	span/24	span/28	span/10
Beams or Joists	span/16	span/18.5	span/21	span/8

required reinforcement, but will reduce the strain in the reinforcing steel in service load condition. Reduced strain in the reinforcement provides reduced deflection. Because using a different grade of reinforcement for a few selected members on a project is not recommended, it is permissible to use Grade 60 steel equal in cross-sectional area to the calculated amount of Grade 40 steel that would be necessary. Values shown in Table 6–6 should be used directly for members with normal-weight concrete and Grade 60 reinforcement. For other conditions the values should be modified as follows:

a. For structural lightweight concrete that has a unit weight in the range of 90–120 pcf, the values shall be multiplied by $\max(1.65 - 0.005\,w_c, 1.09)$, where w_c is the unit weight in pcf.
b. For f_y other than 60,000 psi, the values shall be multiplied by $(0.4 + f_y/100{,}000)$.

6.8 BEAMS AND ONE-WAY SLABS

Figure 6–28 shows a typical beam and one-way slab system. A beam and one-way slab system is an economical choice when the bays are elongated and the superimposed loads are large. The system is especially economical when the structure is subject to large line loads

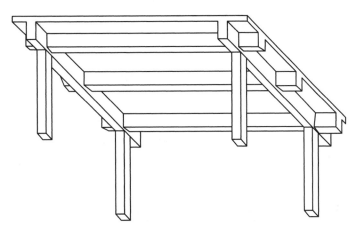

FIGURE 6–28 One-way slabs on beams and girders.

such as heavy partitions. Large openings through the slab can be easily accommodated virtually anywhere in the floor. Beam and one-way slab systems have a larger structural depth than do the other floor systems, and their forming cost is usually higher. These disadvantages are somewhat balanced by savings in concrete and reinforcement usage. This type of system also provides a clear and unambiguous transfer of moments between beams and columns. This is a real advantage in high-wind or seismic zones, where the structural frame resists lateral loads on the building.

Table 6–6 provides information useful for determining the thickness of the slab and the depths of the beams and girders. These values, however, are recommended *minimum* depth values. The following must be considered in selecting the appropriate width for girders (and beams):

a. The width should be enough to lay out the reinforcing in one row in the positive moment regions;
b. The width must provide sufficient cross-sectional area so that $\phi V_c \geq V_u/3$ at least, but preferably $\phi V_c \geq V_u/2$;
c. The width of the girder (or beam) framing into a column should be the same as that of the column face for ease of forming.

6.9   TWO-WAY SLABS ON BEAMS

Figure 6–29 shows a typical two-way slab system. When the aspect ratio (the ratio of the longer span to the shorter span) of a slab that is supported on all four sides is less than about 1.50, the slab exhibits a significant two-way behavior. As discussed in detail in Chapter 2, this means that the slab will carry the loads in both directions. In plan, the load distribution from the slab to the beams may be approximated, as shown on Figure 6–30. The shorter beam supports much less load than the longer span does. So if the aspect ratio is significantly larger

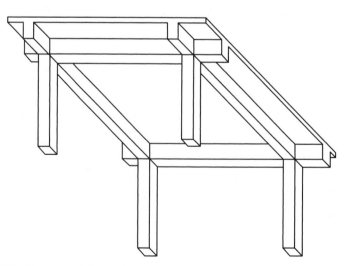

FIGURE 6–29   Two-way slabs on beams.

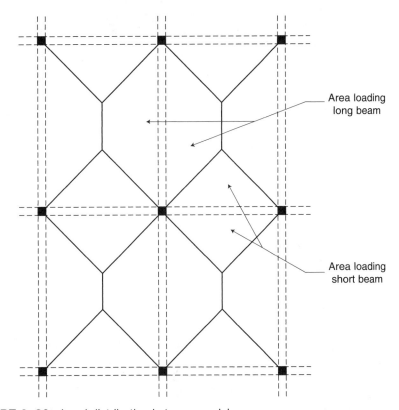

Area loading
long beam

Area loading
short beam

FIGURE 6–30 Load distribution in two-way slabs.

than 1.0, the use of stronger beams in the long direction than in the short direction is recommended. As the aspect ratio approaches 1.0, the load division between the beams is more evenly distributed.

The ACI Code provides recommendations (Section 9.5.3.3) for the minimum thickness of slabs supported on all four sides. These ACI Code formulae are somewhat cumbersome. In the experience of the authors, $h \geq \ell_n/40$ to $\ell_n/45$ is a reasonable value for preliminary design (ℓ_n is the longer clear span of the slab from face of beam to face of beam). The preliminary selection of beams is governed by considerations similar to those for one-way slabs and beams.

6.10 TWO-WAY JOISTS WITH SLAB BAND BEAMS

Figure 6–31 shows a typical two-way joist, or waffle slab, with slab-band beams system. This floor system is an interesting variation of the two-way slabs on beams. Wide beams form a two-way grid of beams (often referred to as *slab bands*) between the columns. The depth of the beams is equal to the depth of the two-way joist system. This arrangement provides a somewhat easier layout of reinforcement in the negative moment regions around the columns. This system may also have a seismic performance better than that of ordinary waffle slabs.

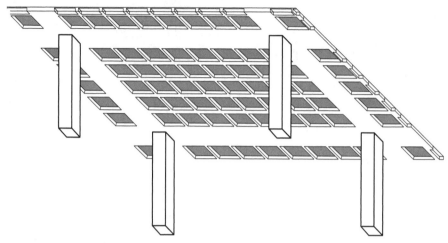

FIGURE 6–31 Waffle slabs (or two-way joists) on beams.

PROBLEMS

6–1 A flat plate reinforced concrete floor system with 16 in. × 16 in. columns is planned on a 25 ft × 25 ft grid. Use $f_c' = 4$ ksi and $f_y = 60$ ksi.

(a) Determine the minimum recommended slab thickness for exterior panels if edge beams are used on the exterior perimeter of the floor.

(b) Determine the minimum recommended slab thickness if no edge beams are planned.

6–2 Repeat Problem 6–1 for $f_y = 40$ ksi reinforcement.

6–3 Repeat Problem 6–1 using drop panels. Determine the minimum required size and thickness of a typical drop panel over an interior column.

6–4 An 8 in.-thick flat plate reinforced concrete floor system with 18 in. × 18 in. columns is planned on a 22 ft × 22 ft grid. The superimposed dead load is 15 psf and the live load is 50 psf. Based on the shear strength of the system around a typical interior column, verify the adequacy of the design. Assume no moment transfer between the slab and the column. Use $f_c' = 4$ ksi, $f_y = 60$ ksi, and $d = 6.5$ in.

6–5 Use the data in Problem 6–4 to calculate the shear in a slab due to the transfer of a factored moment, $M_u = 80$ ft-kip, from the slab to an interior column.

6–6 A flat plate floor system of a reinforced concrete building is shown below. Use $f_c' = 5,000$ psi, $f_y = 60,000$ psi, and concrete cover $= \frac{3}{4}$ in. The superimposed dead load is 20 psf and the live load is 80 psf. Use the ACI direct design method to calculate the moments in the slab.

(a) Determine an appropriate slab thickness for an interior panel. Round the thickness to the nearest $\frac{1}{2}$ in.

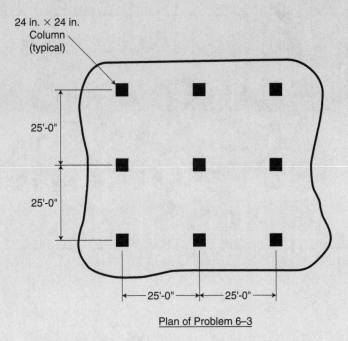

24 in. × 24 in.
Column
(typical)

25'-0"

25'-0"

|← 25'-0" →|← 25'-0" →|

Plan of Problem 6–3

(b) Check the shear around a typical interior column. Assume #6 bars in both direction and use the average d when calculating the shear strength. Assume no moment transfer between the column and the slab.

(c) Calculate the required outer layer reinforcing for (1) positive moment in a column strip, (2) positive moment in a middle strip, (3) negative moment in a column strip, and (4) negative moment in a middle strip.

SELF-EXPERIMENTS

Experiment 1

Make small-scale reinforced concrete models of a one-way joist, a waffle slab, and a flat plate floor system. Place wires for their reinforcements. Record the procedure and your observations. Which system required the least effort in building the forms and placing the reinforcement?

Experiment 2

Identify three different concrete floor systems from local buildings. Record the range of the spans and the bay shapes (square, rectangular, etc.). In addition, record their occupancy types. Write a report that summarizes your findings and includes photos.

7

FOUNDATIONS AND EARTH SUPPORTING WALLS

7.1 INTRODUCTION

Any building structure requires a foundation system in order to transfer the loads to the supporting soil. The strength of concrete typically is 400 to 800 kip per square feet (ksf). Soils typically however, can safely withstand only pressures of 3 to 10 ksf. As a result the foundation system has to spread the load over a large surface area to reduce the pressure when it transfers loads from columns and walls to the supporting soil.

Foundations were constructed of stone and masonry before concrete was used as a building material. Application of concrete has improved the foundation system significantly. Today, virtually all foundations are made of plain or reinforced concrete.

Because the design of foundations requires an understanding of the soil-structure interaction, we must study the different types of soil and their behavior under loading. Therefore, the following sections present an overview of the different types of soil, their classifications, the exploration methods, and the laboratory tests for finding the allowable bearing capacity of soil.

Subsequently, this chapter deals with the different types of foundation systems, with a focus on the design and analysis of wall and column footings. The last part of this chapter discusses the different types of earth supporting walls, with an emphasis on basement walls and cantilever retaining walls.

7.2 TYPES OF SOIL

In general, all subsurface materials fall into one of two groups: rock or soil. But in reality soils are made up mostly from rock eroded by air and water and settled over many millennia. Soils are divided into two main categories: *coarse-grained soils* and *fine-grained soils*.

Coarse-grained soils consist mainly of gravel and sand. The particle sizes are large enough to be seen with the naked eye. Coarse-grained soils are also called *noncohesive soils,* as their grains do not stick to each other when oven-dried. Fine-grained soils are classified as clay or silt. A magnifying glass is needed in order to see their particles. Fine-grained soils are also called *cohesive soils* because their particles stick to each other. Cohesive soils expand when subjected to moisture and shrink when dried.

In addition to these major categories of soil, soils are classified as *organic* or *inorganic* soils. Organic soil consists of decayed vegetable or animal remains. The top soil used to grow plants and vegetations is an organic soil. Inorganic soil, in contrast, is almost completely free of organic materials. Organic soil is not suitable for supporting building structures or even to be used as backfill against basement or retaining walls. If it is encountered in a construction site, it must be replaced with appropriate compacted engineered fill.

7.3 SOIL CLASSIFICATION

A soil classification called the *Unified Soil Classification System (USCS)* has been devised to specify the soil mix and its condition. It is based on the work of Professor Arthur Casagrande at Harvard University. Each designation in this classification consists of two letters. The first letter represents the type of soil: G (gravel), S (sand), M (silt), C (clay), O (organic), and P (peat). The second letter shows the soil condition, for example, W (well-graded) or P (poorly graded). In this classification, soils are divided into 15 types, as shown in Table 7–1. A *well-graded soil* consists of both large and small grains, with the small particles filling the voids between the large ones (sand and gravel). Well graded and compacted sand and gravel are very good substrata. The "poorly graded soil" refers to a soil that does not have the right proportioning of sand and gravel and, as a result, has large voids between adjacent grains.

TABLE 7–1 Unified Soil Classification System

Main Division		Symbol	Description
Coarse-Grained Soils	Gravels	GW	Well-graded gravels, gravel-sand mixtures
		GP	Poorly graded gravels or gravel-sand mixtures
		GM	Silty gravels, gravel-sand-silt mixtures
		GC	Clayey gravels, gravel-sand-clay mixtures
	Sands	SW	Well-graded sands, gravelly sands
		SP	Poorly graded sands or gravelly sands
		SM	Silty sands, silt-sand mixtures
		SC	Clayey sands, sand-clay mixtures
Fine-Grained Soils	Silts and Clays	ML	Inorganic silts and very fine sands, silty or clayey fine sands or clayey silts
		CL	Inorganic clays, gravelly clays, sandy clays, silty clays
		OL	Organic silts and organic silty clays
		MH	Inorganic silts, fine sandy or silty soils
		CH	Inorganic clays
		OH	Organic clays, organic silts
		PT	Peat and other highly organic soils

7.4 TEST BORINGS AND THE STANDARD PENETRATION TEST (SPT)

In order to design foundation systems, we need information about the underlying soil. The most widely used method of exploration drills holes (borings) into the ground at the intended site of the building. The soil is sampled at different depths of borings and standard tests are conducted to obtain information about the soil's properties.

Test borings are distributed so as to obtain insight about the soil under the whole footprint of the building. The spacing and depth of the borings depend mainly on the type of structure to be built and the uniformity of the soil deposit. For low-rise buildings, the spacing of the borings is about 75 to 100 ft. Their depths are about 20 to 30 ft below the foundation level, with one deep boring to search for hidden weak deposits. For high-rise buildings the spacing of borings is closer, around 40 to 50 ft, and the depth often descends to the underlying bedrock.

Boring is performed by an auger drill. A hollow pipe, called the *casing,* is advanced to prevent the soil from collapsing into the borehole. As the bore hole is advanced, the soil is tested in situ at certain locations. This testing is usually performed wherever the driller experiences a different stratum, or at 5-ft intervals within the same stratum. The test used most often is the *Standard Penetration Test* (SPT).

The SPT uses a device called a *split-barrel sampler.* Figure 7–1 shows a schematic drawing of a split-barrel sampler. It is a hollow cylinder two inches in diameter, made up of

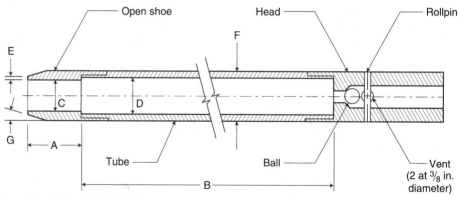

A = 1.0 to 2.0 in. (25 to 50 mm)
B = 18.0 to 30.0 in. (0.457 to 0.762 m)
C = 1.375 ± 0.005 in. (34.93 ± 0.13 mm)
D = 1.50 ± 0.05 − 0.00 in. (38.1 ± 1.3 − 0.0 mm)
E = 0.10 ± 0.02 in. (2.54 ± 0.25 mm)
F = 2.00 ± 0.05 − 0.00 in. (50.8 ± 1.3 − 0.0 mm)
G = 16.0° to 23.0°

The $1\frac{1}{2}$ in. (38 mm) inside diameter split barrel may be used with a 16-gage wall thickness split liner. The penetrating end of the drive shoe may be slightly rounded. Metal or plastic retainers may be used to retain soil samples.

FIGURE 7–1 Schematic drawing of a split-barrel sampler for the Standard Penetration Test (copyright ASTM International. Reprinted with permission).

TABLE 7–2 Relationship Between Soil Condition and Blow Count, N (Data from Terzaghi and Peck, *Soil Mechanics in Engineering Practice,* 2nd ed., 1968)

Sand		Clay	
Condition	**N**	**Condition**	**N**
Very loose	0–4	Very soft	< 2
Loose	4–10	Soft	2–4
Medium	10–30	Medium	4–8
Dense	30–50	Stiff	8–15
Very dense	> 50	Very stiff	15–30
		Hard	> 30

two fitting half cylinders, which are held together by two threaded end-pieces. It is placed at the bottom of the bore hole and driven through the soil by a 140-lb hammer that has a free fall of 30 in. The number of blows needed to move the sampler 3 times six inches into the soil is recorded. The numbers of the blows from the second and third six-in. advancements are added up. This sum gives the so-called N value. The blows from the first six-in. advancement do not reveal the true characteristics of the soil in situ, because the auger tends to leave disturbed soil at the bottom of the hole.

In addition, when the sampler is withdrawn at the end of the test, the device is taken apart and the soil sample contained in the cylinder (part B on Figure 7–1) is placed into a sealed and labeled jar. The sample is then taken to the laboratory for further testing.

The blow count, N, is related to the soil condition. Table 7–2 shows a general classification relating soil condition and the blow count. After careful laboratory analysis of the samples, the geotechnical engineer prepares a boring log of each boring performed at the site. Figure 7–2 shows a sample boring log.

7.5 SOIL FAILURE UNDER FOOTINGS

Soil, like any other material, has a certain load bearing capacity. If the pressure from the footing exceeds this limit, the soil will fail. This would cause the footing to sink into the soil, which may have disastrous consequences on a supported building.

Figure 7–3 shows a simple theoretical failure mechanism for soils under pressure. A wedge is formed directly under the footing (Zone I). This wedge is pushed down by the footing, which in turn pushes Zone II outward. Zone II rotates about a pivot at the top and pushes Zone III sideways and up. The bottom parts of Zones II and III form a shear plane along which the wedges move. This plane provides shear resistance against the movement. The weight of Zone III also provides resistance against the rotation of Zone II. Thus, placing loads on Zone III (surcharge) or moving the footing deeper into the ground will inhibit the movement of the wedges. As a result, the soil will have more bearing capacity.

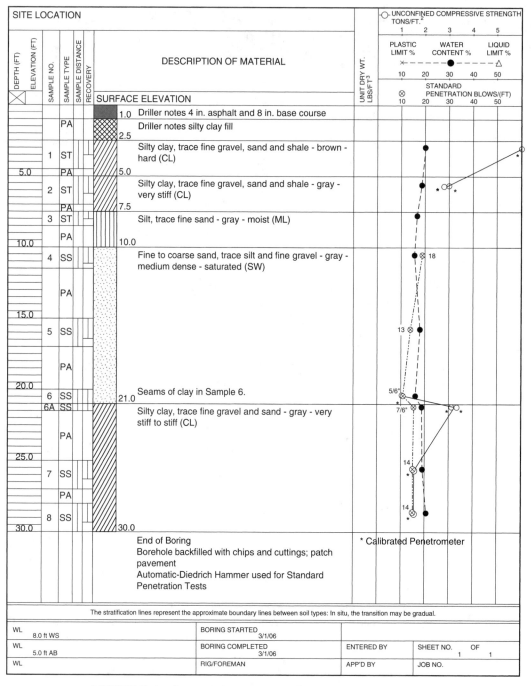

FIGURE 7–2 A sample boring log.

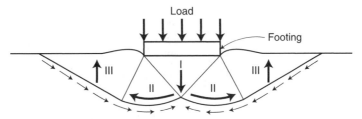

FIGURE 7–3 Bearing capacity failure of soil.

7.6 PRESSURE DISTRIBUTION UNDER FOOTING AND SOIL SETTLEMENT

The pressure at the bottom of the footing propagates through the soil mass. The pressure is most intense directly under the footing, and decreases at increasing horizontal and vertical distances from the footing. This is known as the *pressure bulb effect* as shown in Figure 7–4a for a circular footing. Figure 7–4b shows how the pressure bulb extends in all directions like a balloon. The pressure bulb for a square footing has a shape somewhat similar to the one shown in Figure 7–4. In the case of a continuous strip footing, the pressure distribution extends along the footing with proportions that are cylindrical rather than spherical.

Soils compress into a smaller volume when subjected to pressure. This leads to settlement, or a downward movement of the footings. The amount of settling depends on several factors such as the pressure level under the footing, the size of the footing, and the properties of the soil. The volume of soil affected by the footing is basically the limit of the pressure bulb shown in Figure 7–4. Therefore, settlement also depends on the shape and size of the footing. Settlement can never be completely eliminated unless the footing is directly supported by bedrock. So we design footings to limit the detrimental effects of settlement on the structure.

Soils are composed of three major constituents: solid particles, a system of voids between these particles, and air (gas) or water that fills the voids. When the soil is loose, it has a large void content.

Settlement or consolidation in soils is associated with the squeezing of the moisture or air out of the voids. This permits the solid particles to move closer to each other, resulting in a denser structure.

In granular soils the movement of water (or smaller particles) is easy. Thus, settlement of these soils will take place quickly as the structure is built on top of them and loads are added onto the footings. About 90% of the expected settling will have taken place by the time the building is completed.

In cohesive soils (clays), the movement of moisture is slow, thus the consolidation (settlement) of the soil is also slow. Settlement may take place long after the completion and occupancy of the building.

In general, there are two types of settlement of building foundations:

1. Uniform settlement: This happens when all parts of the entire building settle approximately the same amount. Uniform settlement may damage underground utilities, but usually does not cause any significant structural damage to the building. The Monadnock

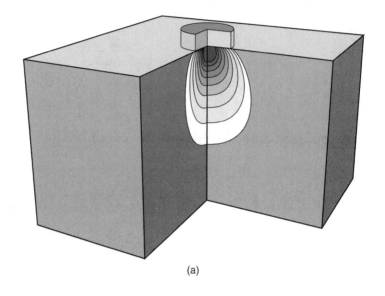

(a)

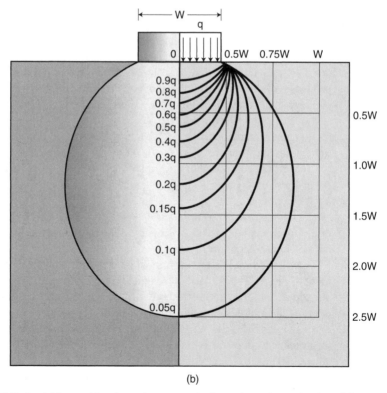

(b)

FIGURE 7–4 (a) Isometric view of pressure bulb under column footing; (b) graph of pressure bulb under column footing.

Building in Chicago, for example, underwent almost two feet of settlement without any damage.

  2. *Differential settlement:* In this case, different footings will experience different amounts of settlement. Differential settlement can cause serious structural distortion and damage, so it is important to design structures and foundations that minimize the effects of differential settlement. It is common practice to try to limit differential settlement to 1/300 of the horizontal distance between adjacent footings.

7.7 ALLOWABLE BEARING SOIL PRESSURE

Because differential settlement can cause severe distortion and structural problems, various methods are used to reduce its effect. One common method is to design the footings so that each applies approximately the same pressure on the soil under the most usual loads. These loads consist of dead loads and an average percentage of the live load depending on the occupancy type.

  There is a limit to the pressure that soils can safely support. This limit is called *allowable soil bearing pressure* or simply *soil bearing capacity,* and is based on two criteria: (1) the soil does not fail, and (2) the settlement is not excessive.

  Building codes recommend soil bearing capacities for specific conditions. These recommended values are generally approximate. Table 7–3 shows approximate soil bearing capacities for each soil type. The proper method of establishing the allowable soil bearing pressure is a soil investigation program conducted by a qualified geotechnical engineer. Usually the structural engineer selects and designs the foundation system based on a soil report from a geotechnical engineer. For smaller projects, however, the structural engineer may use the presumptive bearing capacities recommended by the local building codes such as those shown in Table 7–3 when foundations are supported by soil whose properties are known.

TABLE 7–3 Allowable Soil Bearing Capacities

Soil Type	Bearing Capacity (ksf)
Medium clay	3
Stiff clay	4
Very stiff clay	6
Hard clay	8–10
Medium sand	2–6
Dense sand	6–8
Very dense sand	8–10
Sand and gravel mix	8–12
Soft rock	16
Medium, sound rock	30
Hard rock	40–80
Massive, solid bedrock	200–400

7.8   Types of Foundations

Foundations fall into two main categories: *shallow foundations* and *deep foundations*. Each of these foundation systems consists of different subsystems, as shown in Figure 7–5. Many factors need to be considered in the selection of a foundation system. These include soil strength, soil type, the location of the water table, variation of the soil with depth, and so on. In general, foundations are constructed of plain or reinforced concrete. The typical strength of concrete used in footings is $f'_c = 3,000$ psi. In rare cases, concrete with higher strength may be used to reduce the footing depth and weight.

Shallow Foundations

Shallow foundations are usually located no more than 6 ft below the lowest finished floor. These are the most economical and most common type of foundation. A shallow foundation system generally is used when (1) the soil close to the surface of the ground has sufficient strength, and (2) underlying weaker strata do not result in undue settlement.

Figure 7–5 divides shallow foundations into five major types: wall footings, isolated column spread footings, combined footings, strap footings, and mat or raft foundations.

Depending on the condition and the type of the supporting soil, shallow foundations are cast either into a neat excavation in the soil or into wood forms. It is most economical to cast the concrete into earth forms. But this is only possible with a cohesive soil, such as clay, which remains stable during the concrete placement. If the soil is granular, the concrete is cast using wood side forms. The forms are removed after the concrete gains strength, and the area around the footing is backfilled and compacted.

Wall Footings *Wall footings* support walls made of wood, masonry, or concrete. They are made of plain or reinforced concrete and are continuous under the entire length of the wall. Similar to slabs, structural analysis and design are performed on a 1-ft-long strip of

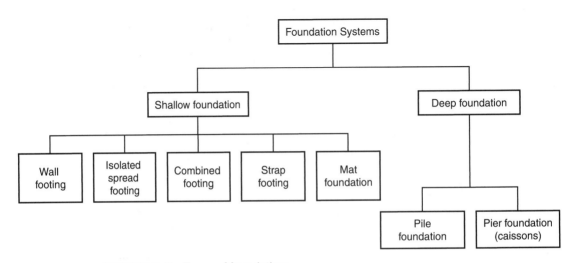

FIGURE 7–5   Types of foundations.

the footing (assuming the wall is evenly loaded). The supported wall is usually placed at the center of the footing to avoid any eccentricity and rotation of the footing. Figure 7–6a shows a typical wall footing.

Isolated Spread Footing An *isolated* or *individual spread* footing supports a single column. Figure 7–6b shows a typical isolated spread footing. Spread footings are usually square shaped, but we can design them in a rectangular shape, if needed. This may be necessary if the footing is close to a neighboring footing or a property line.

Isolated spread footings can be made of plain concrete if they are subjected only to gravity loads and are not located in earthquake-prone areas. Plain concrete footings are usually used only with light loads.

Combined Footings One footing may be used to support two columns when the columns are close to each other and the isolated spread footings for one column would

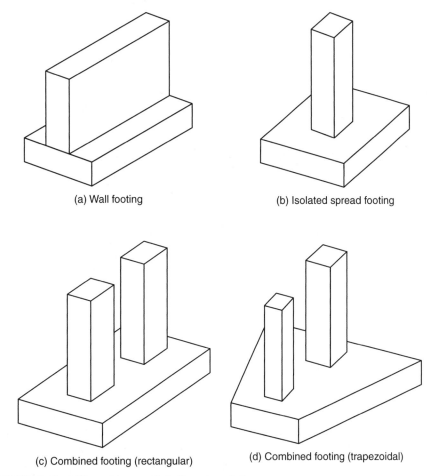

(a) Wall footing

(b) Isolated spread footing

(c) Combined footing (rectangular)

(d) Combined footing (trapezoidal)

FIGURE 7–6   Different types of shallow foundations.

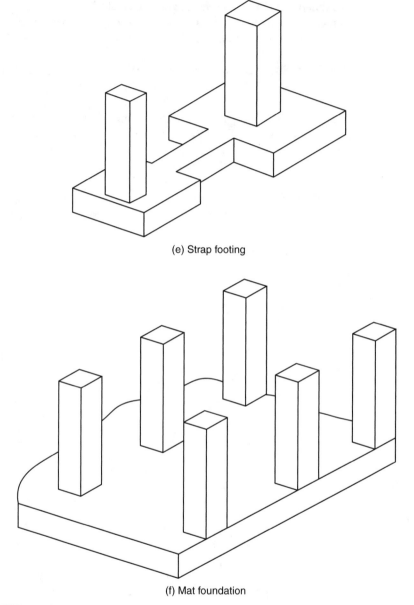

(e) Strap footing

(f) Mat foundation

FIGURE 7–6 (continued)

overlap the other. Such a footing is called a *combined footing*. Figure 7–6c shows a rectangular combined footing. If the column loads are significantly different, we may use a trapezoidal footing such as the one shown in Figure 7–6d. The larger width of the footing will be closer to the column supporting the heavier load.

When designing a combined footing, it is important to size the footing so that it exerts an approximately uniform pressure on the soil. To achieve this, the footing is proportioned so that its centroid is at or near the resultant of the column loads.

Strap Footings A *strap* or *cantilever footing* is a special type of combined footing that uses a "strap" or beam to connect the two footings together. The application of this footing is similar to that of the combined footing. Strap footings may also be useful when underground utility lines prevent the use of rectangular combined footings. Figure 7–6e shows a typical strap footing. The strap acts as a cantilever beam that partially resists the moment from the eccentrically loaded exterior footing. This ensures that the soil pressure is uniform underneath the entire strap footing.

Mat Foundation A *mat* or *raft foundation* consists of a large and thick continuous reinforced concrete slab that supports the entire building. This system is used when the soil bearing capacity is low or column loads are heavy, resulting in more than 50% of the building plan area being required for individual footings. An advantage of a mat foundation is that it drastically reduces differential settlement between columns. Mat foundations are usually made of heavily reinforced concrete slabs at least 24 in. thick. Figure 7–6f shows a typical mat foundation.

Deep Foundations

The use of shallow foundations may not be economical or even possible if the soil bearing capacity close to the surface is too low. Deep foundations are used in these situations to transfer the loads to a strong layer, which may be located at a significant depth below the ground surface. The load is transferred through skin friction and end bearing as shown in Figure 7–7.

There are two main types of deep foundations: *piles* and *piers* (also called *caissons*), as shown in Figure 7–5.

Pile Foundations Piles usually have small cross-section sizes, ranging from 6 to 24 in., and capacities of up to 500 kip. They are made of treated timber, steel, or concrete in different

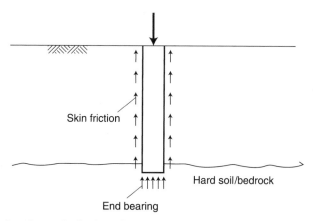

Skin friction

End bearing

Hard soil/bedrock

FIGURE 7–7 Load transfer in deep foundations.

shapes. Piles typically are driven into the ground using pile driving hammers. This process causes noise and vibration, which may disturb sensitive adjacent structures, such as hospitals.

When a pile is driven into the soil, it displaces the soil that is in direct contact with it. The soil around the pile becomes significantly compacted and lateral pressure on the pile increases. This results in friction forces between the soil and the pile.

Timber piles have been used since ancient times. The piles used today are about 25 to 35 ft long although it is possible to splice them for longer length. Timber piles act mainly as friction piles because their end bearing is not significant. These piles have load capacities in the range of 30 to 50 kip.

Steel and precast concrete piles are normally used to carry large loads. For long, slender piles the end bearing on soil is insignificant compared to the resistance from the skin friction. But if the piles are driven to the underlying bedrock, they act as *end-bearing piles* because the end-bearing resistance contributes a large percentage of the total resistance.

Piles are commonly used in groups with a pile cap connecting the tops of the piles and providing a surface area for placing building columns. Figure 7–8a shows a pile group supporting a column. The piles in a group should be separated far enough that the load carrying action of each pile does not affect that of an adjacent one. Typically, at least three piles are used in a group, but two piles are often acceptable in certain conditions. Figure 7–8b shows typical layouts of piles in groups with their associated pile caps.

It is common to use *battered* piles if the piles are to be subjected to lateral loads due to wind or earthquake loads. These piles resist the effects of lateral loads through axial tension and compression forces. Figure 7–8c shows a pile group with battered piles.

Auger-cast piles may be used to alleviate the noise and harmful vibrations associated with pile driving. A long hollow stem screw is drilled into the ground. The auger is withdrawn at the desired depth while, simultaneously, concrete is pumped into the bottom through the hollow stem. The withdrawing auger removes the soil and replaces it with a concrete column. The pile can be reinforced by lowering a wide flange steel section into the fresh concrete immediately after withdrawal of the auger.

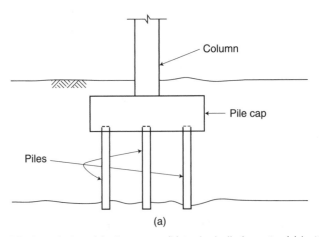

(a)

FIGURE 7–8 Pile foundation: (a) pile group, (b) typical pile layouts, (c) battered piles.

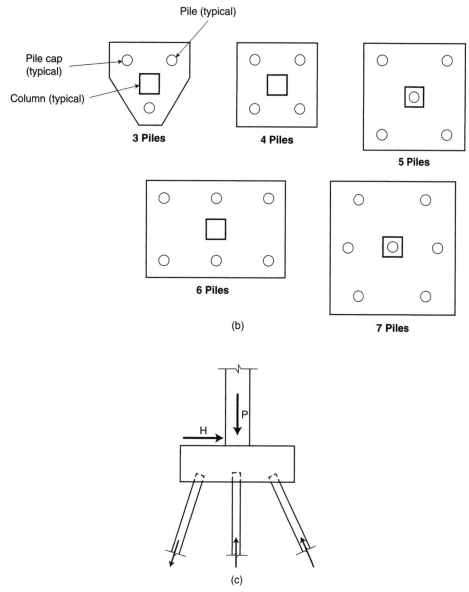

FIGURE 7-8 (continued)

Pier Foundations Piers are typically made by using a large-diameter auger to drill a round hole in the ground and placing concrete into the hole. The drilling process for piers is much less noisy than pile driving. The shaft diameter is usually at least 36 in. This provides enough room to lower a person to inspect and test the soil before placing the concrete. A protective steel cylinder (called a *casing*) is used to prevent the collapse of the sides during drilling.

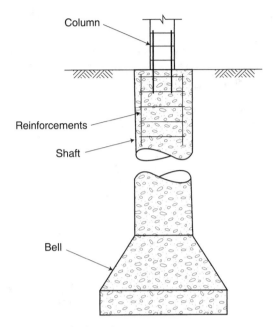

FIGURE 7–9 Pier foundation (caisson).

A drilled pier is sometimes called a *caisson,* which is French for "box." A special device is often used at the bottom of the pier to enlarge the base, creating a belled caisson instead of a straight-shaft caisson. The main purpose of a bell is to increase the bearing area of the caisson. Figure 7–9 shows a typical belled caisson with its components. Bells can be made only in cohesive soils such as clay. It is common practice when designing a caisson to ignore the skin friction between the shaft and soil and use only the end bearing capacity.

Caissons support many tall buildings in Chicago. Each column of the John Hancock Center is supported by a 140-foot-long caisson, which transfers the load to the bedrock. The Sears Tower sits on 114 caissons that are six ft in diameter and over 100 ft long to reach the bedrock.

Considerations for the Placement of Foundations

Several issues must be considered when selecting a footing type. These issues, however, mainly affect the placement of shallow foundations.

Adjacent Property Lines Buildings often have columns or walls close to or right on a property line. Typically building codes and legal considerations do not allow any part of a footing to extend beyond the property line. A good way to avoid this problem is by setting the building's supporting elements away from the property line and letting the supported structure cantilever to the legal limit. But if the design demands supports at or near a property line, the structural designer may need to use an elongated rectangular footing, a combined footing, or a strap footing.

Depth for Frost Penetration The moisture in the soil underneath a footing may freeze during the cold season if the bottom of an exterior footing is located too close to the ground surface. Water expands when it freezes, and the magnitude of the expansion is about 10%. The expansion takes place toward the least resistance, which is usually upward. This phenomenon is called *frost heaving*. This can push the footing upward, which in turn can distort and crack the footing and damage the supported building structure.

This problem is simple to prevent. Exterior footings have to be placed below the *frost line,* as shown in Figure 7–10a. The frost line is the distance measured from

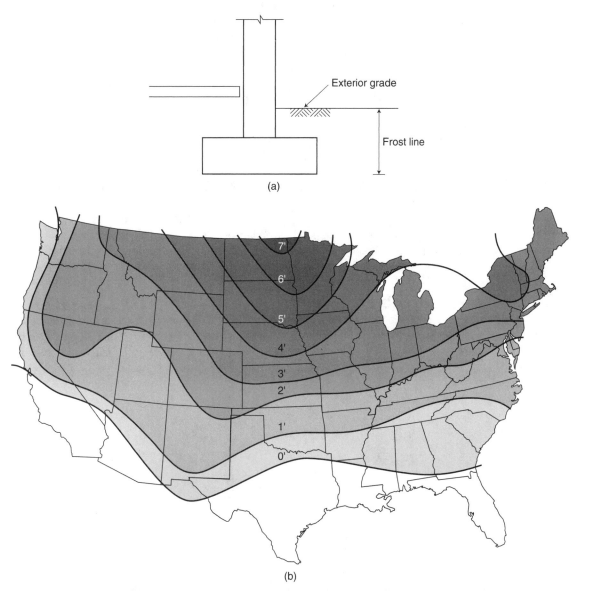

FIGURE 7–10   Frost line (minimum required depth to prevent frost penetration): (a) frost line definition, (b) maximum anticipated depth of freezing from the city building codes.

the finished exterior grade to the bottom of the expected maximum *depth of frost penetration.*

Historic data are available on the depth of the frost line in different locations. For example, 42 in. is a safe depth in most parts of the Midwest. In the northern part of the Great Lakes and in many northern states, exterior footings must be placed at least 60 to 72 in. below the exterior grade. The geotechnical engineer usually provides the necessary information relating to the local frost line. The map in Figure 7–10b shows the variation of the frost line depth for different areas of the United States, based on the values recommended in city building codes. (This map should be used only for general information because it is not necessarily accurate for specific localities.)

In heated buildings, this requirement is mandatory only for exterior footings, as frost does not travel horizontally to affect the interior footings. An exception to this is when the foundations are constructed in the winter, or are left unprotected from freezing during cold spells. In those cases even the interior footings should be placed below the frost line and protected by backfilling or insulating blankets.

Different Elevations of Adjacent Footings   If the elevations of two adjacent footings are different, as shown in Figure 7–11, the pressure on the soil from the upper footing may increase the pressure under the lower footing. Therefore, a limit called the *proximity line* is placed on the slope of the line joining the footings when placing adjacent footings at different elevations.

This slope should preferably be limited to 1:2 if the soil is mainly granular (sand, gravel). If the footings are on good clays, however, the slope may be increased to close to 1:1. The designer should consult the geotechnical engineer regarding the safe elevation difference between neighboring footings.

Presence of Expansive Soil   *Expansive soil* is a type of clay that undergoes significant volumetric changes with moisture variations. For example, a vast area of the southern United States is covered by a clay deposit known as the *Yazoo clay,* which is an expansive soil. Foundations placed directly on expansive soil may experience large upward pressures

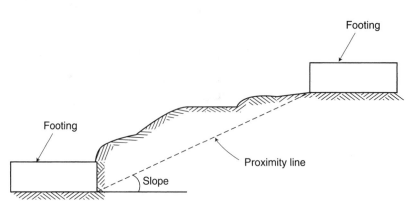

FIGURE 7–11   Adjacent footings at different elevations.

that could cause serious distortion and structural damage throughout the building. The moisture variation affects only the top few feet of an expansive soil; thus, footings are usually placed at an elevation below which the periodic moisture variation is insignificant. Short drilled piers are also commonly used.

Presence of Organic Layers Construction on soils that have significant organic content or underlying layers of organic soils (e.g., peat, marl, etc.) can cause serious problems. Organic matter is highly compressible and in a state of long-term decomposition.

      Two different strategies for dealing with organic soil are available to the designer. One is completely removing the soil to the full extent of the organic layers and replacing it with a so-called "engineered backfill." The second is to use a deep foundation, usually piles. It is also advisable to design the lowest level, which is normally just a slab on grade, as a structural floor. Otherwise, the slab may settle unevenly, cracking and distorting attached nonstructural elements, such as partitions.

7.9   DISTRIBUTION OF SOIL PRESSURE UNDER FOOTINGS

Footings apply pressure on their supporting soil. This pressure has to be limited to a certain allowable level (soil bearing capacity). There is also a reaction from the soil acting on the footing when it presses the soil.

      The true theoretical distribution of the reaction pressure from the soil on the footing depends on the type of supporting soil. Figure 7–12 shows typical pressure distributions for different types of soil. The pressures are larger under the center of footing and smaller along the edges if the soil is sandy. This is because the sand along the edges does not have good lateral support and can easily move laterally (see Figure 7–12a). The shape of the theoretical distribution of pressures in clays is shown in Figure 7–12b. In practice, an average uniform soil pressure distribution, like the one shown in Figure 7–12c, is assumed. This is much simpler and the results have been proven to provide adequate and safe designs.

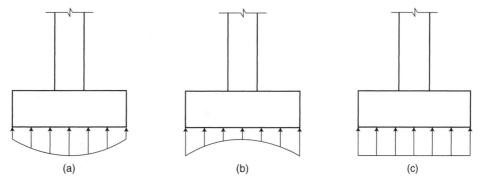

(a)   (b)   (c)

FIGURE 7–12   Soil pressure distributions: (a) sandy soil, (b) clayey soil, (c) design assumption.

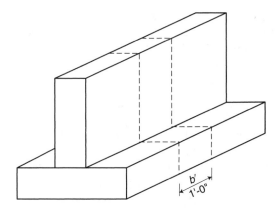

FIGURE 7–13   Wall footings design strip.

7.10   DESIGN OF WALL FOOTINGS

Because wall footings are long continuous members, designers use a 1-ft-long strip of wall ($b' = 12$ in.) and its footing to represent the whole length for design purposes. Figure 7–13 illustrates this concept. Wall footings are made of plain concrete or reinforced concrete. Plain concrete footings are commonly used to support light loads such as residential construction.

Plain Concrete Wall Footings

To design a plain concrete wall footing we need only to determine the depth and width of the footing such that (1) the soil pressure beneath the footing is less than the allowable value (bearing capacity), and (2) the bending and shear strength of the concrete footing is adequate. Generally, the footing width is calculated to satisfy the first requirement, while the depth is computed to satisfy the second requirement.

Plain Concrete Wall Footing Design   We must perform the following steps in order to design a plain concrete wall footing. They are summarized in the flowchart of Figure 7–17.

Step 1   Determine the footing width (b).

Geotechnical reports usually provide the designer with a *net* allowable soil pressure. Net pressure excludes the weight of the footing and the surrounding soil. So the weight of the footing and the weight of the backfill directly above the footing are ignored in the design.

The footing thickness (h) and width (b) are both unknown, so we estimate h and then check the pressure levels to determine how good the estimate was. First we calculate the footing width (b) such that the soil pressure (q_s) is less than the allowable net bearing capacity of soil (q_a), as shown in Figure 7–14.

$$q_s = \frac{w_D + w_L}{b} \le q_a$$

$$b \ge \frac{w_D + w_L}{q_a} \tag{7–1}$$

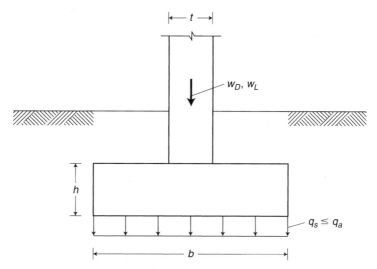

FIGURE 7–14 Wall footing pressure on supporting soil.

In this equation w_D and w_L are the unfactored dead and live loads, respectively. The footing width (b) is usually rounded up to the nearest even inch.

Step 2 Estimate the footing thickness (h).

The rule of thumb for the thickness (h) of the plain concrete wall footing is:

$$h = \frac{b - t}{2} \qquad (7\text{–}2)$$

where t is the wall thickness (see Figure 7–14). The thickness (h) is usually rounded up to the nearest inch.

According to the ACI Code (Section 22.7.4) the wall footing has to have a minimum thickness of 8 in. In practice, we use this value and the thickness of the supported wall, whichever is larger. The footing has to be at least as wide as the wall thickness. In addition, the geotechnical report often states a minimum acceptable footing width, usually at least 16″. Figure 7–15 summarizes these requirements.

Check the footing dimensions against these minimum values. If they are smaller than the minimum, use the minimum width and thickness.

Step 3 Calculate and check the moment.

The bending moment is the critical factor in determining the required thickness (h) for plain concrete wall footings. The critical sections for moment (where moments are the largest) are based on the type of wall being supported. According to the ACI Code (Section 22.7.5) the critical section for moment for a masonry wall (more flexible than a concrete wall) is at a distance $t/4$ to the inside of the wall, as shown in Figure 7–16a. The critical section for a poured concrete wall is at the face of the wall, as shown in Figure 7–16b.

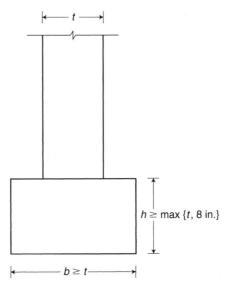

FIGURE 7–15   Minimum dimensions of wall footings.

The weight of soil above the footing and the weight of the footing do not cause any bending or shear in the footing. In this respect, a footing is like a mattress lying flat on a bed. It does not bend; but when you stand on it, you notice how it deforms and bends. Similarly, the only loads that cause bending and shear in the footing are the dead (w_D) and live (w_L) loads. Figure 7–16c shows the bending of the footing subjected to the applied loads. The

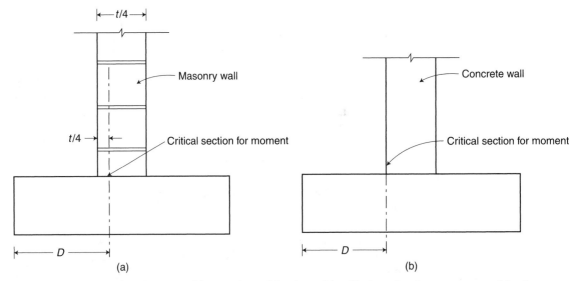

FIGURE 7–16   Moment in wall footings: (a) critical section for masonry wall footing, (b) critical section for concrete wall footing, (c) bending of wall footing, (d) moment at the critical section.

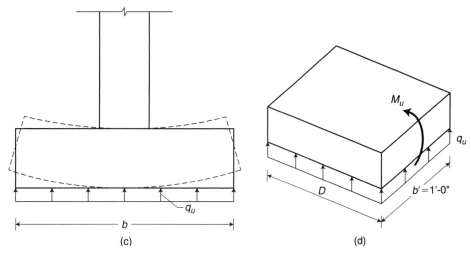

FIGURE 7–16   (continued)

ultimate tensile stress is calculated using factored loads. Therefore, the factored pressure acting from soil on the footing, q_u, is:

$$q_u = \frac{1.2w_D + 1.6w_L}{b} \qquad (7\text{--}3)$$

Figure 7–16d shows the moment at the critical section, which can be found from the equilibrium of the sum of moments:

$$M_u = q_u D \left(\frac{D}{2} \right) = q_u \frac{D^2}{2} \qquad (7\text{--}4)$$

The footing is constructed of plain concrete, so there is no clear definition of the effective depth (d). But ACI Code (Section 22.4.8) recommends a reduction of the overall footing thickness by 2 in. to allow for unevenness of excavation and possible contamination of the concrete adjacent to the soil:

$$d = h - 2 \text{ in.} \qquad (7\text{--}5)$$

The moment, M_u, acts on a section (12 in. $\times$ d). Therefore, the elastic section modulus, S_m, is:

$$S_m = \frac{b'd^2}{6} = \frac{12d^2}{6} \qquad (7\text{--}6)$$

($b' = 12$ in. because we use a 12 in. strip of footing.) The nominal resisting moment of the footing, M_n, is (ACI Code Equation 22–2) therefore:

$$M_n = 5\sqrt{f_c'}\, S_m \qquad (7\text{--}7)$$

The value $5\sqrt{f_c'}$ is the ACI Code-recommended ultimate tensile stress in bending for plain concrete. Then, for the footing section to be acceptable:

$$M_R = \phi M_n \qquad (7\text{--}8)$$

$$M_R \geq M_u \qquad (7\text{--}9)$$

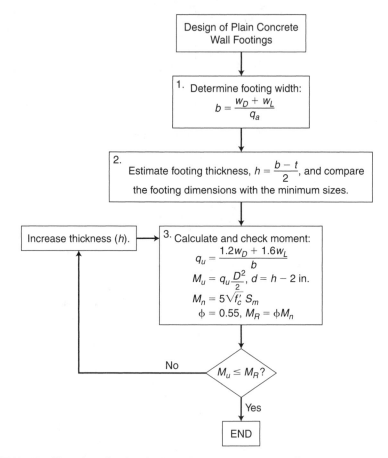

FIGURE 7–17   Flowchart for the design of plain concrete wall footings.

The strength reduction factor for flexure, compression, shear, and bearing of structural plain concrete (ACI Code, Section 9.3.5) is:

$$\phi = 0.55$$

If the footing is not acceptable, we should increase the footing thickness and repeat the process. The thickness can be increased arbitrarily or by solving $M_R = M_u$ for t.

EXAMPLE 7–1

A 12 in. load-bearing CMU (concrete masonry unit) wall supports an outdoor canopy. The wall will support a dead load of 10 kip/ft (including the weight of the wall), and a live load of 5 kip/ft. Design a plain concrete footing for this wall. The compressive strength of the concrete is 3,000 psi. The net bearing capacity of the soil is 3,000 psf. The frost line is four feet from the grade.

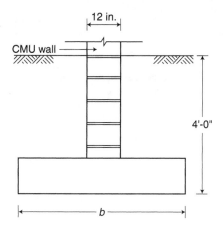

Solution

Step 1   Determine the footing width.

$$\text{Approximate footing width } (b) = \frac{w_D + w_L}{q_a} = \frac{10.0 + 5.0}{3.0} = 5 \text{ ft}$$

Step 2   Estimate the footing thickness.

$$h = \frac{b - t}{2} = \frac{60 - 12}{2} = 24 \text{ in.}$$

The footing dimensions are larger than the minimum sizes.

Step 3   Calculate and check the moment.

We check the moment by comparing the applied moment, M_u, with the resisting moment, M_R. Determine the factored pressure from soil acting on the footing (q_u):

$$q_u = \frac{1.2w_D + 1.6w_L}{b}$$

$$q_u = \frac{1.2(10) + 1.6(5)}{5} = 4.0 \text{ ksf}$$

The critical section for moment, shown in Figure 7–18, is at $t/4$ from the face of the CMU wall. Therefore, the distance, D, from the footing edge to this location is:

$$D = \frac{b - t}{2} + \frac{t}{4}$$

$$D = \frac{(5 \times 12) - 12}{2} + \frac{12}{4} = 27 \text{ in.}$$

$$M_u = q_u \frac{D^2}{2} = (4.0)\frac{\left(\dfrac{27}{12}\right)^2}{2}$$

$$M_u = 10.13 \text{ ft-kip}$$

$$d = h - 2 \text{ in.} = 24 \text{ in.} - 2 \text{ in.} = 22 \text{ in.}$$

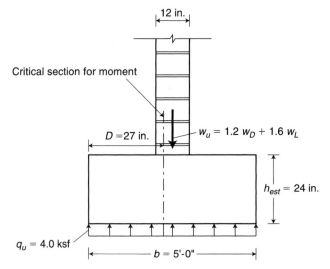

FIGURE 7-18 Plain concrete wall footing of Example 7–1 (checking moment).

$$S_m = \text{elastic section modulus} = \frac{b'd^2}{6}$$

$$S_m = \frac{12(22)^2}{6} = 968 \text{ in}^3.$$

The nominal resisting moment, M_n, from Equation 7–7 is:

$$M_n = 5\sqrt{f'_c}\, S_m$$

$$M_n = 5(\sqrt{3{,}000})\frac{(968)}{12{,}000} \quad \text{Conversion factor for in.-lb to ft-kip}$$

$$M_n = 22.1 \text{ ft-kip}$$
$$M_R = \phi M_n = 0.55(22.1)$$
$$M_R = 12.2 \text{ ft-kip} > M_u = 10.13 \text{ ft-kip} \quad \therefore \text{ ok}$$

Therefore, the footing thickness is enough.

Figure 7–19 shows the final design of this footing.

Reinforced Concrete Wall Footings

Reinforced concrete wall footings usually support larger loads than do their plain concrete counterparts. They are typically reinforced in the short direction.

Reinforced Concrete Wall Footing Design The design of a reinforced concrete wall footing is different from that of a plain concrete footing because the footing is reinforced to develop the required moment. Consequently, the footing can be thinner. As a result the shear also has to be checked in reinforced concrete wall footings. The

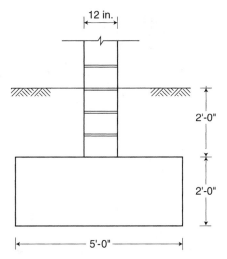

FIGURE 7–19 Final design of wall footing of Example 7–1.

following steps, which are summarized in Figure 7–21, are performed in the design of these footings:

Step 1 Determine the footing width (b).

$$\text{Footing width } (b) = \frac{w_D + w_L}{q_a}$$

Round up b to the nearest even inch, if needed.

Step 2 Estimate the thickness (h).

The footing has no shear reinforcement. Thus, the concrete must be thick enough to have sufficient shear resistance by itself. A reasonable thickness is about 40%–50% of the overhanging width of the footing. Hence, a conservative estimate is:

$$h = 0.5\left(\frac{b - t}{2}\right) \tag{7–10}$$

Round up h to the nearest inch, if needed. Based on ACI Section 15.7, the depth of footing above bottom reinforcement has to be at least 6 in. Considering the cover requirement, the minimum reinforced concrete wall footing depth is 10 in.

Step 3 Calculate and check shear.

Similar to the plain concrete footing, the factored pressure from the soil on the reinforced concrete footing (q_u) is:

$$q_u = \frac{1.2w_D + 1.6w_L}{b}$$

The effective depth of the footing, d, is:

$$d = h - \text{cover} - \text{diameter of bar}/2 \tag{7–11}$$

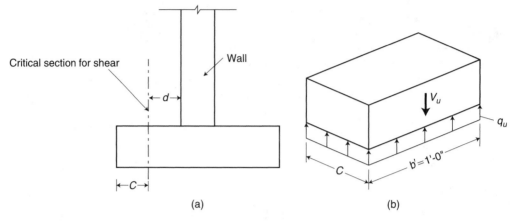

FIGURE 7–20 Shear in wall footings: (a) location of critical section, (b) shear at the critical section.

The minimum cover in footings is 3 in. (ACI Code, Section 7.7.1). Assuming #6 bars:

$$d = h - 3 \text{ in.} - 0.375 = h - 3.38 \text{ in.}$$

According to the ACI Code (Sections 11.1.3.1 and 15.5.2), the critical section for shear is located at a distance d from the face of the wall. Figure 7–20 shows the critical section and the shear at the critical section. To calculate V_u we cut the footing at this location and write the equilibrium of forces (Figure 7–20b):

$$V_u = q_u C \tag{7–12}$$

where

$$C = \frac{b - t}{2} - d \tag{7–13}$$

The nominal shear strength of concrete, V_c, (ACI Equation 11–3) is:

$$V_c = 2\sqrt{f'_c}\, b'd \tag{7–14}$$
$$b' = 12 \text{ in.}$$

According to the ACI Code (Section 9.3.2.3), the strength reduction factor for shear in reinforced concrete (ϕ) is:

$$\phi = 0.75$$

Therefore, for the section to be adequate:

$$\phi V_c \geq V_u \tag{7–15}$$

Otherwise, the footing thickness has to be increased.

Step 4 Determine the required reinforcements.

Based on the ACI Code (Section 15.4.2), the critical section for moment is at the same location as for the plain concrete footing (i.e., $^1/_4$ from the face

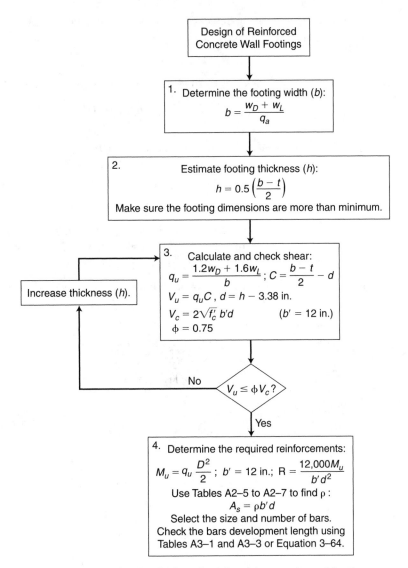

FIGURE 7–21   Flowchart for the design of reinforced concrete wall footings.

of masonry walls and at the face of concrete walls), as shown in Figure 7–16. Calculate the moment at the critical section as follows:

$$M_u = \frac{q_u D^2}{2}$$

Determine the R value:

$$R = \frac{12,000 M_u}{b' d^2} \qquad (b' = 12 \text{ in.}) \qquad (7\text{–}16)$$

Using Tables A2–5 to A2–7, we obtain the steel ratio ρ and calculate the required area of steel:

$$A_s = \rho b' d \qquad (7\text{–}17)$$

Check this value against the minimum area of reinforcement, $A_{s,min}$:

$$A_{s,min} = 0.0018 b' h \qquad (7\text{–}18)$$

Check the bar's development length by using Table A3–3 and the applicable modification factors of Table A3–1. The bar length from the critical section for moment to the edge of the footing has to be more than ℓ_d. Otherwise, the bars have to be hooked at their ends. Therefore,

$$D - 3 \text{ in. (cover)} \geq \ell_d \qquad (7\text{–}19)$$

Note that we can use Equation 3–64 and the corresponding modification factors of Table A3–1 instead of Table A3–3 to obtain the bar development length. This method usually results in a smaller required development length.

Normally, 1 #4 or #5 longitudinal bar is used per foot of width as distributor bars.

EXAMPLE 7–2

Design the wall footing of Example 7–1 using reinforced concrete. Assume $f_y = 60{,}000$ psi.

Solution

Step 1   Determine the required width, b.

$$b = \frac{w_D + w_L}{q_a}$$

$$b = \frac{10 + 5}{3} = 5 \text{ ft}$$

Step 2   Estimate the footing thickness, h.

$$h = 0.5 \left(\frac{b - t}{2} \right) = 0.5 \left(\frac{5 - 1}{2} \right) = 1.0 \text{ ft}$$

$$h = 1'\text{-}0''$$

Step 3   Calculate and check the shear.
Determine the factored pressure on the footing from the soil (q_u):

$$q_u = \frac{1.2 w_D + 1.6 w_L}{b}$$

$$q_u = \frac{1.2(10) + 1.6(5)}{5}$$

$$q_u = 4.0 \text{ ksf}$$

Calculate the effective depth of the footing. Assume #6 bars with 3 in. minimum clear cover. Therefore,

$$d = 12 - 3.38 = 8.62 \text{ in.}$$

The critical section for shear is located at the distance d from the face of wall, as shown in Figure 7–22a; from Equation 7–13:

$$C = \frac{b - t}{2} - d$$

$$C = \frac{5(12) - 12}{2} - 8.62$$

$$C = 15.38 \text{ in.} = 1.28 \text{ ft}$$

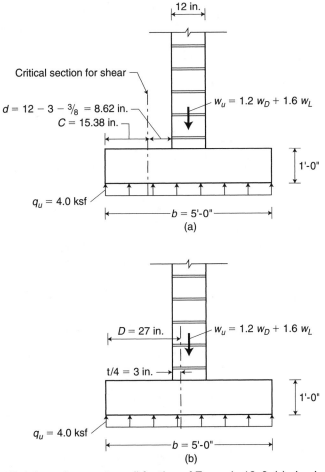

FIGURE 7–22 Reinforced concrete wall footing of Example 10–2: (a) check shear, (b) check moment.

The shear at the critical section (V_u) is:

$$V_u = q_u C$$
$$V_u = 4.0(1.28) = 5.12 \text{ kip}$$

The shear strength of the concrete, ϕV_c, is:

$$\phi V_c = \phi(2\sqrt{f_c'} \, b'd)$$
$$\phi V_c = 0.75 \left[\frac{2\sqrt{3,000}(12)(8.62)}{1,000} \right]$$
$$\phi V_c = 8.50 \text{ kip} > 5.12 \text{ kip}$$

Therefore, the footing thickness is enough to resist the shear.

Step 4   Determine the required reinforcements.

The critical section for moment is at the distance $^1/_4$ from the face of wall, as shown in Figure 7–22b:

$$D = \frac{b - t}{2} + \frac{t}{4}$$
$$D = \frac{5(12) - 12}{2} + \frac{12}{4}$$
$$D = 27 \text{ in.} = 2.25 \text{ ft}$$
$$M_u = \frac{q_u D^2}{2}$$
$$M_u = \frac{4.0(2.25)^2}{2} = 10.13 \text{ ft-kip}$$
$$R = \frac{12,000 M_u}{b'd^2}$$
$$R = \frac{12,000(10.13)}{(12)(8.62)^2} = 136 \text{ psi}$$

From Table A2–6a $\longrightarrow$ $\rho = 0.0026$

$$A_s = \rho b'd = 0.0026(12)(8.62) = 0.27 \text{ in}^2/\text{ft}$$

The minimum required reinforcement is the same as the minimum shrinkage and temperature reinforcement:

$$A_s = 0.0018 b'h = 0.0018(12)(12)$$
$$A_s = 0.26 \text{ in}^2/\text{ft} < 0.27 \text{ in}^2/\text{ft}$$

From Table A2–10 $\longrightarrow$ use #5 @ 13 in. ($A_s = 0.29 \text{ in}^2/\text{ft}$)

Check the reinforcing bars' development length. According to Table A3–2 because clear spacing $= 13 \text{ in.} - 0.625 = 12.375 \text{ in.} > 2(0.625)$ and clear cover $= 3 \text{ in.} > 0.625 \text{ in.}$, therefore condition A is applicable.

From Table A3–3 for #5 bars, $f_y = 60$ ksi, and $f'_c = 3$ ksi:

$$\ell_d = 28 \text{ in.}$$

The ends of the bar must have 3 in. cover. Hence:

$$D - 3 \text{ in.} = 27 \text{ in.} - 3 \text{ in.} = 24 \text{ in.} < 28 \text{ in.} \quad \therefore \text{ N.G.}$$

Therefore, the bars do not satisfy the development length requirements using the simplified expression. We can either bend the bar ends up to create hooks or use the more accurate Equation 3–64 of Chapter 3 to check the required bar development length. From Equation 3–64:

$$\ell_d = \left[\frac{3}{40} \frac{f_y}{\sqrt{f'_c}} \frac{\psi_t \psi_e \psi_s \lambda}{\left(\dfrac{c_b + K_{tr}}{d_b} \right)} \right] d_b \geq 12 \text{ in.}$$

where

$$\frac{c_b + K_{tr}}{d_b} \leq 2.5$$

From Table A3–1:

K_{tr} (no transverse reinforcement) $= 0$

c_b (concrete cover to bar center) $= 3 + \dfrac{0.625}{2} = 3.313$ in.

$\dfrac{(c_b + K_{tr})}{d_b} = \dfrac{(3.313 + 0)}{0.625} = 5.3 \text{ in.} > 2.5 \quad \therefore \text{ Use } 2.5.$

ψ_t (bottom bars) $= 1.0$

ψ_e (coating factor) $= 1.0$

ψ_s (reinforcement size factor) $= 0.80$

λ (normal weight concrete) $= 1.0$

$$\ell_d = \left[\frac{3}{40} \frac{60{,}000}{\sqrt{3{,}000}} \frac{(1.0)(1.0)(0.8)(1.0)}{2.5} \right] d_b \left(\frac{A_{s,\text{required}}}{A_{s,\text{provided}}} \right)$$

$$\ell_d = 26.3 d_b \left(\frac{A_{s,\text{required}}}{A_{s,\text{provided}}} \right) = 26.3(0.625)\left(\frac{0.27}{0.29} \right)$$

$$= 15.3 \text{ in.} > 12 \text{ in.} \quad \therefore \text{ ok}$$

$$D - 3 \text{ in.} = 24 \text{ in.} > 15.3 \text{ in. } \therefore \text{ ok}$$

Therefore, the main reinforcement bar has sufficient development length. Since the footing is 5'-0" wide, we use 5 #4 distributor bars.

Figure 7–23 shows the final design of this footing.

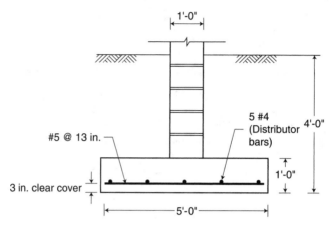

FIGURE 7–23   Final design of wall footing of Example 7–2.

7.11   REINFORCED CONCRETE SQUARE SPREAD FOOTING DESIGN

Square spread footings are the most common type of column footing. Regardless of the material used for the column construction, reinforced concrete (or, rarely, plain concrete) spread footings are used to support columns.

The steps for the design of reinforced concrete square spread footings are summarized in the flowchart of Figure 7–28 and are as follows:

Step 1   Calculate the required area and select the size of the footing.

As shown in Figure 7–24, the footing is sized such that the pressure on the soil (q_s) is less than the soil bearing capacity (q_a).

$$q_s = \frac{P_D + P_L}{A} \le q_a$$

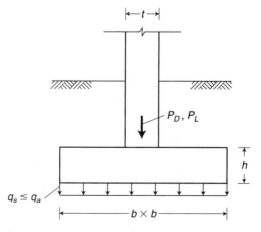

FIGURE 7–24   Footing pressure on supporting soil.

or

$$A_{required} = \frac{P_D + P_L}{q_a} \longrightarrow b_{required} = \sqrt{A_{required}} \qquad (7\text{--}20)$$

where P_D and P_L are the applied service dead and live loads, respectively, and q_a is the allowable soil pressure . The value for b is usually rounded up to the nearest even inch.

Step 2   Estimate the footing thickness.

In a square or rectangular footing, a reasonable estimate of the required thickness is about one-half of the larger overhanging (O. H.) length of the footing. Therefore, for a square footing, the estimated required thickness is:

$$h = 0.5 \, (\text{O. H.}) = 0.5 \left(\frac{b-t}{2} \right) \qquad (7\text{--}21)$$

Round up h to the nearest inch, if needed. The minimum thickness commonly used for column spread footings is 12 in.

Step 3   Calculate and check the shear.

The factored pressure on the footing from the soil, q_u, is:

$$q_u = \frac{1.2P_D + 1.6P_L}{b^2} \qquad (7\text{--}22)$$

Step 3a   Check the two-way (punching) shear.

Typically, the two-way (punching) shear (refer to Chapter 6) is the controlling factor in determining the required thickness.

The critical sections for the two-way shear action are located at the distance $d/2$ from the faces of the concrete column. For a steel column this distance is measured from the midpoint between the face of column and the edge of the base plate, as shown in Figure 7–25.

The footing bends in two perpendicular directions, so it requires reinforcement in the form of a grid. The average effective depth, d, may be taken from the top of footing (which is in compression) to the location between the two layers of bars as follows:

$$d = h - \text{cover} - \text{diameter of bar}$$

The concrete cover for footings is 3 in. Assuming #8 bars, the distance d can be calculated as:

$$d = h - 3 \, \text{in.} - 1 \, \text{in.} = h - 4 \, \text{in.}$$

Compute the length of the critical section, B. Then cut the footing at the critical sections to determine the shear by satisfying the equilibrium of forces in the vertical direction, as shown in Figure 7–26a:

$$B = \text{length of the critical section (one side)}$$

$$B = t + 2 \left(\frac{d}{2} \right) = t + d \qquad (7\text{--}23)$$

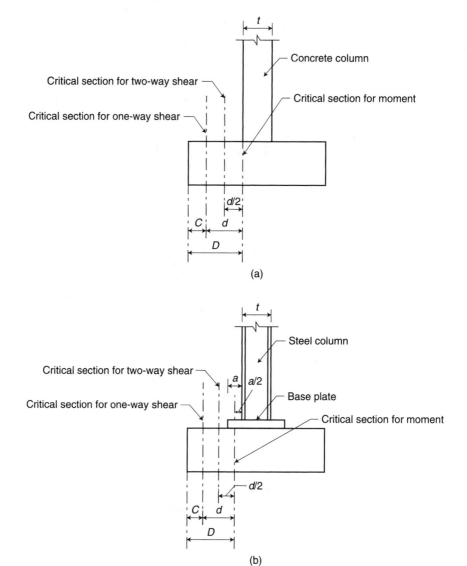

FIGURE 7–25 Critical sections for square spread footings: (a) concrete column, (b) steel column.

The total factored shear acting on the critical shear surface is:

$$V_{u2} = q_u(b^2 - B^2) \qquad (7\text{–}24)$$

According to the ACI Code (Section 11.12.2.1), the nominal shear strength of concrete, V_c, for two-way action is the same as that for slabs:

$$V_{c2} = \min\left\{ \left(2 + \frac{4}{\beta}\right)\sqrt{f_c'}\,b_o d,\ \left(\frac{\alpha_s d}{b_o} + 2\right)\sqrt{f_c'}\,b_o d,\ 4\sqrt{f_c'}\,b_o d \right\} \qquad (7\text{–}25)$$

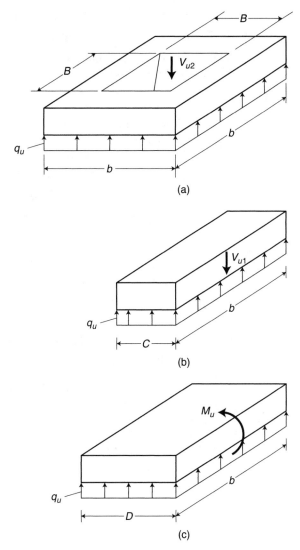

FIGURE 7–26 Shear forces and bending moments at the critical sections: (a) two-way shear, (b) one-way shear, (c) bending moment.

where

$$\beta = \frac{\text{larger dimension of column}}{\text{shorter dimension of column}}$$ ($\beta = 1$ for square columns)

b_o = perimeter of critical section   ($b_o = 4B$ for square columns)

α_s = 40 for columns in the center of footing

  = 30 for columns at an edge of footing

  = 20 for columns at a corner of footing

The footing is adequate in punching shear if:

$$\phi V_{c2} \geq V_{u2} \quad (\phi = 0.75) \tag{7–26}$$

Otherwise, the footing thickness, h, has to be increased, and the process is repeated.

Step 3b   Check one-way or beam shear.

The requirements for one-way (beam action) shear must be satisfied in addition to those for the two-way shear. As shown in Figure 7–25, the critical section for one-way shear is at the distance d from the face of the concrete column. Therefore, cutting the footing at this location and writing the equilibrium equation of forces (see Figure–26b) allows us to calculate the shear as follows:

$$V_{u1} = q_u bC \tag{7–27}$$

where for the case of concrete columns

$$C = \frac{b - t}{2} - d \tag{7–28}$$

The nominal one-way shear strength of concrete is (ACI Code, Section 11.12.1.1):

$$V_{c1} = 2 \sqrt{f'_c}\, bd \tag{7–29}$$

For the footing to be adequate in one-way shear:

$$\phi V_{c1} \geq V_{u1} \quad (\phi = 0.75) \tag{7–30}$$

If the above equation is not satisfied, increase the footing thickness, h, and repeat the process.

Step 4   Determine the required reinforcement.

The footing bends in both directions like a dish when subjected to soil pressure from below. Therefore, we can consider bending of the footing from one side and find the moment at the critical section. Cutting the footing at the critical section for moment, as shown in Figure 7–26c, and setting the sum of moments to zero, we can calculate the moment at the critical section as follows:

$$M_u = q_u b(D)(D/2) = \frac{q_u b D^2}{2} \tag{7–31}$$

The required resistance coefficient, R, is:

$$R = \frac{12,000 M_u}{bd^2}$$

Using the Tables A2–5 to A2–7, we obtain ρ for R. The required area of steel (A_s) then is:

$$A_s = \rho bd$$

Use Table A2–9 to select the number and size of bars. The minimum area of steel is:

$$A_{s,\min} = 0.0018bh$$

Check the bar development length by using Table A3–3 and the applicable modification factors of Table A3–1. The bar length from the critical section for moment has to be more than ℓ_d. Otherwise, the bars have to be hooked at their ends. Thus:

$$D - 3 \text{ in.(cover)} \geq \ell_d$$

Note that we can use Equation 3–64 and the corresponding modification factors of Table A3–1 instead of using Table A3–3 to obtain the bar development length. This method usually results in a smaller required development length.

Step 5   Determine the required dowel bars.

The column at its base transfers the load to the footing on an area equal to the column's cross-sectional area (A_g). This generates a bearing pressure that the footing must resist.

The bearing capacity of the concrete at the column footprint, $N_{bearing}$, is given by the ACI Code (Section 10.17.1):

$$N_1 = \phi(0.85f_c' A_1) \tag{7–32}$$

$$N_2 = \min\left\{ \phi(0.85f_c'A_1)\sqrt{\frac{A_2}{A_1}}, 2\phi(0.85f_c'A_1) \right\} \tag{7–33}$$

$$N_{bearing} = \min\{N_1, N_2\} \tag{7–34}$$

where

ϕ = 0.65 (ACI Code, Section 9.3.2.4)

A_1 = column bearing area, which for a column directly bearing on the footing is equal to A_g of the column

A_2 = area of the part of the footing that is geometrically similar to, and concentric with the column bearing area, A_1 (see Figure 7–27a)

N_1 = bearing capacity of the column

N_2 = bearing capacity of the footing

$N_{bearing}$ = bearing capacity of the concrete at the base of the footing

f_c' = the specified strength of concrete in the column, when evaluating N_1; and the specified strength of concrete in the footing, when evaluating N_2

Because the compressive strength of concrete in columns is usually larger than it is in footings, we must compute both N_1 and N_2 to determine $N_{bearing}$. Dowel bars must resist the difference between the load transferred from the column to the footing, P_u, and the bearing capacity of concrete, $N_{bearing}$. The required area of these bars, A_{sd}, is calculated as follows:

$$P_d = P_u - N_{bearing} \tag{7–35}$$

$$A_{sd} = \max\left\{ \frac{P_d}{f_y}, 0.005 A_g \right\} \tag{7–36}$$

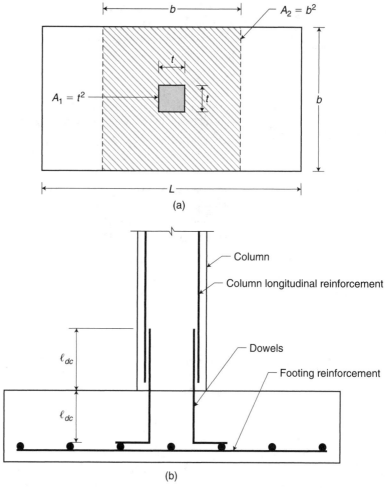

FIGURE 7–27   Dowel reinforcements: (a) bearing areas A_1 and A_2, (b) dowel bars between the column and footing.

The ACI Code, Section 15.8.2.1 requires a minimum amount of dowel reinforcement equal to $0.005A_g$ (A_g is the gross area of column) to transfer the loads from the column to the footing. This is in the form of a minimum of four bars placed at the corners of the column. A minimum development length for the dowels in compression equal to ℓ_{dc} is required. This minimum length has to be provided from the column bearing area extending into the column and the footing, as shown in Figure 7–27b. The dowels are commonly hooked and tied to the footing main reinforcements for ease of construction. The dowels have to be lap spliced in compression based on requirements given in Chapter 3. Use Table A3–6 to obtain the development length for compression bars, ℓ_{dc}, and Table A3–5 to obtain the applicable modification factors.

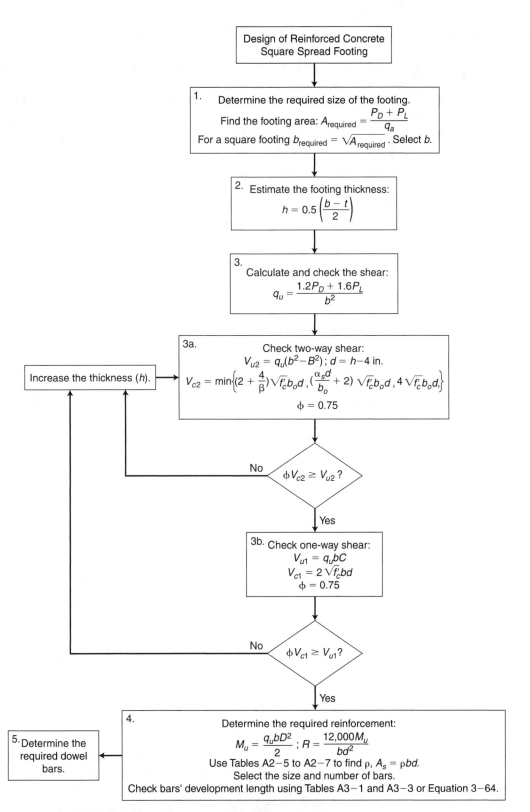

FIGURE 7–28 Flowchart for the design of reinforced concrete square spread footings.

EXAMPLE 7–3

Design a square reinforced concrete footing for the 16 in. square interior concrete column shown below. The dead load is 200 kip and the live load is 100 kip. The allowable net soil pressure (bearing capacity) is 3,500 psf. Use $f'_c = 3,000$ psi for the footing, $f'_c = 4,000$ psi for the column, and $f_y = 60,000$ psi.

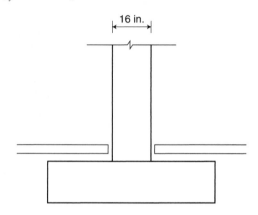

16 in.

Solution

Step 1   Determine the required size of the footing.

$$A_{required} = \frac{P_D + P_L}{q_a} = \frac{200 + 100}{3.5} = 85.7 \text{ ft}^2$$

$$b_{required} = \sqrt{85.7} = 9.26 \text{ ft}$$

Round b to the nearest even inch and select $b = 9'\text{-}4''$.

Step 2   Estimate the footing thickness.

The estimated depth of the footing is:

$$h_{est} = \frac{1}{2}\left[\frac{(9 \times 12 + 4) - 16}{2}\right] = 24 \text{ in.}$$

Step 3   Calculate and check the shear.

The factored pressure on the footing from the soil is:

$$q_u = \frac{1.2P_D + 1.6P_L}{b^2}$$

$$q_u = \frac{1.2(200) + 1.6(100)}{9.33^2}$$

$$q_u = 4.60 \text{ ksf}$$

Step 3a   Check the two-way shear.

The average effective depth, d, is:

$$d = h - \text{cover} - \text{estimated diameter of bar}$$

$$d = 24 - 3 - 1 = 20 \text{ in.}$$

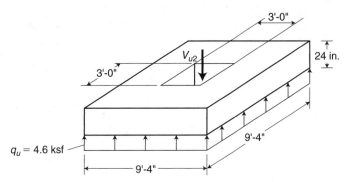

FIGURE 7–29 Two-way shear in the spread footing.

The critical section for the two-way shear is at a distance $d/2$ from the face of the concrete column. Therefore, one side of the critical section, B, is:

$$B = t + d = 16 + 20 = 36 \text{ in.} = 3'\text{-}0''$$

The shear at the critical sections, V_{u2}, is shown in Figure 7–29, and is calculated as follows:

$$V_{u2} = q_u(b^2 - B^2)$$
$$V_{u2} = 4.6(9.33^2 - 3^2)$$
$$V_{u2} = 359 \text{ kip}$$

The nominal two-way shear strength of the concrete, V_{c2}, is:

$$V_{c2} = \min\left\{\left(2 + \frac{4}{\beta}\right)\sqrt{f'_c}\, b_o d,\right.$$

$$\left.\left(\frac{\alpha_s d}{b_o} + 2\right)\sqrt{f'_c}\, b_o d, 4\sqrt{f'_c}\, b_o d\right\}$$

$\beta = 16/16 = 1.0$

$\alpha_s = 40$ (column in the center of footing)

$b_o = 4B = 4 \times 36 = 144$ in.

$$V_{c2} = \min\left\{(2 + 4)\sqrt{3{,}000}\,(144)(20)/1{,}000,\right.$$

$$\left.\left(\frac{40 \times 20}{144} + 2\right)\sqrt{3{,}000}\,(144)(20)/1{,}000, 4\sqrt{3{,}000}\,(144)(20)/1{,}000\right\}$$

$V_{c2} = \min\{946 \text{ kip}, 1{,}192 \text{ kip}, 631 \text{ kip}\} = 631 \text{ kip}$

$\phi V_{c2} = 0.75(631) = 473 \text{ kip} > 359 \text{ kip} \quad \therefore \text{ ok}$

The two-way shear capacity of this footing is acceptable. The estimated depth could be reduced, as the shear capacity is about 30% more than the applied shear force. However, we will conservatively continue with the assumed depth of 24 in.

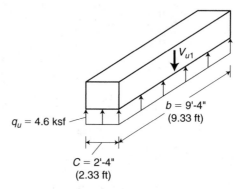

q_u = 4.6 ksf

b = 9'-4"
(9.33 ft)

C = 2'-4"
(2.33 ft)

FIGURE 7–30 One-way shear in the spread footing.

Step 3b Check one-way shear.

The critical section for one-way shear is at a distance, d, from the face of the column:

$$C = \frac{b - t}{2} - d$$

$$C = \frac{9.33 \times 12 - 16}{2} - 20$$

$$C = 28 \text{ in.} = 2'\text{-}4'' = 2.33 \text{ ft}$$

If we cut the footing at the critical section, as shown in Figure 7–30, the shear at this location, V_{u1}, is:

$$V_{u1} = q_u bC$$
$$V_{u1} = 4.60(9.33)(2.33)$$
$$V_{u1} = 100.0 \text{ kip}$$

The nominal one-way shear strength of concrete, V_{c1}, is:

$$V_{c1} = 2\sqrt{f_c'}\, bd$$
$$V_{c1} = 2\sqrt{3,000}\,(9.33 \times 12)(20)/1,000$$
$$V_{c1} = 245.3 \text{ kip}$$
$$\phi V_{c1} = 0.75(245.3) = 184 \text{ kip} > 100 \text{ kip} \quad \therefore \text{ ok}$$

Therefore, the footing has enough capacity against the one-way shear.

Step 4 Determine the required reinforcement.

To calculate the required reinforcement, we first calculate the bending moment at the critical section for moment (the face of the column):

$$D = \frac{b - t}{2} = \frac{9.33 \times 12 - 16}{2}$$

$$D = 48 \text{ in.} = 4.00 \text{ ft}$$

The moment at the critical section as shown in Figure 7–31 is:

$$M_u = q_u bD^2/2$$
$$M_u = 4.6(9.33)(4.0)^2/2$$

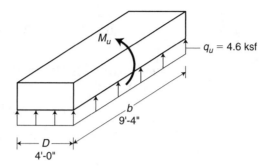

FIGURE 7–31   Bending moment in the spread footing.

$$M_u = 343.3 \text{ ft-kip}$$

$$R = \frac{12{,}000 M_u}{bd^2}$$

$$R = \frac{12{,}000(343.3)}{(9.33 \times 12)(20)^2} = 92.0 \text{ psi}$$

From Table A2–6a ($f_c' = 3{,}000$ psi, $f_y = 60{,}000$ psi) $\longrightarrow$ $\rho = 0.0018$ (conservatively) and the required area of steel, A_s, is :

$$A_s = \rho bd$$

$$A_s = 0.0018(9.33 \times 12)(20)$$

$$A_s = 4.0 \text{ in}^2$$

$$A_{s,\min} = 0.0018\, bh$$

$$A_{s,\min} = 0.0018(9.33 \times 12)(24)$$

$$A_{s,\min} = 4.84 \text{ in}^2 > 4.0 \text{ in}^2$$

Therefore, the required area of steel is:

$$A_s = 4.84 \text{ in}^2$$

Table A2–9 $\longrightarrow$ $\therefore$ Use 9 #7 each way ($A_s = 5.40$ in^2)
Check the bars development length.

From Table A3–2, because cover > 0.875 in. and clear space $> 2(0.875$ in.), condition A is applicable, and from Table A3–3:

$$\ell_d = 48 \text{ in. (\#7 bars)}$$

From Table A3–1, ℓ_d can be reduced by $\dfrac{A_{s,\text{required}}}{A_{s,\text{provided}}}$, therefore:

$$\ell_d = 48 \times \frac{4.00}{5.40} = 35.5 \text{ in.}$$

The bar length measured from the critical section for moment is:

$$D - 3 \text{ in.} = 48 \text{ in.} - 3 \text{ in.} = 45 \text{ in.} > 35.5 \text{ in.} \quad \therefore \text{ ok}$$

Step 5   Determine the required dowel bars.

$$N_1 = \phi(0.85 f_c' A_1); \ (f_c')_{\text{column}} = 4.0 \text{ ksi}$$

$$N_1 = 0.65[0.85(4.0)(16 \times 16)] = 565.8 \text{ kip}$$

$$N_2 = \min\left\{ \phi(0.85f_c'A_1)\sqrt{\frac{A_2}{A_1}},\ 2\phi(0.85f_c'A_1) \right\};\ (f_c')_{\text{footing}} = 3.0\,\text{ksi}$$

$$N_2 = \min\left\{ 0.65\left[0.85(3.0)(16 \times 16)\sqrt{\frac{112 \times 112}{16 \times 16}} \right], \right.$$

$$\left. 2(0.65)[0.85(3.0)(16 \times 16)] \right\}$$

$$N_2 = \min\{2,970, 848.6\} = 848.6\,\text{kip}$$

$$N_{\text{bearing}} = \min\{N_1, N_2\} = 565.8\,\text{kip}$$

$$P_u = 1.2(200) + 1.6(100) = 400\,\text{kip} < 565.8\,\text{kip}$$

$$\therefore \text{ Use minimum area for dowels.}$$

$$A_{sd} = 0.005A_g = 0.005(16 \times 16) = 1.28\,\text{in}^2$$

(Table A2–9 $\longrightarrow$ use 4 #6 ($A_s = 1.76\,\text{in}^2$))

It has to be noted that for practical purposes the dowel bar size is usually selected to match the column main reinforcements, which for a 16 in. × 16 in. column is expected to be larger that #6 bars. We have, however, selected #6 bars here for consistency and clarity in the solution.

The required development length in the footing from Table A3–6 (compression bars) for $f_c' = 3,000$ psi is:

$$\ell_{dc} = 17\,\text{in. (#6 bars)}$$

Adjusting the dowel length using Table A3–5:

$$\ell_{\text{dowel}} = \ell_{dc}\left(\frac{A_{s,\text{required}}}{A_{s,\text{provided}}}\right) = 17\left(\frac{1.28}{1.76}\right) = 13\,\text{in.}$$

Conservatively, we use the same development length in the column.

Figure 7–32 shows the final design of this footing.

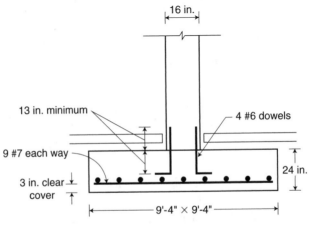

16 in.

13 in. minimum

4 #6 dowels

9 #7 each way

24 in.

3 in. clear
cover

9'-4" × 9'-4"

FIGURE 7–32 Final design of the spread footing of Example 7–3.

EXAMPLE 7–4

Design the square reinforced concrete footing shown in Figure 7–33 for the interior column of Example 5–3. The bearing capacity of the soil is 8,000 psf, $f'_c = 3,000$ psi for the footing, and $f_y = 60,000$ psi.

Solution

From Example 5–3:

$$\text{Column} = 16 \text{ in.} \times 16 \text{ in.}$$
$$P_D = 387 \text{ kip}$$
$$P_L = 117 \text{ kip}$$
$$f'_c = 4,000 \text{ psi}$$

Step 1 Determine the required footing size.

$$A_{required} = \frac{387 + 117}{8.0} = 63.0 \text{ ft}^2$$

$$b_{required} = \sqrt{63.0} = 7.94 \text{ ft} \quad \text{Round to } 8'\text{-}0''.$$

Step 2 Estimate footing thickness.

$$h_{est} = 0.5\left(\frac{b - t}{2}\right) = 0.5\left(\frac{8.0 \times 12 - 16}{2}\right) = 20 \text{ in.}$$

Step 3 Calculate and check shear.

$$q_u = \frac{1.2P_D + 1.6P_L}{b^2} = \frac{1.2(387.0) + 1.6(117.0)}{(8.0)^2}$$

$$q_u = 10.2 \text{ ksf}$$

Step 3a Check two-way shear.

$$d = 20 - 4 = 16 \text{ in.}$$

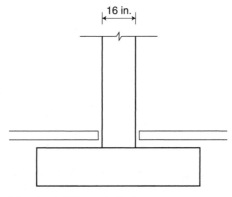

FIGURE 7–33 Spread footing of Example 7–4.

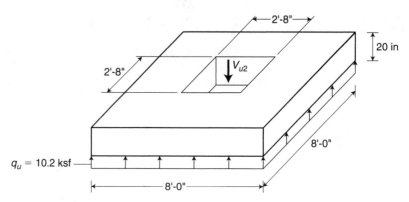

FIGURE 7-34 Two-way shear action.

The critical sections, as shown in Figure 7–34, are at distance $d/2$ from the face of the column.

$$B = t + d = 16 + 16 = 32 \text{ in.} = 2'\text{-}8'' = 2.67 \text{ ft}$$

$$b_o = 4 \times 32 = 128 \text{ in.}$$

From Figure 7–34:

$$V_{u2} = q_u(b^2 - B^2) = 10.2(8.0^2 - 2.67^2)$$
$$V_{u2} = 580 \text{ kip}$$

The nominal shear capacity of the concrete for the two-way action is:

$$V_{c2} = \min\left\{ \left(2 + \frac{4}{\beta}\right)\sqrt{f_c'}\, b_o d, \left(\frac{\alpha_s d}{b_o} + 2\right)\sqrt{f_c'}\, b_o d, 4\sqrt{f_c'}\, b_o d \right\}$$

$$\beta = \frac{16}{16} = 1, \quad \alpha_S = 40 \text{ (column at the center of footing)}$$

$$V_{c2} = \min\left\{ \left(2 + \frac{4}{1}\right)\sqrt{3,000}\,(128)(16)/1,000, \right.$$

$$\left(\frac{(40 \times 16)}{128} + 2\right)\sqrt{3,000}\,(128)(16)/1,000,$$

$$\left. 4\sqrt{3,000}(128)(16)/1,000 \right\}$$

$$V_{c2} = \min\{673, 785, 449\}$$
$$V_{c2} = 449 \text{ kip}$$
$$\phi V_{c2} = 0.75\,(449) = 337 \text{ kip} < V_{u2} = 580 \text{ kip} \quad \therefore \text{ N.G.}$$

Therefore, we need to increase the footing thickness, which is usually done through a trial-and-error process. The difference between the shear capac-

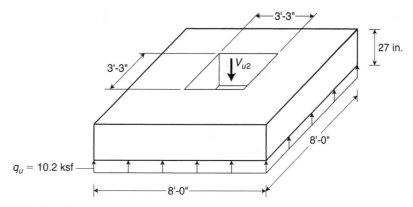

FIGURE 7–35 Two-way shear action (second trial).

ity and the shear demand is quite large, so we try to increase the footing thickness by 7 in.

$$\therefore \text{Try } h = 27 \text{ in.} = 2'\text{-}3''$$

Step 3R (Repeat) Calculate and check shear.
Step 3R (a) Check two-way shear.
As shown in Figure 7–35:

$$q_u = 10.2 \text{ ksf}$$
$$d = 27 - 4 = 23 \text{ in.}$$

$$B = t + d = 16 + 23$$

$$= 39 \text{ in.} = 3'\text{-}3'' = 3.25 \text{ ft}$$

$$b_o = 4(39) = 156 \text{ in.}$$

Using Figure 7–35:

$$V_{u2} = q_u(b^2 - B^2) = 10.2(8.0^2 - 3.25^2)$$
$$V_{u2} = 545 \text{ kip}$$

Again, calculating the shear capacity of concrete:

$$V_{c2} = \min\left\{ \left(2 + \frac{4}{1}\right)\sqrt{3{,}000}(156)(23)/1{,}000, \right.$$

$$\left(\frac{(40 \times 23)}{156} + 2\right)\sqrt{3{,}000}(156)(23)/1{,}000,$$

$$\left. 4\sqrt{3{,}000}(156)(23)/1{,}000 \right\}$$

$$V_{c2} = \min\{1{,}179 \text{ kip}, 1{,}552 \text{ kip}, 786 \text{ kip}\}$$
$$V_{c2} = 786 \text{ kip}$$
$$\phi V_{c2} = 0.75(786) = 590 \text{ kip} > 545 \text{ kip} \quad \therefore \text{ ok}$$

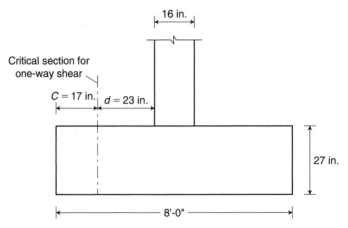

FIGURE 7-36 Critical section for one-way shear.

The new footing is deep enough, and we continue with checking the one-way shear action.

Step 3b One-way shear.

The critical section for the one-way shear, as shown in Figure 7–36, is at a distance d(23 in.) from the face of the column.

$$C = \frac{8.0 \times 12 - 16}{2} - 23 = 17\,\text{in.} = 1.42\,\text{ft}$$

Figure 7–37 shows the shear acting on the section. Thus:

$$V_{u1} = q_u bC = 10.2(8.0)(1.42)$$
$$V_{u1} = 116\,\text{kip}$$

The one-way shear capacity of the concrete, V_{c1} is:

$$V_{c1} = 2\sqrt{f_c'}\,bd$$
$$= 2\sqrt{3{,}000}(8.0 \times 12)(23)/1{,}000$$
$$V_{c1} = 242\,\text{kip}$$
$$\phi V_{c1} = 0.75(242) = 182\,\text{kip} > 116\,\text{kip} \quad \therefore \text{ ok}$$

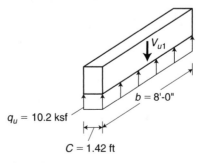

FIGURE 7-37 One-way shear force.

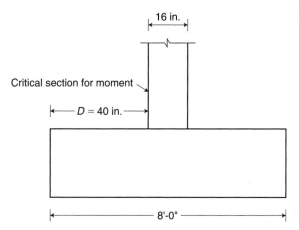

FIGURE 7–38 Critical section for bending moment.

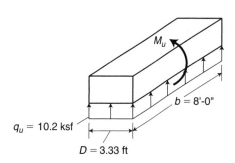

FIGURE 7–39 Bending moment at the critical section.

Step 4 Determine the required reinforcements.

The critical section for moment, as shown in Figure 7–38, is at the face of the column:

$$D = \frac{b - t}{2}$$

$$D = \frac{8.0 \times 12 - 16}{2} = 40 \text{ in.} = 3.33 \text{ ft}$$

Figure 7–39 shows the moment acting on the critical section. Thus:

$$M_u = q_u b D \left(\frac{D}{2}\right)$$

$$M_u = 10.2(8.0)(3.33)\left(\frac{3.33}{2}\right)$$

$$M_u = 452 \text{ ft-kip}$$

$$R = \frac{12,000 M_u}{bd^2}$$

$$R = \frac{12,000 \times 452}{(8.0 \times 12)(23)^2} = 107 \text{ psi}$$

From Table A2–6a ($f_c' = 3,000 \text{ psi}, f_y = 60,000 \text{ psi}$) $\longrightarrow$ $\rho = 0.0021$

$$A_s = \rho b d = 0.0021(8.0 \times 12)(23)$$
$$A_s = 4.64 \text{ in}^2$$
$$A_{s,min} = 0.0018 \, bh$$
$$A_{s,min} = 0.0018(8.0 \times 12)(27) = 4.67 \text{ in}^2 > 4.64 \text{ in}^2$$

From Table A2–9 $\therefore$ Use 8 #7 ($A_s = 4.80 \text{ in}^2$)

This reinforcement is required for both direction.

Check the bars' development length:

From Table A3–2, because cover $> d_b = 0.875$ in. and clear space $> 2d_b = 2(0.875 \text{ in.})$, we use condition A. From Table A3–3, the required development length is:

$$\ell_d = 48 \text{ in. (\#7 bars)}$$

From Table A3–1, ℓ_d can be reduced by $\dfrac{A_{s,\text{required}}}{A_{s,\text{provided}}}$ to get the required length (ℓ_{req}). Therefore,

$$\ell_{\text{req}} = 48 \times \frac{4.64}{4.80} = 46 \text{ in.}$$

The bar length measured from the critical section for moment is:

$$D - 3 \text{ in.} = (3.33 \times 12) - 3 = 37 \text{ in.} < 46 \text{ in.} \quad \therefore \; N.\,G.$$

Use Equation 3–64 to calculate the more accurate required bar development length:

$$c_b = 3 + \frac{0.875}{2} = 3.44$$

$$K_{tr} = 0 \quad \text{(conservatively)}$$

$$\frac{c_b + K_{tr}}{d_b} = \frac{3.44}{0.875} = 3.93 > 2.5 \quad \therefore \; \text{Use 2.5}$$

$$\ell_d = \left[\frac{3}{40} \frac{f_y}{\sqrt{f_c'}} \frac{\psi_t \psi_e \psi_s \lambda}{\left(\dfrac{c_b + K_{tr}}{d_b} \right)} \right] d_b \left(\frac{A_{s,\text{required}}}{A_{s,\text{provided}}} \right) \geq 12 \text{ in.}$$

$$= \left[\frac{3}{40} \frac{60{,}000}{\sqrt{3{,}000}} \frac{(1.0)(1.0)(1.0)(1.0)}{2.5} \right] d_b \left(\frac{4.64}{4.80} \right) = 31.77\, d_b$$

$$= 31.77(0.875) = 28 \text{ in.} > 12 \text{ in.}$$

$$D - 3 = 37 \text{ in.} > 28 \text{ in.} \quad \therefore \; \text{The bar development length is adequate.}$$

Step 5 Determine the required dowel bars.

Using Equations 7–32 to 7–36:

$$N_1 = \phi(0.85 f_c' A_1)$$

$$N_1 = 0.65[0.85(4.0)(16 \times 16)] = 565.8 \text{ kip}$$

$$N_2 = \min\left\{ \phi(0.85 f_c' A_1)\sqrt{\frac{A_2}{A_1}},\; 2\phi(0.85 f_c' A_1) \right\}$$

$$N_2 = \min\left\{0.65[0.85(3.0)(16 \times 16)\sqrt{\frac{96 \times 96}{16 \times 16}}\,],\right.$$

$$\left.2(0.65)[0.85(3.0)(16 \times 16)]\right\}$$

$$N_2 = \min\{2{,}546\,\text{kip}, 848.6\,\text{kip}\} = 848.6\,\text{kip}$$

$$N_{\text{bearing}} = \min\{N_1, N_2\} = \min\{565.8\,\text{kip}, 848.6\,\text{kip}\} = 565.8\,\text{kip}$$

$$P_u = 1.2P_D + 1.6P_L = 1.2(387) + 1.6(117) = 651.6\,\text{kip} > 565.8\,\text{kip}$$

$$P_d = P_u - N_{\text{bearing}} = 651.6 - 565.8 = 85.8\,\text{kip}$$

$$A_{sd} = \max\left\{\frac{P_d}{f_y}, 0.005A_g\right\}$$

$$= \max\left\{\frac{85.8}{60}, 0.005(16 \times 16)\right\}$$

$$= \max\{1.43\,\text{in}^2, 1.28\,\text{in}^2\} = 1.43\,\text{in}^2$$

Table A2–9 $\longrightarrow$ Use 4 #6 ($A_s = 1.76\,\text{in}^2$)

Refer to the comment regarding the selection of size of dowels in the solution for Example 7–3.

From Table A3–6 the required development length of #6 bars in the footing for $f_c' = 3{,}000$ psi is:

$$\ell_{dc} = 17\,\text{in.}$$

$$\ell_{\text{dowel}} = \ell_{dc}\left(\frac{A_{s,\text{required}}}{A_{s,\text{provided}}}\right) = 17\left(\frac{1.43}{1.76}\right) = 13.8\,\text{in.}$$

Therefore, use 14 in. minimum dowel length in the footing. The compression lap splice in the column for $f_y \leq 60{,}000$ is (see Chapter 3):

$$= 0.0005f_y\,d_b \geq 12\,\text{in.}$$

$$= 0.0005(60{,}000)(0.75) = 22.5\,\text{in.}$$

Use a 23 in. splice in the column.

Figure 7–40 is a sketch of the final design.

7.12 RECTANGULAR REINFORCED CONCRETE FOOTING

A square footing is sometimes impractical due to space limitations. For example, if a building column is located close to a property line, the designer has to size the footing to keep it within the property boundaries. A rectangular footing may be used in such cases. The design method for rectangular footings is similar to that for square footings. There are a few differences, however. The steps for the design of rectangular reinforced concrete footings follow and are also summarized in the flowchart of Figure 7–45.

Step 1   Determine the required area of the footing:

$$A_{\text{required}} = \frac{P_D + P_L}{q_a}$$

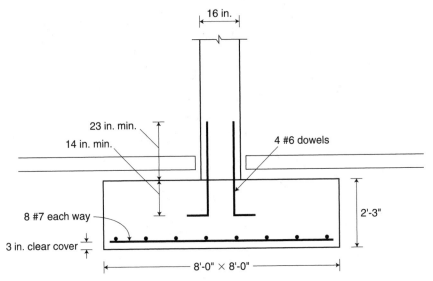

FIGURE 7–40 Final design of the spread footing of Example 7–4.

If the footing area is $b_\ell \times b_s$ (b_ℓ is the longer dimension, and b_s the shorter, as shown in Figure 7–41) and the side (b_s) is limited to a certain known value, then $b_\ell = \dfrac{A}{b_s}$. Round up b_ℓ to the nearest even inch and calculate the footing contact area (A):

$$A = b_\ell \times b_s$$

Step 2   Estimate the footing thickness.
A reasonable preliminary thickness is about 50% of the overhanging length for a square footing of equivalent area. Thus:

$$b = \sqrt{A}$$

$$h_{est} = 0.5\left(\frac{b - t}{2}\right)$$

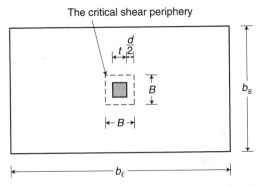

FIGURE 7–41 Two-way shear in rectangular reinforced concrete footings.

where t is the column width. Round up the thickness to the nearest inch, if necessary.

Step 3   Calculate and check shear.

The factored pressure on the footing from the soil is:

$$q_u = \frac{1.2P_D + 1.6P_L}{A}$$

Step 3a   Check two-way (punching) shear.

The two-way shear requirements for a rectangular footing are similar to those of a square footing. The critical sections are at a distance $^d\!/_2$ from the concrete column face ($d = h - 3$ in. $- 1$ in.).

Figure 7–41 shows the critical two-way shear perimeter for a rectangular footing with a square column. Cut the footing at the critical sections and obtain the resulting shear force:

$$B = t + 2\left(\frac{d}{2}\right) = t + d$$

$$V_{u2} = q_u(A - B^2)$$

The nominal two-way shear capacity of concrete, V_{c2}, is:

$$V_{c2} = \min\left\{\left(2 + \frac{4}{\beta}\right)\sqrt{f_c'}\,b_o d, \left(\frac{\alpha_s d}{b_o} + 2\right)\sqrt{f_c'}\,b_o d, 4\sqrt{f_c'}\,b_o d\right\}$$

which is the same equation (Equation 7–25) as for square footings. (Refer to the section on square footings for the definitions of the parameters.)

In order to satisfy the ACI Code's requirements, the shear capacity of the concrete has to be greater than the applied shear force:

$$\phi V_{c2} \geq V_{u2}$$

If this condition is not satisfied, we need to increase the footing depth, h, and repeat the process.

Step 3b   Check one-way shear.

The one-way shear requirements for rectangular footings are also similar to those for square footings. But, the rectangular shape of the footing places the critical section for the one-way shear at the distance d from the face of the column in the long direction, as shown in Figure 7–42a. Therefore, the distance, C, from the edge of the footing to the critical section is:

$$C = \frac{b_\ell - t}{2} - d$$

Figure 7–42b shows the applied loads at the critical section for shear, V_{u1}, which can be calculated as:

$$V_{u1} = q_u b_s C$$

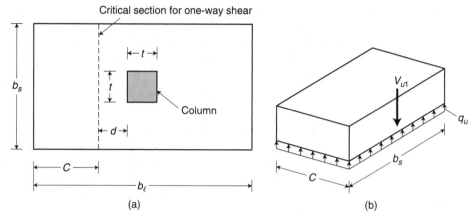

FIGURE 7–42 One-way shear in rectangular reinforced concrete footings: (a) plan view, (b) cut at the critical section.

The one-way shear strength of the footing is:

$$\phi V_{c1} = \phi(2\sqrt{f_c'}\,bd) = \phi(2\sqrt{f_c'}\,b_s d)$$

For the footing to be adequate in one-way shear:

$$\phi V_{c1} \geq V_{u1}$$

If the above relationship is not satisfied, we increase the footing depth and repeat the process.

Step 4   Determine required reinforcement.

Finding the required area of steel reinforcements for rectangular footings is much different than for square footings, as the bending moments in the footing are different in each direction. The critical sections for moments are at the face of the column in each direction, as shown in Figure 7–43a.

Therefore, the distance in the long direction (D_ℓ) from the edge of the footing to the critical section is:

$$D_\ell = \frac{b_\ell - t}{2}$$

and the moment at the critical section, $M_{u\ell}$, shown in Figure 7–43b, is:

$$M_{u\ell} = q_u b_s D_\ell \left(\frac{D_\ell}{2}\right)$$

$$M_{u\ell} = q_u b_s \frac{D_\ell^2}{2}$$

The coefficient of resistance, R, can be calculated as follows:

$$R = \frac{12{,}000 M_{u\ell}}{bd^2} = \frac{12{,}000 M_{u\ell}}{b_s d^2}$$

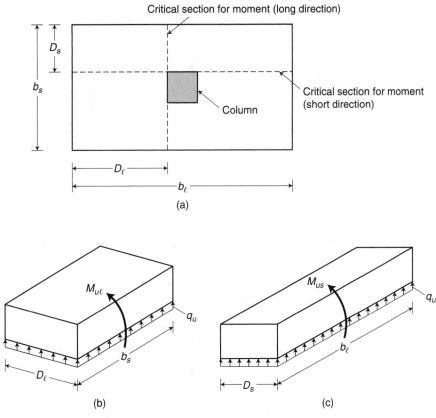

FIGURE 7–43   (a) Plan view of critical sections for moment; (b) moment at the critical section in the long direction; (c) moment at the critical section in the short direction.

Use Tables A2–5 to A2–7 to obtain ρ, and calculate the required area of steel in the long direction:

$$A_s = \rho b_s d$$

This reinforcing is distributed uniformly across the width (b_s) of the footing. For the short direction, the location of the critical section, D_s, is:

$$D_s = \frac{b_s - t}{2}$$

and the moment at the critical section, M_{us}, shown in Figure 7–43c is:

$$M_{us} = q_u b_\ell D_s \left(\frac{D_s}{2}\right)$$

$$M_{us} = q_u b_\ell \frac{D_s^2}{2}$$

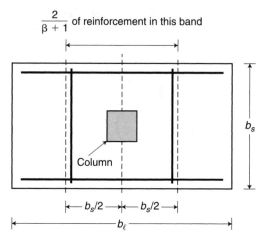

$\dfrac{2}{\beta + 1}$ of reinforcement in this band

FIGURE 7–44   Rectangular footing plan for reinforcement distribution.

The coefficient of resistance, R, can be calculated as follows:

$$R = \frac{12{,}000 M_{us}}{bd^2} = \frac{12{,}000 M_{us}}{b_\ell d^2}$$

Using Tables A2–5 to A2–7, we obtain ρ. Then, the required area of steel in the short direction is:

$$A_s = \rho b_\ell d$$

The reinforcement is not distributed uniformly in the short direction. Figure 7–44 shows how the reinforcement is distributed in the short direction. According to the ACI Code (Section 15.4.4.2), a portion of the total reinforcement equal to $\gamma_s A_s$ should be uniformly distributed over a bandwidth equal to the footing width (b_s) under the column. The remainder of the reinforcement $[(1 - \gamma_s)A_s]$ should be distributed uniformly outside this bandwidth. γ_s is defined as:

$$\gamma_s = \frac{2}{\beta + 1} \tag{7–37}$$

where

$$\beta = \frac{\text{long side of footing}}{\text{short side of footing}} = \frac{b_\ell}{b_s} \tag{7–38}$$

We must check the development length of the reinforcements in both directions, measured from the critical section for moment. The procedure is similar to that used for square footings.

Step 5   Determine the required dowel bars.

The dowel requirements are the same as those for square footings.

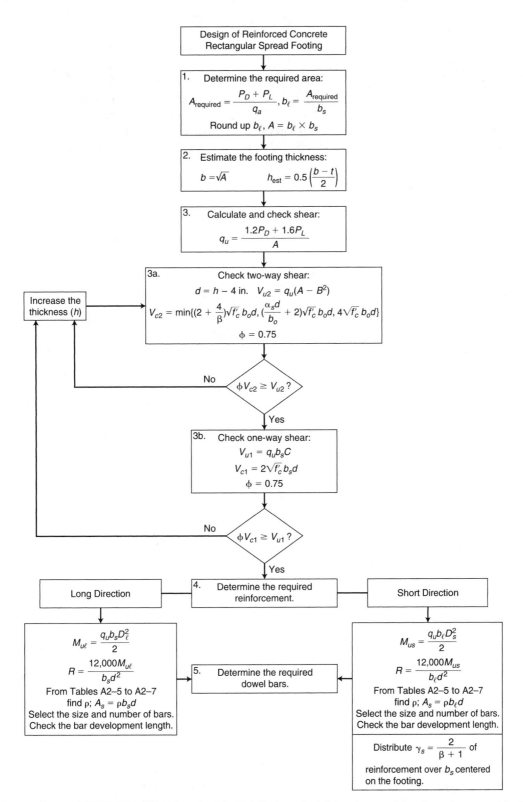

FIGURE 7-45 Flowchart for the design of reinforced concrete rectangular spread footing.

EXAMPLE 7–5

Design a rectangular reinforced concrete footing for the 16 in. square reinforced concrete exterior column shown in Figure 7–46. The dead load is 100 kip and the live load is 75 kip. The soil bearing capacity is 3,500 psf, $f'_c = 3,000$ psi for the footing and the column, and $f_y = 60,000$ psi. The width of the footing is limited to 6'-0" due to its proximity to the property line.

Solution

Step 1   Determine the required area.

$$A = \frac{P_D + P_L}{q_a} = \frac{100 + 75}{3.5} = 50 \text{ ft}^2$$

The long dimension, b_ℓ is:

$$b_\ell = \frac{A}{b_s} = \frac{50}{6} = 8.33 \text{ ft}$$

Select an 8'-4" × 6'-0" footing

$$A = 8.33 \times 6 = 50 \text{ ft}^2$$

Step 2   Estimate the footing thickness.

$$b = \sqrt{A} = \sqrt{50} = 7.07 \text{ ft}$$

$$h_{est} = 0.5\left(\frac{b - t}{2}\right) = 0.5\left(\frac{(7.07 \times 12) - 16}{2}\right) = 17.2 \text{ in.}$$

$$\therefore h = 18 \text{ in.}$$

Step 3   Calculate and check shear.

The factored pressure acting on the footing from the soil, q_u, is:

$$q_u = \frac{1.2P_D + 1.6P_L}{A}$$

$$q_u = \frac{1.2(100) + 1.6(75)}{50}$$

$$q_u = 4.8 \text{ ksf}$$

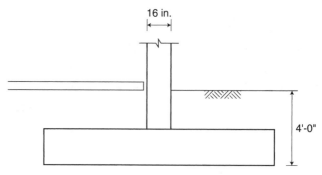

FIGURE 7–46   Rectangular spread footing of Example 7–5.

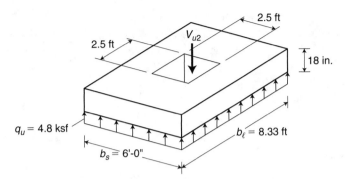

FIGURE 7–47 Two-way shear in the rectangular spread footing.

Step 3a Check two-way shear.

$$d = h - 4 \text{ in.} = 18 - 4 = 14 \text{ in.}$$

The length of one side of the critical section for two-way shear, B, is:

$$B = t + d = 16 + 14 = 30 \text{ in.} = 2.5 \text{ ft}$$

Figure 7–47 shows the forces acting at the critical two-way shear sections, V_{u2}, which can be calculated as follows:

$$V_{u2} = q_u(A - B^2)$$
$$V_{u2} = 4.8[50 - (2.5)^2]$$
$$V_{u2} = 210 \text{ kip}$$

The two-way shear capacity of the concrete, V_{c2}, is:

$$V_{c2} = \min\left\{ \left(2 + \frac{4}{\beta}\right)\sqrt{f_c'}\,b_o d, \left(\frac{\alpha_s d}{b_o} + 2\right)\sqrt{f_c'}b_o d, 4\sqrt{f_c'}\,b_o d \right\}$$

$$\beta = \frac{16}{16} = 1.0; \quad \alpha_s = 40 \text{ (column at the center of the footing)}$$

$$b_o = 4B = 4 \times 30 = 120 \text{ in.}$$

$$V_{c2} = \min\left\{ \left(2 + \frac{4}{1}\right)\sqrt{3{,}000}\,\frac{(120)(14)}{1{,}000}, \left(\frac{40 \times 14}{120} + 2\right) \right.$$
$$\left. \sqrt{3{,}000}\,\frac{(120)(14)}{1{,}000}, 4\sqrt{3{,}000}\,\frac{120 \times 14}{1{,}000} \right\}$$

$$V_{c2} = \min\{552, 613, 368\} = 368 \text{ kip}$$
$$\phi V_{c2} = 0.75(368) = 276 \text{ kip} > 210 \text{ kip} \quad \therefore \text{ ok}$$

Therefore, the footing thickness is adequate for the two-way shear action.

Step 3b Check one-way shear.

The critical section for one-way shear, as shown in Figure 7–48, is at a distance d from the face of the column in the long direction:

$$C = \frac{b_\ell - t}{2} - d$$

$$C = \frac{8.33 \times 12 - 16}{2} - 14 = 28 \text{ in.} = 2.33 \text{ ft}$$

$$V_{u1} = q_u b_s C$$

$$V_{u1} = 4.8(6.0)(2.33)$$

$$V_{u1} = 67.1 \text{ kip}$$

$$V_{c1} = 2\sqrt{f_c'}\, b_s d$$

$$V_{c1} = 2\sqrt{3{,}000}\, \frac{(6 \times 12)(14)}{1{,}000}$$

$$V_{c1} = 110.4 \text{ kip}$$

$$\phi V_{c1} = 0.75(110.4) = 82.8 \text{ kip} > 67.1 \text{ kip} \quad \therefore \text{ ok}$$

Therefore, the one-way shear is ok, and the footing thickness is adequate.

Step 4   Calculate the required reinforcement.

Long Direction

The location of the critical section for moment, shown in Figure 7–49, is:

$$D_\ell = \frac{b_\ell - t}{2} = \frac{8.33 \times 12 - 16}{2} = 42 \text{ in.} = 3.50 \text{ ft}$$

$$M_{u\ell} = q_u b_s \frac{D_\ell^2}{2}$$

$$M_{u\ell} = 4.8(6) \frac{(3.50)^2}{2} = 176.4 \text{ ft-kip}$$

$$R = \frac{12{,}000(176.4)}{(6 \times 12)(14)^2} = 150 \text{ psi}$$

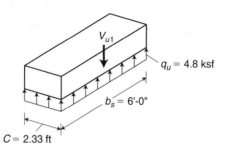

FIGURE 7–48 One-way shear in the rectangular spread footing.

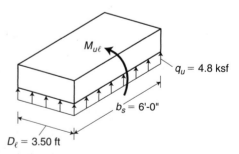

FIGURE 7–49 Moment at the critical section in the long direction.

From Table A2–6a (f'_c = 3,000 psi, f_y = 60,000 psi) $\longrightarrow$ ρ = 0.0029 (rounded up)

$$A_s = \rho b_s d$$
$$A_s = 0.0029(6 \times 12)(14)$$
$$A_s = 2.92 \text{ in}^2$$
$$A_{s,min} = 0.0018 b_s h$$
$$A_{s,min} = 0.0018(6 \times 12)(18)$$
$$A_{s,min} = 2.33 \text{ in}^2 < 2.92 \text{ in}^2$$

From Table A2–9 select 7 #6 (long direction).

$$(A_s = 3.08 \text{ in}^2)$$

Short Direction

The location of the critical section for moment, shown in Figure 7–50, is:

$$D_s = \frac{b_s - t}{2} = \frac{(6 \times 12) - 16}{2} = 28 \text{ in.} = 2.33 \text{ ft}$$

The moment at the critical section (see Figure 7–50) is:

$$M_{us} = q_u b_\ell \frac{D_s^2}{2}$$

$$M_{us} = 4.8(8.33) \frac{(2.33)^2}{2} = 109 \text{ ft-kip}$$

$$R = \frac{12,000 M_{us}}{b_\ell d^2}$$

$$R = \frac{12,000(109)}{(8.33 \times 12)(14)^2} = 67 \text{ psi}$$

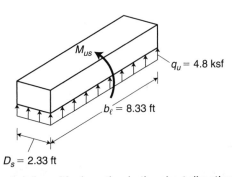

FIGURE 7–50 Moment at the critical section in the short direction.

From Table A2–6a $\longrightarrow$ $\rho = 0.0013$

$$A_s = \rho b_\ell d = 0.0013(8.33 \times 12)(14)$$
$$A_s = 1.82 \text{ in}^2$$
$$A_{s,\text{min}} = 0.0018 b_\ell h$$
$$A_{s,\text{min}} = 0.0018(8.33 \times 12)(18)$$
$$A_{s,\text{min}} = 3.24 \text{ in}^2 > 1.82 \text{ in}^2$$

Therefore, we need the minimum required reinforcement.
From Table A2–9 $\longrightarrow$ use 8 #6 ($A_s = 3.52 \text{ in}^2$)
For the distribution of reinforcement in the short direction:

$$\beta = \frac{b_\ell}{b_s} = \frac{8.33}{6} = 1.39$$

We must place a portion of reinforcement equal to γ_s of the total in a band centered on the column and having a width equal to b_s:

$$\gamma_s = \frac{2}{\beta + 1} = \frac{2}{1.39 + 1} = 0.84$$

Therefore, the number of bars to be distributed in this band is:

$$\gamma_s A_s = 0.84(8) = 6.7$$

Use 7 #6 bars in the 6'-0" center bandwidth and one #6 bars on each side.
This results in 9 #6 bars ($A_s = 3.96 \text{ in}^2$).
Check the bars' development length:
    Bar spacing in the long direction

$$= \frac{6(12) - 2(3) - (0.75)}{6} = 10.9 \text{ in.}$$

From Table A3–2, because cover > 0.75 in. and clear space $>$ $2(0.75 \text{ in.})$, use condition A. From Table A3–3:

$$\ell_d = 33 \text{ in. (#6 bar)}$$

From Table A3–1, ℓ_d can be reduced by $\dfrac{A_{s,\text{required}}}{A_{s,\text{provided}}}$:

For the long direction:

$$\ell_d = 33 \times \frac{2.92}{3.08} = 31 \text{ in.}$$

The provided bar length is:

$$D_\ell - 3 \text{ in.} = 3.5(12) - 3 = 39 \text{ in.} > 31 \text{ in.} \quad \therefore \text{ ok}$$

For the short direction:

$$\ell_d = 33 \times \frac{1.82}{3.96} = 15.2 \text{ in.}$$

The provided bar length in the short direction is:

$$D_s - 3 \text{ in.} = 2.33(12) - 3 = 25 \text{ in.} > 15.2 \text{ in.} \quad \therefore \text{ ok}$$

We demonstrate how to use Equation 3–64 and Table A3–1 to calculate the development length more accurately:

$$c_b = 3 + \frac{0.75}{2} = 3.38 \text{ in.}$$

$$K_{tr} = 0$$

$$\frac{c_b + K_{tr}}{d_b} = \frac{3.38 \text{ in.}}{0.75} = 4.5 > 2.5 \quad \therefore \text{ use } 2.5$$

$$\ell_d = \left[\frac{3}{40} \frac{f_y}{\sqrt{f_c'}} \frac{\psi_t \psi_e \psi_s \lambda}{\left(\frac{c_b + K_{tr}}{d_b} \right)} \right] d_b \left(\frac{A_{s,\text{required}}}{A_{s,\text{provided}}} \right) \geq 12 \text{ in.}$$

$$= \left[\frac{3}{40} \frac{60{,}000}{\sqrt{3{,}000}} \frac{(1.0)(1.0)(0.8)(1.0)}{2.5} \right] d_b \left(\frac{1.82}{3.96} \right) = 12.08 \, d_b$$

$$= 12.08(0.75) = 9 \text{ in.} < 25 \text{ in.} \quad \therefore \text{ ok}$$

$$\therefore \text{ The bars are long enough.}$$

Step 5 Determine the required dowel bars.
Using Equations 7–32 to 7–36:

$$N_1 = \phi(0.85 f_c' A_1)$$
$$N_1 = 0.65[0.85(3)(16 \times 16)] = 424.3 \text{ kip}$$
$$N_2 = \min\left\{ \phi(0.85 f_c' A_1) \sqrt{\frac{A_2}{A_1}}, \, 2\phi(0.85 \, f_c' A_1) \right\}$$
$$N_2 = \min\left\{ 0.65 \left[0.85(3.0)(16 \times 16) \sqrt{\frac{72 \times 72}{16 \times 16}} \right], \, 2(424.3) \right\}$$
$$N_2 = \min\{1{,}909, \, 848.6\} = 848.6 \text{ kip}$$
$$N_{\text{bearing}} = \min\{N_1, N_2\} = \min\{424.3, 848.6\} = 424.3 \text{ kip}$$
$$P_u = 1.2(100) + 1.6(75) = 240 \text{ kip} < 424.3 \text{ kip}$$
$$\therefore \text{ Use the minimum area for dowels.}$$
$$A_{sd} = 0.005 A_g = 0.005(16 \times 16) = 1.28 \text{ in}^2$$

Refer to the comment regarding the selection of size of dowels in the solution for Example 7–3.
Table A2–9 $\longrightarrow$ Use 4 #6 ($A_s = 1.76 \text{ in}^2$).
The required development length from Table A3–6 is:

$$\ell_d = 17 \text{ in.}$$

$$\ell_{\text{dowel}} = \ell_d \left(\frac{A_{s,\text{required}}}{A_{s,\text{provided}}} \right) = 17 \left(\frac{1.28}{1.76} \right) = 12.4 \text{ in.}$$

$$\therefore \text{ Use 13 in. minimum.}$$

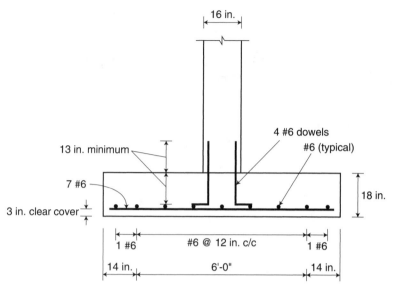

FIGURE 7–51   Final design of the rectangular footing of Example 7–5.

Figure 7–51 shows the final design of the footing.

7.13 EARTH SUPPORTING WALLS

Basement walls and retaining walls are two common concrete (plain or reinforced) structural systems. Sometimes, however, they are made of concrete masonry units. These structural elements have to resist lateral soil pressure. Therefore, it is important to understand the action of soil on them. This section briefly explains lateral soil pressure, then discusses the different aspects of the design and analysis of basement and retaining walls.

Lateral Earth Pressure

Soil that is retained on one side of a wall is confined on the higher grade and prevented from moving freely. Figure 7–52 shows a vertical section of a retaining wall. A wedge-shaped part of the soil in this vertical cut is pulled downward by gravity and tries to slide down along a plane of rupture. The wall, however, prevents these downward and outward movements, resulting in a lateral pressure on the wall.

Frictional resistance occurs along the plane of rupture as the grains of the soil try to slide by one another. If the soil has clay content, cohesion increases this sliding resistance. The plane also supports part of the weight (W) of the wedge. The combination of the weight support and the sliding resistance results in the force R.

The earth pressure (E) on the back of the wall is the resultant of the soil friction on the wall and the lateral earth pressure. The reaction to this force, E', acts on the soil wedge, which is in equilibrium with the weight of the wedge (W) and the R force. The inclination

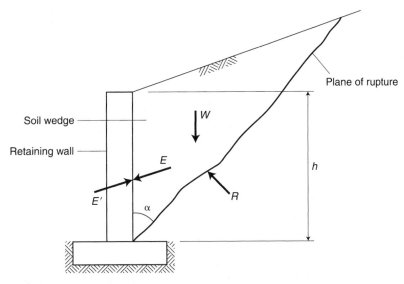

FIGURE 7–52　The retained soil wedge.

of force E is due to the frictional resistance the wall offers to the sliding wedge. If the wall surface is practically frictionless (e.g., smooth waterproofing on a basement wall), then the earth pressure on the wall is horizontal.

The magnitude and the distribution of the earth pressure on the back of the wall depend on many factors. The most important of these are the type of the retained soil (granular or cohesive), its moisture content, and the slope of the backfill (if any). In addition, there may be loads on the upper surface, called *surcharge loads*. These are caused by stored materials, traffic, or permanent installations such as neighboring building foundations. All these factors increase the gravity loads on the sliding wedge, which in turn increase the pressure on the wall.

A detailed discussion of the theory of lateral earth pressure is beyond the scope of this book, as it belongs to the field of soil mechanics. Our intention is to familiarize you sufficiently with the results of the theory.

There are three different types of earth pressure, distinguished from each other by their pressure coefficients. The first type is *earth pressure at rest*. This is the theoretical pressure on an essentially immovable object. The second type is the so-called *active earth pressure*. This occurs when the wall moves ever so slightly. This very slight movement activates the sliding resistance along the plane of rupture, which in turn reduces the pressure on the back of the wall. The *active* earth pressure, which occurs in the direction of wall movement, is significantly less than the *at-rest* earth pressure. The third type is the *passive earth pressure*. This happens when the wall is moving *against* the soil.

This section of the text is concerned mainly with the *active* earth pressure. We use Rankine theory here, which is the easiest and simplest theoretical solution for calculating active earth pressure. The theory assumes that the rupture plane is a straight line and the backfill material is cohesionless. It also assumes that the frictional resistance at the back of

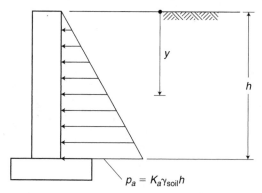

FIGURE 7–53   Pressure distribution on the back of a wall (no surcharge).

the wall is nonexistent (i.e., the wall is smooth). Thus, our primary concern is with the horizontal component of the pressure that is exerted on the wall.

The theory assumes that the distribution of the pressure on the back of the wall is triangular, as shown in Figure 7–53. In the absence of surcharge loads, the pressure is zero at the top and increases linearly with depth.

Level Backfill, No Surcharge   The pressure at any depth (y) can be expressed as:

$$p_{a,y} = K_a \gamma_{soil} y \qquad (7\text{--}39)$$

where
$K_a =$ the coefficient of the active pressure
$\gamma_{soil} =$ the unit weight of the soil in pcf
$y =$ the depth measured from the surface

The maximum pressure at the base is:

$$p_a = K_a \gamma_{soil} h \qquad (7\text{--}40)$$

The value of K_a depends on the angle of internal friction within the soil (ϕ). This relationship is:

$$K_a = \tan^2 \left(45° - \frac{\phi}{2} \right) \qquad (7\text{--}41)$$

As the value of ϕ increases, K_a decreases. Conversely, when the angle of internal friction decreases, the value of K_a increases. Thus, if water is present in the soil, the friction between the solid particles is reduced, and K_a increases. Hence, it is important to have good drainage in the backfill. Footing drains (a drain tile system at the bottom of the basement walls) and weep holes in retaining walls can provide this drainage.

The value of K_a changes within narrow limits of about 0.27 to 0.34 for level backfill when it is evaluated for well-drained granular soils using the values listed in Table 7–4.

TABLE 7-4	Angle of Internal Friction for Drained Granular Soils
Soil Type	ϕ **(Degrees)**
Gravel and coarse sand	33–36
Medium to fine sand	29–32
Silty sand	27–30

Sloping Backfill   When the backfill slopes, as shown in Figure 7–54, the formula for K_a is more involved. The equation is presented here for completeness. Using the Rankine formula:

$$K_a = \cos \beta \frac{\cos \beta - \sqrt{\cos^2 \beta - \cos^2 \phi}}{\cos \beta + \sqrt{\cos^2 \beta - \cos^2 \phi}} \tag{7-42}$$

The maximum pressure at the base is:

$$p_{max} = K_a \gamma_{soil} h$$

The Effect of Surcharge   Any additional load surcharge atop the surface increases the gravity force on the sliding wedge. This in turn increases the lateral pressure on the back of the wall, as illustrated in Figure 7–55. The lateral pressure at any depth is $p_a = K_a \gamma_{soil} y$,

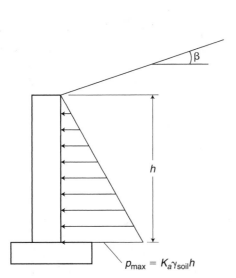

FIGURE 7-54   Horizontal component of the pressure distribution, sloping granular drained backfill.

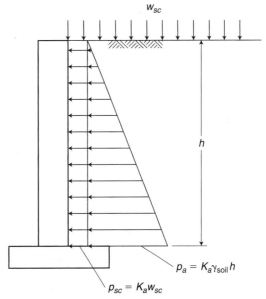

FIGURE 7-55   Additional lateral pressure from surcharge.

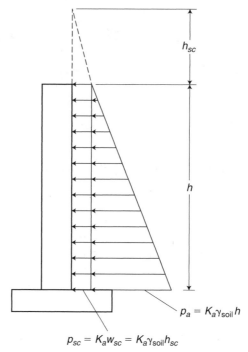

$$p_{sc} = K_a w_{sc} = K_a \gamma_{\text{soil}} h_{sc}$$

FIGURE 7–56   Representation of surcharge by additional backfill height.

where the product $\gamma_{\text{soil}} y$ is the weight of the soil above level y, so the increased lateral pressure from a distributed surcharge load (w_{sc}) will be

$$p_{a,y} = K_a(\gamma_{\text{soil}} y + w_{sc}) = K_a \gamma_{\text{soil}} y + K_a w_{sc} \qquad (7\text{–}43)$$

The second part of the equation represents the increased lateral pressure from the surcharge, which is independent of the depth. The surcharge may also be conceptualized as having an additional height (h_{sc}) of soil atop the finish surface. If we express the surcharge with the unit weight of the soil as

$$w_{sc} = h_{sc} \gamma_{\text{soil}} \qquad (7\text{–}44)$$

then this fictitious height, as illustrated in Figure 7–56, is:

$$h_{sc} = \frac{w_{sc}}{\gamma_{\text{soil}}} \qquad (7\text{–}45)$$

Equivalent Fluid Pressure   The triangular lateral earth pressure is similar to a liquid pressure. The pressure in a liquid (like water) at a given depth is uniform in every direction and is equal to the unit weight of the liquid multiplied by the depth. So if the unit weight of a fictitious liquid is $\gamma_a = K_a \gamma_{\text{soil}}$, and we substitute this value into Equations 7–39 and 7–40 that express the lateral pressure, we obtain what is referred to as *equivalent fluid pressure*. Geotechnical engineers usually make their recommendations regarding lateral pressures on walls in terms of the equivalent fluid density (γ_a).

The unit weight of compacted granular backfill is between 105 to 115 pcf. From the average values of K_a the calculated equivalent fluid density for granular backfills is between 30 and 40 pcf, with a mean value of about 35 pcf.

Note that these values hold for soils that are well drained. Clay soils or saturated soils may produce much higher pressure values. The designer should always consult with a geotechnical engineer to verify the most likely equivalent fluid density prior to designing retaining structures.

Basement Walls

Basement walls are earth retaining walls that are supported laterally by the first floor construction at their top and by the basement slab on grade at their bottom. In addition, they are vertically supported on wall footings.

Figure 7–57 shows a schematic section through a basement wall. The wall will be stable only after the first floor construction is complete, so no backfill (or only a very limited height of backfill) should be placed against the wall until after the first floor is in place. The backfill should be a compacted granular fill that will drain well into the footing drain. The footing drain (drain tile) is made of perforated tiles or plastic drain pipes. This drain tile is then connected into either the storm drains or a sump pit out of which the water is pumped. This prevents water from accumulating behind the basement wall, and thereby prevents the increase of the lateral pressure.

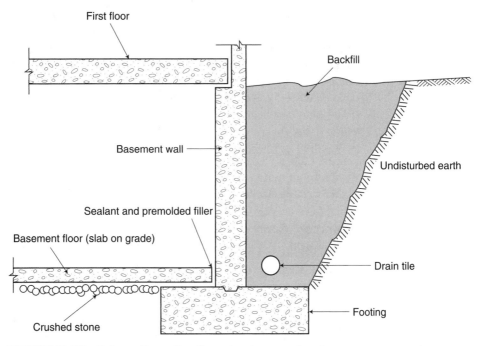

FIGURE 7–57 Schematic section through a basement wall.

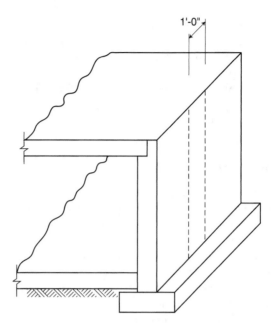

FIGURE 7–58 Basement wall design is based on a 1-ft strip.

Basement walls are usually made of concrete (either reinforced or unreinforced). In residential construction they are sometimes built using concrete masonry units (CMU) and hence are called *CMU walls*. These also may be reinforced or unreinforced.

Design of Basement Walls The structural behavior of a basement wall is similar to that of a simply-supported one-way slab spanning vertically between the slab on grade at the base and the first floor at the top. Similar to slabs, only a 1-ft-long strip of the wall, as shown in Figure 7–58, is considered in the design of these walls.

The minimum thickness commonly used for unreinforced concrete basement walls is 10 inches. It is difficult to properly consolidate the concrete within the forms for thinner walls. In addition, basement walls have to be thick enough to provide width for placement of members such as stud walls, brick veneer, and so on. In unreinforced concrete walls, it is advisable to use vertical control joints at a maximum spacing of 20 ft. The control joints prevent the random cracking of the wall due to volumetric changes.

Unreinforced Concrete Basement Wall Design The steps to design unreinforced (plain) concrete basement walls are as follows. They are summarized in Figure 7–61:

Step 1   Calculate the maximum moment.
          To help with the analysis, Figures 7–59 and 7–60 show the four most common loading cases with closed-form solutions. The ACI Code considers the lateral earth pressure to be live load. Hence, we must multiply the moments that are calculated from unfactored pressures by a load factor of 1.6 to get $M_{u,max}$.

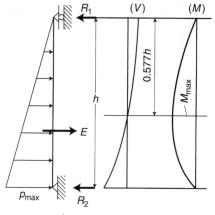

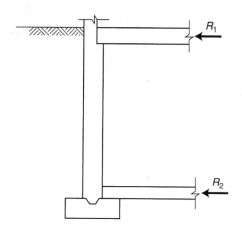

$$E = p_{max}\frac{h}{2} \quad V_{max} = R_2$$

$$R_1 = \frac{E}{3} \qquad R_2 = \frac{2E}{3}$$

$$M_{max} = (@0.577h \text{ from grade}) = 0.128\ Eh$$

(a)

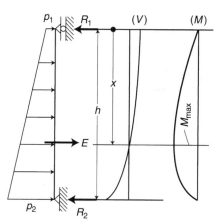

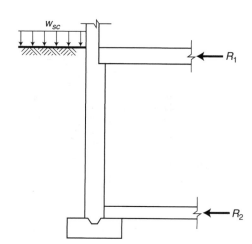

$$p_1 = K_a w_{sc} \quad p_2 = p_1 + K_a \gamma_{soil} h \quad V_{max} = R_2$$

$$E = (p_1 + p_2)\frac{h}{2}$$

$$R_1 = (2p_1 + p_2)\frac{h}{6} \quad R_2 = E - R_1$$

M_{max} (@ x from grade)

$$= \frac{x}{6h}[h^2(2p_1 + p_2) - 3hp_1x - (p_2 - p_1)x^2]$$

x is found by solving:

$$\frac{(p_2 - p_1)}{h}x^2 + 2p_1x - 2R_1 = 0$$

(b)

FIGURE 7–59 Shear force and bending moment in basement wall with full height backfill: (a) without surcharge, (b) with surcharge.

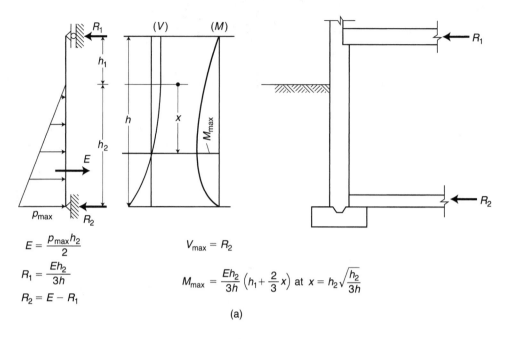

$$E = \frac{p_{max}h_2}{2}$$

$$R_1 = \frac{Eh_2}{3h}$$

$$R_2 = E - R_1$$

$$V_{max} = R_2$$

$$M_{max} = \frac{Eh_2}{3h}\left(h_1 + \frac{2}{3}x\right) \text{ at } x = h_2\sqrt{\frac{h_2}{3h}}$$

(a)

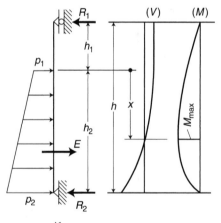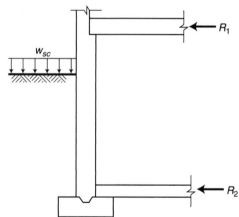

$$p_1 = K_a w_{sc}$$

$$E = (p_1 + p_2)\frac{h_2}{2}$$

$$R_1 = \frac{h_2^2(2p_1 + p_2)}{6h}$$

$$R_2 = E - R_1$$

$V_{max} = R_2$

M_{max} (@ x from grade)

$$= \frac{h_2^2}{6h}(2p_1 + p_2)(h_1 + x) - \frac{x^2}{6h_2}[3h_2p_1 + x(p_2 - p_1)]$$

x is found from solving:

$$\frac{(p_2 - p_1)}{h_2}x^2 + 2p_1x - 2R_1 = 0$$

(b)

FIGURE 7-60 Shear force and bending moment in basement wall with partial backfill: (a) without surcharge, (b) with surcharge.

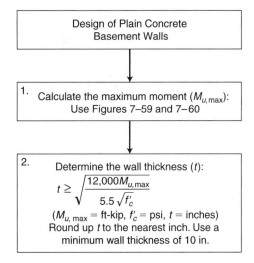

FIGURE 7–61   Flowchart for the design of plain concrete basement walls.

Step 2   Determine the wall thickness (t).

To determine the wall thickness, set the wall resisting moment, $M_R = \phi M_n$, equal to the maximum moment, $M_{u,\max}$, calculated in step 1. ACI Code Equation 22–2 gives the nominal resisting moment of a plain concrete section as:

$$M_n = \left(5\sqrt{f_c'}\right)S_m$$

where S_m is the elastic section modulus of 1-ft-long wall, or:

$$S_m = \frac{bt^2}{6}, \quad b = 12 \text{ in.}$$

$$M_R = \phi M_n \geq M_{u,\max}$$

where for an unreinforced concrete wall, $\phi = 0.55$ (ACI Code Section 9.3.5). Substituting ϕ and S_m into the above equation:

$$0.55\left(5\sqrt{f_c'}\right)\frac{12t^2}{6} \geq 12{,}000 M_{u,\max} \tag{7–46}$$

Solving for t:

$$t \geq \sqrt{\frac{12{,}000 M_{u,\max}}{5.5\sqrt{f_c'}}} \tag{7–47}$$

In this equation, $M_{u,\max}$ is in ft-kip, f_c' is in psi, and t is in inches. There is no need to check for shear. Shear is never critical in the design of basement walls subject to lateral earth pressure.

EXAMPLE 7–6

Design the plain concrete basement wall shown in Figure 7–62. The backfill is made of granular material with a unit weight of $\gamma_{\text{soil}} = 120$ pcf and the coefficient of active soil pressure,

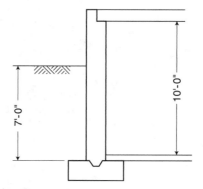

FIGURE 7–62   Basement wall of Example 7–6.

$K_a = 0.33$. Consider two cases: (a) without a surcharge; (b) with a surcharge of 100 psf acting on the backfill. Use $f_c' = 4,000$ psi.

Solution

(a) Without Surcharge

Step 1   Calculate the maximum moment.

This is Case (a) on Figure 7–60, a basement wall with partial backfill and without surcharge. The lateral soil pressure and resulting maximum moments are:

Equivalent fluid density $= \gamma_a = K_a \gamma_{soil} = 0.33(120) = 40\,\text{pcf}$

The pressure at the base of the wall, p_{max}, is:

$$p_{max} = \gamma_a h_2 = 40 \times 7 = 280\,\text{lb/ft}^2$$

Use the factored pressure, which is obtained by multiplying the actual pressure by the soil pressure load factor (H) of 1.6, instead of using the actual pressure. Then all of the results will be factored values.

$$p_u = 1.6 p_{max} = 1.6 \times 280 = 448\,\text{lb/ft}^2/1000 = 0.45\,\text{kip/ft}^2$$

Figure 7–63 shows the wall with the factored pressures.

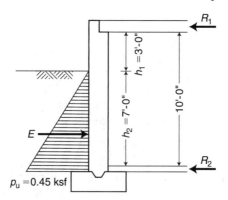

FIGURE 7–63   Factored pressure distribution on basement wall in Example 7–6.

Using the equations for reactions and the maximum moment from Figure 7–60:

$$E_u = \frac{p_u h_2}{2} = \frac{0.45(7)}{2} = 1.58 \text{ kip/ft of wall}$$

$$R_{u1} = \frac{E_u h_2}{3h} = \frac{1.58(7)}{3(10)} = 0.37 \text{ kip/ft}$$

$$R_{u2} = E_u - R_{u1} = 1.58 - 0.37 = 1.21 \text{ kip/ft}$$

and the location of the maximum moment, $x = h_2 \sqrt{\dfrac{h_2}{3h}}$:

$$x = 7.0 \sqrt{\frac{7.0}{3(10)}} = 3.38 \text{ ft}$$

$$M_{u,\max} = \frac{E_u h_2}{3h}\left(h_1 + \frac{2}{3}x\right)$$

$$M_{u,\max} = \frac{1.58(7)}{3(10)}\left(3 + \frac{2}{3} \times 3.38\right)$$

$$M_{u,\max} = 1.94 \text{ kip-ft/ft}$$

Step 2   Determine the wall thickness.
From Equation 7–47:

$$t \geq \sqrt{\frac{12{,}000 M_{u,\max}}{5.5\sqrt{f_c'}}}$$

$$t \geq \sqrt{\frac{12{,}000(1.94)}{5.5\sqrt{4{,}000}}}$$

$$t \geq 8.2 \text{ in.}$$

$$\therefore \text{ Use } t = 10 \text{ in. (minimum wall thickness)}$$

(b) With Surcharge

Step 1   Calculate the maximum moment.
This is Case (b) in Figure 7–60. Use the formulae from this figure to calculate the lateral loads from the retained soil and the surcharge, and the resulting maximum moment:

$$p_a = \gamma_a h = 40 \times 7 = 280 \text{ lb/ft}^2$$

Surcharge pressure $= p_s = K_a w_{sc} = 0.33(100) = 33 \text{ lb/ft}^2$.
The factored pressures from the soil, p_u, and the surcharge, p_{su}, are:

$$p_u = 1.6\, p_a = 1.6 \times 280 = 448 \text{ lb/ft}^2/1000 = 0.45 \text{ ksf}$$

$$p_{su} = 1.6\, p_s = 1.6 \times 33 = 53 \text{ lb/ft}^2/1000 = 0.053 \text{ ksf}$$

Obtain the equations for the reactions and the maximum moment from Figure 7–60:

$$p_{u1} = p_{su} = 0.053 \text{ ksf}$$

$$p_{u2} = p_u + p_{su} = 0.45 + 0.053 = 0.503 \text{ ksf}$$

$$E_u = (p_{u1} + p_{u2})\frac{h_2}{2} = (0.053 + 0.503)\frac{7}{2} = 1.95 \text{ kip/ft}$$

$$R_{u1} = \frac{h_2^2(2p_{u1} + p_{u2})}{6h} = \frac{(7.0)^2(2 \times 0.053 + 0.503)}{6 \times 10} = 0.50 \text{ kip/ft}$$

$$R_{u2} = E_u - R_{u1} = 1.95 - 0.50 = 1.45 \text{ kip/ft}$$

To determine the location of the maximum moment (x), solve the quadratic equation shown in Figure 7–60, Case (b):

$$\frac{(p_{u2} - p_{u1})}{h_2}x^2 + 2p_{u1}x - 2R_{u1} = 0$$

$$\frac{(0.503 - 0.053)}{7.0}x^2 + 2(0.053)x - 2(0.50) = 0$$

$$0.0643x^2 + 0.106x - 1.0 = 0$$

$$x = \frac{-0.106 + \sqrt{(0.106)^2 + 4(0.0643)(1.0)}}{2(0.0643)}$$

$$x = 3.20 \text{ ft}$$

$$M_{u,\max} = \frac{h_2^2}{6h}(2p_{u1} + p_{u2})(h_1 + x) - \frac{x^2}{6h_2}[3h_2\,p_{u1} + x(p_{u2} - p_{u1})]$$

$$M_{u,\max} = \frac{(7.0)^2}{6(10)}(2 \times 0.053 + 0.503)(3 + 3.20)$$

$$- \frac{(3.20)^2}{6(7.0)}[3(7.0)(0.053) + 3.20(0.503 - 0.053)]$$

$$M_{u,\max} = 2.46 \text{ kip-ft/ft}$$

Step 2   Determine the wall thickness.

Use the formula (Equation 7–47) developed for calculating the necessary wall thickness for a plain concrete basement wall.

$$t \geq \sqrt{\frac{12,000M_{u,\max}}{5.5\sqrt{f_c'}}}$$

$$t \geq \sqrt{\frac{12,000(2.46)}{5.5\sqrt{4,000}}}$$

$$t \geq 9.2 \text{ in.}$$

$$\therefore \text{ Use } t = 10 \text{ in.}$$

Reinforced Concrete Basement Wall Design   The design of reinforced concrete basement walls is often dictated by considerations other than the absolute minimum wall thickness required by flexure. It is difficult to place and consolidate concrete into the forms when the design contains at least two layers of reinforcing (vertical and horizontal) near the inside face, especially with thin walls. To make matters more difficult, often both faces of the wall may need reinforcement to better control cracking induced by shrinkage and temperature changes.

Architectural requirements also influence the selection of an appropriate wall thickness. The support of the exterior wall finish (e.g., a brick ledge), in addition to providing for adequate support for the first floor construction, often results in much thicker walls than would be required by strict structural considerations only.

Hence, the thickness of reinforced concrete basement walls is usually preselected by the designer, and the wall is strengthened by providing the needed amount of reinforcement. An absolute minimum thickness in a reinforced concrete wall is 8 in. As with plain concrete basement walls, only flexure needs to be considered; the shear stresses in normal basement walls are never excessive.

The steps in the design are as follows and are summarized in the flowchart of Figure 7–64.

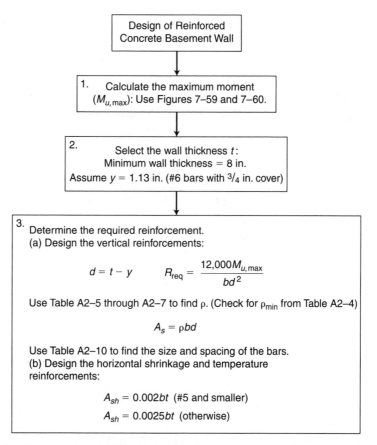

FIGURE 7–64 Flowchart for the design of reinforced concrete basement walls.

Step 1   Calculate the maximum moment.

Use the formulae listed in Figures 7–60 and 7–61 to calculate the factored pressures, reactions, and the maximum moment, $M_{u,\,max}$.

Step 2   Select an appropriate wall thickness (t).

Use a minimum wall thickness of 8 in. Keep in mind, however, that you may need at least 10 in. or more thickness in some situations to provide enough width at the top to place studs, brick veneer, and so on.

Step 3   Determine the required area of vertical reinforcement.

The primary reinforcing will be located near the inside face of the wall, and the ACI Code requires a minimum concrete cover of $^3/_4$ in. Assuming #6 bars for the vertical reinforcing, the effective depth (d) is:

$$d = t - 0.75 - 0.75/2 = t - 1.13 \text{ in.}$$

The required resistance coefficient is:

$$R_{\text{req}} = \frac{12,000 M_u}{bd^2}$$

Use the appropriate f'_c and f_y in Tables A2–5 through A2–7 to obtain the steel ratio, ρ. Then the required area of vertical reinforcement, A_s is:

$$A_s = \rho bd$$

According to the ACI Code (Section 10.5.1), the minimum area of the vertical flexural reinforcements is:

$$A_{s,\,min} = \rho_{min}bd = \max\left\{\frac{3\sqrt{f'_c}}{f_y}, \frac{200}{f_y}\right\}bd \qquad (7\text{–}48)$$

Table A2–4 lists the corresponding ρ_{min} values.

Use Table A2–10 to select the size and spacing of the vertical reinforcements. The horizontal shrinkage and temperature reinforcement is specified in the ACI Code (Section 14.3.3) as:

$$A_{sh} = 0.002bt \quad \text{(when \#5 and smaller bars are used)} \qquad (7\text{–}49)$$
$$A_{sh} = 0.0025bt \quad \text{(when larger bars are used)} \qquad (7\text{–}50)$$

The bar spacing for the vertical and horizontal reinforcements (ACI Code, Section 14.3.5) is limited to:

$$s \leq \min\{3t, 18 \text{ in.}\} \qquad (7\text{–}51)$$

For basement walls with a thickness of 10 in. or less, shrinkage and temperature horizontal reinforcement typically is placed on only one face. For thicker walls, we distribute the required horizontal reinforcing evenly between the inside and the outside faces.

EXAMPLE 7-7

Design the reinforced concrete basement wall shown in Figure 7–65. The unit weight of backfill is $\gamma_{soil} = 115$ pcf, and the coefficient of active soil pressure is $K_a = 0.33$. The surcharge on the backfill is 150 psf. Use $f_c' = 4,000$ psi and $f_y = 60,000$ psi.

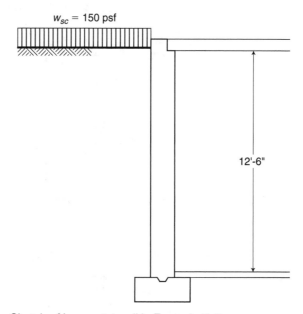

$w_{sc} = 150$ psf

12'-6"

FIGURE 7-65   Sketch of basement wall in Example 7–7.

Solution

Step 1   Calculate the maximum moment.

This basement wall is subjected to full backfill with surcharge. This is Case (b) in Figure 7–59.

The equivalent fluid density and pressure are:

$$\gamma_a = K_a \gamma_{soil} = 0.33(115) = 38 \text{ pcf}$$
$$p_a = \gamma_a h = 38(12.5) = 475 \text{ psf}$$

The pressure from the surcharge is:

$$p_s = K_a w_{sc} = 0.33(150) = 50 \text{ psf}$$

The factored pressures are:

$$p_{au} = 1.6 p_a = 1.6 \times 475 = 760 \text{ psf}/1,000 = 0.76 \text{ ksf}$$
$$p_{su} = 1.6 p_s = 1.6 \times 50 = 80 \text{ psf}/1,000 = 0.08 \text{ ksf}$$

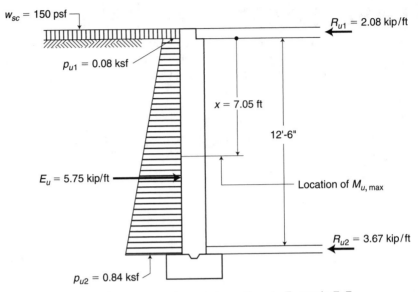

w_{sc} = 150 psf

R_{u1} = 2.08 kip/ft

p_{u1} = 0.08 ksf

x = 7.05 ft

12'-6"

E_u = 5.75 kip/ft

Location of $M_{u,\,max}$

R_{u2} = 3.67 kip/ft

p_{u2} = 0.84 ksf

FIGURE 7–66   Factored pressure values and reactions in Example 7–7.

Figure 7–66 shows the wall with the factored pressures.
Using the equations in Case (b) of Figure 7–59, we calculate reactions and maximum moment.

$$p_{u1} = p_{su} = 0.08 \text{ ksf}$$

$$p_{u2} = p_{au} + p_{su} = 0.76 + 0.08 = 0.84 \text{ ksf}$$

$$E_u = (p_{u1} + p_{u2})\frac{h}{2} = (0.08 + 0.84)\frac{12.5}{2} = 5.75 \text{ kip/ft}$$

$$R_{u1} = (2p_{u1} + p_{u2})\frac{h}{6} = (2 \times 0.08 + 0.84)\frac{12.5}{6} = 2.08 \text{ kip/ft}$$

$$R_{u2} = E_u - R_{u1} = 5.75 - 2.08 = 3.67 \text{ kip/ft}$$

To determine the location of the maximum moment, x, we solve the following equation:

$$\frac{(p_{u2} - p_{u1})}{h}x^2 + 2p_{u1}x - 2R_{u1} = 0$$

$$\frac{(0.84 - 0.08)}{12.5}x^2 + 2(0.08)x - 2(2.08) = 0$$

$$0.061x^2 + 0.16x - 4.16 = 0$$

$$x = \frac{-0.16 + \sqrt{(0.16)^2 + 4(0.061)(4.16)}}{2(0.061)}$$

$$x = 7.05 \text{ ft}$$

The maximum factored moment is:

$$M_{u,max} = \frac{x}{6h}\left[h^2(2p_{u1} + p_{u2}) - 3hp_{u1}x - (p_{u2} - p_{u1})x^2\right]$$

$$M_{u,max} = \frac{7.05}{6(12.5)}\left[(12.5)^2(2 \times 0.08 + 0.84) - 3(12.5)(0.08)(7.05)\right.$$
$$\left. - (0.84 - 0.08)(7.05)^2\right]$$

$$M_{u,max} = 9.15 \text{ kip-ft/ft}$$

Step 2 Select a wall thickness.
Assume that there are no particular architectural requirements for the thickness of the wall. Select $t = 8$ in.

Step 3 Design the required reinforcement.
(a) Design the vertical reinforcement.
Assume $\frac{3}{4}$ in. cover and #6 bars; then

$$d = 8 - 1.13 = 6.87 \text{ in.}$$

Calculate the required coefficient of R:

$$R_{req} = \frac{12,000M_u}{bd^2} = \frac{12,000(9.15)}{12(6.87)^2} = 194 \text{ psi}$$

Obtain the required reinforcement ratio from Table A2–6b:

$$\rho_{req} = 0.0038$$

Check for the minimum flexural reinforcement required from Table A2–4:

$$\rho_{min} = 0.0033 < 0.0038 \quad \therefore \text{ ok}$$
$$A_s = (0.0038)(12)(6.87) = 0.31 \text{ in}^2/\text{ft}$$

From Table A2–10 select #5 @ 12 in. c/c ($A_s = 0.31$ in²/ft).
(b) Design the horizontal reinforcements.
The horizontal reinforcement required for shrinkage and temperature, based on the ACI Code (Section 14.3.3) is:

$$A_{sh} = 0.0020bt \quad \text{(assuming #5 or smaller bars)}$$
$$A_{sh} = 0.0020(12)(8) = 0.19 \text{ in}^2/\text{ft}$$

From Table A2–10 select #4 @ 12 in. ($A_s = 0.20$ in²/ft).
Check for the maximum permitted bar spacing:

$$s_{max} = \min\{3t, 18 \text{ in.}\} = \min\{3(8), 18\} = 18 \text{ in.}$$

With 12 in. bar spacing, the requirement is satisfied.

Figure 7–67 shows the sketch of the final design.

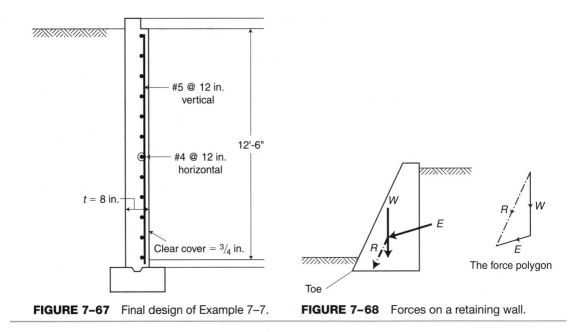

FIGURE 7–67   Final design of Example 7–7.   **FIGURE 7–68**   Forces on a retaining wall.

Retaining Walls

The behavior of retaining walls is very different from that of basement walls. Basement walls are vertical simply-supported slabs bending between two supports (i.e., the basement floor and the first floor). Unlike basement walls, retaining walls are not supported at the top. They must have substantial weight to prevent toppling over from the earth pressure.

Figure 7–68 shows the acting forces on a retaining wall. In addition to the weight (W), the earth pressure (E) is applied to the back side of the wall. If the back of the wall is smooth (i.e., frictionless), the E force is horizontal. The two forces, W and E, are combined into the resultant (R).

The E force "wants" to *overturn* the wall, or to pivot it around the toe point. In Figure 7–68 E exerts a counterclockwise moment on the toe. The W force (i.e., the weight of the wall) "wants" to prevent the overturning by applying a clockwise moment about the toe. As long as the resisting moment, M_r, of W is greater than the overturning moment, M_{ot}, from E, the wall will be stable. Another way to express the same concept is that the wall is stable as long as the resultant force, R, intercepts the base of the wall inside the bottom width, as shown in Figure 7–68. On the other hand, the wall will tip over if the R force intercepts the base line outside the bottom width. Figure 7–69 shows a typical gravity retaining wall, and the applied forces (assuming a smooth wall), with their corresponding application locations. The overturning moment, M_{ot}, can be calculated as follows:

$$M_{ot} = E\frac{h}{3} \qquad (7\text{--}52)$$

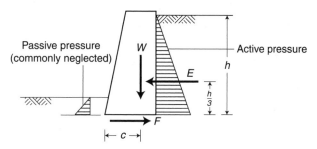

FIGURE 7-69   Forces on a retaining wall.

And the resisting moment, M_r, is:

$$M_r = Wc \qquad (7\text{--}53)$$

Stability against overturning requires that:

$$M_r \geq FS_{ot} \times M_{ot} \qquad (7\text{--}54)$$

where FS_{ot} is the factor of safety against overturning. A recommended minimum factor of safety against *overturning* is

$$FS_{ot,\,min} = 2.0$$

A second failure mode besides overturning exists. The earth pressure (E) tries to push the wall (from right to left in Figure 7–69) to make it slide along its base. Resistance against *sliding* comes from two sources. The first is frictional resistance (F) at the bottom of the wall. The magnitude of this force is the weight of the wall (W) multiplied by the coefficient of friction (μ) between the two materials (i.e., the wall and the soil). The larger the weight, the larger is the frictional resistance. The second force that resists sliding is the *passive* earth pressure in the front of the wall. The bottom of the retaining wall, as with any other footing, is usually placed below the frost line. The fill at the front provides passive resistance, which can be considerable when the fill is there. Sometimes, however, this fill in the front is removed for one reason or another. Thus, its continuous presence is not a given, and most designers disregard it.

To ensure safety against sliding:

$$F = \mu W \geq FS_s \times E \qquad (7\text{--}55)$$

where

μ = the coefficient of friction between the bottom of the wall and the soil
FS_s = the factor of safety against sliding. (The recommended minimum safety factor against sliding is 1.5.)

All retaining walls in essence are gravity walls, although only one type is designated as such. They differ only in the way we provide the mass needed to safely retain the soil at the upper elevation. These differences in design, however, increase the diversity of structural behavior *within* the wall structures themselves. Figure 7–70 shows various types of retaining walls, and discussed as follows.

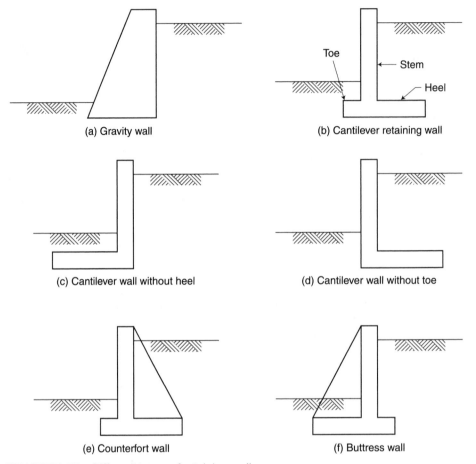

FIGURE 7–70   Different types of retaining walls.

1. Gravity walls These walls are constructed of plain concrete, stone, or brick masonry. Their extensive use of material and labor costs limit their economy to relatively low heights of about 8 ft. above the low grade (Figure 7–70a).

2. Cantilever retaining walls These are by far the most common type of retaining wall. They are constructed with reinforced concrete or reinforced masonry, and are economical to use for heights up to about 20 ft. Figure 7–70b shows a typical cantilever retaining wall. The backfill above the heel provides much of the weight needed for the stability of the wall. There are, however, variations of these walls such as cantilever walls without heel or toe as shown in Figures 7–70c and 7–70d. They are used when property lines or other limitations prevent the footing from extending beyond one side of the wall. These walls are not as efficient as the typical cantilever retaining walls.

3. Counterfort walls The wall stem will be subjected to very large bending moments if a cantilever retaining wall is higher than 20 ft. In such cases it may be economical to construct the wall with counterforts (walls perpendicular to the stem, spaced about 12 to

15 ft apart) that attach the stem to the heel. The reinforced counterforts act like tension members supporting the stem. They also greatly increase the bending strength of the stem. Figure 7–70e shows a counterfort wall.

4. *Buttress walls* These walls are similar to counterfort walls, except that the buttresses, which attach the stem to the toe, are located in front of the wall (see Figure 7–70f).

Vertical Soil Pressure Under the Base of a Retaining Wall So far we have discussed only wall and column footings that are concentrically loaded (i.e., the load acts at the centroid of the footing, and the distribution of the pressures on the soil is uniform). In general, concentric loading cannot be achieved under retaining walls. The resultant force (R), as shown in Figure 7–68, does not intercept the footing at its center, but rather is *eccentric* to it.

Figure 7–71 shows three different possibilities of pressure distribution under a footing. In Case I the load is concentrically applied to the footing. In Case II the load is applied at a small eccentricity (e), which is less than $b/6$. In Case III the load is applied at an eccentricity larger than $b/6$. Because a 1-ft-long strip is considered in a wall footing:

$$A = b \times 1 = b \text{ ft}^2 \quad \text{and} \quad S_m = \frac{1 \times b^2}{6} \text{ ft}^3$$

The pressures for the individual cases then can be found as:

$$f_{max} = \frac{P}{b} \quad \bigg| \quad f_{max} = \frac{P}{b}\left(1 + \frac{6e}{b}\right) \quad \bigg| \quad f_{max} = \frac{2P}{3c} \quad (7\text{–}56)$$

$$f_{min} = \frac{P}{b}\left(1 - \frac{6e}{b}\right)$$

(Case I)   (Case II)   (Case III)

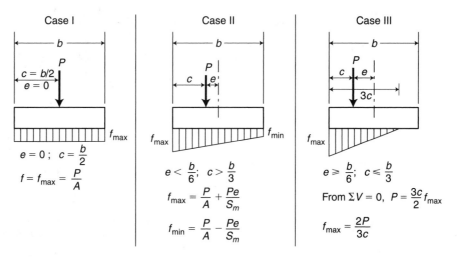

FIGURE 7–71   Eccentric pressures under wall footings.

If $\left(1 - \dfrac{6e}{b} \right) < 0$ in Case II, the expression for f_{min} becomes negative, which indicates that theoretically there is tension between the footing and the soil. But tension cannot develop between the bottom of the footing and the soil, as a gap would appear. This is an impossible and inadmissible situation, so Case III applies. With a straight-line pressure distribution, the pressure volume under the footing must be in equilibrium with the P force. Hence, the resultant of the reaction pressure must be colinear with the P force. Then the neutral axis (i.e., where the pressures become zero) must be located at a distance $3c$ from the toe (see Figure 7–71).

Shear keys and weep holes We can easily increase inadequate *sliding resistance* in a retaining wall by using a *shear key* at the bottom of the footing, as shown in Figure 7–72. The passive earth pressure in front of the key provides a sure and economical resistance.

As with basement walls, it is important to prevent water saturation of the backfill behind retaining walls. Saturated backfill increases the earth pressure dramatically and may endanger the stability of the wall. Providing *weep holes* in the wall at regular spacing is the easiest and safest way to drain the backfill. A typical retaining wall weep hole is shown in Figure 7–72. Figure 7–73 shows a flowchart for the stability analysis and design of cantilever retaining walls.

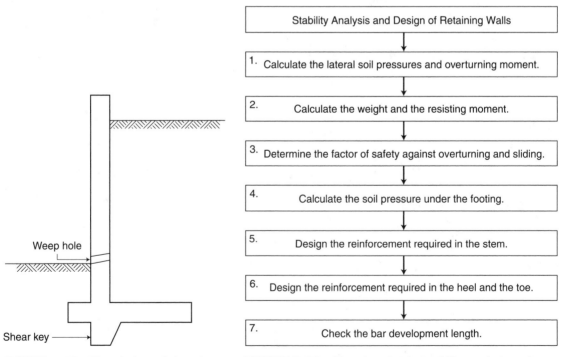

FIGURE 7–72 Weep hole and shear key in retaining walls.

FIGURE 7–73 Flowchart for the stability analysis and design of retaining walls.

EXAMPLE 7–8

Analyze the stability of the gravity retaining wall shown in Figure 7–74. The backfill is sandy gravel, γ_{soil} = 120 pcf; the coefficient of the lateral active earth pressure is K_a = 0.32; and the coefficient of friction at the base is μ = 0.52. The wall is constructed of concrete, which weighs 150 pcf. A surcharge load of 150 psf exists on the upper elevation. Calculate the toe pressure on the soil in addition to the stability analysis. Disregard the passive pressure in front of the toe in the analysis.

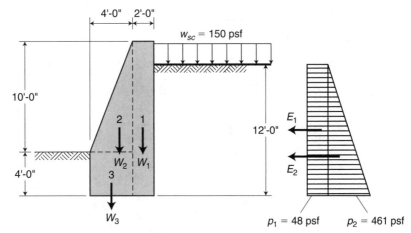

FIGURE 7–74   Sketch for Example 7–8.

Solution

Step 1   Calculate the lateral soil pressures and the overturning moment.

$$p_1 = K_a w_{sc} = 0.32 \times 150 = 48 \text{ psf}$$
$$E_1 = 48 \times 12 = 576 \text{ lb/ft}$$

applied at half the soil height.

$$p_2 = K_a \gamma_{soil} h = 0.32 \times 120 \times 12 = 461 \text{ psf}$$
$$E_2 = 461 \times 12/2 = 2{,}766 \text{ lb/ft}$$

applied at one-third of the soil height.

$$\sum E = 576 + 2{,}766 = 3{,}342 \text{ lb/ft}$$

The overturning moment about the toe is the sum of the moments caused by E_1 and E_2:

$$M_{ot} = 576 \times 12/2 + 2{,}766 \times 12/3 = 14{,}520 \text{ ft-lb/ft}$$

Step 2   Calculate the weight and the resisting moment.

We divide the area of the wall into three component parts, as shown in Figure 7–74, and do our calculations in a tabulated form.

Part No.		W	x (Distance of W from Toe)	Wx
1	$2 \times 14 \times 150 =$	4,200 lb/ft	5.00 ft	21,000 ft-lb/ft
2	$(4 \times 10/2) \times 150 =$	3,000 lb/ft	2.67 ft	8,000 ft-lb/ft
3	$4 \times 4 \times 150 =$	2,400 lb/ft	2.00 ft	4,800 ft-lb/ft
		$\Sigma = 9,600$ lb/ft		33,800 ft-lb/ft

Step 3   Determine the safety factor against overturning and sliding:

$$FS_{ot} = \frac{M_r}{M_{ot}} = \frac{33,800}{14,520} = 2.33 \longrightarrow 2.33 > 2.0 \quad \therefore \text{ ok}$$

Calculate the factor of safety against sliding:
The friction force

$$F = \mu W = 0.52 \times 9,600 = 4,992 \text{ lb/ft}$$

$$FS_s = \frac{F}{\Sigma E} = \frac{4,992}{3,342} = 1.49 \longrightarrow 1.49 \approx 1.5 \quad \therefore \text{ ok}$$

Step 4   Calculate the soil pressure under the footing.

Figure 7–75 shows all the forces acting on the wall. Determine the location where the R force (the reaction of the resultant of W and E) intercepts

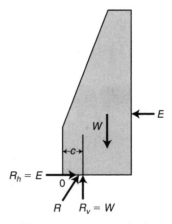

FIGURE 7–75 The R force and its components at the base.

the base of the wall. The vertical component of the resultant is $R_v = W$, and the horizontal component is $R_h = E$; (W, E, R_h, and R_v are the forces acting on the wall). The moment of a resultant about any point must equal the sum of the moments of the composing forces about the same point.

The moments of the composing forces are already known. The moment of W about the toe is the resisting moment (M_r). The moment of E is the overturning moment (M_{ot}).

Hence, from the equilibrium of forces acting on the wall:

$$\sum M_o = 0$$
$$-Wc + M_r - M_{ot} = 0$$
$$Wc = M_r - M_{ot}$$
$$c = \frac{M_r - M_{ot}}{W} \qquad (7\text{--}57)$$

Substituting the calculated values, we obtain c:

$$c = \frac{33{,}800 - 14{,}520}{9{,}600} = 2.0 \text{ ft}$$

Because $b = 6.0$ ft, this is Case III $\left(c = \dfrac{b}{3} \right)$. Use Case III from Figure 7–71, and find the maximum pressure, f_{max}:

$$f_{max} = \frac{2 \times 9{,}600}{3 \times 2.0} = 3{,}200 \text{ psf}$$

Figure 7–76 shows the resulting soil pressure distribution.

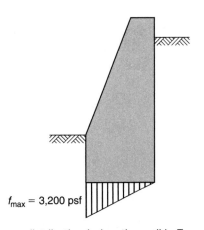

$f_{max} = 3{,}200$ psf

FIGURE 7–76   Soil pressure distribution below the wall in Example 7–8.

EXAMPLE 7–9

Analyze the stability of the reinforced concrete cantilever retaining wall shown in Figure 7–77. Calculate the reinforcement required in the wall and the footing. Disregard the passive resistance of the soil in front of the toe. Assume $K_a = 0.32$, $\gamma_{soil} = 115$ pcf, and $\mu = 0.50$. Use $f_c' = 3,000$ psi and $f_y = 60,000$ psi.

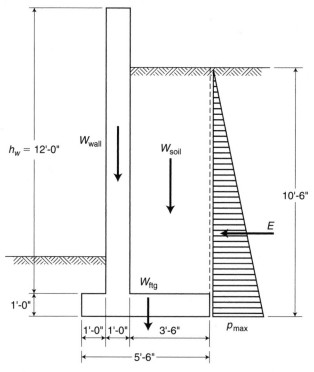

FIGURE 7–77 Sketch of the wall in Example 7–9.

Solution

Step 1 Calculate the lateral soil pressure and the overturning moment:

$$p_{max} = K_a \gamma_{soil} h = 0.32 \times 115 \times 10.5 = 386.4 \text{ psf}$$

$$E = \frac{p_{max} h}{2} = \frac{386.4 \times 10.5}{2} = 2,029 \text{ lb/ft}$$

$$M_{ot} = E\frac{h}{3} = 2,029 \times \frac{10.5}{3} = 7,102 \text{ ft-lb/ft}$$

Step 2 Calculate the weight and the resisting moment. Include the weight of the backfill atop the heel of the wall and treat that as an integral part of the retaining wall.

Item	W				x (From Toe)	Wx (Moment to Toe)
Wall	12	×	1 × 150 =	1,800 lb/ft	1.50	2,700 ft-lb/ft
Footing	5.5 ×		1 × 150 =	825 lb/ft	2.75	2,269 ft-lb/ft
Soil	3.5 × 9.5 × 115 =			3,824 lb/ft	3.75	14,340 ft-lb/ft
			Σ =	6,449 lb/ft		19,309 ft-lb/ft

Step 3   Determine the factors of safety against overturning and sliding:

$$FS_{ot} = \frac{M_r}{M_{ot}} = \frac{19,309}{7,102} = 2.72 > 2.0 \quad \therefore \text{ ok}$$

$$FS_s = \frac{\mu W}{E} = \frac{0.50 \times 6,449}{2,029} = 1.59 > 1.5 \quad \therefore \text{ ok}$$

Step 4   Calculate the soil pressure under the footing. Determine the location at which the resultant force intersects the bottom of the footing and calculate the resulting soil pressures.

$$c = \frac{M_r - M_{ot}}{W} = \frac{19,309 - 7,102}{6,449} = 1.89 \text{ ft from the toe}$$

So the eccentricity is

$$e = \frac{b}{2} - c = \frac{5.50}{2} - 1.89 = 0.86 \text{ ft}$$

Because

$$e = 0.86 < \frac{b}{6} = \frac{5.5}{6} = 0.92 \text{ ft}$$

Therefore, Case II (see Figure 7–71) is applicable. The soil pressures under the footing are:

$$f_{max} = \frac{W}{b}\left(1 + \frac{6e}{b}\right) = \frac{6,449}{5.50}\left(1 + \frac{6 \times 0.86}{5.50}\right) = 2,273 \text{ psf}$$

$$f_{min} = \frac{W}{b}\left(1 - \frac{6e}{b}\right) = \frac{6,449}{5.50}\left(1 - \frac{6 \times 0.86}{5.50}\right) = 72 \text{ psf}$$

Step 5   Design the required reinforcement in the stem. Calculate the factored bending moment in the stem, as illustrated in Figure 7–78. The maximum pressure at the base of the stem is:

$$p_a = 0.32 \times 115 \times 9.5 = 349.6 \text{ psf}$$
$$p_u = 1.6 \times 349.6 = 559.4 \text{ psf}$$

The factored design moment at the bottom of the stem is:

$$M_u = \left(\frac{p_u h_{stem}}{2}\right)\left(\frac{h_{stem}}{3}\right) = \frac{559.4 \times 9.5}{2} \times \frac{9.5}{3}$$
$$= 8,414 \text{ ft-lb/ft} = 8.41 \text{ ft-kip/ft}$$

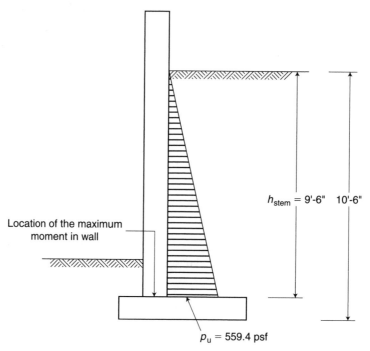

FIGURE 7–78 Pressures on the stem.

The minimum concrete cover required for #6 or larger bars (ACI Code, Section 7.7.1) is 2 in., as the back of the wall is exposed to the soil. Thus (assuming #6 bars):

$$d = 12 - 2 - (0.75/2) = 9.63 \text{ in.}$$

For a 1-ft length of the wall:

$$R = \frac{12,000 M_u}{bd^2} = \frac{12,000(8.41)}{(12)(9.63)^2} = 91 \text{ psi}$$

Using $f_c' = 3,000$ psi concrete and $f_y = 60,000$ psi steel from Table A2–6a:

$$\rho_{req} = 0.0018 \quad \text{(conservatively)}$$

From Table A2–4:

$$\rho_{min} = 0.0033 > 0.0018 \quad \therefore \text{ use } \rho_{min} = 0.0033$$
$$A_{s,req} = 0.0033 \times 12 \times 9.63 = 0.38 \text{ in.}^2/\text{ft}$$

From Table A2–10 select

$$\#5 @ 9 \text{ in. c/c} \quad (A_s = 0.41 \text{ in.}^2/\text{ft})$$

or

$$\#6 @ 12 \text{ in. c/c} \quad (A_s = 0.44 \text{ in.}^2/\text{ft})$$

We place some vertical reinforcements to support the horizontal bars on the exterior face of the wall. Use #4 @ 18 in. for walls with a height, $h_w \le 14$ ft,

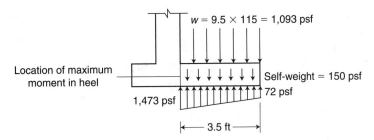

FIGURE 7-79 Forces on the heel in Example 7-9.

and use #5 @ 18 in. where $h_w > 14$ ft. Therefore, here we use #4 @ 18 in. since $h_w = 12$ ft. The horizontal shrinkage and temperature reinforcement required in the stem and footing is:

$$A_{sh} = 0.002bt = 0.002 \times 12 \times 12 = 0.288 \text{ in.}^2/\text{ft}$$

From Table A2–10 select #5 @ 12 in. c/c ($A_s = 0.31$ in.^{2}/ft) for the footing. Use #4 @ 16 in. on each face of the stem, as walls thicker than 10 in. require two layers of reinforcement (total $A_s = 2 \times 0.15 = 0.30$ in.^{2}/ft).

Step 6   Design the reinforcement required in the heel and toe.

The heel acts like a cantilever from the back of the stem to the end of the heel, as shown in Figure 7–79. The loads acting on it are the weight of the soil from above, its self-weight, and the upward reaction pressures at its bottom (found in step 4):

$$p_1 = 72 \text{ psf}$$

$$p_2 = 72 + \frac{2{,}273 - 72}{5.5}(3.5)$$

$$p_2 = 1{,}473 \text{ psf}$$

The moment at the intersection of the heel and the stem is:

$$M = [(1{,}093 + 150) \times 3.5] \times \frac{3.5}{2} - 72 \times 3.5 \times \frac{3.5}{2}$$

$$- (1{,}473 - 72) \times \frac{3.5}{2} \times \frac{3.5}{3} = 4{,}312 \text{ ft-lb/ft}$$

The factored moment is:

$$M_u = 1.6 \times 4{,}312 = 6{,}900 \text{ ft-lb/ft} = 6.9 \text{ ft-kip/ft}$$

The reinforcement will be placed at the top of the heel. Thus, 2 in. cover (as in the case of stem) is required, therefore, $d = 9.63$ in. For a 1-ft length of the heel:

$$R = \frac{12{,}000M_u}{bd^2} = \frac{12{,}000(6.9)}{(12)(9.63)^2} = 74 \text{ psi}$$

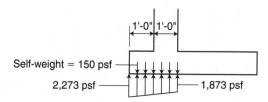

FIGURE 7–80 Forces on the toe in Example 7–9.

From Table A2-6a:

$$\rho_{req} = 0.0014 \longrightarrow A_s = 0.0014(12)(9.63) = 0.16 \text{ in.}^2/\text{ft}$$

Per the ACI Code, Sections 10.5.4 and 7.12 the minimum reinforcement is equal to the required shrinkage and temperature reinforcements.

$$A_{s,min} = 0.0018bt = 0.0018bt = 0.0018 \times 12 \times 12$$
$$= 0.26 \text{ in.}^2/\text{ft} > 0.16 \text{ in.}^2/\text{ft}$$

From Table A2–10 select   #5 @ 14 in. c/c ($A_s = 0.27$ in.^{2}/ft).

$$S_{max} = \min\{3h, 18\} = \min\{3 \times 12, 18\} = 18 \text{ in.} > 14 \text{ in.} \quad \therefore \text{ ok}$$

The toe is only 1'-0" long, so the reinforcement required will not be significant. For the sake of thoroughness, however, we will also determine the reinforcement in the toe.

$$p_1 = 2,273 \text{ psf}$$

$$p_2 = 72 + \frac{2,273 - 72}{5.5}(4.5) = 1,873 \text{ psf}$$

Neglecting, conservatively, the soil on the toe, the moment at the intersection of the toe and the stem, as shown in Figure 7–80, is:

$$M = (1,873 - 150)(1.0)\left(\frac{1.0}{2}\right) + (2,273 - 1,873)\left(\frac{1.0}{2}\right)\left(\frac{2}{3} \times 1.0\right)$$
$$= 995 \text{ ft-lb/ft}$$

$$M_u = 1.6(995) = 1,592 \text{ ft-lb} = 1.6 \text{ ft-kip/ft}$$

$$d = 12 - 3(\text{footing cover}) - \frac{0.75}{2} = 8.62 \text{ in. (assuming #6 bars)}$$

For a 1-ft length of the toe:

$$R = \frac{12,000(1.6)}{12(8.62)^2} = 22 \text{ psi} \longrightarrow \text{Table A2-6a} \longrightarrow \rho_{req} < 0.0010$$

$$\therefore \text{ use minimum steel}$$

$$A_s = 0.0018(12 \times 12) = 0.26 \text{ in./ft}$$

Table A2-10 $\longrightarrow$ Use #6 @ 12 in. c/c to dowel bars from the footing into the stem ($A_s = 0.44$ in.^{2}/ft). These bars will also serve for reinforcement in the toe.

Step 7   Check the bar development length.

a.   *Stem Reinforcement*   For the reinforcements in the stem, we must lap splice the dowels in the footing to the main reinforcement. From Table A3–3, for #6 bars, $\ell_d = 33$ in., which can be reduced by $\dfrac{A_{s,\,required}}{A_{s,\,provided}}$.

Therefore

$$\ell_{required} = 33 \times \frac{0.38}{0.44} = 29 \text{ in.}$$

Using a Class B lap splice according to the requirements discussed in Chapter 3:

$$\text{Required lap splice} = 1.3\ell_d = 1.3(29) = 38 \text{ in.}$$

The footing is too small to provide the above length, so the bars are hooked into the toe. The development length (per Equation 3–65) is:

$$\ell_{dh} = \left(\frac{0.02 \, \psi_e \lambda f_y}{\sqrt{f_c'}} \right) d_b$$

$$\ell_{dh} = \left(\frac{0.02(1.0)(1.0)(60,000)}{\sqrt{3,000}} \right) d_b = 21.9 d_b$$

$$\ell_{dh} = 21.9(0.75) = 16 \text{ in.}$$

From Table A3–4, we use an adjustment factor of 0.70 because the cover is more than 2 in. Therefore, the required length is:

$$\ell_{required} = 16 \times 0.7 \times \frac{0.38}{0.44} = 9.7 \text{ in.} > \min\{8(0.75), 6 \text{ in.}\} \quad \therefore \text{ ok}$$

$$\ell_{required} = 9.7 \text{ in.} < \ell_{provided} = 24 - (3 + 2) = 19 \text{ in.} \quad \therefore \text{ ok}$$

↗   ↖

Toe cover   Stem cover

We could also use a shear key with the dowels extended into the key to provide the required bar length if needed (see Figure 7–81).

b.   *Heel Reinforcement*   The development length for #5 bars, per Table A3–3, is 28 in. The required length is:

$$\ell_d = 28 \times \frac{0.26}{0.27} = 27 \text{ in.}$$

The provided reinforcement length in the heel is (cover = 3 in.):

$$3.5(12) - 3 = 39 \text{ in.} > 27 \text{ in.} \quad \therefore \text{ ok}$$

Hooks are required in the toe area, as the toe is not long enough.

$$\ell_{dh} = \left(\frac{0.02\,\psi_e\lambda f_y}{\sqrt{f_c'}}\right)d_b = \frac{0.02(1.0)(1.0)(60,000)}{\sqrt{3,000}}(0.625) = 13.7 \text{ in.}$$

$$\ell_{required} = 13.7\left(\frac{0.26}{0.27}\right) = 13 \text{ in.} > \min\{8(0.625), 6 \text{ in.}\}$$

$$\ell_{provided} = 24 \text{ in.} - 3 \text{ in.} = 21 \text{ in.} > 13 \text{ in.} \quad \therefore \text{ ok}$$

c. *Toe Reinforcement* The dowels from the stem are used as reinforcement for the toe. From Equation 3–65, the required length of bars with hook for #6 bars is:

$$\ell_{dh} = \left(\frac{0.02\psi_e\lambda f_y}{\sqrt{f_c'}}\right)d_b$$

$$\ell_{dh} = \left(\frac{0.02(1.0)(1.0)(60,000)}{\sqrt{3,000}}\right)0.75 = 16.4 \text{ in.}$$

Using the adjustment factors of Table A3–4 (cover > 2.5 in.):

$$\ell_{required} = 16.4 \times 0.7 \times \frac{0.26}{0.44} = 6.8 \text{ in.} \geq \min\{8(0.75), 6 \text{ in.}\}$$

$$\ell_{provided} = 12 \text{ in.} - 3 \text{ in.} = 9 \text{ in.} > 6.8 \text{ in.} \quad \therefore \text{ ok}$$

Figure 7–81 shows the retaining wall and the details of the reinforcements.

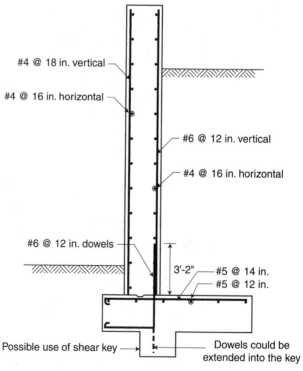

#4 @ 18 in. vertical

#4 @ 16 in. horizontal

#6 @ 12 in. vertical

#4 @ 16 in. horizontal

#6 @ 12 in. dowels

3'-2"

#5 @ 14 in.
#5 @ 12 in.

Possible use of shear key

Dowels could be extended into the key

FIGURE 7–81 Final reinforcement results for Example 7–9.

PROBLEMS

7–1   Design a plain concrete wall footing to support a 12 in. thick concrete wall. The dead load, including the weight of the wall, is 5.0 kip/ft, and the live load is 6.0 kip/ft. The bearing capacity of the soil is 2,500 psf and $f'_c = 3,000$ psi.

7–2   Redesign the footing of Problem 7–1 for a soil bearing capacity of 6,000 psf.

7–3   Rework Problem 7–1 for a reinforced concrete wall footing. Use $f_y = 60,000$ psi.

7–4   The following figure shows a partial section of a four-story office building. It is constructed of 8 in.-thick precast hollow core planks for roof and floors, supported by 12 in. block walls. The planks weigh 55 psf, and the block wall weighs 80 psf. The floor superimposed dead load is 25 psf, and the floor live load is 50 psf. The roofing weighs 15 psf, and the roof snow load is 30 psf. The soil bearing capacity is 3,500 psf. Use $f'_c = 3,000$ psi and $f_y = 60,000$ psi. Design a reinforced concrete footing for the interior walls shown.

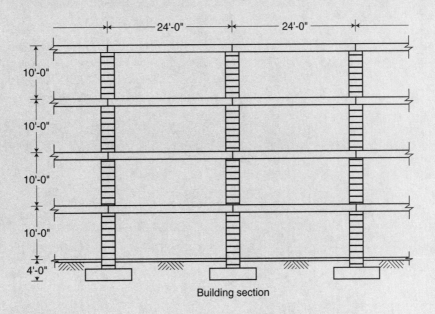

Building section

7–5   A 16 in. × 16 in. reinforced concrete column supports 150 kip dead load and 75 kip live load. The allowable soil bearing pressure is 4,000 psf. Design a square footing to support the column. Use $f'_c = 3,000$ psi for the column and the footing and $f_y = 60,000$ psi.

7–6   Design a square reinforced concrete spread footing for the interior columns of Problem 5–11. The soil bearing capacity is 6,000 psf. Use $f'_c = 3,000$ psi for the footing, $f'_c = 4,000$ psi for the column, and $f_y = 60,000$ psi.

7–7   Design a square reinforced concrete spread footing to support a 24 in. × 24 in. column carrying a 600 kip dead load and a 400 kip live load. The soil bearing capacity is 10,000 psf. Use $f_c' = 3,000$ psi for the footing, $f_c' = 4,000$ psi for the column, and $f_y = 60,000$ psi.

7–8 Redesign the footing of Example 7–3, if one of the horizontal dimensions of the footing is limited to 7'-0" due to the proximity of an adjacent property line.

7–9 Redesign the footing of Problem 7–7 if one of the horizontal dimensions of the footing is limited to 8'-0".

7–10 Determine the thickness of the unreinforced basement wall shown below for the following cases. The unit weight of the backfill material is 120 pcf, and the coefficient of active pressure $K_a = 0.33$. Show the soil lateral pressure and draw the shear force and bending moment diagrams for the applied loads. Use $f_c' = 4,000$ psi.

(a) $h_2 = 8'-0"$, no surcharge $(w_{sc} = 0)$ (c) $h_2 = 7'-0"$, $w_{sc} = 200$ psf
(b) $h_2 = 6'-0"$, no surcharge   (d) $h_2 = 10'-0"$, $w_{sc} = 200$ psf

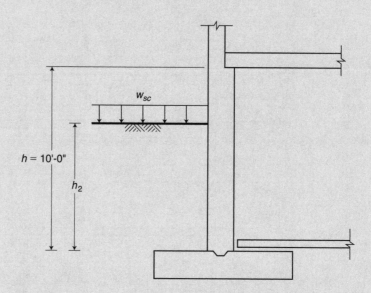

7–11 Rework Problem 7–10 for $h = 12'-0"$. For case d, use $h_2 = 12'-0"$.

7–12 Design the plain concrete basement wall shown. The equivalent fluid active density of the backfill material is 36 pcf. The unit weight of the soil is 120 pcf. Consider two cases: (a) without surcharge, and (b) a surcharge of 150 psf acting on the backfill. Use $f_c' = 3,000$ psi.

7–13 Redesign the basement wall of Problem 7–12 using reinforced concrete. Use $f_y = 60,000$ psi.

7–14 Check the adequacy of the 10-in.-thick reinforced concrete basement wall shown below. Use $f_c' = 4,000$ psi and $f_y = 60,000$ psi. The clear cover is $\frac{3}{4}$ in. The unit weight of the backfill is 100 pcf, and the coefficient of active soil pressure $K_a = 0.30$.

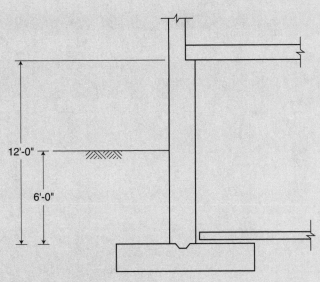

FIGURE FOR PROBLEM 7-12

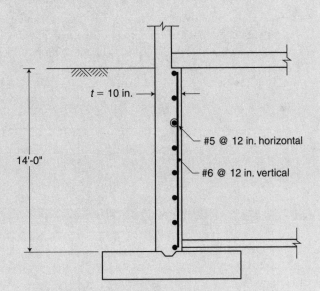

FIGURE FOR PROBLEM 7-14

7-15 What is the maximum allowable surcharge that can be placed on the outside grade of the basement wall of Problem 7-14?

7-16 Check the stability of the concrete gravity retaining wall shown below. Also, determine the soil pressure distribution on the base. The unit weight of the backfill is 120 pcf, the coefficient of active soil pressure is 0.30, and the coefficient of friction at the base is 0.50. The unit weight of the concrete is 150 pcf. The applied surcharge on the backfill is 100 psf. Disregard the passive pressure action on the wall.

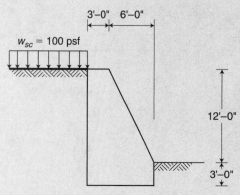

FIGURE FOR PROBLEM 7-16

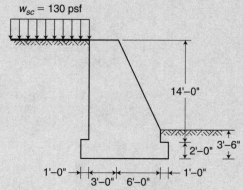

FIGURE FOR PROBLEM 7-17

7-17 Check the stability of the concrete gravity retaining wall shown. Also, determine the soil pressure distribution on the base. The unit weight of the backfill is 115 pcf, the coefficient of active soil pressure is 0.33, and the coefficient of friction at the base is 0.45. The unit weight of the concrete is 150 pcf. The applied surcharge on the backfill is 130 psf. Disregard the passive pressure action on the wall.

7-18 Check the stability of the reinforced concrete cantilever retaining wall shown below. Disregard the passive resistance of the soil in front of the toe. Assume $K_a = 0.3$, $\gamma_{soil} = 120 \, pcf$, and $\mu = 0.52$. The unit weight of the concrete is 150 pcf.

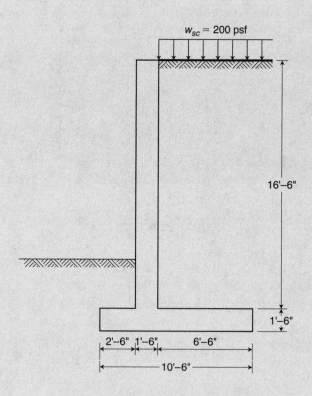

7–19 Design the cantilever retaining wall of Problem 7–18. Use $f_c' = 3,000$ psi and $f_y = 60,000$ psi. Use the ACI Code-recommended minimum covers.

SELF-EXPERIMENTS

In this self-experiment we study the behavior of square spread column footings. Include all the details of your tests such as sizes, times, concrete and ingredient proportions, problems you encountered, and so on, together with images showing the steps of the tests in your report.

Experiment 1

To study the behavior of spread footings, we will use a square piece of rubber (about $\frac{1}{2}$ in. thick) or any other flexible material. Put the rubber on some soft soil or sand. Place a Styrofoam column on the center of the rubber. Press the column down, as shown in Figure SE 7–1, and observe how the rubber mat reacts. Document all findings and observations.

Experiment 2

To gain a better understanding of the behavior of different types of soil under a foundation, we repeat Experiment 1 in the following order:

1. Fill a dish with dry sand. Compact the sand by gently pounding it with the bottom of a bottle. Smooth the top, and place a wood block (representing a square footing) on the sand. Load the block with an increasing load. Note what happens to the sand around the loaded block.

2. Repeat step 1 using wet sand.

3. Repeat step 1 using wet clay. The clay must be wet enough to be moldable.

4. Repeat step 3 after letting the clay dry for a few days.

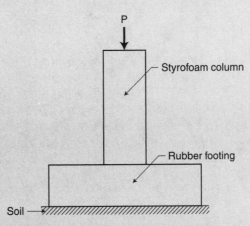

FIGURE SE 7–1   Spread footing under concentrated load.

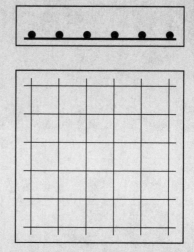

FIGURE SE 7–2 Reinforced spread footing.

Experiment 3

Form and cast a reinforced concrete square spread column footing. We will use wires to represent the two required sets of reinforcement, as illustrated in Figure SE 7–2. Document all your problems and findings in casting the footing.

Experiment 4

Using the concrete footing of Experiment 3, repeat Experiment 1 by adding a square reinforced concrete column at the center of the footing. Remember that you need dowel bars to tie the column to the footing. Document all problems and your findings in the construction of the column and footing model.

What is the approximate capacity of the footing if it is placed on a soil with a bearing capacity of 2,000 psf?

8

OVERVIEW OF PRESTRESSED CONCRETE

8.1   INTRODUCTION

Concrete has a considerable compressive strength, but its tensile strength is quite limited. Thus, designers use reinforcement in conjunction with concrete to make useful elements in buildings.

Early on, researchers realized that tensile stresses could be eliminated in concrete structures by adding sufficient compressive stresses to balance them out. Then the element would have a stress distribution throughout that consisted of compression only.

Figure 8–1 shows a simply-supported beam with an applied distributed load $w = 800$ lb/ft. The cross section of the beam is $b = 12$ in. and $h = 18$ in. If this beam is made of a homogeneous elastic material, the stresses can be calculated as follows:

$$M_{\text{max}} = \frac{0.8(20)^2}{8} = 40 \text{ kip-ft}$$

The elastic section modulus is:

$$S_m = \frac{12(18)^2}{6} = 648 \text{ in}^3$$

The maximum stress at the location of the largest moment is:

$$f_{\text{max}} = \frac{40 \times 12,000}{648} = 741 \text{ psi}$$

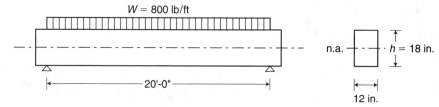

FIGURE 8–1   Bending of a simply-supported beam.

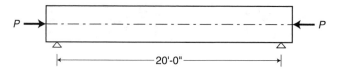

FIGURE 8–2   Simply-supported beam subject to concentric axial load.

This value represents the compression at the top edge of the section, and a similar magnitude of tension at the bottom edge. Based on the discussions on the tensile strength of concrete in Chapter 1, we conclude that a plain (unreinforced) concrete beam would fail under this load.

Now imagine that the beam is precompressed (prestressed) before it is loaded with the distributed load. Figure 8–2 shows this beam with a pair of forces acting on the centerline.

The force P would produce uniform compressive stresses over the cross section. The magnitude of the force P needed to eliminate any tension on this beam could be determined as follows:

$$f = \frac{P}{A} = 741 \text{ psi}$$

Then

$$P = 741 \times (12 \times 18) = 160,000 \text{ lb}$$

Figure 8–3 shows a graphical representation of the superimposed stresses caused by the axial load and the distributed load at the midspan of the beam.

As just mentioned, these are the stresses at midspan. At the bottom edge of the section, the compression from the force P and the tension from the maximum moment will exactly balance each other. At the top edge, the compression from the moment is added to

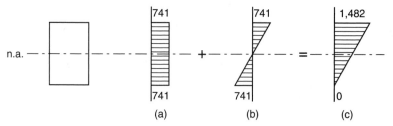

FIGURE 8–3   Stresses in the beam: (a) from force P, (b) from bending, (c) superimposed.

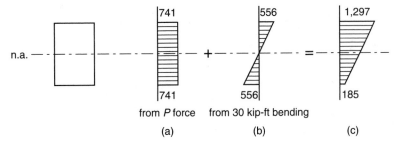

from *P* force   from 30 kip-ft bending

(a)   (b)   (c)

FIGURE 8–4 Stresses at 5'-0" from the support: (a) from force *P*, (b) from 30 ft-kip moment, (c) superimposed.

that from the force *P*. The rest of the beam will have compressions at both the top and the bottom. For example, at 5'-0" from the support, the moment is only $M = 30$ kip-ft. Figure 8–4 shows the stresses at that location.

   Now consider what will happen if the initial prestressing force is moved downward at an eccentricity *e* from the centroid. As discussed in Chapter 5, an eccentric force has the same effect as a concentric force plus a moment. See Figure 8–5. The force *P* still causes a uniform compression on the section; however, the $M = P \times e$ moment will cause tension at the top edge and compression at the bottom edge. So how large an eccentricity is needed so that the tension from the moment and the compression from the force *P* at the top edge cancel out each other? To answer this question we must first algebraically make the stresses due to the force equal to the stresses due to the moment. Mathematically this is expressed as follows:

$$\frac{P}{A} - \frac{M}{S_m} = 0 \quad \text{or} \quad \frac{P}{bd} - \frac{P \times e}{\frac{bd^2}{6}} = 0$$

Then we solve for *e*:

$$e = \frac{d}{6}$$

Returning to the numerical problem: Assume $e = 3$ in. and determine the amount of eccentric prestress force, *P*, is needed to have zero tension after the application of uniform loads of 800 lb/ft to the beam. The equation expressing this condition is:

$$\frac{P}{A} + \frac{P \times e}{S_m} - \frac{M_{\text{max}} \times 12,000}{S_m} = 0$$

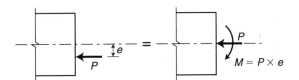

FIGURE 8–5 Moment due to an eccentric axial force, *P*.

where $\frac{P}{A}$ is the compressive stress caused by the force P, $\frac{P \times e}{S_m}$ is the compressive stresses caused by the eccentricity of the force P, and $\frac{M_{\max} \times 12,000}{S_m}$ is the tensile stress caused by the maximum moment. P is assumed to be in lb, and $M_{\max}$ is in ft-kip. (Note that a positive value designates compressive stress and a negative value designates tensile stress.)

Substituting $e = 3$ in., $A = 12 \times 18 = 216$ in^2, $S_m = 648$ in^3, and $M_{\max} = 40$ kip-ft and solving for P:

$$P = 80,000 \text{ lb}$$

This value is only half that needed to achieve the same result when the P force was concentrically applied. So, providing a well-selected eccentricity to the prestressing force can drastically reduce the magnitude of the force and still have the same effect as a concentrically applied force.

Although the tensile strength of concrete is small compared to its compressive strength, the ACI Code (Section 18.4.1) allows the section initially (at the time the prestressing force is applied) to have tensile stress equal to $f_t = 3\sqrt{f'_{ci}}$. Here f'_{ci} is the specified compressive strength of the concrete at the time the prestress is applied. This value is usually smaller than the final design strength of the concrete, as prestressing is usually accomplished before the concrete is completely cured. Similarly, the ACI Code permits a tensile stress of $f_t = 7.5\sqrt{f'_c}$ under full service load condition in most applications.

For the final introductory example we examine the use of the allowable tensile stress at the bottom of the section at midspan. Assume that $f'_c = 5,000$ psi. Then the beam can have $f_t = 7.5\sqrt{5,000} = 530.3$ psi of tension in the final service load condition. Using $e = 3$ in., we obtain the following equation for the service load condition at midspan after the substitutions:

$$\frac{P}{216} + \frac{P \times 3}{648} - \frac{40 \times 12,000}{648} = -530.3$$

(Note: The negative sign in front of "530.3" indicates tension.) Solving for P:

$$P = 22,728 \text{ lb}$$

This is the amount of prestress force that the beam needs at service load condition.

Some further refinements to these foregoing introductory examples will be discussed later in this chapter.

8.2   ADVANTAGES OF PRESTRESSED CONCRETE STRUCTURES

One major advantage of prestressing is that it prevents cracks in the concrete structure by either limiting or completely eliminating tensile stresses in the structure. But prestressing has another very important advantage. Prestressed structural elements can be much shallower than ordinary reinforced concrete elements for the same span and loading conditions,

TABLE 8–1 Recommended Maximum Span-Depth Ratios for Prestressed Floor Structures with Moderate Live Loads

	Single Span (Floor)	Continuous Spans (Floor)
Prestressed hollow core slabs	36	N.A.
Prestressed double tees	32	N.A.
Posttensioned one-way solid slabs	44	48
Posttensioned solid slab cantilevers	18	N.A.
Posttensioned flat plates (supported on columns)	N.A.	45
Posttensioned waffle slabs (supported on columns)	N.A.	35

while still maintaining good span/deflection ratios. The shallower depths in turn result in lighter structural elements, thus providing considerable savings in the dead loads the structure must carry. The savings extend to reduced floor-to-floor heights and lighter column and foundation loads as well. The reduced floor-to-floor height in multistory buildings results in large savings in nonstructural building elements such as walls. The reduced building volume also lessens the energy needed for heating and cooling.

It is somewhat difficult to give precise span-depth ratios for prestressed concrete structural elements, so you should use the values listed in Table 8–1 only as a recommendation for preliminary selection of structural depth. In the authors' experience, using these values as limits will result in structural elements that perform well and without excessive camber, deflection, or bothersome vibration.

The ratios may safely be exceeded by about 10% for roof structures. Thus, prestressed hollow core slabs used for roofs will perform well with span-depth ratios of 40.

Compare the values in Table 8–1 to those recommended in foregoing chapters for elements using normal reinforced concrete, and note the depth and weight savings the prestressing offers. For example, the 10-in. flat plate in Example 6–3 can be designed with a thickness of 7.5 to 8 in. if prestressing is used, reducing the dead loads by 20%–25%.

8.3 TYPES OF PRESTRESSING

Pretensioning

In *pretensioning,* as the expression illustrates, the prestressing strands are tensioned before the casting of the element by stretching and fixing them against two bulkheads, as illustrated in Figure 8–6. The bulkheads are very strong and designed to take the large forces from

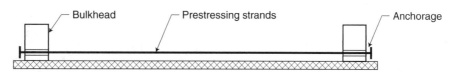

FIGURE 8–6 Prestressing process (step 1: placing strands).

FIGURE 8–7 Prestressing process (step 2: placing concrete).

the initial stretching of the prestressing strands. The strands are anchored at one end (the "dead" end) and then pulled from the other end (the "live" end) one by one with a specially designed hydraulic jack. The force in the strands can be measured directly on the jack or determined from the amount of elongation. Elongation is directly related to the stress in the strand, so we can readily determine the force in the strand if we know the cross-sectional area of the strand. The strands are anchored at the live end as well once the appropriate force has been reached.

In the next step we set up the forms around the stressed strands and cast the concrete, as illustrated in Figure 8–7.

After the concrete has gained sufficient strength, the strands are released, as illustrated in Figure 8–8. The bond established between the strands and the cured concrete transfers the force in the strands into the concrete element. The tension in the strands now becomes compression on the concrete. This transfer occurs within a few inches from the ends of the members.

This method is applicable for precast and prestressed elements produced in manufacturing plants. The production technique often involves the casting of elements in long (up to 600 ft) casting beds, which permits the simultaneous fabrication of many elements with a single tensioning of the strands. A large saw is used to cut each element to its individual required length. Accelerated curing techniques permit the release of the strands in only 16 to 18 hours after the placement of the concrete; thus, a 24-hour manufacturing cycle can be maintained. The production is highly mechanized and provides great quality control. Many standardized profiles are available in catalog form for spans and loading capacities. These can be readily called out and specified by the designer. Figures 8–9 through 8–13 show some typical profiles that are popular in building construction.

Hollow core decks are a popular precast and prestressed building element. They are typically available in standard depths of 6 in., 8 in., 10 in., and 12 in., and are manufactured mostly in 4'-0" width, although some manufacturers may supply them in 2'-0" or

FIGURE 8–8 Prestressing process (step 3: releasing strands).

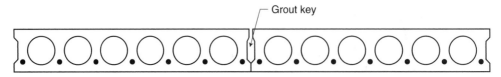

FIGURE 8–9 Hollow core decks (without concrete topping).

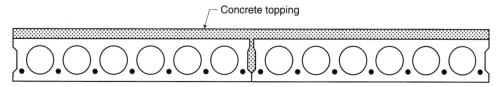

FIGURE 8–10   Hollow core decks (with concrete topping).

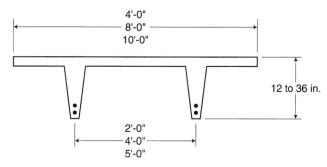

FIGURE 8–11   Double tee section.

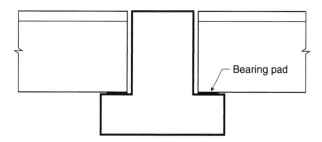

FIGURE 8–12   Inverted T-beam supporting double tees.

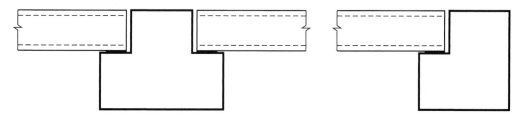

FIGURE 8–13   Inverted T-beam and L-beam supporting hollow core decks.

8'-0" widths. The shape of the cores may also differ from the circles shown on Figure 8–9, as different patented manufacturing processes are used to form them. Not all building designs can use the standard widths, so narrower filler panels are made by slicing the panels lengthwise. Panels can also be cut at an angle to accommodate supporting beams or girders that are not perpendicular to the span of the panels.

A *grout key* is formed at the sides of the individual decks. As the name indicates, the formed keyway is grouted solid after the erection of the panels. The keyways prevent individual panels from deflecting differently after the grout cures. Thus, a kind of lateral load transfer occurs if one of the floor panels is loaded much more than its neighbors.

The top surface of hollow core decks is not smooth enough for floor structures, which receive finish materials like tiles or carpets. Thus, in those types of applications, the decks usually receive a 2 in.-thick (nominal) concrete topping (see Figure 8–10) that can be finished to the desired flatness and smoothness. The concrete topping bonds to the surface of the decks and becomes a composite part of the whole. It also makes the floor thicker, and consequently considerably stiffer than one made of untopped decks.

The topping thickness is a *nominal* thickness. Decks usually have an upward deflection or camber when they are erected. This is a natural result of the pretensioning process, in which the strands compress the bottom and sometimes also cause tension on the top. Thus, the top elongates and the bottom shortens, resulting in an upward curvature (i.e., camber). Calculation of the camber is a somewhat complicated process. It requires a reasonable estimate of the concrete's modulus of elasticity, as the concrete continues to cure even after the prestress has been applied. It also requires a knowledge of the rate of shrinkage and creep deformation that take place as the concrete ages. For example, a typical 8-in.-thick hollow core deck, 20 to 25 ft long, may exhibit a $\frac{3}{4}$ in. camber at the time of its erection. In order to eliminate midspan humps from the finished structure, engineers may use only $1\frac{1}{2}$ in. topping at the center of the span and $2\frac{1}{4}$ in. at the supports, or some similar combination, to make the finished floor as flat as possible.

Another popular precast and prestressed building element is the *double tee,* which is shown in Figure 8–11. These are also standard elements, although some manufacturers may make them in only one width, or may not provide them in depths beyond a certain dimension.

Double tees are quite light; the top slab is typically 2 in. thick only at the outer edges. They are very economical for covering large spans (spans of 100 ft or more are not uncommon). Such long elements, however, are difficult to transport in a tight urban environment. Double tees are rather flexible, so their camber or deflection under load can be significant, especially on longer spans. So the designer must carefully plan the interface of double tees with other building elements. A double tee on an 80-ft or 90-ft span element may have an initial camber of 3 to 4 in. (or more), with perhaps a similar magnitude of deflection under loads. That much movement requires very careful consideration of details, such as when the design contemplates a window-wall parallel with the span of a double tee.

Figures 8–12 and 8–13 show some other typical precast and prestressed elements that are frequently used to support floor elements in building construction. The shapes shown here are just a few of the many different shapes readily available to the designer.

Posttensioning

Posttensioning is a technique used to prestress concrete structures on the job site after the concrete has been cast into the forms and cured. It differs from pretensioning, which is

FIGURE 8–14 Draped strands in a single span beam.

typically used to manufacture building elements *away* from the building site. The posttensioning technique places flexible hollow metal or plastic tubes into the formwork to form ducts. Tendons are inserted through the ducts after the concrete has cured. Other techniques place plastic sheeted tendons into the formwork. The plastic sheet prevents the tendons from bonding with the cast concrete. The strand or tendon is anchored at one end to a device or plate cast into the concrete (the "dead" end). Portable hydraulic jacks from the "live" end provide tensioning. The jack leans against the concrete surface while pulling on the tendon. A calibrated gage on the stressing jack shows the amount of force in the tendon, while the elongation of the tendon is also measured to ensure quality control. After the design force has been reached, the stretched tendon is anchored to the concrete.

The space between the duct and the tendon is pressure grouted in certain applications. The grouting not only creates a continuous bond between the tendon and the duct, which is bonded to the concrete, but also provides enhanced corrosion protection for the tendon. In other applications, the tendons are left ungrouted and must rely on the continuing performance of the end anchorages throughout the service life of the structure. These tendons are usually pregreased inside the ducts. Greasing helps to minimize the frictional losses on curved tendons. (See discussion later in this chapter.)

Posttensioning has one vast advantage over pretensioning: It usually uses curved tendons. This enables the designer to change the location of the prestressing force from section to section along the length of the structure.

Figure 8–14 shows a single span beam with the strands draped in a parabolic form. After tensioning, the eccentricity of the prestressing force to the neutral axis is zero at the ends and maximum at the center. The moment resulting from the prestressing causes compression at the bottom and tension at the top. The maximum of these forces occurs at midspan, as illustrated in Figure 8–14. In fact, the forces closely balance the effects of the gravity loads on the beam.

Posttensioning also enables designers to use prestressing on continuous spans. Pretensioned members, because of their straight strands, are used only as simply-supported single spans, although short cantilevers can also be accommodated by adding conventional reinforcing. Posttensioning, however, can use draped strands, as shown in Figure 8–15, to follow the requirements of the bending moments from the gravity loads. The strands are near the bottom of the section where the bending moments are positive, and are near the top of the section where the bending moments are negative.

FIGURE 8–15 Draped strands in continuous beams or slabs.

8.4 PRESTRESSED CONCRETE MATERIALS

Concrete

Prestressed concretes generally are high-strength concretes with $f_c' = 5{,}000$ to 6,000 psi. These high-strength concretes are better for many reasons. Chief among these reasons is that these concretes have a smaller amount of shrinkage and creep, which lessens the loss of prestress. Another reason is that in both pre- and posttensioning, very highly stressed regions arise in compression at the anchorages; high-strength concretes are needed to withstand these stresses.

Prestressing Steel

The earliest experiments with prestressing failed mainly because they were performed with ordinary steels (yield strength in the range of 36 to 40 ksi). All, or almost all, of the prestress was lost with the passage of time, due to a series of contributory reasons. (Loss of prestress in normal applications may amount to 25 to 35 ksi, or higher.)

The use of very high-strength steel wires helped to solve this problem, as these wires, even after the considerable prestress losses, retained sufficient stress levels. Although many proprietary prestressing (mostly post-tensioning) systems use large-diameter, high-strength bars, most systems employ prestressing strands manufactured from cold-drawn wires conforming to ASTM A421. Usually six wires are wound tightly around a seventh (and usually slightly larger-diameter) wire into a uniform pitch helix, as illustrated in Figure 1–21. The pitch is 12 to 16 times the diameter of the wires. After manufacture, the strands are put through a stress-relieving heat treatment to make them conform to the requirements of ASTM A216, "Standard Specifications for Uncoated Seven-Wire Stress-Relieved Strand for Prestressed Concrete." They are also prestretched to increase their apparent modulus of elasticity. The strands may be manufactured in Grade 250 or in Grade 270, the numbers referring to the minimum ultimate strength of the strand in ksi. Table 8–2 shows the properties of Grade 270 strands.

Figure 8–16 shows a typical stress-strain curve for prestressing strands. The strands, unlike normal reinforcing steel, do not have a defined yield. They remain elastic up to about 85% of their ultimate strength. An arbitrary yield point is often used for specification purposes. ASTM A216 requires a minimum value of $0.85f_{pu}$ at 1% extension (or strain), where f_{pu} is the minimum guaranteed ultimate strength. By this definition, the yield for Grade 270 K strand may be taken as $0.85 \times 270 = 230$ ksi.

TABLE 8–2 Properties of Grade 270 Strands

Nominal Diameter of Strand (in.)	Breaking (Ultimate) Strength of Strand (kip)	Nominal Steel Area of Strand (in²)	Nominal Weight of Strand (lb/ft)
3/8	23.0	0.085	0.29
7/16	31.0	0.115	0.39
1/2	41.3	0.153	0.52
3/5	58.6	0.217	0.74

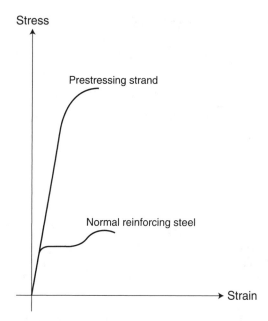

FIGURE 8–16   Stress-strain curve of prestressing strands versus reinforcing steel.

8.5 LOSS OF PRESTRESSING

The final service level stresses in the prestressing strands will be significantly lower than they were at the time of the initial stressing. The contributory causes for this prestress loss are numerous. The five major ones are discussed below.

Elastic Shortening of Concrete

When prestress is applied to concrete (i.e., when the prestressing strands are released in the pretensioning type of application), the concrete shortens due to the compressive stresses transferred to it. As the concrete shortens, so do the strands bonded inside the concrete. This shortening lessens the stress in the steel, and correspondingly lessens the compression on the concrete. How much the concrete member shortens depends on the concrete's modulus of elasticity, which in turn depends on the concrete's strength at the time the prestress was applied. The higher the strength of the concrete, the lesser the loss due to elastic shortening. In posttensioning, however, very little loss is caused by the elastic shortening; as the stressing and the shortening of the concrete take place simultaneously; so when the force in the strand is measured, the change in the concrete's length has already taken place. (This is not exactly true, because as the tendons are pulled one by one, each stressed and anchored tendon will lose some of its stress when its neighboring tendon is stressed.)

Shrinkage of Concrete

If the concrete shrinks due to loss of moisture after prestressing, the shrinkage will shorten the member. Correspondingly, the stretched strands will also shorten by the same amount. This shortening leads to a loss of stress in the strand. Nearly 80% of the shrinkage takes place in the first year of life of the structure. The magnitude of shrinkage depends on many variables, but it can be estimated to a reasonable degree of accuracy based on experimental data.

Creep of Concrete

Sustained compression shortens concrete over time. Creep, therefore, is a time-dependent deformation. The magnitude of creep depends on many variables. The most important of these are the strength of the concrete, the age of concrete at the time of prestressing, and the average compressive stress in the concrete.

Relaxation of the Prestressing Steel

By definition, *relaxation* is the change in stress in a material held at a constant strain. This phenomenon is a very complex, time-dependent characteristic of prestressing wires and strands that are subject to high stresses. Relaxation contributes less than shrinkage or creep of concrete to the total sum of the losses, but it still must be considered.

Friction Losses in Curved Tendons

Friction loss occurs only in posttensioning. Figures 8–14 and 8–15 show typical paths of curved tendons. Sometimes posttensioned flat plates use tendons that curve in the horizontal direction to accommodate floor openings or ducts. As a result, when a tendon is pulled from one end, it leans against the duct. Figure 8–17 illustrates the problem.

If the tendon is pulled from the right, P_2 will be less than P_1. The difference will be the loss due to the frictional resistance at the contact surfaces. The loss depends on the radius of curvature and the friction coefficient between the tendon and the duct. Sharper curves and larger friction coefficients will result in larger loss of prestress. In other words, the force in the tendon at locations away from the live end will be less than that measured at the stressing jack. In addition to the intentional curving of the ducts, an unintentional curving also takes place during the concreting operation. This is referred to as *wobble*. The so-called wobble coefficient accounts for this curving, but is a rather vague value. For example, the ACI Code recommends a value between 0.0003 and 0.0020 for 7-wire pregreased strands. The recommended value for the curvature friction coefficient for similar strands is between 0.05 and 0.15.

Equation 8–1 gives the formula for calculating frictional losses (ACI Code Equation 18–1):

$$P_{px} = P_{pj} e^{-(K\ell_{px} + \mu_p \alpha_{px})} \tag{8–1}$$

FIGURE 8–17   Friction between the tendon and the duct in posttensioning.

where
K = wobble friction coefficient per foot of tendon
ℓ_{px} = distance in feet from jacking end of prestressing steel element to the point under consideration (point x)
μ_p = post-tensioning curvature friction coefficient
α_{px} = the total angular change in radians of the tendon profile from the jacking end to the point x
P_{pj} = prestressing force at jacking end in pounds
P_{px} = prestressing force evaluated at distance ℓ_{px} from the jacking end in pounds

EXAMPLE 8–1

Figure 8–18 shows a curved tendon between two inflection points in an 8-in.-thick slab. The inflection points are 15 ft apart. Between these two points the tendon rises 2.75 in. Assume $K = 0.001$ and $\mu_p = 0.10$. Find the prestress loss within these two points.

Solution

We use the given data to calculate the radius of curvature, which is 122.8 ft. The value of α_{px} is then equal to $7° = 0.122$ radians. Substituting these values into Equation 8–1 we obtain the prestress loss due to friction between the two points as a function of the prestressing force.

$$P_{px} = P_{pj}\,e^{-(0.001 \times 15 + 0.10 \times 0.122)}$$
$$P_{px} = P_{pj} \times 0.9731$$

This means that the loss of force along the curve is 2.69%. In multiple curves the losses will combine, and may become very significant depending on the number of spans.

Several techniques can be used to mitigate the frictional losses. One of them is to stress the tendons from both ends. After the tendon has been stressed from the "live" end, that end is anchored and the tendon is restressed by pulling from the former "dead" end. Other techniques involve stressing very long tendons one section at a time, coupling the next section to the already stressed segment.

Total Losses

Calculating the total losses of prestress is a very complex problem. Even with the best available research information, we can obtain only approximate values. The values of the parameters that influence the loss from any of the major sources are only approximate ones. The current ACI Code frowns on using lump sum values for estimating prestress losses, but the authors feel that the values listed in Table 8–3 quite closely approximate the true ones (at least in building construction).

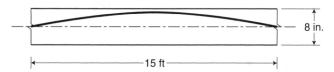

FIGURE 8–18   Part of a posttensioned slab for Example 8–1.

	f_{pu} (ksi)	Initial Prestress (ksi)	Loss (ksi) (w/o Friction)	Remnant (ksi)
TABLE 8–3 Average Prestressing Losses				
Pretensioning	270	216	40–42	175
Posttensioning	270	189	28–30	160

For typical 270 K strands that are initially stressed to the ACI Code-recommended value of $0.7f_{pu}$ for post-tensioning applications and $0.8f_{pu}$ for pretensioning applications, a good estimate of losses from volumetric changes for average conditions (not including friction losses) is 18%–20% for pretensioning, and 15%–16% for posttensioning. This translates to the average loss values shown in Table 8–3.

8.6   ULTIMATE STRENGTH

Prestressed elements are designed to limit stresses at service load conditions. The ACI Code, however, also requires that prestressed elements satisfy *ultimate strength* requirements as well.

The ultimate moment strength is calculated using equations similar to those used with ordinary reinforcement. But these familiar equations substitute a calculated (somewhat fictitious) yield value for f_y in the calculations. This yield value, f_{ps}, can be calculated using Equations 18–3, 18–4, and 18–5 of the ACI Code.

This introductory chapter on prestressing includes only the formulae that deal with unbonded tendons. The reader should consult the ACI Code for further information on the ultimate strength calculations for bonded tendons.

Thus, Equation 8–2 (ACI Code Equation 18–4) can be used to calculate the yield strength for members with unbonded tendons and with a span/depth ratio of 35 or less:

$$f_{ps} = f_{se} + 10,000 + \frac{f_c'}{100\rho_p} \tag{8–2}$$

and Equation 8–3 (ACI Code Equation 18–5) is applicable for members with unbonded tendons and with a span/depth ratio greater than 35:

$$f_{ps} = f_{se} + 10,000 + \frac{f_c'}{300\rho_p} \tag{8–3}$$

where

f_{se} is the effective stress in the tendons after losses in psi

f_c' is the specified compressive strength of the concrete in psi

ρ_p is the ratio of the prestressed reinforcement, $\dfrac{A_{ps}}{bd_p}$

A_{ps} is the area of the prestressing steel

d_p is the distance from the compression edge to the centroid of the prestressing tendons

If the design uses unbonded tendons, the ACI Code (Section 18.9) requires the addition of a minimum amount of bonded normal reinforcement. This minimum amount is given below:

$$A_s = 0.004A_{ct}$$

where A_{ct} is the area of the concrete section between the tension face and the centroid of the gross concrete section.

The ACI Code also permits the use of nonprestressed reinforcing to help with the required ultimate strength. The detailed discussion of this subject is beyond the scope of this text.

8.7 THE CONCEPT OF LOAD BALANCING

In the simple numerical examples given above, the analyses of prestressed sections were conducted by finding a sufficiently large concentric, or eccentric, prestressing force that eliminated, or greatly reduced, unwanted large tensile stresses from applied loads. This procedure is sometimes referred to as *superposition* (i.e., adding axial compressive stresses to the ones caused by flexure).

In posttensioning there is a much more easily visualized method of analysis. The *load balancing method,* introduced by T. Y. Lin in 1963, is the most widely used and most powerful analytical tool for the design of prestressed structures. This method provides prestressing by using a system of stressed tendons selected to impose loads on the element *in opposition* to the acting gravity loads; hence, the name *load balancing method.*

This concept is illustrated in Figure 8–19. If the path of the tendon is a parabola, its effect, after stressed on the simple span beam, is equivalent to an upward acting uniform load. The balancing loads, w_{bal}, can be calculated using simple statics as follows. The moment caused by the horizontal component of the prestressing force, as shown in

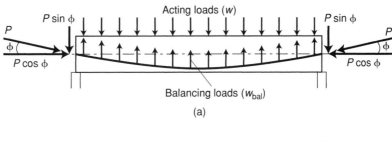

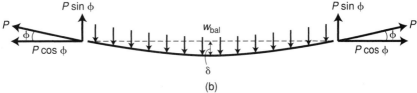

FIGURE 8–19   (a) Balancing uniform load, (b) free body diagram of the tendon.

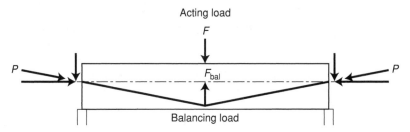

FIGURE 8–20 Balancing a concentrated load at the center of the span.

Figure 8–19b, is $(P\cos\phi)\delta$. This *prestress moment* must balance the moment caused by acting loads. Mathematically this is expressed as:

$$(P\cos\phi)\delta \cong P\delta = \frac{w_{bal}\ell^2}{8}$$

Thus

$$P = \frac{w_{bal}\ell^2}{8\delta} \tag{8–4}$$

where δ is the sag of the tendon.

A balanced beam theoretically has uniform compression, $f = P/A$, throughout. The balanced load, w_{bal}, can be equal to the acting loads, or only a part of the load the designer wants to balance. This is normally the case when a certain amount of tension is permitted at service load levels.

The loads on the beam do not have to be uniformly distributed in order to apply the load balancing concept. This is illustrated in Figures 8–20 and 8–21. If we use similar notations for the loading conditions shown in Figures 8–20 and 8–21 and change the tendon's path from a parabola to those shown, the balancing forces for the midspan point load and one-third span point load conditions are $P = \dfrac{F_{bal}\ell}{4\delta}$ and $P = \dfrac{F_{bal}\ell}{3\delta}$, respectively. We can easily extend the load balancing concept to continuous spans as well as to flat plates or flat slabs. The detailed discussion of the intricacies involved, however, are beyond the scope of this text.

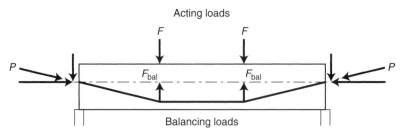

FIGURE 8–21 Balancing two equal loads at the third points of the span.

EXAMPLE 8–2

The 30-ft-long, 12 in. × 16 in. simply-supported beam shown in Figure 8–22 will support a superimposed distributed load of 400 lb/ft. A single parabolic tendon is used to posttension the beam. Determine the required force in the tendon using the load balancing method. Use $f'_c = 5,000$ psi, and $f'_{ci} = 3,750$ psi and the unit weight of the concrete is 150 pcf.

Solution

For efficiency select the largest sag permitted by the beam section, $\delta = 6$ in. = 0.5 ft. The self-weight of the beam is:

$$w_{sw} = 150 \times \frac{12 \times 16}{144} = 200 \text{ lb/ft}$$

To balance all the dead loads and one-half the live loads, w_{bal} is

$$w_{bal} = 200 + \frac{1}{2} \times 400 = 400 \text{ lb/ft}$$

and from Equation 8–4

$$P = \frac{w_{bal}\ell^2}{8\delta} = \frac{(0.4 \text{ kip/ft}) \times 30^2}{8 \times (0.5 \text{ ft})} = 90 \text{ kip}$$

Only the self-weight of the beam acts on the beam at the time of the transfer of the prestressing force. Hence, the upward loads from the prestressing tendon and the self-weight of the beam yield a *net* result of 200 lb/ft upward. So we calculate the initial stresses due to prestress at midspan as follows:

$$M = \frac{0.2 \times 30^2}{8} = 22.5 \text{ ft-kip}$$

The elastic section modulus of the beam's cross section is:

$$S_m = \frac{bh^2}{6} = \frac{12 \times 16^2}{6} = 512 \text{ in}^3$$

and the area is $A = 12 \times 16 = 192 \text{ in}^2$.

$$f = \frac{M}{S_m} = \frac{22.5 \times 12,000}{512} = 527 \text{ psi}$$

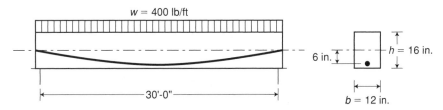

FIGURE 8–22   Elevation and section of the beam in Example 8–2.

tension at the top and compression at the bottom. (The net difference of w_{bal} and the self-weight acts upward.)

From the 90-kip axial compression load at the time of transfer of the prestress (transferred at the anchorages at the ends):

$$f = \frac{P}{A} = \frac{90 \times 1,000}{192} = 469 \text{ psi (compression)}$$

Combining the axial compression (469 psi) and the flexural stresses due to the net 200 lb/ft upward load, the initial stresses at midspan are:

$$f_{top} = 469 - 527 = -58 \text{ psi (tension)}$$
$$f_{bottom} = 469 + 527 = 996 \text{ psi (compression)}$$

Because these resulting stresses at the time of the load transfer are less than the ACI Code (Section 18.4.1) permits:

$$3\sqrt{f'_{ci}} = 3\sqrt{3,750} = 184 \text{ psi (tension)}$$

and

$$0.6 f'_{ci} = 2,250 \text{ psi (compression)}$$

respectively. The stress values at the time of transfer are acceptable.

From the 400 lb/ft = 0.4 kip/ft superimposed load:

$$M_{S.I.} = \frac{0.4 \times 30^2}{8} = 45 \text{ ft-kip}$$

$$f = \frac{M}{S_m} = \frac{45 \times 12,000}{512} = 1,055 \text{ psi}$$

compression at the top and tension at the bottom. Combining these stresses due to total live load with the initial stresses gives us:

$$f_{top} = -58 + 1,055 = 997 \text{ psi (compression)}$$
$$f_{bottom} = 997 - 1,055 = -58 \text{ psi (tension)}$$

Now check the stress after long-term losses have taken place.
Assume long-term prestressing losses of 15%.
The final remaining prestressing force is only:

$$P_{final} = 0.85 \times 90,000 = 76,500 \text{ lb}$$

and the adjusted upward-acting balancing load is:

$$w_{bal} = \frac{8P\delta}{\ell^2} = \frac{8 \times 76,500 \times 0.5}{30^2} = 340 \text{ lb/ft}$$

rather than the full 400 lb/ft that was first used.

The *net* difference between the downward-acting gravity loads (self-weight of the beam plus superimposed loads) and the upward-acting balancing loads is:

$$200 + 400 - 340 = 260 \text{ lb/ft} \quad \text{(downward)}$$

The moment caused by this load is:

$$M_{net} = \frac{0.26 \times 30^2}{8} = 29.25 \text{ ft-kip}$$

This moment causes compression at the top and tension at the bottom. These stresses, combined with the compression stresses from the remaining posttensioning force, will give the final stresses:

$$f_{top} = \frac{76,500}{192} + \frac{29.25 \times 12,000}{512} = 1,084 \text{ psi} \quad \text{(compression)}$$

$$f_{bottom} = \frac{76,500}{192} - \frac{29.25 \times 12,000}{512} = -287 \text{ psi} \quad \text{(tension)}$$

These stresses are lower than those permitted by the ACI Code (Sections 18.3.3 and 18.4.2) for service load stage.

$$0.45 f_c' = 0.45 \times 5,000 = 2,250 \text{ psi compression} > 1,084 \text{ psi}$$
$$7.5\sqrt{f_c'} = 7.5\sqrt{5,000} = 530 \text{ psi tension} > 287 \text{ psi}$$

Hence, the selected amount of prestressing is satisfactory.

PROBLEMS

8–1 A plain concrete beam has a width of 14 in. and a total depth of 24 in., and is simply supported with a span of 24 ft. What is the maximum tensile stress acting on the beam due to its weight?

8–2 The beam of Problem 8–1 is prestressed with a straight tendon at the centroid of the section to produce a prestressing force of 200 kip.

1. What will be the maximum stresses on the beam at midspan (a) at the top, and (b) at the bottom?
2. How much uniformly distributed load may be placed on the beam if no tension is permitted in the beam?

8–3 Assume that the straight prestressing tendon of Problem 8–2 is placed 4 in. from the bottom of the beam.

1. What are the maximum stresses on the beam at midspan (a) at the top, and (b) at the bottom, when only the beam's self-weight acts?
2. How much uniformly distributed load may be placed on the beam if 424 psi maximum tension is permitted?

8–4 The rectangular prestressed beam (16 in. × 32 in.) shown below is subject to a total dead and live load of 2.2 kip/ft (including the beam weight). The parabolic tendon will have a sag of 10 in., as shown, to provide an upward balancing load of 1.6 kip/ft.

1. Calculate the required final prestressing force (after losses).
2. Calculate the final stresses in the beam at midspan and at the ends (a) at the top of the section, and (b) at the bottom of the section.

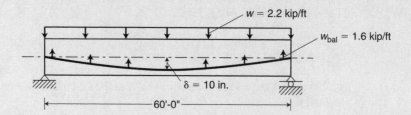

SELF-EXPERIMENTS

In this self-experiment, you will study the behavior of prestressed and posttensioned beams. Record all the details of the tests and include photos showing different stages of the experiments in your report.

Experiment 1

In this experiment we study the behavior of prestressed beams using a styrofoam beam. Place the beam between two supports. Apply a predetermined load (a few pounds) on the beam and record the magnitude of the beam deflection.

Make a hole at the bottom of the beam and pass a few plastic strings through it. Anchor the strings at one end. Pull the strings from the other side and anchor them as shown in Figure SE 8–1. Place the beam on the same supports and apply the same load. Determine how much the beam deflects. Compare the results with those of the previous test.

Experiment 2

In this experiment, we study the behavior of prestressed beams using concrete models. Cast two concrete beams of the same size. Reinforce one beam with regular wires, and thread the ends of the wire for the other. Before placing concrete for the prestressed beam, pull the wires from the two sides as shown in Figure SE 8–2. Compare the behavior of the two beams by placing them on two supports and gradually loading them. Which one deflects more? Why? Discuss your observations.

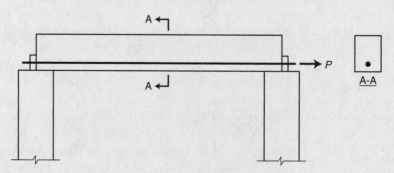

FIGURE SE 8–1   Prestressing a styrofoam beam.

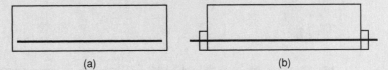

(a)   (b)

FIGURE SE 8–2   (a) Reinforced concrete beam, (b) prestressed concrete beam.

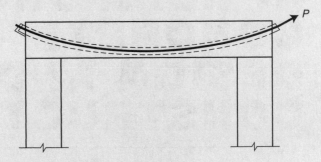

FIGURE SE 8–3   Posttensioned styrofoam beam.

Experiment 3

Here we use a styrofoam beam similar to that of Experiment 1 to study the behavior of post-tensioned beams. Make a hole of the same size as that of Experiment 1 on the side of the beam using a hot wire. The hole should have a curved shape, as shown in Figure SE 8–3. Pass a few plastic strings through this hole, anchor them to the beam from one end, and pull and anchor to the other end. Now place the beam on the two supports and apply the same load as before. Record the beam deflection at the midspan. Compare the results with those of Experiment 1. Which case resulted in less deflection?

Experiment 4

This experiment involves the use of a posttensioned concrete beam. As in Experiment 2, cast two beams. One will use regular wire, and the other will be posttensioned. For the posttensioned beam, place a plastic tube inside the beam and cast the concrete. After the concrete is set (72 hours), insert steel wires, anchor them to the beam from one end, and pull and anchor to the other end (Figure SE 8–4). Place the two beams on the two supports and gradually load them. Compare their behavior. Which one deflects more? Why? Any other observations?

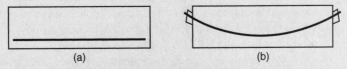

(a)   (b)

FIGURE SE 8–4   (a) Reinforced concrete beam, (b) posttensioned concrete beam.

9

METRIC SYSTEM
IN REINFORCED
CONCRETE
DESIGN AND
CONSTRUCTION

9.1 INTRODUCTION

Efforts to change the U.S. measurement units to the metric system have been under way for quite a while. This chapter briefly discusses this matter as it relates to reinforced concrete structure design and construction. We present a few examples using this system of units so that you will better understand how to make the conversions.

9.2 BRIEF HISTORY OF METRIC SYSTEM ADOPTION IN THE UNITED STATES

Historically, the United States has used the British system of measurements. Most other countries, however, use variations of the metric system. To conform with the rest of the world, and to increase the international competitiveness, productivity, and quality of U.S. industry, the U.S. Congress enacted the Metric Conversion Act of 1975. A version of the metric system called *Le Système International d'Unites* (International System of Units), or the SI system, was adopted. Furthermore, in 1988, the U.S. Congress passed the *Omnibus Trade and Competitiveness Act,* which resulted in the formation of the *Construction Metrication Council.* This council is part of the National Institute of Building Sciences (NIBS) located in Washington, D.C. The Council publishes a newsletter, *Construction Metrication,* which provides the latest efforts on system conversions. You can obtain a copy of this newsletter from the NIBS Web site (www.nibs.org).

The American Concrete Institute has published an equivalent metric version of the ACI Code since 1983. The current metric version of the ACI code is ACI 318M-05 ("M" stands for

TABLE 9–1 Equivalent Soft Metric Designation for Rebars

Bar Size	Equivalent Soft Metric Designation
#3	#10
#4	#13
#5	#16
#6	#19
#7	#22
#8	#25
#9	#29
#10	#32
#11	#36
#14	#43
#18	#57

metric). Adopting the metric system has two major ramifications: (1) using metric units for structural calculations, and (2) changing the physical sizes of products based on the metric system of units. The first task can be accomplished with relative simplicity, as this chapter will show. But the manufacture of construction material for concrete structures, in particular reinforcing bars, has been one of the major obstacles in the adoption of the SI system of units.

To prevent the costly maintenance of two different inventories of steel reinforcement (in British and SI units), the producers of reinforcing bars adopted a *soft metric* conversion in 1997. This conversion allows mills to produce reinforcing bars in the customary British unit sizes, but to designate them with their equivalent metric values instead of multiples of 100 mm^2 as required in the hard metric conversion. As a result, nearly all reinforcing bars currently produced are marked with the soft metric equivalent sizes. Table 9–1 shows the equivalent soft metric bar size designations for the customary British unit sizes.

9.3 CONVERSION TO SI UNITS

A familiarity with the SI units is required to convert British units to their equivalent SI units. Table 9–2 shows the main SI units along with the most common prefixes used in the design of building structures.

Two important quantities that we need to understand well are *mass* and *force.* The SI unit of mass is the *kilogram,* kg, which is used as the unit of force in other versions of the metric system. The SI unit used for force is the *newton,* which is equal to 1 kg-m/sec^2. Force is equal to mass (m) multiplied by the gravitational acceleration (g). Thus,

$$F = m \times g$$
$$F = (1.0 \text{ kg})(9.80665 \text{ m/s}^2)$$
$$F = 9.80665 \text{ kg-m/s}^2 \approx 9.81 \text{ newtons (N)}$$

Therefore, a one-kilogram mass generates 9.81 newtons (N) of force.

TABLE 9–2 Principal SI Units and the Common Prefixes		
Main SI Units		
Quantity	**Unit**	**Symbol (Expression)**
length	meter	m
mass	kilogram	kg
time	second	s
force	newton	N (kg-m/s^2)
stress/pressure	pascal	Pa (N/m^2)
SI Prefixes		
Prefix	**Symbol**	**Value**
micro	μ	10^{-6}
milli	m	10^{-3}
kilo	k	10^3
mega	M	10^6

The British units of *pound-mass* (lbm) and *pound-force* (lbf) are also defined. These are related to each other as follows:

$$F = m \times g$$
$$\text{lbf} = (\text{lbm})(32.174 \text{ ft/s}^2)$$
$$\text{lbf} = 32.174 \text{ lbm-ft/s}^2 \approx 32.2 \text{ lbm-ft/s}^2$$

The relationship between lbm and kg is:

$$1 \text{ lbm} = 0.45359 \text{ kg} \approx 0.454 \text{ kg}$$

The relationship between ft and m is:

$$1 \text{ ft} = 0.3048 \text{ m}$$

Therefore, substituting:

$$1 \text{ lbf} = 32.174 \, (0.45359)(0.3048) \text{ kg-m/s}^2$$
$$1 \text{ lbf} = 4.448 \text{ kg-m/s}^2$$
$$1 \text{ lbf} = 4.448 \text{ N}$$

Note that weight is defined in units of mass. To use weight as load we need to consider the gravitational acceleration of 9.81 m/s^2. The following is an important conversion:

$$1 \frac{\text{lbf}}{\text{ft}^3} = \frac{4.448 \text{ N}}{(0.3048)^3 \text{ m}^3} = 157.1 \text{ N/m}^3$$

This is in units of weight. In the SI units, however, it is defined in units of mass. Therefore:

$$1 \frac{\text{lbm}}{\text{ft}^3} = \frac{0.45359 \text{ kg}}{(0.3048 \text{ m})^3} = 16.02 \text{ kg/m}^3$$

TABLE 9–3   Conversion Factors Between the SI and the British System of Units

Unit	Multiply	By	To Get:
Length	inch (in.)	25.4	millimeter (mm)
	foot (ft)	0.3048	meter (m)
	millimeter (mm)	0.03937	inch (in.)
	meter (m)	3.281	foot (ft)
Area	square inch (in^2)	645.2	square millimeter (mm^2)
	square foot (ft^2)	0.0929	square meter (m^2)
	square millimeter (mm^2)	0.00155	square inch (in^2)
	square meter (m^2)	10.764	square foot (ft^2)
Volume	cubic inch (in^3)	16,387	cubic millimeter (mm^3)
	cubic foot (ft^3)	0.028317	cubic meter (m^3)
	cubic millimeter (mm^3)	0.000061024	cubic inch (in^3)
	cubic meter (m^3)	35.315	cubic foot (ft^3)
Mass	pound-mass (lbm)	0.454	kilogram (kg)
	kilogram (kg)	2.205	pound
Density	pound per cubic foot (lb/ft^3)	16.02	kilogram per cubic meter (kg/m^3)
	kilogram per cubic meter (kg/m^3)	0.06243	pound per cubic foot (lb/ft^3)
Force	pound-force (lbf)	4.448	newton (N)
	kip	4,448	newton (N)
	pound per foot (lb/ft)	14.594	newton per meter (N/m)
	kip per foot (kip/ft)	14.594	kilonewton per meter (kN/m)
	newton (N)	0.2248	pound-force (lbf)
	newton (N)	0.0002248	kip
	newton per meter (N/m)	0.06852	pound per foot (lb/ft)
	kilonewton per meter (kN/m)	0.06852	kip per foot (kip/ft)
Moment of Inertia	inch^4 (in^4)	416,231	millimeter^4 (mm^4)
	millimeter^4 (mm^4)	0.000002403	inch^4 (in^4)
Bending Moment	pound-inch (lb-in.)	0.113	newton-meter (N-m)
	pound-foot (lb-ft)	1.356	newton-meter (N-m)
	kip-inch (kip-in.)	0.113	kilonewton-meter (kN-m)
	kip-foot (kip-ft)	1.356	kilonewton-meter (kN-m)
	newton-meter (N-m)	8.851	pound-inch (lb-in)
	newton-meter (N-m)	0.738	pound-foot (lb-ft)
	kilonewton-meter (kN-m)	8.851	kip-inch (kip-in.)
	kilonewton-meter (kN-m)	0.738	kip-foot (kip-ft)
Pressure, Stress	pound per square inch (psi)	6,895	pascal (Pa)
	kip per square inch (ksi)	6,895	kilopascal (kPa)
		6.895	megapascal (MPa)
	pound per square foot (psf)	47.88	pascal (Pa)
	kip per square foot (ksf)	47.88	kilopascal (kPa)
	pascal (Pa)	0.000145	pound per square inch (psi)
	kilopascal (kPa)	0.14503	pound per square inch (psi)
	megapascal (MPa)	0.14503	kip per square inch (ksi)
	pascal (Pa)	0.020886	pound per square foot (psf)

TABLE 9–4 ASTM Standard Metric Reinforcing Bars

Bar Size Designation	Nominal Dimensions		
	Area (mm²)	Weight (kg/m)	Diameter (mm)
#10	71	0.560	9.5
#13	129	0.994	12.7
#16	199	1.522	15.9
#19	284	2.235	19.1
#22	387	3.042	22.2
#25	510	3.973	25.4
#29	645	5.060	28.7
#32	819	6.404	32.3
#36	1,006	7.907	35.8
#43	1,452	11.38	43.0
#57	2,581	20.24	57.3

For example, the unit mass of concrete (normal weight) is:

$$150\,\text{lbm/ft}^3 = 150(16.02) = 2,400\,\text{kg/m}^3$$

Table 9–3 shows the complete set of conversion factors between the SI and the British systems of units. The following examples solve problems posed by examples in previous chapters using the equivalent SI units. Since we must use the soft metric reinforcing bar sizes, Table 9–4 shows their designations along with their properties.

EXAMPLE 9–1 (SI Version of Example 1–2)

A 75 mm × 150 mm, 2.70 m-long plain concrete beam was simply supported at its ends and tested to determine the modulus of rupture of the concrete. Two concentrated loads, P, were placed at the third points. The beam failed at $P = 670\,\text{N}$. The specified compressive strength of the concrete is $f_c' = 28\,\text{MPa}$. The concrete weight (mass) is 2,400 kg/m³. Determine the modulus of rupture of the concrete using (a) the results of the test, and (b) the ACI Code approximate equation.

Solution

(a) Test Results

Determine the loads acting on the beam shown in Figure 9–1:

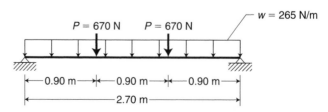

FIGURE 9–1 Example 9–1.

$$w = (2,400) \frac{(75)(150)}{(1,000)(1,000)} = 27.0 \text{ kg/m} \times 9.81 \text{ m/s}^2 = 265 \text{ N/m}$$

$$M_{total} = \frac{w\ell^2}{8} + \frac{P\ell}{3}$$

$$M_{total} = \frac{265(2.70)^2}{8} + \frac{670(2.70)}{3}$$

$$M_{total} = 242 + 603 = 845 \text{ N-m} = 845,000 \text{ N-mm}$$

The maximum tensile stress at the bottom of the beam (f_r) is:

$$f_r = \frac{Mc}{I} = \frac{M}{S_m}$$

$$S_m = \frac{bh^2}{6} = \frac{75(150)^2}{6} = 281,250 \text{ mm}^3$$

$$f_r = \frac{M}{S_m} = \frac{845 \times 1,000}{281,250} = 3.00 \text{ N/mm}^2$$

$$= 3.00 \times 10^6 \text{ N/m}^2 = 3.00 \text{ MPa}$$

(b) ACI Approximate Equation

From Equation 9–10 of ACI 318M-05:

$$f_r = 0.70\sqrt{f_c'}$$

$$f_r = 0.70\sqrt{28} = 3.70 \text{ MPa}$$

EXAMPLE 9–2 (SI Version of Example 2–8)

Calculate M_R for the reinforced concrete section shown in Figure 9–2. Use $f_y = 420$ MPa, and $f_c' = 28$ MPa. $A_s = 6\,\#32 = 4,914 \text{ mm}^2$.

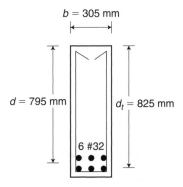

FIGURE 9–2   Sketch of Example 9–2.

Solution

Step 1

$$\rho = \frac{A_s}{bd} = \frac{4,914}{305 \times 795} = 0.0203$$

From Table A2–4:

$$\rho_{min} = 0.0033 < 0.0203 \quad \therefore \text{ ok}$$

From Table A2–3:

$$\rho_{max} = 0.0207 > 0.0203 \quad \therefore \text{ ok}$$

Step 2

$$a = \frac{A_s f_y}{0.85 f'_c b} = \frac{4,914(420)}{0.85 \times 28 \times 305} = 284 \text{ mm}$$

Step 3

$$c = \frac{a}{\beta_1} = \frac{284}{0.85} = 334 \text{ mm}$$

Step 4

$$\frac{c}{d_t} = \frac{334}{825} = 0.405 > 0.375$$

Because $0.405 > 0.375$, the section is in the transition zone:

$$\phi = A_2 + \frac{B_2}{\dfrac{c}{d_t}}$$

$$\phi = 0.233 + \frac{0.25}{0.405}$$

$$\phi = 0.850$$

Step 5

$$\overset{\text{mm}^2 \quad \text{N/mm}^2 \text{(MPa)} \quad \text{mm}}{M_R = \phi A_s f_y \left(d - \frac{a}{2} \right) = (0.850)(4,914)(420) \left(795 - \frac{284}{2} \right)}$$

$$M_R = 1,145,556,594 \text{ N-mm}/10^6 = 1,146 \text{ kN-m}$$

EXAMPLE 9–3 (SI Version of Example 4–2)

Determine the spacing of #10 strirrups at the critical section for a reinforced concrete beam with $b_w = 380$ mm, $h = 610$ mm, and $\overline{V}_u = 270$ kN. Use $f'_c = 21$ MPa, and $f_{yt} = 420$ MPa.

Solution

$$d_{est} = h - 65 = 610 - 65 = 545 \text{ mm}$$

Using Equation 11–3 of ACI 318M-05:

$$V_c = \frac{\sqrt{f_c'}}{6} b_w d$$

$$V_c = \frac{\sqrt{21}}{6}(380)(545)$$

$$V_c = 158{,}175 \text{ N} = 158.2 \text{ kN}$$

The shear force to be resisted by the stirrups at the critical section, $\overline{V}_s$, is:

$$\overline{V}_s = \frac{\overline{V}_u}{\phi} - V_c$$

$$\overline{V}_s = \frac{270}{0.75} - 158.2$$

$$\overline{V}_s = 201.8 \text{ kN}$$

So the required spacing of the bars is:

$$s = \frac{A_v f_{yt} d}{V_s}$$

$$\overline{s} = \frac{(2 \times 71)(420)(545)}{201.8 \times 1000} = 161 \text{ mm}$$

Use $\overline{s} = 160$ mm.

EXAMPLE 9–4 (SI Version of Example 5–4)

Design a short square tied column to carry an axial dead load of 1,300 kN and a live load of 900 kN. Assume that the applied moments on the column are negligible. Use $f_c' = 28$ MPa, $f_y = 420$ MPa, and a concrete clear cover of 40 mm.

Solution

Step 1   The factored load, P_u, is:

$$P_u = 1.2P_D + 1.6P_L$$
$$P_u = 1.2(1{,}300) + 1.6(900)$$
$$P_u = 3{,}000 \text{ kN}$$

Step 2   Assuming $\rho_g = 0.03$, the required area of column, A_g, is:

$$A_g = \frac{P_u}{0.8\phi[0.85f_c'(1 - \rho_g) + f_y\rho_g]}$$

$$A_g = \frac{3{,}000 \times 1000}{0.8(0.65)[0.85(28)(1 - 0.03) + 420(0.03)]}$$

$$A_g = 161{,}667 \text{ mm}^2$$

Step 3   The column size, h, is:

$$h = \sqrt{A_g} = \sqrt{161,667}$$

$$h = 402 \text{ mm} \quad \therefore \text{ Use } h = 400 \text{ mm}$$

Therefore, the column is 400 mm $\times$ 400 mm, and the column gross area, A_g, is:

$$A_g = 400 \times 400 = 160,000 \text{ mm}^2$$

Step 4   The required area of reinforcement, A_{st}, is:

$$A_{st} = \frac{P_u - 0.8\phi(0.85f_c'A_g)}{0.8\phi(f_y - 0.85f_c')}$$

Conversion from MPa to kPa

$$= \frac{3,000 - 0.8 \times 0.65(0.85 \times 28 \times 10^{-3} \times 160,000)}{0.8 \times 0.65(420 \times 10^{-3} - 0.85 \times 28 \times 10^{-3})}$$

$$A_{st} = 4,950 \text{ mm}^2$$

Step 5   Using Table 9–4, select 8 #29 bars ($A_s = 8 \times 645 = 5,160 \text{ mm}^2$).

Step 6   Using #10 for the ties, the maximum spacing, s_{max}, (ACI Code, Section 7.10.5.2) is:

$$s_{max} = \min\{16d_b, 48d_t, b_{min}\}$$

$$s_{max} = \min\{16(28.7), 48(9.5), 400\}$$

$$s_{max} = \min\{459, 456, 400\}$$

$$s_{max} = 400 \text{ mm}$$

Therefore, the ties are #10 @ 400 mm.

Use Figure 5–12 to check the arrangement of the ties. Determine the clear space between the longitudinal bars:

Cover   #10 Ties   #29 Bars

$$\text{Clear space} = \frac{400 - 2(40) - 2(9.5) - 3(28.7)}{2}$$

Clear space = 107 mm < 150 mm

Therefore, one tie per set is enough, as shown in Figure 9–3.

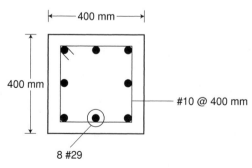

8 #29

FIGURE 9–3   Final design of Example 9–4.

EXAMPLE 9–5 (SI Version of Example 7–1)

A 300 mm load-bearing CMU wall supports an outdoor canopy. The dead load is 150 kN/m (including the wall weight) and the live load is 75 kN/m. Design the plain concrete footing shown in Figure 9–4 to support this wall. The compressive strength of the concrete is 21 MPa, and the net bearing capacity of the soil is 150 kPa. The frost line is at 1.20 m from the outside grade.

Solution

Step 1 Determine the footing width (b).

$$\text{Approximate footing width } (w) = \frac{w_D + w_L}{q_a}$$

$$= \frac{(150 + 75)}{150} = 1.50 \text{ m} \quad \nearrow \text{ kN/m}$$

$$\nwarrow \text{ kPa} = \text{kN/m}^2$$

$$\therefore b = 1.5 \text{ m}$$

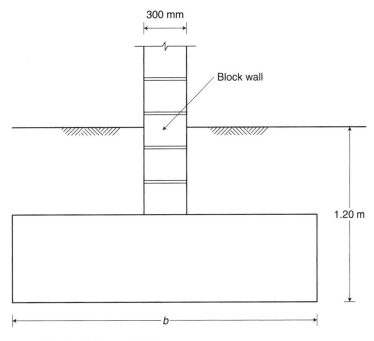

300 mm

Block wall

1.20 m

b

FIGURE 9–4 Sketch of Example 9–5.

Step 2   Estimate the footing thickness (h).

$$h = \frac{b - t}{2} = \frac{1.5 - 0.30}{2} = 0.60 \text{ m}$$

$$\therefore h = 0.60 \text{ m (600 mm)}$$

Step 3   Calculate and check the moment.

$$q_u = \frac{1.2 w_D + 1.6 w_L}{b}$$

$$q_u = \frac{1.2 \times 150 + 1.6 \times 75}{1.5} = 200 \text{ kN/m}^2 \text{ (kPa)}$$

The distance from the edge of the footing to the critical section for moment (D) is:

$$D = \frac{b - t}{2} + \frac{t}{4}$$

$$D = \frac{1.5 \times 1,000 - 300}{2} + \frac{300}{4}$$

$$D = 675 \text{ mm}$$

$$M_u = q_u \frac{D^2}{2} = (200) \frac{\left(\dfrac{675}{1,000}\right)^2}{2}$$

$$M_u = 45.6 \text{ kN-m}$$

$$d = h - 50 = 600 - 50 = 550 \text{ mm}$$

Considering a 1-m (1,000 mm) strip of footing:

$$S_m = \frac{b d^2}{6}$$

$$S_m = \frac{1,000(550)^2}{6} = 50.42 \times 10^6 \text{ mm}^3$$

The nominal resisting moment, M_n, (ACI 318M-05 Equation 22–2) is:

$$M_n = \left(\frac{5}{12}\right) \sqrt{f_c'}\, S_m$$

$$M_n = \frac{5}{12} \sqrt{21} (50.42 \times 10^6)$$

$$M_n = 96.3 \times 10^6 \text{ N-mm}/10^6 = 96.3 \text{ kN-m}$$

$$M_R = \phi M_n = 0.55(96.3)$$

$$M_R = 53.0 \text{ kN-m} > 45.6 \text{ kN-m} \quad \therefore \text{ ok}$$

Figure 9–5 shows the final design of this footing.

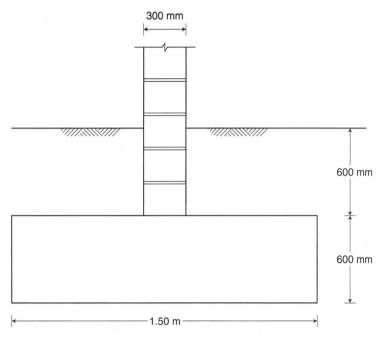

300 mm

600 mm

600 mm

1.50 m

FIGURE 9–5   Final design of Example 9–5.

PROBLEMS

9–1   (SI Version of Problem 1–7) Draw the bending moment and shear force diagrams for a 300 mm × 600 mm concrete beam made of lightweight concrete with a unit weight (mass) of 1,800 kg/m^3 and subjected to a uniformly distributed load of 15 kN/m. Assume that the beam is simply-supported and has a 3.0-m span.

9–2   (SI Version of Problem 1–10) Determine the maximum span for a 200 mm × 300 mm simply-supported plain concrete beam constructed of normal-weight concrete with a unit weight (mass) of 2400 kg/m^3 and loaded by a uniformly distributed load of 30 kN/m just before it fails. The specified compressive strength of the concrete is 28 MPa. Use the ACI Code–recommended value for the modulus of rupture.

9–3   (SI Version of Problem 2–7) The rectangular reinforced concrete beam shown below is subjected to a dead load moment of 250 kN-m and a live load moment of 125 kN-m. Determine whether this beam is adequate for the applied moment using the Method I. Use $f'_c = 28$ MPa and $f_y = 420$ MPa. The stirrups are #10 and the cover is 40 mm.

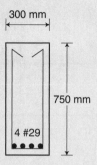

9–4 (SI Version of Problem 2–11) Determine the moment capacity, M_R, of the reinforced concrete section shown below if it is subjected to a negative moment. Use the Method I. The stirrups are #10 and the cover is 40 mm. Use $f'_c = 28$ MPa and $f_y = 420$ MPa.

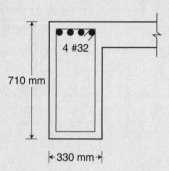

9–5 (SI Version Problem 4–2) A beam is subjected to a uniformly distributed load and has a maximum shear of 270 kN at the face of its supports. The beam clear span is 9.0 m, $b_w = 300$ mm, and $d = 600$ mm. Use $f'_c = 28$ MPa, and $f_{yt} = 420$ MPa. Determine the shear at the critical section. Determine the spacing of #10 stirrups at the critical section.

9–6 (SI Version of Problem 4–5) The shear force at the critical section, $\overline{V}_u$, of a reinforced concrete beam is 265 kN. If the beam has $b_w = 360$ mm, $f'_c = 21$ MPa, and $f_{yt} = 420$ MPa, what is the required effective depth, d, such that the minimum spacing of #10 stirrups is 230 mm?

9–7 (SI Version of Problem 5–4) The square reinforced concrete tied column shown below is subjected to a dead load of 900 kN and a live load of 1,000 kN. Determine whether this column is adequate. The clear cover is 40 mm and the load eccentricity is negligible. Use $f'_c = 28$ MPa and $f_y = 420$ MPa. Checking the ties is not required.

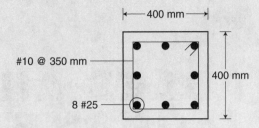

9–8   (SI Version of Problem 5–8) Design a square tied reinforced concrete column subjected to a dead load of 1,100 kN and a live load of 1,300 kN. The moments due to the loads are negligible. Use $f_c' = 28$ MPa, $f_y = 420$ MPa, and 40 mm clear cover.

9–9   (SI Version of Problem 7–1) Design a plain concrete wall footing to support a 300 mm thick concrete wall. The dead load, including the weight of wall, is 70 kN/m, and the live load is 90 kN/m. The bearing capacity of the soil is 120 kPa, and $f_c' = 21$ MPa.

9–10 (SI Version of Problem 7–3) Rework Problem 9–9 for a reinforced concrete wall footing. Use $f_y = 420$ MPa.

A
TABLES AND DIAGRAMS

TABLE A1–1   Mechanical Properties of Steel Reinforcing Bars

Type of Steel	Grade	f_y (ksi)	ϵ_y
Billet, A615	40	40	0.00138
	60	60	0.00207
	75	75	0.00259
Rail, A996	50	50	0.00172
	60	60	0.00207
Axle, A996	40	40	0.00138
	60	60	0.00207
Low Alloy, A706	60	60	0.00207

TABLE A1–2   Steel Bar Sizes

Bar Size	#3	#4	#5	#6	#7	#8	#9	#10	#11	#14	#18
Diameter (in.)	0.375	0.500	0.625	0.750	0.875	1.000	1.128	1.270	1.410	1.693	2.257
Area (in^2)	0.11	0.20	0.31	0.44	0.60	0.79	1.00	1.27	1.56	2.25	4.00

TABLE A2–1 ACI Approximate Design Moments and Shears for Beams and One-Way Slabs

Positive Moment	End spans	
	Discontinuous end unrestrained	$w_u\ell_n^2/11$
	Discontinuous end integral with support	$w_u\ell_n^2/14$
	Interior spans	$w_u\ell_n^2/16$
Negative Moment	At exterior face of the first interior support	
	Two spans	$w_u\ell_n^2/9$
	More than two spans	$w_u\ell_n^2/10$
	At other faces of interior supports	$w_u\ell_n^2/11$
	At the face of all supports for SLABS with spans not exceeding 10 ft; and BEAMS where ratio of sum of column stiffnesses to beam stiffness exceeds 8 at each end of the span	$w_u\ell_n^2/12$
	At interior face of exterior support for members built integrally with supports	
	Where support is a spandrel beam	$w_u\ell_n^2/24$
	Where support is a column	$w_u\ell_n^2/16$
Shear	In end members at the face of the first interior support	$1.15\,w_u\ell_n/2$
	At face of all other supports	$w_u\ell_n/2$

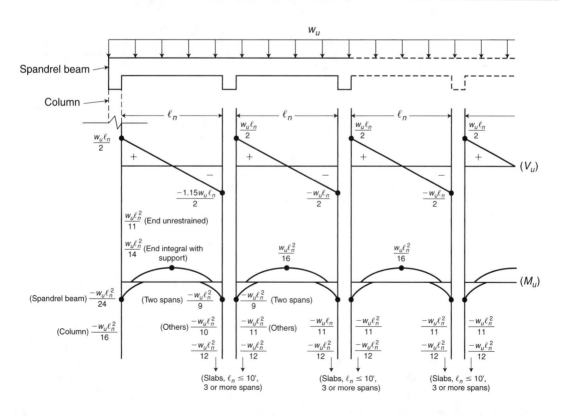

TABLE A2–2a Values of A_1 and B_1 for Commonly Used Reinforcing Steels

f_y (psi)	ϵ_y	A_1	B_1
40,000	0.00138	0.555	69.1
60,000	0.00207	0.473	85.3
75,000	0.00259	0.381	103.7

TABLE A2–2b Values of A_2 and B_2 for Commonly Used Reinforcing Steels

f_y (psi)	d_t/c_b	c_b/d_t	A_2	B_2
40,000	1.460	0.685	0.345	0.208
60,000	1.690	0.592	0.233	0.250
75,000	1.863	0.537	0.067	0.312

TABLE A2–3 ρ_{max} and ρ_{tc} for Common Grades of Steel and Compressive Strength of Concrete (Single Layer of Steel, i.e., $d = d_t$)

f_y (psi)	ρ_{max} ($\epsilon_t = 0.004$)			ϕ
	$f_c' = 3,000$ psi	$f_c' = 4,000$ psi	$f_c' = 5,000$ psi	
40,000	0.0232	0.0310	0.0364	0.83
60,000	0.0155	0.0207	0.0243	0.81
75,000	0.0124	0.0165	0.0194	0.80

f_y (psi)	ρ_{tc} ($\epsilon_t = 0.005$)			ϕ
	$f_c' = 3,000$ psi	$f_c' = 4,000$ psi	$f_c' = 5,000$ psi	
40,000	0.0203	0.0270	0.0318	0.90
60,000	0.0135	0.0180	0.0212	0.90
75,000	0.0108	0.0144	0.0169	0.90

Note: For multiple layers of reinforcements, multiply the table values by $\dfrac{d_t}{d}$.

TABLE A2–4 Minimum Steel Ratio (ρ_{min})

f_y (psi)	ρ_{min}			
	$f_c' = 3,000$ psi	$f_c' = 4,000$ psi	$f_c' = 5,000$ psi	$f_c' = 6,000$ psi
40,000	0.0050	0.0050	0.0053	0.0058
60,000	0.0033	0.0033	0.0035	0.0039
75,000	0.0027	0.0027	0.0028	0.0031

TABLE A2–5a Resistance Coefficient R (in psi) Versus Reinforcement Ratio (ρ); $f'_c = 3{,}000$ psi, $f_y = 40{,}000$ psi (for Beams $\rho_{min} = 0.005$)

ρ	R	ρ	R	ρ	R	ρ	R	ρ	R	ϕ	ρ	R	ϕ
0.001	36	0.0051	176	0.0092	307	0.0133	429	0.0174	541		0.0215	643	0.87
0.0011	39	0.0052	180	0.0093	310	0.0134	432	0.0175	544		0.0216	646	0.87
0.0012	43	0.0053	183	0.0094	313	0.0135	435	0.0176	546		0.0217	648	0.86
0.0013	46	0.0054	186	0.0095	317	0.0136	437	0.0177	549		0.0218	651	0.86
0.0014	50	0.0055	189	0.0096	320	0.0137	440	0.0178	551		0.0219	653	0.86
0.0015	53	0.0056	193	0.0097	323	0.0138	443	0.0179	554		0.022	655	0.86
0.0016	57	0.0057	196	0.0098	326	0.0139	446	0.018	557		0.0221	658	0.86
0.0017	60	0.0058	199	0.0099	329	0.014	449	0.0181	559		0.0222	660	0.85
0.0018	64	0.0059	203	0.01	332	0.0141	451	0.0182	562		0.0223	662	0.85
0.0019	67	0.006	206	0.0101	335	0.0142	454	0.0183	564		0.0224	665	0.85
0.002	71	0.0061	209	0.0102	338	0.0143	457	0.0184	567		0.0225	667	0.85
0.0021	74	0.0062	212	0.0103	341	0.0144	460	0.0185	569		0.0226	669	0.84
0.0022	78	0.0063	216	0.0104	344	0.0145	463	0.0186	572		0.0227	672	0.84
0.0023	81	0.0064	219	0.0105	347	0.0146	465	0.0187	574		0.0228	674	0.84
0.0024	85	0.0065	222	0.0106	350	0.0147	468	0.0188	577		0.0229	676	0.84
0.0025	88	0.0066	225	0.0107	353	0.0148	471	0.0189	580		0.023	679	0.84
0.0026	92	0.0067	229	0.0108	356	0.0149	474	0.019	582		0.0231	681	0.83
0.0027	95	0.0068	232	0.0109	359	0.015	476	0.0191	585		0.0232	683	0.83
0.0028	99	0.0069	235	0.011	362	0.0151	479	0.0192	587				
0.0029	102	0.007	238	0.0111	365	0.0152	482	0.0193	590				
0.003	105	0.0071	241	0.0112	368	0.0153	485	0.0194	592				
0.0031	109	0.0072	245	0.0113	371	0.0154	487	0.0195	595				
0.0032	112	0.0073	248	0.0114	374	0.0155	490	0.0196	597				
0.0033	116	0.0074	251	0.0115	377	0.0156	493	0.0197	600				
0.0034	119	0.0075	254	0.0116	380	0.0157	496	0.0198	602				
0.0035	123	0.0076	257	0.0117	383	0.0158	498	0.0199	605				
0.0036	126	0.0077	260	0.0118	385	0.0159	501	0.02	607				
0.0037	129	0.0078	264	0.0119	388	0.016	504	0.0201	610				
0.0038	133	0.0079	267	0.012	391	0.0161	506	0.0202	612				
0.0039	136	0.008	270	0.0121	394	0.0162	509	0.0203	614	ρ_{tc}			
0.004	139	0.0081	273	0.0122	397	0.0163	512	0.0204	617	0.90			
0.0041	143	0.0082	276	0.0123	400	0.0164	514	0.0205	619	0.90			
0.0042	146	0.0083	279	0.0124	403	0.0165	517	0.0206	622	0.89			
0.0043	150	0.0084	282	0.0125	406	0.0166	520	0.0207	624	0.89			
0.0044	153	0.0085	286	0.0126	409	0.0167	522	0.0208	627	0.89			
0.0045	156	0.0086	289	0.0127	412	0.0168	525	0.0209	629	0.88			
0.0046	160	0.0087	292	0.0128	415	0.0169	528	0.021	631	0.88			
0.0047	163	0.0088	295	0.0129	417	0.017	530	0.0211	634	0.88			
0.0048	166	0.0089	298	0.013	420	0.0171	533	0.0212	636	0.88			
0.0049	170	0.009	301	0.0131	423	0.0172	536	0.0213	639	0.87			
0.005	173	0.0091	304	0.0132	426	0.0173	538	0.0214	641	0.87			

TABLE A2–5b Resistance Coefficient R (in psi) Versus Reinforcement Ratio (ρ); $f'_c = 4{,}000$ psi, $f_y = 40{,}000$ psi (for Beams $\rho_{min} = 0.005$)

ρ	R	ρ	R	ρ	R	ρ	R	ρ	R
0.001	36	0.0051	178	0.0092	313	0.0133	441	0.0174	562
0.0011	39	0.0052	181	0.0093	316	0.0134	444	0.0175	565
0.0012	43	0.0053	185	0.0094	320	0.0135	447	0.0176	568
0.0013	46	0.0054	188	0.0095	323	0.0136	450	0.0177	571
0.0014	50	0.0055	192	0.0096	326	0.0137	453	0.0178	574
0.0015	54	0.0056	195	0.0097	329	0.0138	456	0.0179	577
0.0016	57	0.0057	198	0.0098	332	0.0139	459	0.018	579
0.0017	61	0.0058	202	0.0099	336	0.014	462	0.0181	582
0.0018	64	0.0059	205	0.01	339	0.0141	465	0.0182	585
0.0019	68	0.006	208	0.0101	342	0.0142	468	0.0183	588
0.002	71	0.0061	212	0.0102	345	0.0143	471	0.0184	591
0.0021	75	0.0062	215	0.0103	348	0.0144	474	0.0185	594
0.0022	78	0.0063	218	0.0104	351	0.0145	477	0.0186	596
0.0023	82	0.0064	222	0.0105	355	0.0146	480	0.0187	599
0.0024	85	0.0065	225	0.0106	358	0.0147	483	0.0188	602
0.0025	89	0.0066	228	0.0107	361	0.0148	486	0.0189	605
0.0026	92	0.0067	232	0.0108	364	0.0149	489	0.019	608
0.0027	96	0.0068	235	0.0109	367	0.015	492	0.0191	610
0.0028	99	0.0069	238	0.011	370	0.0151	495	0.0192	613
0.0029	103	0.007	242	0.0111	374	0.0152	498	0.0193	616
0.003	106	0.0071	245	0.0112	377	0.0153	501	0.0194	619
0.0031	110	0.0072	248	0.0113	380	0.0154	504	0.0195	621
0.0032	113	0.0073	252	0.0114	383	0.0155	507	0.0196	624
0.0033	116	0.0074	255	0.0115	386	0.0156	510	0.0197	627
0.0034	120	0.0075	258	0.0116	389	0.0157	513	0.0198	630
0.0035	123	0.0076	261	0.0117	392	0.0158	516	0.0199	633
0.0036	127	0.0077	265	0.0118	395	0.0159	519	0.02	635
0.0037	130	0.0078	268	0.0119	398	0.016	522	0.0201	638
0.0038	134	0.0079	271	0.012	402	0.0161	525	0.0202	641
0.0039	137	0.008	274	0.0121	405	0.0162	528	0.0203	644
0.004	141	0.0081	278	0.0122	408	0.0163	531	0.0204	646
0.0041	144	0.0082	281	0.0123	411	0.0164	533	0.0205	649
0.0042	147	0.0083	284	0.0124	414	0.0165	536	0.0206	652
0.0043	151	0.0084	287	0.0125	417	0.0166	539	0.0207	654
0.0044	154	0.0085	291	0.0126	420	0.0167	542	0.0208	657
0.0045	158	0.0086	294	0.0127	423	0.0168	545	0.0209	660
0.0046	161	0.0087	297	0.0128	426	0.0169	548	0.021	663
0.0047	165	0.0088	300	0.0129	429	0.017	551	0.0211	665
0.0048	168	0.0089	304	0.013	432	0.0171	554	0.0212	668
0.0049	171	0.009	307	0.0131	435	0.0172	557	0.0213	671
0.005	175	0.0091	310	0.0132	438	0.0173	559	0.0214	673

(continued)

TABLE A2–5b *(continued)*

ρ	R	ρ	R	φ	ρ	R	φ
0.0215	676	0.0256	783		0.0297	835	0.85
0.0216	679	0.0257	785		0.0298	835	0.85
0.0217	681	0.0258	788		0.0299	836	0.85
0.0218	684	0.0259	790		0.03	837	0.85
0.0219	687	0.026	793		0.0301	837	0.84
0.022	690	0.0261	795		0.0302	838	0.84
0.0221	692	0.0262	798		0.0303	838	0.84
0.0222	695	0.0263	800		0.0304	839	0.84
0.0223	697	0.0264	803		0.0305	839	0.84
0.0224	700	0.0265	805		0.0306	840	0.84
0.0225	703	0.0266	808		0.0307	840	0.84
0.0226	705	0.0267	810		0.0308	841	0.83
0.0227	708	0.0268	813		0.0309	841	0.83
0.0228	711	0.0269	815		0.031	842	0.83
0.0229	713	0.027	818	ρ_{tc}			
0.023	716	0.0271	820	0.90			
0.0231	719	0.0272	821	0.90			
0.0232	721	0.0273	821	0.90			
0.0233	724	0.0274	822	0.89			
0.0234	726	0.0275	822	0.89			
0.0235	729	0.0276	823	0.89			
0.0236	732	0.0277	824	0.89			
0.0237	734	0.0278	824	0.89			
0.0238	737	0.0279	825	0.88			
0.0239	739	0.028	825	0.88			
0.024	742	0.0281	826	0.88			
0.0241	745	0.0282	826	0.88			
0.0242	747	0.0283	827	0.88			
0.0243	750	0.0284	828	0.87			
0.0244	752	0.0285	828	0.87			
0.0245	755	0.0286	829	0.87			
0.0246	757	0.0287	829	0.87			
0.0247	760	0.0288	830	0.87			
0.0248	763	0.0289	830	0.87			
0.0249	765	0.029	831	0.86			
0.025	768	0.0291	832	0.86			
0.0251	770	0.0292	832	0.86			
0.0252	773	0.0293	833	0.86			
0.0253	775	0.0294	833	0.86			
0.0254	778	0.0295	834	0.85			
0.0255	780	0.0296	834	0.85			

TABLE A2–5c Resistance Coefficient R (in psi) Versus Reinforcement Ratio (ρ); $f'_c = 5{,}000$ psi, $f_y = 40{,}000$ psi (for Beams $\rho_{min} = 0.0053$)

ρ	R	ρ	R	ρ	R	ρ	R	ρ	R
0.001	36	0.0051	179	0.0092	317	0.0133	449	0.0174	575
0.0011	39	0.0052	183	0.0093	320	0.0134	452	0.0175	578
0.0012	43	0.0053	186	0.0094	323	0.0135	455	0.0176	581
0.0013	47	0.0054	189	0.0095	327	0.0136	458	0.0177	584
0.0014	50	0.0055	193	0.0096	330	0.0137	461	0.0178	587
0.0015	54	0.0056	196	0.0097	333	0.0138	465	0.0179	590
0.0016	57	0.0057	200	0.0098	337	0.0139	468	0.018	593
0.0017	61	0.0058	203	0.0099	340	0.014	471	0.0181	596
0.0018	64	0.0059	207	0.01	343	0.0141	474	0.0182	599
0.0019	68	0.006	210	0.0101	346	0.0142	477	0.0183	602
0.002	71	0.0061	213	0.0102	350	0.0143	480	0.0184	605
0.0021	75	0.0062	217	0.0103	353	0.0144	483	0.0185	608
0.0022	78	0.0063	220	0.0104	356	0.0145	486	0.0186	611
0.0023	82	0.0064	223	0.0105	359	0.0146	489	0.0187	614
0.0024	85	0.0065	227	0.0106	363	0.0147	493	0.0188	617
0.0025	89	0.0066	230	0.0107	366	0.0148	496	0.0189	620
0.0026	92	0.0067	234	0.0108	369	0.0149	499	0.019	623
0.0027	96	0.0068	237	0.0109	372	0.015	502	0.0191	626
0.0028	99	0.0069	240	0.011	376	0.0151	505	0.0192	629
0.0029	103	0.007	244	0.0111	379	0.0152	508	0.0193	632
0.003	106	0.0071	247	0.0112	382	0.0153	511	0.0194	635
0.0031	110	0.0072	250	0.0113	385	0.0154	514	0.0195	638
0.0032	113	0.0073	254	0.0114	388	0.0155	517	0.0196	641
0.0033	117	0.0074	257	0.0115	392	0.0156	520	0.0197	643
0.0034	120	0.0075	260	0.0116	395	0.0157	523	0.0198	646
0.0035	124	0.0076	264	0.0117	398	0.0158	527	0.0199	649
0.0036	127	0.0077	267	0.0118	401	0.0159	530	0.02	652
0.0037	131	0.0078	270	0.0119	404	0.016	533	0.0201	655
0.0038	134	0.0079	274	0.012	408	0.0161	536	0.0202	658
0.0039	138	0.008	277	0.0121	411	0.0162	539	0.0203	661
0.004	141	0.0081	280	0.0122	414	0.0163	542	0.0204	664
0.0041	145	0.0082	284	0.0123	417	0.0164	545	0.0205	667
0.0042	148	0.0083	287	0.0124	420	0.0165	548	0.0206	670
0.0043	152	0.0084	290	0.0125	424	0.0166	551	0.0207	673
0.0044	155	0.0085	294	0.0126	427	0.0167	554	0.0208	676
0.0045	159	0.0086	297	0.0127	430	0.0168	557	0.0209	678
0.0046	162	0.0087	300	0.0128	433	0.0169	560	0.021	681
0.0047	165	0.0088	304	0.0129	436	0.017	563	0.0211	684
0.0048	169	0.0089	307	0.013	439	0.0171	566	0.0212	687
0.0049	172	0.009	310	0.0131	443	0.0172	569	0.0213	690
0.005	176	0.0091	314	0.0132	446	0.0173	572	0.0214	693

(continued)

TABLE A2–5c *(continued)*

ρ	R	ρ	R	ρ	R	φ	ρ	R	φ
0.0215	696	0.0256	811	0.0297	920		0.0338	987	0.87
0.0216	699	0.0257	813	0.0298	922		0.0339	988	0.87
0.0217	701	0.0258	816	0.0299	925		0.034	989	0.87
0.0218	704	0.0259	819	0.03	928		0.0341	989	0.86
0.0219	707	0.026	821	0.0301	930		0.0342	990	0.86
0.022	710	0.0261	824	0.0302	933		0.0343	991	0.86
0.0221	713	0.0262	827	0.0303	935		0.0344	991	0.86
0.0222	716	0.0263	830	0.0304	938		0.0345	992	0.86
0.0223	719	0.0264	832	0.0305	940		0.0346	992	0.86
0.0224	721	0.0265	835	0.0306	943		0.0347	993	0.86
0.0225	724	0.0266	838	0.0307	946		0.0348	994	0.85
0.0226	727	0.0267	840	0.0308	948		0.0349	994	0.85
0.0227	730	0.0268	843	0.0309	951		0.035	995	0.85
0.0228	733	0.0269	846	0.031	953		0.0351	995	0.85
0.0229	736	0.027	848	0.0311	956		0.0352	996	0.85
0.023	738	0.0271	851	0.0312	958		0.0353	997	0.85
0.0231	741	0.0272	854	0.0313	961		0.0354	997	0.84
0.0232	744	0.0273	857	0.0314	963		0.0355	998	0.84
0.0233	747	0.0274	859	0.0315	966		0.0356	998	0.84
0.0234	750	0.0275	862	0.0316	968		0.0357	999	0.84
0.0235	752	0.0276	865	0.0317	971		0.0358	1000	0.84
0.0236	755	0.0277	867	0.0318	973	ρ_{tc}	0.0359	1000	0.84
0.0237	758	0.0278	870	0.0319	976	0.90	0.036	1001	0.84
0.0238	761	0.0279	873	0.032	976	0.90	0.0361	1001	0.84
0.0239	764	0.028	875	0.0321	977	0.90	0.0362	1002	0.83
0.024	766	0.0281	878	0.0322	977	0.89	0.0363	1002	0.83
0.0241	769	0.0282	880	0.0323	978	0.89	0.0364	1003	0.83
0.0242	772	0.0283	883	0.0324	979	0.89			
0.0243	775	0.0284	886	0.0325	979	0.89			
0.0244	778	0.0285	888	0.0326	980	0.89			
0.0245	780	0.0286	891	0.0327	981	0.89			
0.0246	783	0.0287	894	0.0328	981	0.88			
0.0247	786	0.0288	896	0.0329	982	0.88			
0.0248	789	0.0289	899	0.033	983	0.88			
0.0249	791	0.029	902	0.0331	983	0.88			
0.025	794	0.0291	904	0.0332	984	0.88			
0.0251	797	0.0292	907	0.0333	984	0.88			
0.0252	800	0.0293	909	0.0334	985	0.87			
0.0253	802	0.0294	912	0.0335	986	0.87			
0.0254	805	0.0295	915	0.0336	986	0.87			
0.0255	808	0.0296	917	0.0337	987	0.87			

TABLE A2–6a Resistance Coefficient R (in psi) Versus Reinforcement Ratio (ρ); $f'_c = 3{,}000$ psi, $f_y = 60{,}000$ psi (for Beams $\rho_{min} = 0.0033$)

ρ	R	ρ	R	ρ	R	ρ	R	ϕ
0.001	53	0.0051	259	0.0092	443	0.0133	606	
0.0011	59	0.0052	264	0.0093	447	0.0134	610	
0.0012	64	0.0053	268	0.0094	451	0.0135	613	ρ_{tc}
0.0013	69	0.0054	273	0.0095	456	0.0136	615	0.90
0.0014	74	0.0055	278	0.0096	460	0.0137	615	0.89
0.0015	80	0.0056	282	0.0097	464	0.0138	616	0.89
0.0016	85	0.0057	287	0.0098	468	0.0139	616	0.88
0.0017	90	0.0058	292	0.0099	472	0.014	616	0.88
0.0018	95	0.0059	296	0.01	476	0.0141	616	0.87
0.0019	100	0.006	301	0.0101	481	0.0142	616	0.87
0.002	105	0.0061	306	0.0102	485	0.0143	617	0.86
0.0021	111	0.0062	310	0.0103	489	0.0144	617	0.86
0.0022	116	0.0063	315	0.0104	493	0.0145	617	0.86
0.0023	121	0.0064	320	0.0105	497	0.0146	617	0.85
0.0024	126	0.0065	324	0.0106	501	0.0147	617	0.85
0.0025	131	0.0066	329	0.0107	505	0.0148	618	0.84
0.0026	136	0.0067	333	0.0108	509	0.0149	618	0.84
0.0027	141	0.0068	338	0.0109	513	0.015	618	0.83
0.0028	146	0.0069	342	0.011	517	0.0151	618	0.83
0.0029	151	0.007	347	0.0111	521	0.0152	618	0.83
0.003	156	0.0071	351	0.0112	525	0.0153	619	0.82
0.0031	161	0.0072	356	0.0113	529	0.0154	619	0.82
0.0032	166	0.0073	360	0.0114	533	0.0155	619	0.81
0.0033	171	0.0074	365	0.0115	537			
0.0034	176	0.0075	369	0.0116	541			
0.0035	181	0.0076	374	0.0117	545			
0.0036	186	0.0077	378	0.0118	549			
0.0037	191	0.0078	383	0.0119	553			
0.0038	196	0.0079	387	0.012	557			
0.0039	201	0.008	391	0.0121	560			
0.004	206	0.0081	396	0.0122	564			
0.0041	211	0.0082	400	0.0123	568			
0.0042	216	0.0083	404	0.0124	572			
0.0043	220	0.0084	409	0.0125	576			
0.0044	225	0.0085	413	0.0126	580			
0.0045	230	0.0086	417	0.0127	583			
0.0046	235	0.0087	422	0.0128	587			
0.0047	240	0.0088	426	0.0129	591			
0.0048	245	0.0089	430	0.013	595			
0.0049	249	0.009	435	0.0131	598			
0.005	254	0.0091	439	0.0132	602			

TABLE A2–6b Resistance Coefficient R (in psi) Versus Reinforcement Ratio (ρ); $f'_c = 4,000$ psi, $f_y = 60,000$ psi (for Beams $\rho_{min} = 0.0033$)

ρ	R	ρ	R	ρ	R	ρ	R	ρ	R	ϕ
0.001	54	0.0051	263	0.0092	456	0.0133	634	0.0174	795	
0.0011	59	0.0052	268	0.0093	461	0.0134	638	0.0175	799	
0.0012	64	0.0053	273	0.0094	465	0.0135	642	0.0176	803	
0.0013	69	0.0054	278	0.0095	470	0.0136	646	0.0177	807	
0.0014	75	0.0055	283	0.0096	474	0.0137	650	0.0178	810	
0.0015	80	0.0056	287	0.0097	479	0.0138	654	0.0179	814	
0.0016	85	0.0057	292	0.0098	483	0.0139	659	0.018	818	ρ_{tc}
0.0017	90	0.0058	297	0.0099	488	0.014	663	0.0181	820	0.90
0.0018	96	0.0059	302	0.01	492	0.0141	667	0.0182	820	0.89
0.0019	101	0.006	307	0.0101	497	0.0142	671	0.0183	820	0.89
0.002	106	0.0061	312	0.0102	501	0.0143	675	0.0184	821	0.89
0.0021	111	0.0062	316	0.0103	506	0.0144	679	0.0185	821	0.88
0.0022	116	0.0063	321	0.0104	510	0.0145	683	0.0186	821	0.88
0.0023	122	0.0064	326	0.0105	514	0.0146	687	0.0187	821	0.88
0.0024	127	0.0065	331	0.0106	519	0.0147	691	0.0188	822	0.87
0.0025	132	0.0066	336	0.0107	523	0.0148	695	0.0189	822	0.87
0.0026	137	0.0067	340	0.0108	528	0.0149	699	0.019	822	0.87
0.0027	142	0.0068	345	0.0109	532	0.015	703	0.0191	822	0.86
0.0028	147	0.0069	350	0.011	536	0.0151	707	0.0192	822	0.86
0.0029	153	0.007	355	0.0111	541	0.0152	711	0.0193	823	0.86
0.003	158	0.0071	359	0.0112	545	0.0153	715	0.0194	823	0.85
0.0031	163	0.0072	364	0.0113	549	0.0154	719	0.0195	823	0.85
0.0032	168	0.0073	369	0.0114	554	0.0155	723	0.0196	823	0.85
0.0033	173	0.0074	374	0.0115	558	0.0156	726	0.0197	823	0.84
0.0034	178	0.0075	378	0.0116	562	0.0157	730	0.0198	824	0.84
0.0035	183	0.0076	383	0.0117	567	0.0158	734	0.0199	824	0.84
0.0036	188	0.0077	388	0.0118	571	0.0159	738	0.02	824	0.83
0.0037	193	0.0078	392	0.0119	575	0.016	742	0.0201	824	0.83
0.0038	198	0.0079	397	0.012	579	0.0161	746	0.0202	824	0.83
0.0039	203	0.008	402	0.0121	584	0.0162	750	0.0203	825	0.82
0.004	208	0.0081	406	0.0122	588	0.0163	754	0.0204	825	0.82
0.0041	213	0.0082	411	0.0123	592	0.0164	757	0.0205	825	0.82
0.0042	218	0.0083	415	0.0124	596	0.0165	761	0.0206	825	0.82
0.0043	223	0.0084	420	0.0125	601	0.0166	765	0.0207	825	0.81
0.0044	228	0.0085	425	0.0126	605	0.0167	769			
0.0045	233	0.0086	429	0.0127	609	0.0168	773			
0.0046	238	0.0087	434	0.0128	613	0.0169	777			
0.0047	243	0.0088	438	0.0129	617	0.017	780			
0.0048	248	0.0089	443	0.013	621	0.0171	784			
0.0049	253	0.009	447	0.0131	626	0.0172	788			
0.005	258	0.0091	452	0.0132	630	0.0173	792			

TABLE A2–6c   Resistance Coefficient R (in psi) Versus Reinforcement Ratio (ρ); $f_c' = 5,000$ psi, $f_y = 60,000$ psi (for Beams $\rho_{min} = 0.0035$)

ρ	R	ρ	R	ρ	R	ρ	R	ρ	R	ϕ	ρ	R	ϕ
0.001	54	0.0051	265	0.0092	465	0.0133	651	0.0174	824		0.0215	976	0.89
0.0011	59	0.0052	270	0.0093	469	0.0134	655	0.0175	828		0.0216	976	0.89
0.0012	64	0.0053	275	0.0094	474	0.0135	660	0.0176	832		0.0217	977	0.89
0.0013	70	0.0054	280	0.0095	479	0.0136	664	0.0177	836		0.0218	977	0.88
0.0014	75	0.0055	285	0.0096	483	0.0137	668	0.0178	840		0.0219	977	0.88
0.0015	80	0.0056	290	0.0097	488	0.0138	673	0.0179	844		0.022	978	0.88
0.0016	85	0.0057	295	0.0098	493	0.0139	677	0.018	848		0.0221	978	0.87
0.0017	91	0.0058	300	0.0099	497	0.014	681	0.0181	853		0.0222	978	0.87
0.0018	96	0.0059	305	0.01	502	0.0141	686	0.0182	857		0.0223	978	0.87
0.0019	101	0.006	310	0.0101	507	0.0142	690	0.0183	861		0.0224	979	0.86
0.002	106	0.0061	315	0.0102	511	0.0143	694	0.0184	865		0.0225	979	0.86
0.0021	112	0.0062	320	0.0103	516	0.0144	699	0.0185	869		0.0226	979	0.86
0.0022	117	0.0063	325	0.0104	520	0.0145	703	0.0186	873		0.0227	980	0.86
0.0023	122	0.0064	330	0.0105	525	0.0146	707	0.0187	877		0.0228	980	0.85
0.0024	127	0.0065	335	0.0106	530	0.0147	711	0.0188	880		0.0229	980	0.85
0.0025	133	0.0066	340	0.0107	534	0.0148	716	0.0189	884		0.023	980	0.85
0.0026	138	0.0067	345	0.0108	539	0.0149	720	0.019	888		0.0231	981	0.85
0.0027	143	0.0068	350	0.0109	543	0.015	724	0.0191	892		0.0232	981	0.84
0.0028	148	0.0069	354	0.011	548	0.0151	728	0.0192	896		0.0233	981	0.84
0.0029	153	0.007	359	0.0111	552	0.0152	733	0.0193	900		0.0234	981	0.84
0.003	159	0.0071	364	0.0112	557	0.0153	737	0.0194	904		0.0235	982	0.83
0.0031	164	0.0072	369	0.0113	562	0.0154	741	0.0195	908		0.0236	982	0.83
0.0032	169	0.0073	374	0.0114	566	0.0155	745	0.0196	912		0.0237	982	0.83
0.0033	174	0.0074	379	0.0115	571	0.0156	750	0.0197	916		0.0238	982	0.83
0.0034	179	0.0075	384	0.0116	575	0.0157	754	0.0198	920		0.0239	983	0.82
0.0035	184	0.0076	388	0.0117	580	0.0158	758	0.0199	924		0.024	983	0.82
0.0036	189	0.0077	393	0.0118	584	0.0159	762	0.02	928		0.0241	983	0.82
0.0037	195	0.0078	398	0.0119	589	0.016	766	0.0201	931		0.0242	983	0.82
0.0038	200	0.0079	403	0.012	593	0.0161	771	0.0202	935		0.0243	984	0.81
0.0039	205	0.008	408	0.0121	598	0.0162	775	0.0203	939		0.0244	984	0.81
0.004	210	0.0081	412	0.0122	602	0.0163	779	0.0204	943				
0.0041	215	0.0082	417	0.0123	607	0.0164	783	0.0205	947				
0.0042	220	0.0083	422	0.0124	611	0.0165	787	0.0206	951				
0.0043	225	0.0084	427	0.0125	615	0.0166	791	0.0207	954				
0.0044	230	0.0085	431	0.0126	620	0.0167	795	0.0208	958				
0.0045	235	0.0086	436	0.0127	624	0.0168	800	0.0209	962				
0.0046	240	0.0087	441	0.0128	629	0.0169	804	0.021	966				
0.0047	245	0.0088	446	0.0129	633	0.017	808	0.0211	970				
0.0048	250	0.0089	450	0.013	638	0.0171	812	0.0212	973	ρ_{tc}			
0.0049	255	0.009	455	0.0131	642	0.0172	816	0.0213	976	0.90			
0.005	260	0.0091	460	0.0132	646	0.0173	820	0.0214	976	0.90			

TABLE A2–7a   Resistance Coefficient R (in psi) Versus Reinforcement Ratio (ρ); $f_c' = 3{,}000$ psi, $f_y = 75{,}000$ psi (for Beams $\rho_{min} = 0.0027$)

ρ	R	ρ	R	ρ	R	ϕ
0.001	67	0.0051	318	0.0092	537	
0.0011	73	0.0052	324	0.0093	542	
0.0012	80	0.0053	330	0.0094	547	
0.0013	86	0.0054	336	0.0095	552	
0.0014	93	0.0055	341	0.0096	557	
0.0015	99	0.0056	347	0.0097	561	
0.0016	105	0.0057	352	0.0098	566	
0.0017	112	0.0058	358	0.0099	571	
0.0018	118	0.0059	364	0.01	576	
0.0019	125	0.006	369	0.0101	580	
0.002	131	0.0061	375	0.0102	585	
0.0021	137	0.0062	380	0.0103	590	
0.0022	144	0.0063	386	0.0104	595	
0.0023	150	0.0064	391	0.0105	599	
0.0024	156	0.0065	397	0.0106	604	
0.0025	163	0.0066	402	0.0107	609	
0.0026	169	0.0067	408	0.0108	613	ρ_{tc}
0.0027	175	0.0068	413	0.0109	615	0.90
0.0028	181	0.0069	418	0.011	614	0.89
0.0029	187	0.007	424	0.0111	613	0.88
0.003	194	0.0071	429	0.0112	613	0.87
0.0031	200	0.0072	435	0.0113	612	0.87
0.0032	206	0.0073	440	0.0114	611	0.86
0.0033	212	0.0074	445	0.0115	611	0.85
0.0034	218	0.0075	450	0.0116	610	0.85
0.0035	224	0.0076	456	0.0117	609	0.84
0.0036	230	0.0077	461	0.0118	609	0.83
0.0037	236	0.0078	466	0.0119	608	0.83
0.0038	242	0.0079	471	0.012	608	0.82
0.0039	248	0.008	476	0.0121	607	0.81
0.004	254	0.0081	482	0.0122	606	0.81
0.0041	260	0.0082	487	0.0123	606	0.80
0.0042	266	0.0083	492	0.0124	605	0.80
0.0043	272	0.0084	497			
0.0044	278	0.0085	502			
0.0045	284	0.0086	507			
0.0046	289	0.0087	512			
0.0047	295	0.0088	517			
0.0048	301	0.0089	522			
0.0049	307	0.009	527			
0.005	313	0.0091	532			

TABLE A2–7b Resistance Coefficient R (in psi) Versus Reinforcement Ratio (ρ); $f'_c = 4,000$ psi, $f_y = 75,000$ psi (for Beams $\rho_{min} = 0.0027$)

ρ	R	ρ	R	ρ	R	ρ	R	ϕ
0.001	67	0.0051	325	0.0092	558	0.0133	766	
0.0011	73	0.0052	331	0.0093	563	0.0134	771	
0.0012	80	0.0053	337	0.0094	569	0.0135	776	
0.0013	86	0.0054	343	0.0095	574	0.0136	780	
0.0014	93	0.0055	349	0.0096	579	0.0137	785	
0.0015	100	0.0056	355	0.0097	585	0.0138	790	
0.0016	106	0.0057	361	0.0098	590	0.0139	794	
0.0017	113	0.0058	366	0.0099	595	0.014	799	
0.0018	119	0.0059	372	0.01	601	0.0141	804	
0.0019	126	0.006	378	0.0101	606	0.0142	808	
0.002	132	0.0061	384	0.0102	611	0.0143	813	
0.0021	138	0.0062	390	0.0103	616	0.0144	818	ρ_{tc}
0.0022	145	0.0063	396	0.0104	621	0.0145	820	0.90
0.0023	151	0.0064	402	0.0105	627	0.0146	819	0.89
0.0024	158	0.0065	407	0.0106	632	0.0147	818	0.89
0.0025	164	0.0066	413	0.0107	637	0.0148	818	0.88
0.0026	170	0.0067	419	0.0108	642	0.0149	817	0.87
0.0027	177	0.0068	425	0.0109	647	0.015	816	0.87
0.0028	183	0.0069	430	0.011	652	0.0151	816	0.86
0.0029	189	0.007	436	0.0111	658	0.0152	815	0.86
0.003	196	0.0071	442	0.0112	663	0.0153	815	0.85
0.0031	202	0.0072	447	0.0113	668	0.0154	814	0.85
0.0032	208	0.0073	453	0.0114	673	0.0155	813	0.84
0.0033	215	0.0074	459	0.0115	678	0.0156	813	0.84
0.0034	221	0.0075	464	0.0116	683	0.0157	812	0.83
0.0035	227	0.0076	470	0.0117	688	0.0158	811	0.83
0.0036	233	0.0077	476	0.0118	693	0.0159	811	0.82
0.0037	240	0.0078	481	0.0119	698	0.016	810	0.82
0.0038	246	0.0079	487	0.012	703	0.0161	809	0.82
0.0039	252	0.008	492	0.0121	708	0.0162	809	0.81
0.004	258	0.0081	498	0.0122	713	0.0163	808	0.81
0.0041	264	0.0082	503	0.0123	718	0.0164	808	0.80
0.0042	270	0.0083	509	0.0124	723	0.0165	807	0.80
0.0043	276	0.0084	514	0.0125	727			
0.0044	283	0.0085	520	0.0126	732			
0.0045	289	0.0086	525	0.0127	737			
0.0046	295	0.0087	531	0.0128	742			
0.0047	301	0.0088	536	0.0129	747			
0.0048	307	0.0089	542	0.013	752			
0.0049	313	0.009	547	0.0131	756			
0.005	319	0.0091	553	0.0132	761			

TABLE A2–7c Resistance Coefficient R (in psi) Versus Reinforcement Ratio (ρ); f'_c = 5,000 psi, f_y = 75,000 psi (for Beams ρ_{min} = 0.0028)

ρ	R	ρ	R	ρ	R	ρ	R	ϕ	ρ	R	ϕ
0.001	67	0.0051	329	0.0092	571	0.0133	792		0.0174	973	0.88
0.0011	74	0.0052	335	0.0093	576	0.0134	798		0.0175	973	0.88
0.0012	80	0.0053	341	0.0094	582	0.0135	803		0.0176	972	0.87
0.0013	87	0.0054	347	0.0095	587	0.0136	808		0.0177	971	0.87
0.0014	93	0.0055	353	0.0096	593	0.0137	813		0.0178	971	0.86
0.0015	100	0.0056	359	0.0097	599	0.0138	818		0.0179	970	0.86
0.0016	106	0.0057	365	0.0098	604	0.0139	823		0.018	970	0.85
0.0017	113	0.0058	371	0.0099	610	0.014	828		0.0181	969	0.85
0.0018	120	0.0059	378	0.01	615	0.0141	833		0.0182	969	0.85
0.0019	126	0.006	384	0.0101	621	0.0142	838		0.0183	968	0.84
0.002	133	0.0061	390	0.0102	627	0.0143	843		0.0184	967	0.84
0.0021	139	0.0062	396	0.0103	632	0.0144	848		0.0185	967	0.83
0.0022	146	0.0063	402	0.0104	638	0.0145	854		0.0186	966	0.83
0.0023	152	0.0064	408	0.0105	643	0.0146	859		0.0187	966	0.82
0.0024	159	0.0065	414	0.0106	649	0.0147	864		0.0188	965	0.82
0.0025	165	0.0066	420	0.0107	654	0.0148	869		0.0189	965	0.82
0.0026	171	0.0067	426	0.0108	660	0.0149	874		0.019	964	0.81
0.0027	178	0.0068	431	0.0109	665	0.015	878		0.0191	963	0.81
0.0028	184	0.0069	437	0.011	670	0.0151	883		0.0192	963	0.81
0.0029	191	0.007	443	0.0111	676	0.0152	888		0.0193	962	0.80
0.003	197	0.0071	449	0.0112	681	0.0153	893		0.0194	962	0.80
0.0031	204	0.0072	455	0.0113	687	0.0154	898				
0.0032	210	0.0073	461	0.0114	692	0.0155	903				
0.0033	216	0.0074	467	0.0115	697	0.0156	908				
0.0034	223	0.0075	473	0.0116	703	0.0157	913				
0.0035	229	0.0076	479	0.0117	708	0.0158	918				
0.0036	235	0.0077	484	0.0118	714	0.0159	923				
0.0037	242	0.0078	490	0.0119	719	0.016	928				
0.0038	248	0.0079	496	0.012	724	0.0161	932				
0.0039	254	0.008	502	0.0121	730	0.0162	937				
0.004	260	0.0081	508	0.0122	735	0.0163	942				
0.0041	267	0.0082	513	0.0123	740	0.0164	947				
0.0042	273	0.0083	519	0.0124	745	0.0165	952				
0.0043	279	0.0084	525	0.0125	751	0.0166	956				
0.0044	285	0.0085	531	0.0126	756	0.0167	961				
0.0045	292	0.0086	536	0.0127	761	0.0168	966				
0.0046	298	0.0087	542	0.0128	766	0.0169	971	ρ_{tc}			
0.0047	304	0.0088	548	0.0129	772	0.017	975	0.90			
0.0048	310	0.0089	554	0.013	777	0.0171	975	0.90			
0.0049	316	0.009	559	0.0131	782	0.0172	974	0.89			
0.005	323	0.0091	565	0.0132	787	0.0173	974	0.89			

TABLE A2–8 b_{min} and b_{max} for Reinforced Concrete Beams (in.)

Number of Bars in Single Layer	b_{min}								b_{max}
	#3 or #4	#5	#6	#7	#8	#9	#10	#11	
2	6.0	6.0	6.5	7.0	7.0	7.5	8.0	8.5	14.0
3	7.5	8.0	8.0	9.0	9.0	10.0	10.5	11.0	24.0
4	9.0	9.5	10.0	10.5	11.0	12.0	13.0	14.0	34.0
5	10.5	11.0	11.5	12.5	13.0	14.5	15.5	17.0	44.0
6	12.0	12.5	13.5	14.5	15.0	16.5	18.0	19.5	54.0
7	13.5	14.5	15.0	16.5	17.0	19.0	20.5	22.5	64.0
8	15.0	16.0	17.0	18.0	19.0	21.0	23.0	25.5	74.0
9	16.5	17.5	18.5	20.0	21.0	23.5	26.0	28.0	84.0
10	18.0	19.0	20.5	22.0	23.0	25.5	28.5	31.0	94.0

TABLE A2–9 Areas of Multiple Reinforcing Bars (in^2)

Number of Bars	Bar Size								
	#3	#4	#5	#6	#7	#8	#9	#10	#11
1	0.11	0.20	0.31	0.44	0.60	0.79	1.00	1.27	1.56
2	0.22	0.40	0.62	0.88	1.20	1.58	2.00	2.54	3.12
3	0.33	0.60	0.93	1.32	1.80	2.37	3.00	3.81	4.68
4	0.44	0.80	1.24	1.76	2.40	3.16	4.00	5.08	6.24
5	0.55	1.00	1.55	2.20	3.00	3.95	5.00	6.35	7.80
6	0.66	1.20	1.86	2.64	3.60	4.74	6.00	7.62	9.36
7	0.77	1.40	2.17	3.08	4.20	5.53	7.00	8.89	10.92
8	0.88	1.60	2.48	3.52	4.80	6.32	8.00	10.16	12.48
9	0.99	1.80	2.79	3.96	5.40	7.11	9.00	11.43	14.04
10	1.10	2.00	3.10	4.40	6.00	7.90	10.00	12.70	15.60
11	1.21	2.20	3.41	4.84	6.60	8.69	11.00	13.97	17.16
12	1.32	2.40	3.72	5.28	7.20	9.48	12.00	15.24	18.72
13	1.43	2.60	4.03	5.72	7.80	10.27	13.00	16.51	20.28
14	1.54	2.80	4.34	6.16	8.40	11.06	14.00	17.78	21.84
15	1.65	3.00	4.65	6.60	9.00	11.85	15.00	19.05	23.40
16	1.76	3.20	4.96	7.04	9.60	12.64	16.00	20.32	24.96
17	1.87	3.40	5.27	7.48	10.20	13.43	17.00	21.59	26.52
18	1.98	3.60	5.58	7.92	10.80	14.22	18.00	22.86	28.08
19	2.09	3.80	5.89	8.36	11.40	15.01	19.00	24.13	29.64
20	2.20	4.00	6.20	8.80	12.00	15.80	20.00	25.40	31.20

TABLE A2–10   Areas of Reinforcement in One-Foot-Wide Sections

Areas of steel are given in square inches for one-foot-wide sections of concrete (slabs, walls, footings) for various center-to-center spacings of reinforcing bars.

Spacing (inches)	Bar Sizes								
	#3	#4	#5	#6	#7	#8	#9	#10	#11
3	0.44	0.80	1.24	1.76	2.40	3.16	4.00	5.08	6.24
4	0.33	0.60	0.93	1.32	1.80	2.37	3.00	3.81	4.68
5	0.26	0.48	0.74	1.06	1.44	1.90	2.40	3.05	3.74
6	0.22	0.40	0.62	0.88	1.20	1.58	2.00	2.54	3.12
7	0.19	0.34	0.53	0.75	1.03	1.35	1.71	2.18	2.67
8	0.17	0.30	0.47	0.66	0.90	1.19	1.50	1.91	2.34
9	0.15	0.27	0.41	0.59	0.80	1.05	1.33	1.69	2.08
10	0.13	0.24	0.37	0.53	0.72	0.95	1.20	1.52	1.87
11	0.12	0.22	0.34	0.48	0.65	0.86	1.09	1.39	1.70
12	0.11	0.20	0.31	0.44	0.60	0.79	1.00	1.27	1.56
13	0.10	0.18	0.29	0.41	0.55	0.73	0.92	1.17	1.44
14	0.09	0.17	0.27	0.38	0.51	0.68	0.86	1.09	1.34
15	0.09	0.16	0.25	0.35	0.48	0.63	0.80	1.02	1.25
16	0.08	0.15	0.23	0.33	0.45	0.59	0.75	0.95	1.17
17	0.08	0.14	0.22	0.31	0.42	0.56	0.71	0.90	1.10
18	0.07	0.13	0.21	0.29	0.40	0.53	0.67	0.85	1.04

TABLE A3–1   Description of Factors Used in Embedment Length Formulae

Symbol	Name	Condition	Value
ℓ_d	Development length		As calculated, but not less than $12d_b$
ψ_t	Reinforcement location factor	Horizontal reinforcement placed so that more than 12 in. of fresh concrete is cast in the member below the development length or splice	1.3
		Other reinforcement	1.0
ψ_e	Coating factor	Epoxy-coated bars or wires with cover less than $3d_b$ or clear spacing less than $6d_b$	1.5
		All other epoxy-coated bars or wires	1.2
		Uncoated reinforcement	1.0
ψ_s	Reinforcement size factor	#6 and smaller bars and wires	0.8
		#7 and larger bars	1.0
λ	Lightweight aggregate concrete factor	When lightweight aggregate concrete is used	1.3
		When f_{ct} is specified, however, λ shall be permitted to be taken as	$6.7\sqrt{f'_c}/f_{ct} > 1.0$
		When normal-weight concrete is used	1.0
c_b	Spacing or cover dimension, in.	Use the smaller of either distance from the center of the bar to the nearest concrete surface, or one-half of the center-to-center spacing of the bars being developed.	
K_{tr}	Transverse reinforcement index	It is permitted to use $K_{tr} = 0$ as a design simplification, even if transverse reinforcement is present.	$\dfrac{A_{tr}f_{yt}}{1500\,sn}$
	Excess reinforcement	Reinforcement in a flexural member is in excess of that required by analysis.	$\dfrac{A_{s,required}}{A_{s,provided}}$

Note: The development length used may not be less than 12 in.

TABLE A3–2 Simplified Expression of Development Length, ℓ_d, for Bars in Tension Based on ACI Code Section 12.2.2

Conditions	#6 and Smaller Bars and Deformed Wires	#7 and Larger Bars
A. Clear spacing of bars being developed or spliced not less than d_b, clear cover not less than d_b, and stirrups or ties throughout ℓ_d not less than the ACI Code minimum; or clear spacing of bars being developed or spliced not less than $2d_b$ and clear cover not less than d_b	$\left(\dfrac{f_y\psi_t\psi_e\lambda}{25\sqrt{f_c'}}\right)d_b$	$\left(\dfrac{f_y\psi_t\psi_e\lambda}{20\sqrt{f_c'}}\right)d_b$
B. Other cases	$\left(\dfrac{3f_y\psi_t\psi_e\lambda}{50\sqrt{f_c'}}\right)d_b$	$\left(\dfrac{3f_y\psi_t\psi_e\lambda}{40\sqrt{f_c'}}\right)d_b$

Note: The development length used may not be less than 12 in.

TABLE A3–3 Development Length for Tension Bars (ℓ_d) with $f_y = 60$ ksi ($\psi_e = \psi_t = \lambda = 1.0$) [in.]

Bar Size	$f_c' = 3$ ksi Condition A	Condition B	$f_c' = 4$ ksi Condition A	Condition B
#3	17	25	15	22
#4	22	33	19	29
#5	28	41	24	36
#6	33	50	29	43
#7	48	72	42	63
#8	55	83	48	72
#9	62	93	54	81
#10	70	105	61	91
#11	78	116	67	101

Note: Conditions A and B are based on Table A3–2.

TABLE A3–4   Applicable Reduction Factors Permitted by the ACI Code, Section 12.5.3, for the Development Length of Hook Terminated Bars

Condition	Reduction Factor
For #11 bar and smaller hooks with side cover (normal to the plane of the hook) not less than 2.5 in., and for 90-degree hooks with cover on bar extension beyond hook not less than 2 in.	0.7
For 90-degree hooks of #11 and smaller bars that are either enclosed within ties or stirrups perpendicular to the bar being developed, spaced not greater than $3d_b$ along ℓ_{dh}; or enclosed within ties or stirrups parallel to the bar being developed, spaced not greater than $3d_b$ along the length of the tail extension of the hook plus bend	0.8
For 180-degree hooks of #11 and smaller bars enclosed within stirrups or ties perpendicular to the bar being developed, spaced not greater than $3d_b$ along ℓ_{dh}	0.8
Where anchorage or development for f_y is not specifically required, reinforcement in excess of that required by analysis	$A_{s,\text{required}}/A_{s,\text{provided}}$

Note: The development length used may not be less than the smaller of $8d_b$ or 6 in.

TABLE A3–5   Permitted Reduction Factors for the Development Length of Bars in Compression Based on ACI Code Section 12.3.3

Condition	Reduction Factor
When the reinforcement provided is in excess of that required by analysis	$A_{s,\text{required}}/A_{s,\text{provided}}$
When reinforcement is enclosed within spiral reinforcement not less than $\frac{1}{4}$ in. diameter and not more than 4 in. pitch; or within #4 ties in conformance with ACI Code, Section 7.10.5, and spaced at not more than 4 in. on center	0.75

Note: The development length used may not be less than 8 in.

TABLE A3–6   Development Length for Compression Bars (ℓ_{dc}) with $f_y = 60$ ksi and Various f'_c Values (in.)

Bar Size	$f'_c = 3$ ksi	$f'_c = 4$ ksi	$f'_c \geq 5$ ksi
#3	9	8	7
#4	11	10	9
#5	14	12	12
#6	17	15	14
#7	20	17	16
#8	22	19	18
#9	25	22	21
#10	28	24	23
#11	31	27	26

TABLE A4–1a Values of V_c in kips ($f'_c = 3,000$ psi) [$d = h - 2.5$ in.($+/-$)]

$f'_c = 3,000$ psi

h (in.)	bw (in.)												
	6	8	10	12	14	16	18	20	22	24	26	28	30
10	4.9	6.6	8.2	9.9	11.5	13.1	14.8	16.4	18.1	19.7	21.4	23.0	24.6
12	6.2	8.3	10.4	12.5	14.6	16.7	18.7	20.8	22.9	25.0	27.1	29.1	31.2
14	7.6	10.1	12.6	15.1	17.6	20.2	22.7	25.2	27.7	30.2	32.8	35.3	37.8
16	8.9	11.8	14.8	17.7	20.7	23.7	26.6	29.6	32.5	35.5	38.5	41.4	44.4
18	10.2	13.6	17.0	20.4	23.8	27.2	30.6	34.0	37.4	40.8	44.1	47.5	50.9
20	11.5	15.3	19.2	23.0	26.8	30.7	34.5	38.3	42.2	46.0	49.8	53.7	57.5
22	12.8	17.1	21.4	25.6	29.9	34.2	38.5	42.7	47.0	51.3	55.5	59.8	64.1
24	14.1	18.8	23.6	28.3	33.0	37.7	42.4	47.1	51.8	56.5	61.2	65.9	70.7
26	15.4	20.6	25.7	30.9	36.0	41.2	46.3	51.5	56.6	61.8	66.9	72.1	77.2
28	16.8	22.3	27.9	33.5	39.1	44.7	50.3	55.9	61.5	67.0	72.6	78.2	83.8
30	18.1	24.1	30.1	36.1	42.2	48.2	54.2	60.2	66.3	72.3	78.3	84.3	90.4
32	19.4	25.9	32.3	38.8	45.2	51.7	58.2	64.6	71.1	77.6	84.0	90.5	96.9
34	20.7	27.6	34.5	41.4	48.3	55.2	62.1	69.0	75.9	82.8	89.7	96.6	103.5
36	22.0	29.4	36.7	44.0	51.4	58.7	66.1	73.4	80.7	88.1	95.4	102.8	110.1
38	23.3	31.1	38.9	46.7	54.4	62.2	70.0	77.8	85.6	93.3	101.1	108.9	116.7
40	24.6	32.9	41.1	49.3	57.5	65.7	73.9	82.2	90.4	98.6	106.8	115.0	123.2
42	26.0	34.6	43.3	51.9	60.6	69.2	77.9	86.5	95.2	103.8	112.5	121.2	129.8

TABLE A4–1b Values of V_c in kips ($f'_c = 4,000$ psi) [$d = h - 2.5$ in.($+/-$)]

$f'_c = 4,000$ psi

h (in.)	bw (in.)												
	6	8	10	12	14	16	18	20	22	24	26	28	30
10	5.7	7.6	9.5	11.4	13.3	15.2	17.1	19.0	20.9	22.8	24.7	26.6	28.5
12	7.2	9.6	12.0	14.4	16.8	19.2	21.6	24.0	26.4	28.8	31.2	33.6	36.0
14	8.7	11.6	14.5	17.5	20.4	23.3	26.2	29.1	32.0	34.9	37.8	40.7	43.6
16	10.2	13.7	17.1	20.5	23.9	27.3	30.7	34.2	37.6	41.0	44.4	47.8	51.2
18	11.8	15.7	19.6	23.5	27.4	31.4	35.3	39.2	43.1	47.1	51.0	54.9	58.8
20	13.3	17.7	22.1	26.6	31.0	35.4	39.8	44.3	48.7	53.1	57.6	62.0	66.4
22	14.8	19.7	24.7	29.6	34.5	39.5	44.4	49.3	54.3	59.2	64.1	69.1	74.0
24	16.3	21.8	27.2	32.6	38.1	43.5	49.0	54.4	59.8	65.3	70.7	76.1	81.6
26	17.8	23.8	29.7	35.7	41.6	47.6	53.5	59.5	65.4	71.3	77.3	83.2	89.2
28	19.4	25.8	32.3	38.7	45.2	51.6	58.1	64.5	71.0	77.4	83.9	90.3	96.8
30	20.9	27.8	34.8	41.7	48.7	55.7	62.6	69.6	76.5	83.5	90.4	97.4	104.4
32	22.4	29.9	37.3	44.8	52.2	59.7	67.2	74.6	82.1	89.6	97.0	104.5	111.9
34	23.9	31.9	39.8	47.8	55.8	63.8	71.7	79.7	87.7	95.6	103.6	111.6	119.5
36	25.4	33.9	42.4	50.8	59.3	67.8	76.3	84.7	93.2	101.7	110.2	118.6	127.1
38	26.9	35.9	44.9	53.9	62.9	71.8	80.8	89.8	98.8	107.8	116.8	125.7	134.7
40	28.5	37.9	47.4	56.9	66.4	75.9	85.4	94.9	104.4	113.8	123.3	132.8	142.3
42	30.0	40.0	50.0	60.0	69.9	79.9	89.9	99.9	109.9	119.9	129.9	139.9	149.9

TABLE A4–1c Values of V_c in kips ($f'_c = 5{,}000$ psi) [$d = h - 2.5$ in.(+/−)]

$f'_c = 5{,}000$ psi

h (in.)	bw (in.)												
	6	8	10	12	14	16	18	20	22	24	26	28	30
10	6.4	8.5	10.6	12.7	14.8	17.0	19.1	21.2	23.3	25.5	27.6	29.7	31.8
12	8.1	10.7	13.4	16.1	18.8	21.5	24.2	26.9	29.6	32.2	34.9	37.6	40.3
14	9.8	13.0	16.3	19.5	22.8	26.0	29.3	32.5	35.8	39.0	42.3	45.5	48.8
16	11.5	15.3	19.1	22.9	26.7	30.5	34.4	38.2	42.0	45.8	49.6	53.5	57.3
18	13.2	17.5	21.9	26.3	30.7	35.1	39.5	43.8	48.2	52.6	57.0	61.4	65.8
20	14.8	19.8	24.7	29.7	34.6	39.6	44.5	49.5	54.4	59.4	64.3	69.3	74.2
22	16.5	22.1	27.6	33.1	38.6	44.1	49.6	55.2	60.7	66.2	71.7	77.2	82.7
24	18.2	24.3	30.4	36.5	42.6	48.6	54.7	60.8	66.9	73.0	79.1	85.1	91.2
26	19.9	26.6	33.2	39.9	46.5	53.2	59.8	66.5	73.1	79.8	86.4	93.1	99.7
28	21.6	28.8	36.1	43.3	50.5	57.7	64.9	72.1	79.3	86.5	93.8	101.0	108.2
30	23.3	31.1	38.9	46.7	54.4	62.2	70.0	77.8	85.6	93.3	101.1	108.9	116.7
32	25.0	33.4	41.7	50.1	58.4	66.8	75.1	83.4	91.8	100.1	108.5	116.8	125.2
34	26.7	35.6	44.5	53.5	62.4	71.3	80.2	89.1	98.0	106.9	115.8	124.7	133.6
36	28.4	37.9	47.4	56.9	66.3	75.8	85.3	94.8	104.2	113.7	123.2	132.7	142.1
38	30.1	40.2	50.2	60.2	70.3	80.3	90.4	100.4	110.5	120.5	130.5	140.6	150.6
40	31.8	42.4	53.0	63.6	74.2	84.9	95.5	106.1	116.7	127.3	137.9	148.5	159.1
42	33.5	44.7	55.9	67.0	78.2	89.4	100.6	111.7	122.9	134.1	145.2	156.4	167.6

TABLE A4–2a   Values of V_s in kips, with 2 Legs of #3 Stirrups

#3 Stirrups—2 Legs

h	Spacing s (in.)										
(in.)	2	3	4	5	6	8	10	12	14	16	18
10	49.5	33.0	24.8								
12	62.7	41.8	31.4	25.1							
14	75.9	50.6	38.0	30.4	25.3						
16	89.1	59.4	44.6	35.6	29.7						
18	102.3	68.2	51.2	40.9	34.1	25.6					
20	115.5	77.0	57.8	46.2	38.5	28.9					
22	128.7	85.8	64.4	51.5	42.9	32.2	25.7				
24	141.9	94.6	71.0	56.8	47.3	35.5	28.4				
26	155.1	103.4	77.6	62.0	51.7	38.8	31.0	25.9			
28	168.3	112.2	84.2	67.3	56.1	42.1	33.7	28.1			
30	181.5	121.0	90.8	72.6	60.5	45.4	36.3	30.3	25.9		
32	194.7	129.8	97.4	77.9	64.9	48.7	38.9	32.5	27.8		
34	207.9	138.6	104.0	83.2	69.3	52.0	41.6	34.7	29.7	26.0	
36	221.1	147.4	110.6	88.4	73.7	55.3	44.2	36.9	31.6	27.6	
38	234.3	156.2	117.2	93.7	78.1	58.6	46.9	39.1	33.5	29.3	26.0
40	247.5	165.0	123.8	99.0	82.5	61.9	49.5	41.3	35.4	30.9	27.5

Note: Multiply table values by 2 for #3 stirrups with 4 legs.

TABLE A4–2b   Values of V_s in kips, with 2 Legs of #4 Stirrups

#4 Stirrups—2 Legs

h	Spacing s (in.)										
(in.)	2	3	4	5	6	8	10	12	14	16	18
10	90.0	60.0	45.0								
12	114.0	76.0	57.0	45.6							
14	138.0	92.0	69.0	55.2	46.0						
16	162.0	108.0	81.0	64.8	54.0						
18	186.0	124.0	93.0	74.4	62.0	46.5					
20	210.0	140.0	105.0	84.0	70.0	52.5					
22	234.0	156.0	117.0	93.6	78.0	58.5	46.8				
24	258.0	172.0	129.0	103.2	86.0	64.5	51.6				
26	282.0	188.0	141.0	112.8	94.0	70.5	56.4	47.0			
28	306.0	204.0	153.0	122.4	102.0	76.5	61.2	51.0			
30	330.0	220.0	165.0	132.0	110.0	82.5	66.0	55.0	47.1		
32	354.0	236.0	177.0	141.6	118.0	88.5	70.8	59.0	50.6		
34	378.0	252.0	189.0	151.2	126.0	94.5	75.6	63.0	54.0	47.3	
36	402.0	268.0	201.0	160.8	134.0	100.5	80.4	67.0	57.4	50.3	
38	426.0	284.0	213.0	170.4	142.0	106.5	85.2	71.0	60.9	53.3	47.3
40	450.0	300.0	225.0	180.0	150.0	112.5	90.0	75.0	64.3	56.3	50.0

Note: Multiply table values by 2 for #4 stirrups with 4 legs.

TABLE A5–1 Maximum Number of Bars in Columns

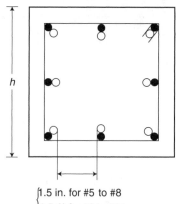 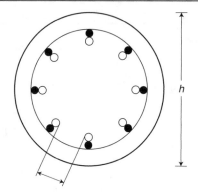

$\begin{cases}1.5 \text{ in. for \#5 to \#8} \\ 1.5d_b^* \text{ for \#9 to \#11}\end{cases}$ 1.5 in. for #5 to #8
1.5d_b^* for #9 to #11

(d_b^* = diameter of longitudinal bars.)

h (in)	Square Tied Column							Round Spiral Column						
	#5	#6	#7	#8	#9	#10	#11	#5	#6	#7	#8	#9	#10	#11
10	8	4	4	4	4	—	—	6	—	—	—	—	—	—
11	8	8	8	4	4	4	4	7	6	—	—	—	—	—
12	12	8	8	8	4	4	4	8	7	6	6	—	—	—
13	12	12	8	8	8	4	4	10	9	8	7	—	—	—
14		12	12	12	8	8	4		10	9	8	7	—	—
15		12	12	12	8	8	8		12	10	9	8	6	—
16			16	12	12	8	8		12	11	9	7	6
17			16	16	12	12	8		13	12	10	8	7
18				16	16	12	12		14	13	11	9	8
19				20	16	12	12			14	12	10	9
20				20	16	16	12			16	13	11	10
21				20	20	16	12				15	12	11
22					20	16	16				16	13	12
23					20	20	16				17	14	13
24						20	16				18	15	13
25						20	20					16	14
26						20	20					17	15
27							20					18	16
28							20						17
29							24						18
30							24						19

Note: Values are based on 1½ in. cover, #4 ties or spirals, with clear space of 1½ in. for #5 to #8, and 1.5 times bar diameter for #9 to #11.

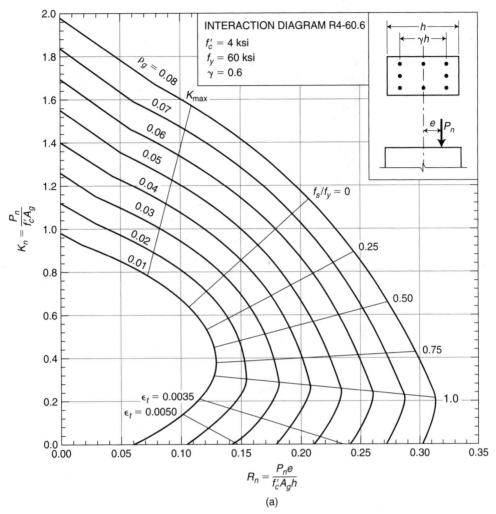

FIGURE A5–1   (a) ACI column interaction diagram [SP-17(97)], *Courtesy of American Concrete Institute.* (b) K_n versus ϕ relationship.

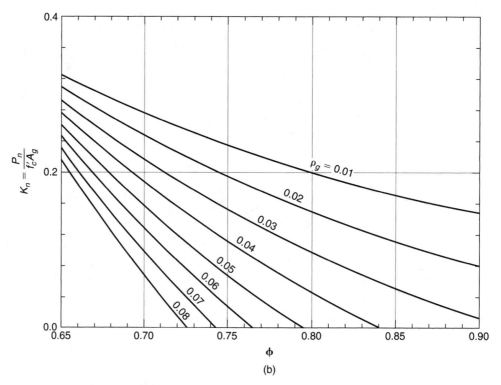

(b)

FIGURE A5–1   (*continued*)

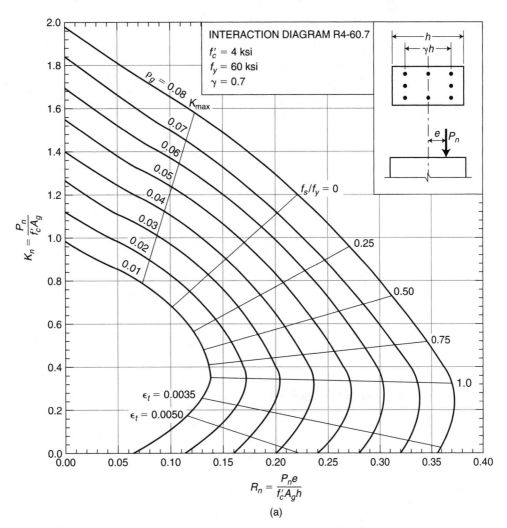

FIGURE A5–2   (a) ACI column interaction diagram [SP-17(97)], *Courtesy of American Concrete Institute.* (b) K_n versus ϕ relationship.

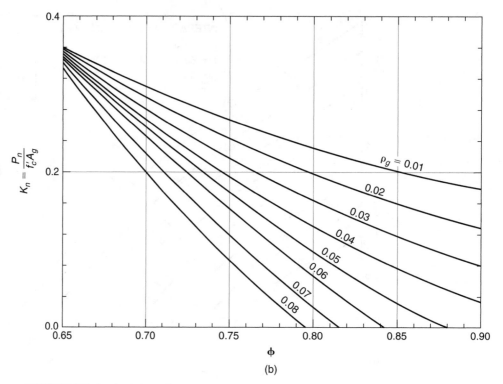

FIGURE A5–2 (*continued*)

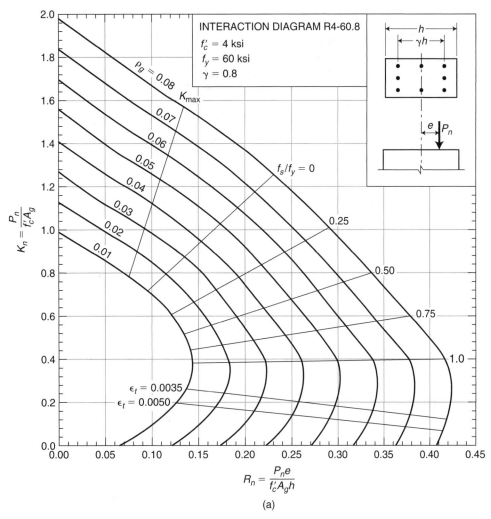

FIGURE A5–3   (a) ACI column interaction diagram [SP-17(97)], *Courtesy of American Concrete Institute.* (b) K_n versus ϕ relationship.

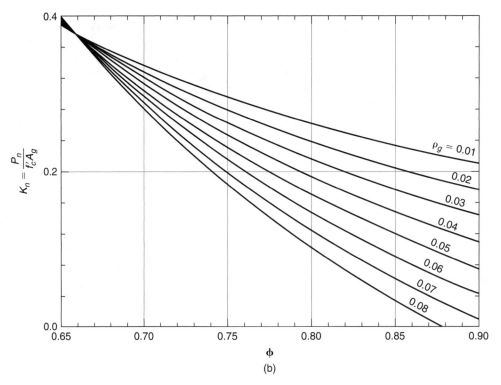

FIGURE A5–3   (*continued*)

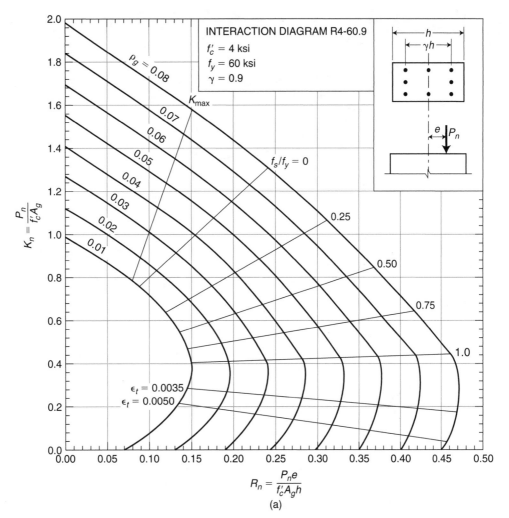

FIGURE A5–4 (a) ACI column interaction diagram [SP-17(97)], *Courtesy of American Concrete Institute.* (b) K_n versus ϕ relationship.

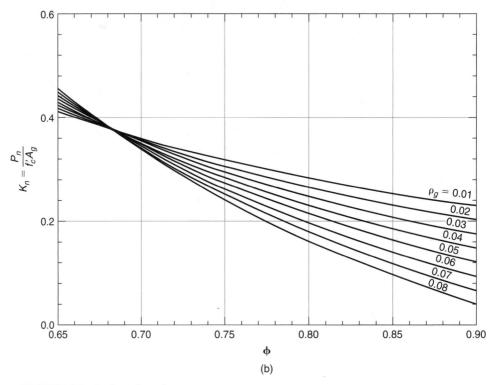

FIGURE A5–4 (continued)

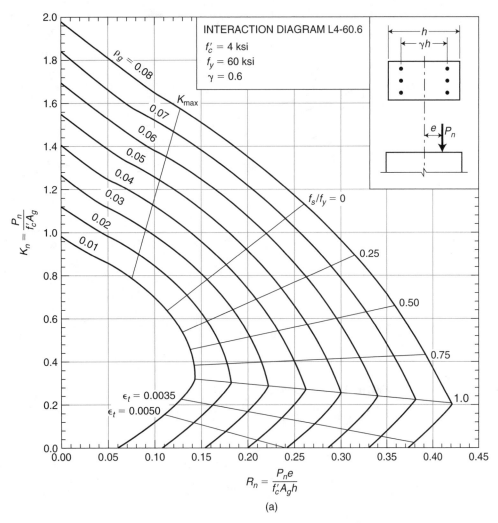

FIGURE A5–5 (a) ACI column interaction diagram [SP-17(97)], *Courtesy of American Concrete Institute.* (b) K_n versus ϕ relationship.

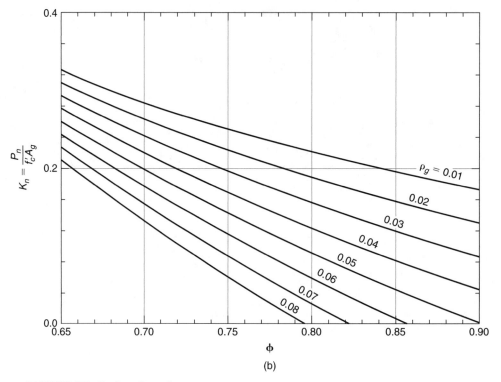

(b)

FIGURE A5–5 (*continued*)

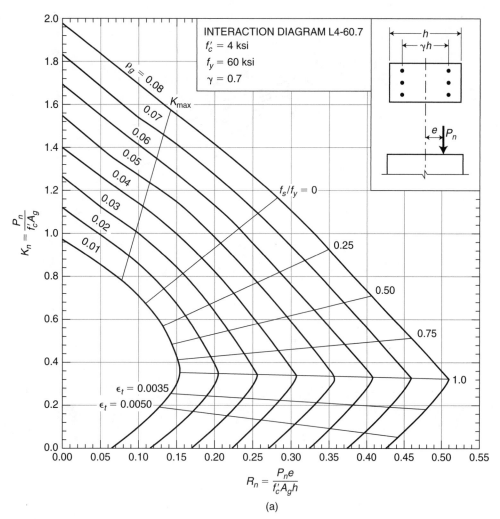

FIGURE A5–6   (a) ACI column interaction diagram [SP-17(97)], *Courtesy of American Concrete Institute.* (b) K_n versus ϕ relationship.

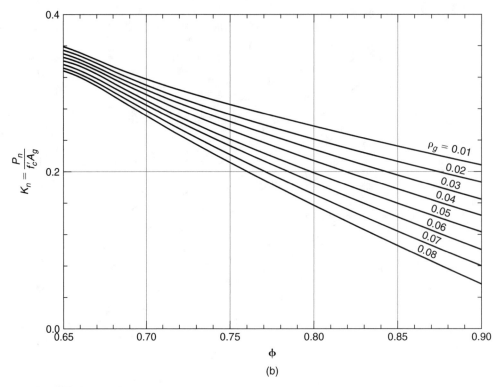

(b)

FIGURE A5–6 (*continued*)

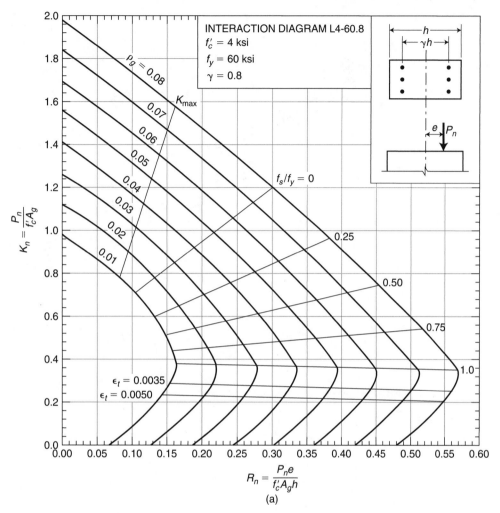

(a)

FIGURE A5–7 (a) ACI column interaction diagram [SP-17(97)], *Courtesy of American Concrete Institute.* (b) K_n versus ϕ relationship.

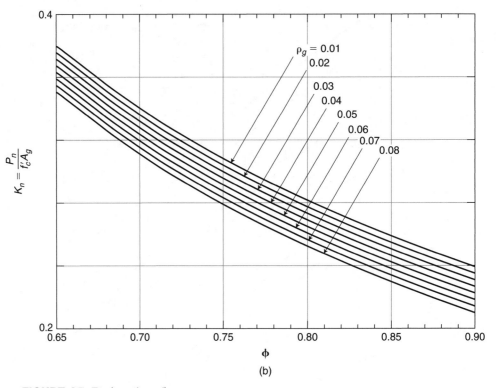

(b)

FIGURE A5–7   (*continued*)

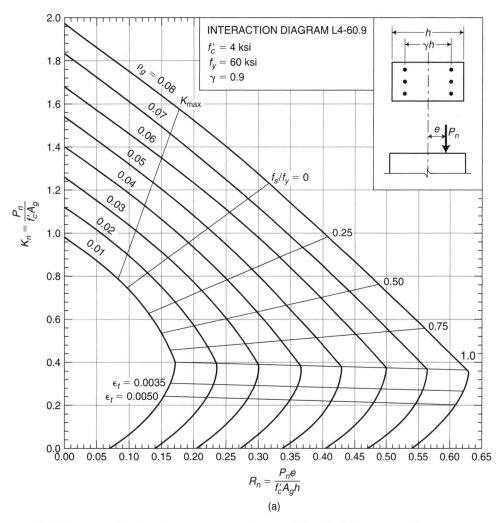

FIGURE A5–8   (a) ACI column interaction diagram [SP-17(97)], *Courtesy of American Concrete Institute.* (b) K_n versus ϕ relationship.

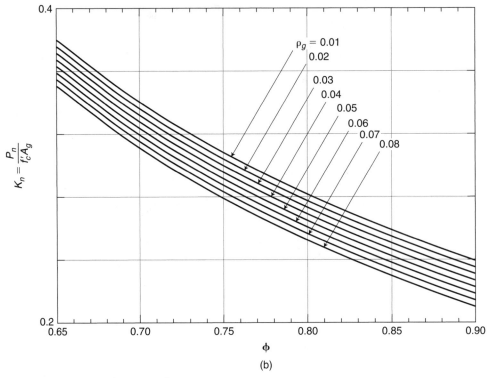

(b)

FIGURE A5–8   (*continued*)

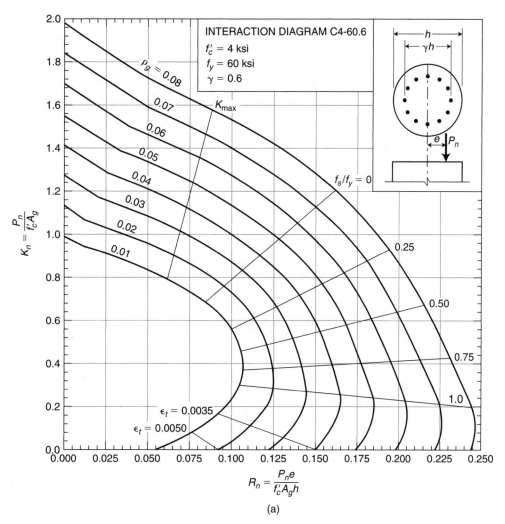

FIGURE A5–9 (a) ACI column interaction diagram [SP-17(97)], *Courtesy of American Concrete Institute.* (b) K_n versus ϕ relationship.

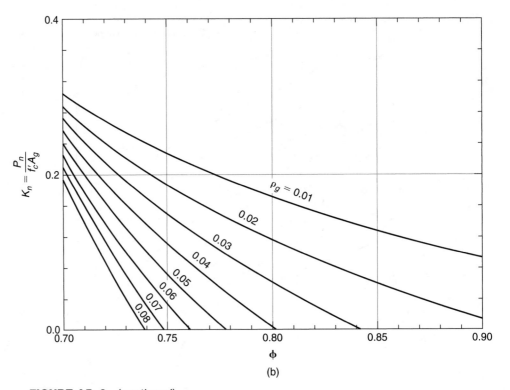

(b)

FIGURE A5–9 (*continued*)

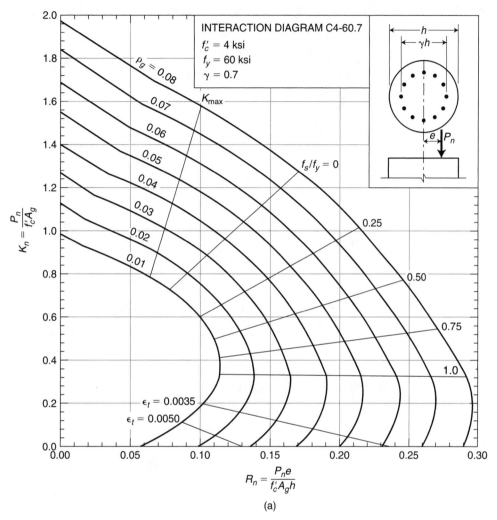

FIGURE A5–10 (a) ACI column interaction diagram [SP-17(97)], *Courtesy of American Concrete Institute.* (b) K_n versus ϕ relationship.

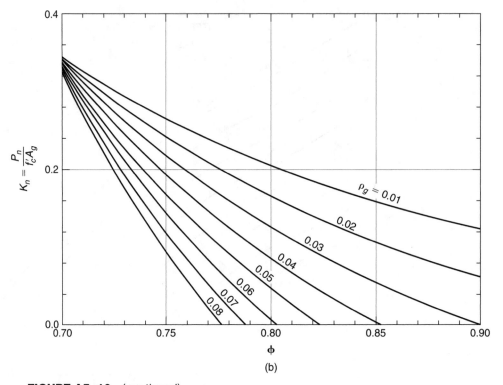

FIGURE A5–10 (*continued*)

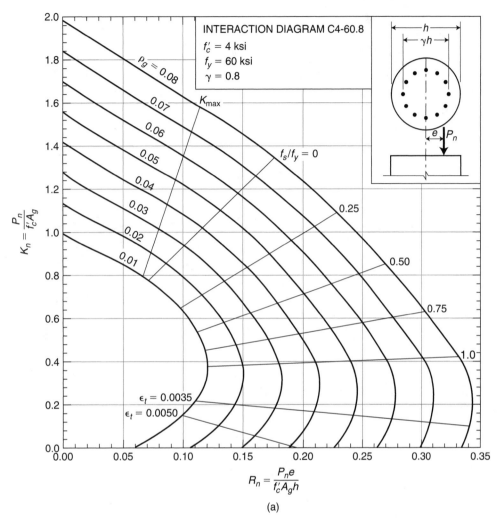

FIGURE A5–11   (a) ACI column interaction diagram [SP-17(97)], *Courtesy of American Concrete Institute.* (b) K_n versus ϕ relationship.

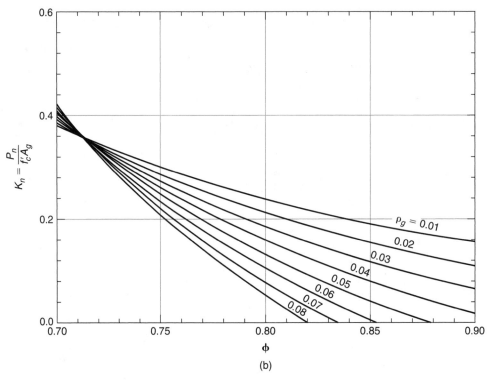

FIGURE A5–11   (*continued*)

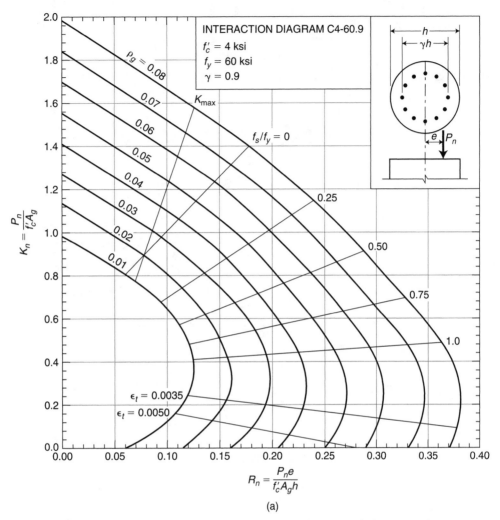

FIGURE A5–12 (a) ACI column interaction diagram [SP-17(97)], *Courtesy of American Concrete Institute.* (b) K_n versus ϕ relationship.

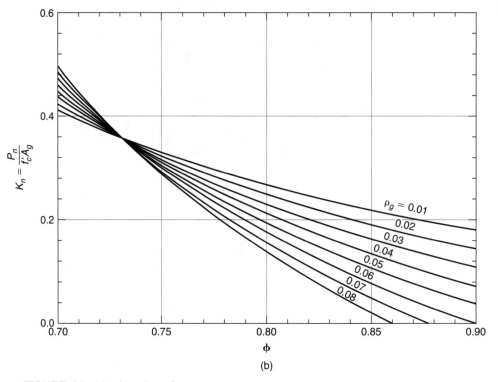

(b)

FIGURE A5–12   (continued)

B
STANDARD ACI NOTATIONS

a	=	depth of equivalent rectangular stress block
A_{ch}	=	cross-sectional area of a structural member measured out-to-out of transverse reinforcements
A_{ct}	=	area of that part of the cross section between the flexural tension face and the center of gravity of the gross section
A_g	=	gross area of concrete section. For a hollow section, A_g is the area of the concrete only and does not include the area of the void(s)
A_{ps}	=	area of prestressing steel in flexural tension zone
A_s	=	area of nonprestressed longitudinal tension reinforcement
A_s'	=	area of longitudinal compression reinforcement
$A_{s,\min}$	=	minimum area of flexural reinforcement
A_{st}	=	total area of nonprestressed longitudinal reinforcements
A_v	=	area of shear reinforcement within spacing s
$A_{v,\min}$	=	minimum area of shear reinforcement within spacing s
A_1	=	loaded area
A_2	=	area of the lower base of the largest frustum of a pyramid, cone, or tapered wedge, contained wholly within the support and having for its upper base the loaded area, and having side slopes of 1 (vertical) and 2 (horizontal)
b	=	width of the compression face of a member
b_o	=	perimeter of the critical section for shear in slabs and footings
b_w	=	web width, or diameter of a circular section
c	=	distance from extreme compression fiber to the neutral axis

c_b = smaller of (a) the distance from the center of a bar to the nearest concrete surface, and (b) one-half the center-to-center spacing of the bars being developed

c_c = clear cover of reinforcement

C = cross-sectional constant to define torsional properties of slab and beam

d = distance from extreme compression fiber to the centroid of longitudinal tension reinforcement

d' = distance from extreme compression fiber to the centroid of longitudinal compression reinforcement

d_b = nominal diameter of a bar, wire, or prestressing strand

d_p = distance from extreme compression fiber to centroid of prestressing steel

d_t = distance from extreme compression fiber to the centroid of the extreme layer of longitudinal tension steel

D = dead loads, or related internal moments and forces

E = load effects of earthquake, or related internal moments and forces

E_c = modulus of elasticity of concrete

E_{cb} = modulus of elasticity of beam concrete

E_{cs} = modulus of elasticity of slab concrete

E_s = modulus of elasticity of reinforcement and structural steel

f'_c = specified compressive strength of concrete

f'_{ci} = specified compressive strength of concrete at time of initial prestress

f_{ct} = average splitting tensile strength of lightweight concrete

f_{ps} = stress in prestressing steel at nominal flexural strength

f_{pu} = specified tensile strength of prestressing steel

f_r = modulus of rupture of concrete

f_s = calculated stress in reinforcement at service loads

f_{se} = effective stress in prestressing steel (after allowance for all prestress losses)

f_t = extreme fiber stress in tension in the precompressed tensile zone calculated at service loads using gross section properties

f_y = specified yield strength of reinforcement

f_{yt} = specified yield strength, f_y, of transverse reinforcement

F = loads due to weight and pressures of fluids with well-defined densities and controllable maximum heights, or related internal moments and forces

h = overall thickness or height of member

H = loads due to weight and pressure of soil, water in soil, or other materials, or related internal moments and forces

I = moment of inertia of section about the centroidal axis

I_b = moment of inertia of gross section of beam about the centroidal axis

I_{cr} = moment of inertia of cracked section transformed to concrete

I_e = effective moment of inertia for computation of deflection

I_g = moment of inertia of gross concrete section about the centroidal axis, neglecting reinforcement

I_s = moment of inertia of gross section of slab about the centroidal axis defined for calculating α_f and β_t

k = effective length factor for compression members

K = wobble friction coefficient per foot of tendon

K_{tr} = transverse reinforcement index

ℓ	=	span length of beam or one-way slab, clear projection of cantilever
ℓ_n	=	length of clear span measured face-to-face of supports
ℓ_{px}	=	distance from the jacking end of a prestressing steel element to the point under consideration
ℓ_u	=	unsupported length of a compression member
ℓ_1	=	length of span in the direction that moments are being determined, measured center-to-center of supports
ℓ_2	=	length of span in the direction perpendicular to ℓ_1, measured center-to-center of supports
L	=	live loads, or related internal moments and forces
L_r	=	roof live load, or related internal moments and forces
M_a	=	maximum unfactored moment in a member at the stage deflection is computed
M_{cr}	=	cracking moment
M_n	=	nominal flexural strength at section
M_o	=	total factored static moment
M_u	=	factored moment at section
M_1	=	smaller factored end moment on a compression member; taken as positive if member is bent in single curvature, and negative if bent in double curvature
M_2	=	larger factored end moment on a compression member, always positive
N_u	=	factored axial force normal to a cross section occurring simultaneously with V_u or T_u; taken as positive for compression and negative for tension
P_b	=	nominal axial strength at balanced strain conditions
P_n	=	nominal axial strength of cross section
$P_{n,\max}$	=	maximum allowable value of P_n
P_o	=	nominal axial strength at zero eccentricity
P_{pj}	=	prestressing force at jacking end
P_{px}	=	prestressing force evaluated at distance ℓ_{px} from the jacking end
P_u	=	factored axial force; taken as positive for compression and negative for tension
q_u	=	factored load per unit area
r	=	radius of gyration of cross section of a compression member
R	=	rain load, or related internal moments and forces
s	=	center-to-center spacing of items, such as longitudinal reinforcement, transverse reinforcement, prestressing tendons, wires, or anchors
S	=	snow load, or related internal moments and forces
S_m	=	elastic section modulus
t	=	wall thickness of a hollow section
T	=	cumulative effect of temperature, creep, shrinkage, differential settlement, and shrinkage-compensating concrete
U	=	strength required to resist factored loads or related internal moments and forces
v_n	=	nominal shear stress
V_c	=	nominal shear strength provided by concrete
V_n	=	nominal shear strength
V_s	=	nominal shear strength provided by shear reinforcement
V_u	=	factored shear force at section
w_c	=	unit weight of concrete
w_u	=	factored load per unit length of beam or one-way slab

W = wind load, or related internal moments and forces

x = shorter overall dimension of rectangular part of a cross section

y = longer overall dimension of rectangular part of a cross section

y_t = distance from the centroidal axis of the gross section, neglecting reinforcement, to a tension face

α_f = ratio of the flexural stiffness of a beam section to the flexural stiffness of a width of slab bounded laterally by centerlines of adjacent panels (if any) on each side of the beam

α_{f1} = α_f in direction of ℓ_1

α_{f2} = α_f in direction of ℓ_2

α_{px} = total angular change of tendon profile from the tendon's jacking end to the point under consideration in radians

α_s = constant used to compute V_c in slabs and footings

β = ratio of long to short dimensions: clear spans for two-way slabs; sides of column, concentrated load, or reaction area

β_t = ratio of the torsional stiffness of an edge beam section to the flexural stiffness of a width of slab equal to the span length of the beam, center-to-center of supports

β_1 = factor relating the depth of the equivalent rectangular compressive stress block to the neutral axis depth

γ_s = factor used to determine the portion of reinforcement located in the center band of a footing

ε_t = net tensile strain in extreme tension steel at nominal strength, excluding strains due to effective prestress, creep, shrinkage, and temperature

λ = development length modification factor related to the unit weight of the concrete

λ_Δ = multiplier for additional deflection due to long-term effects

μ_p = posttensioning curvature friction coefficient

ξ = time-dependent factor for sustained load

ρ = ratio of A_s to bd

ρ' = ratio of A_s' to bd

ρ_b = ratio of A_s to bd producing balanced strain conditions

ρ_g = ratio of A_{st} to A_g

ρ_p = ratio of A_{ps} to bd_p

ρ_s = ratio of the volume of spiral reinforcement to the total volume of core confined by the spiral (measured out-to-out of spirals)

ρ_w = ratio A_s to $b_w d$

ψ_e = factor used to modify development length based on reinforcement coating

ψ_s = factor used to modify development length based on reinforcement size

ψ_t = factor used to modify development length based on reinforcement location

ϕ = strength reduction factor

ω = tension reinforcement index

INDEX